MARE
TRANQUILLITATIS
44
Arago D
4300
ARAGO
(1800)
LAMONT
Arago C
Maskelyne G
Sabine B
Sabine D
Sabine E
Sabine C
Tranquillity Base
470
MOLTKE
(1070)
3308
Censor
Moltke A
Moltke B
Torricelli G
Hypatia G
Hypatia D
Hypatia DA
HYPATIA
Torricelli K
Hypatia H
Hypatia B
4000
TORRICELLI
3200
Hypatia A
(2520)

A Man on the Moon

A Man on the Moon

The Voyages of the Apollo Astronauts

ANDREW CHAIKIN

MICHAEL JOSEPH
an imprint of
PENGUIN BOOKS

MICHAEL JOSEPH

UK | USA | Canada | Ireland | Australia
India | New Zealand | South Africa

Michael Joseph is part of the Penguin Random House group of companies
whose addresses can be found at global.penguinrandomhouse.com

Penguin
Random House
UK

First published in Great Britain by Michael Joseph 1994
Published in Penguin Books 1995
Reissued with a Foreword 1998
Reissued in this edition 2018
001

Grateful acknowledgement is made for permission to reprint
excerpts from the following copyrighted works:
Carrying the Fire by Michael Collins. Copyright © 1974 by Michael Collins.
Reprinted by permission of Farrar, Straus & Giroux, Inc.;
Extract taken from the song 'How Great Thou Art' by Stuart K. Hine (1899–1989), copyright © 1953
Stuart K. Hine/Kingsway's Thankyou Music, PO Box 75, Eastbourne, East Sussex BN23 6NT, UK.
Worldwide excluding USA, Canada. Used by permission;
'Fly Me to the Moon', words and music by Bart Howard.
Copyright © 1954 Almanac Music, Incorporated, New York, NY.
TRO Essex Music Limited, SW10 0SZ. All rights reserved;
'Spacewalk' and 'Quietly, Like a Night Bird' from *Hello Earth, Greetings from Endeavour*
by Alfred M. Worden (Nash Publishing Company, 1974).
Reprinted by permission of the author.

Printed and bound in Great Britain by Clays Ltd, Elcograf S.p.A.

A CIP catalogue record for this book is available from the British Library

HARDBACK ISBN: 978–0–241–36382–9
OM PAPERBACK ISBN: 978–0–241–36384–3

www.greenpenguin.co.uk

MIX
Paper from
responsible sources
FSC
www.fsc.org
FSC® C018179

Penguin Random House is committed to a
sustainable future for our business, our readers
and our planet. This book is made from Forest
Stewardship Council® certified paper.

For my father and mother

Contents

Foreword

Tom Hanks

I was captured by the spirit of the Apollo program when I was twelve years old. I had been following the space program throughout the Mercury and Gemini flights, building model kits and watching the launches from my mom's house in northern California. We had an above-ground pool in the backyard, and I would put a brick in the back of my swim trunks to hold me down on the bottom, suck in air through a garden hose, and lay there with my arms and legs adrift, pretending I was walking in space. I was, of course, eagerly anticipating the Apollo missions to the moon, because that would give me more models to build. But it wasn't until the lunar orbit flight of Apollo 8 at Christmastime, 1968, that I really recognized what was happening. Mankind was leaving Earth, for the first time in history. I thought about the three astronauts traveling all that distance, and wondered what they felt and what they must have seen. I'll never forget the telecast they made as they circled the moon Christmas Eve, and their reading from the book of Genesis. It was a perfect moment that captured the romantic, epic, historic adventure of it all.

I returned to that adventure in 1994 for the filming of *Apollo 13*. In preparing for the role of astronaut Jim Lovell (who, appropriately, had also been aboard Apollo 8), I realized there was a great deal about the Apollo program that had never been brought to light, things that I did not know. I wanted to understand the events that enabled Neil Armstrong and Buzz Aldrin to make the first landing on the moon. I also wanted to know what went on when other men, like Pete Conrad, Al Bean, Dave Scott, Jack Schmitt, and Gene Cernan, made their footprints in the lunar soil in the five landings that came afterward. I wanted the whole story of mankind's exploration of the moon.

I found it in Andy Chaikin's impressive and illuminating book, *A Man on the Moon*. It had just come out; I picked it up and read it straight through. It was fascinating, just great stuff. What I enjoyed most was the way Andy captured the personalities of the men themselves—not just what they did, but who they were and what made them take on the job of flying to the moon. At the same time, he did a great job of describing the nuts and bolts of how we got there. I am, quite frankly, not a rocket scientist. (I only played one in a movie.) I had been confused by the technology of Apollo. But here it was, explained in a way that I could understand and that was, above all, entertaining. Even the moon itself—full of mystery and spectacular scenery—was an ongoing character in a grand scientific detective story.

A Man on the Moon showed me that each of the missions was fraught with peril, and each accomplished things that were of stellar importance to humanity. Each was an example of the best part of us all. The only thing that is disappointing about the Apollo story is that it comes to an end; our work on the moon was halted after the Apollo 17 mission in December 1972. (I have the secret hope that the story is only Volume One, with Volume Two coming soon.)

Late in the winter of 1995, after *Apollo 13* had finished filming, I read Andy's book again, this time for sheer pleasure. By then, I had met many of the people I was reading about, and that gave the stories new meaning and immediacy. And I began playing with the idea of bringing the entire Apollo saga to television, in a dramatic, episodic event—something like a combination of *Brideshead Revisited* and *The Civil War* by way of all those official NASA mission re-cap films. Not long afterward I called Andy, whom I had met on the set of *Apollo 13*. By summer, work on the series of *HBO From the Earth to the Moon* had begun.

As my colleagues became immersed in the task of putting Apollo on film, excerpts from *A Man on the Moon* were the first things in our reference binders. The big challenge we faced was to make each new mission as interesting as the one before, to give each episode of the series its own unique dramatic narrative and theme. Often these themes came right out of what Andy writes about, or hints at, in the book's pages. When there was a question of accuracy or accomplishment, someone on the staff always asked the most obvious question: "What does Chaikin say about this?"

What I hope to convey with *From the Earth to the Moon* is what Andy's captured so well in this book—just how magnificent an undertaking Apollo really was. That going to the moon was not just a technological endeavor, but an artistic one, like Michelangelo's frescoes on the Sistine Chapel ceiling. The same kind of imagination that allowed Michelangelo to produce the crowning achievement of his era helped NASA's engineers build their moonships. Just as Michelangelo needed faith in his own abilities to sustain him during the long years of his effort, so faith was at the heart of what it took to put men, and their shoes and socks, and pictures of their children, on the surface of the moon.

Above all, Apollo was a voyage of inspiration. The thing that still fuels me in my day-to-day life, and what I want to convey to my children, and to the audience, is that if mankind can figure out a way to put twelve men on the moon, then, honestly, we can solve anything. That's why I believe the six Apollo landings are six of the seven greatest stories ever told.

Preface

The sixties were a time of cultural earthquakes: the horror of the Kennedy and King assassinations, the arrival of four mop-topped singers from Liverpool, the flower-fragrant Summer of Love, the din of protest, and—most of all—the violence of the war in Vietnam. And something else extraordinary happened: on the night of July 20, 1969, two Americans walked on the moon. In what seemed like a miracle of technology, we witnessed it live on television. Across the country, 400,000 people who had worked to make it happen celebrated their triumph. TV commentators and editorial writers proclaimed that five hundred years from now our century would be remembered for those footsteps, when human beings left their home planet to explore the universe.

But in our own time their impact was fleeting. Even as astronauts returned to the moon for bolder and more ambitious explorations, our attention was diverted. While the nation was absorbed in the war, the environment, and unrest at home, NASA was expanding the reach of human beings on another world. On Apollo 11, the first lunar landing, Neil Armstrong and Buzz Aldrin had spent little more than a day on a bland acre of moonscape. In their single moonwalk—which lasted a bit longer than a feature-length film—they never ventured more than a couple of hundred feet from their lander. By 1972, the final pair of moonwalkers, Gene Cernan and Jack Schmitt, were living for three full days on the moon, exploring a lunar valley ringed by spectacular mountains whose rocks would provide a key to the origin of the solar system. During three moonwalks, each lasting more than seven hours, they drove more than ten miles in a battery-powered car. Their activities were broadcast live, in color, and with remarkable clarity, by a TV camera that was remotely controlled from earth. But we had stopped watching.

We have never really come to terms with what we saw on that summer night twenty-five years ago, or with the events that followed. In large measure, it is because we never really understood what had happened. TV showed us what the astronauts did on the moon, but could not transmit the immensity of the venture. The astronauts knew this, and when they returned to earth, they struggled to describe their experiences. But astronauts are not communicators, and with rare exceptions, their words could not bridge the gap between the high-tech realm of spaceflight and everyday experience. The real impact of Apollo—the experiences of the first men to visit another world—remained, like the moon itself, beyond our grasp.

I was thirteen when Armstrong and Aldrin landed on the moon; like countless other space-struck teenagers, I kept a daily vigil in front of the TV, surrounded by press kits, lunar maps, and scale models of rockets and spacecraft. That passion stayed with me when I embarked on a science journalism career, writing articles on astronomy and space exploration as an editor of *Sky & Telescope* magazine. The Space Shuttle era had arrived, but for me, it was Apollo that held special fascination. By 1984 I had interviewed Gene Kranz, the flight director who had been in the trenches of mission control during the first lunar landing, and Harvard geologist Clifford Frondell, who had been present when the first box of moon rocks was opened inside the windowless expanse of NASA's Lunar Receiving Laboratory. For an article on space motion sickness I had experienced weightlessness aboard a special NASA cargo plane used for astronaut training. I had tried on a lunar space suit and flown the Apollo command module simulator. Without realizing it, I had been preparing myself to become a vicarious participant in the Apollo adventure.

By the summer of 1984, the idea of a book about the moon

experience began to take shape. To write it, I would need to talk at length to each of the twenty-three surviving moon voyagers (Apollo 13 astronaut Jack Swigert had died in 1982). The following year I conducted the first interviews for the project, with the crew of Apollo 12, Pete Conrad, Dick Gordon, and Alan Bean. In 1986, I began to work on the book full-time, and my house became a museum of lunar exploration: panoramas of the moon on the walls, mission transcripts and debriefings on the bookshelves, Apollo videotapes in the VCR.

At first a number of the astronauts were reluctant to get involved; some of the interviews came about only after years of perseverance. Not until 1992, seven years after I began, did the last of the lunar veterans, Apollo 16's Ken Mattingly, sit down with me to tell his stories. The encounter was typical of the interviews on this project: we talked for six hours, in which time Mattingly vividly recounted his own unique lunar experiences.

During the past eight years, when people asked me what my book was about, I would say that it was the story of the lunar voyages that the astronauts never wrote. These men are such loners that the thought of all of them sitting down to write a book together seems more far-fetched than the idea that they have been to the moon. I hope that my efforts will serve in their stead, to tell the story of a unique handful of men who have been to the edge of human experience. If we can know what it was like for them—if we can sense the men inside the space suits—then *we* can look back and see what really began on that July night twenty-five years ago: We touched the face of another world, and became a people without limits.

Andrew Chaikin
December 1993

Acknowledgments

This book would not have been possible without the generous assistance of the NASA history offices at the Johnson Space Center and at NASA Headquarters. My debt to Janet Kovacevich and her staff in Houston is immeasurable. In Washington, Lee Saegesser and his colleagues were tirelessly helpful. Nor can I give enough thanks to Diana Ryan and Peter Nubile in JSC's sound department, who provided me with the audiotapes I used in my research, and who were as valuable to me as the astronauts themselves. For Apollo photographs, Mike Gentry and his staff were always helpful, always professional.

Outside NASA, many friends and colleagues made essential contributions. Mark Washburn was a voice of experience and encouragement. Tony Reichhardt sustained me during the darkest days of the project with thoughtful readings of the manuscript and many helpful discussions. Don Wilhelms guided me through the intricacies of Apollo lunar science and lent his peerless expertise to the reading of several chapters. Eric Jones, Apollo historian and kindred spirit, was an invaluable sounding board in the final year of the project. My former colleagues at *Sky & Telescope*, and especially Kelly Beatty, were most generous with the use of their offices, research materials, and laser printer, throughout the project. I thank Richard Maurer for his inspiration and for his perspectives on Apollo as history, a key contribution to this book.

Gregg Linebaugh provided videotapes of Apollo television transmissions. At Brown University's planetary data center, Deborah Glavin was generous in her support. Video Vision Associates supplied a laser disc of Apollo mission photography. Judy Mintz of XyQuest introduced me to the word-processing joys of XyWrite. Marcia Bartusiak, Marsha Cohen, David Cooper, Douglas Dinsmoor, Tom Finn, Steve O'Meara, and Donna Donovan O'Meara were sources of inspiration and insight. Kelly Beatty, Hank Bonney, Chip Cohen, Rick Friedman, Holly Hanson, Richard Maurer, Timothy McCall, Mark Washburn, and Frank White provided helpful comments on draft chapters. Lisa Clark was unfailing in her advice and encouragement.

At Viking, Dan Frank saw the potential of this project and gave me the chance to make it a reality. Al Silverman was a steady and supportive helmsman in bringing the book to completion. I owe my greatest debt to Connie Roosevelt: quite simply the best editor anyone could ask for. This book is what it is because of her wisdom, her enthusiasm, and her patience.

Of course, I am grateful to the astronauts, and everyone whom I interviewed, for their time and recollections. Many of them read draft chapters and gave helpful feedback. Most of all, they allowed me to tell their stories.

Any author, especially a first-time one, finds the task of writing a book to be bigger than he or she expected. By the time this one was finished my family and friends—who had tolerated my ups and downs and, hardest of all, cajoled me away from the keyboard from time to time—were as glad to see the ordeal finished as I was. To them, I give my heartfelt thanks: for their patience, their support, and their love.

Prologue

Wednesday, September 12, 1962

Morning
Rice University, Houston, Texas

John Kennedy stood at the podium at Rice University stadium in the heat of a Texas sun. He had come to Rice, on the outskirts of Houston, to help dedicate the National Aeronautics and Space Administration's new Manned Spacecraft Center, 22 miles away. NASA's burgeoning facilities were proof that America's space program was being transformed from an experimental venture into a bold reach a quarter of a million miles across the void. And it was happening because Kennedy had said it should.

Nearly sixteen months earlier, in May 1961, the young president had stood before a joint session of Congress, and in the calmest of words, had stated the unthinkable: "I believe this nation should commit itself, before this decade is out, to landing a man on the moon and returning him safely to the earth." Whatever Kennedy's words lacked in emotion, their impact was immediate. Human beings had barely taken their first toddling steps off the planet. NASA had just lofted Alan Shepard on a fifteen-minute suborbital flight. Now Kennedy was giving them less than nine years to get to the moon. But the stakes were high: Space was the new battleground of the Cold War, and the Soviet Union was winning. In April the Soviets had sent the first man into space; for Kennedy that embarrassment was compounded a week later by the Bay of Pigs fiasco. Suddenly nothing mattered more than beating the Soviets in space, and Kennedy's advisers told him only a race to land on the moon offered the U.S. a chance of victory. Within days after Shepard's flight, he had made his decision.

In the past sixteen months, as Kennedy's vision had materialized, so had the clarity of his purpose. To those who questioned this audacious venture, he would now give his answer.

"Why choose this as our goal?" Kennedy asked his audience. "And they may well ask, Why climb the highest mountain? Why, thirty-five years ago, fly the Atlantic?"

"Why," he added without missing a beat, "does Rice play Texas?"

The crowd sat quietly in the heat, fanning themselves, mopping their brows, as Kennedy spelled out the technological hurdles that would have to be cleared to build the Apollo spacecraft and its Saturn V booster. With no hint of unease he laid out the staggering costs of the venture: already the space budget had increased to forty cents per person per week for every man, woman, and child in the United States—more than the allocations of the previous eight years combined—and soon it would be more than fifty cents. The effort would spawn new jobs, new knowledge, new technology, Kennedy said, but ultimately, the first voyages to another world would be "in some measure an act of faith and vision, for we do not know what benefits await us." And if the road to the moon seemed long in 1962, there were men and women in this audience who would dedicate themselves to traveling it, and Kennedy's words were fuel for their young, ambitious hearts.

"We choose to go to the moon! We choose to go to the moon in this decade and do the other things—*not* because they are easy, but because they are *hard.*" His voice rang with energy and confidence; his words soared above the sound of applause. "Because that goal will serve to organize and measure the best of our abilities and skills, because that challenge is one that we are willing to accept, one we are unwilling to postpone, and one which we intend to win . . ."

Afternoon
Miramar Naval Air Station, San Diego, California

The howl of jet engines enveloped Pete Conrad as he climbed out of his Phantom supersonic fighter, the energy of the flight still inside him. It was past four in the afternoon, and Conrad planned to change out of his flight gear and head straight home. But when he reached the ready room there was a message waiting for him from NASA's Deke Slayton in Houston. Conrad wouldn't let himself believe the thought that flashed through his mind, that Slayton's call would change his life.

At age thirty-two, navy lieutenant Charles "Pete" Conrad had already made a name for himself as one of the most gifted test pilots in the country. He'd come to Miramar from the navy's Flight Test School at Patuxent River, Maryland, where he was known as a superb stick-and-rudder man and a fine instructor. Recently, he'd requalified for night carrier landings, the hairiest, most demanding flying in the navy or anywhere else, for his money. It was carrier flying that prompted navy fliers to call themselves "aviators" instead of just pilots. Now Conrad was cleared for combat operations; soon he and the rest of Fighter Squadron 96 would deploy with the aircraft carrier *Ranger* on a training mission in the Pacific.

Since joining the squadron here at Miramar, Conrad had begun to think about leaving the navy. Mostly it came down to money. He had four boys to put through college someday, and the pay for test pilots was better in industry. Back in June, he'd sent inquiring letters to several aerospace companies, but in 1962 there were few industry jobs for test pilots. His only offer came from North American Aviation in Columbus, Ohio, where they were building the new A-3J Vigilante attack plane. Conrad thought hard about the job and decided it was too risky. If he went to Columbus and the Vigilante program didn't come through, he'd end up in pilot purgatory, flying a twin Beechcraft for some corporate president for the rest of his days. In the end, he'd figured, the navy wasn't such a bad deal after all. For one thing, there was job security; and the flying—that was wonderful. Nothing in life mattered more to Conrad than flying.

If anyone had lived the made-to-order childhood of a fighter pilot, Pete Conrad had. At the age of five his father took him to a country fair in Ambler, Pennsylvania, where someone was selling rides in a small Waco cabin plane. The boy badgered his father into paying for a ride, then scrambled into the front seat ahead of everyone else. When the old Waco climbed into the blue Conrad felt as if he belonged there. From then on, flying became his obsession. Growing up in the Main Line suburbs of Philadelphia, he would sit for hours in cockpits he made from grocery crates and upended chairs, pretending to be Lindbergh flying the Atlantic, or Eddie Rickenbacker in a World War I dogfight. As a teenager he hung around at the local airport, doing odd jobs in exchange for flying lessons—a little too often, in fact, for his parents, who shipped him off to boarding school for a year to keep him out of the sky. But he returned in time to earn his private pilot's license on his seventeenth birthday. As an undergraduate at Princeton, Conrad entered the navy with an eye toward becoming a naval aviator. After graduation he had started down the road that brought him here, to the ranks of the top military pilots in the country. But that wasn't as high as a pilot could go, not any more, and therein lay the one disappointment in Conrad's past.

In 1959, Conrad had been one of sixty-nine young fliers who received secret orders to report to Washington, where they were told about Project Mercury, the country's high-priority effort to put a man in space. Thirty-seven of the pilots decided not to volunteer. The remaining thirty-two, including Conrad, went on to the Lovelace Clinic in Albuquerque, New Mexico, for a week of exhaustive medical tests. Conrad ran the treadmill. He breathed into a bag while pedaling a stationary bike. His body was subjected to every conceivable invasion and indignity. They measured everything from his IQ to the length of his lower intestine. Conrad began to think NASA was looking for superb medical specimens even more than piloting skills. And his body did not fail him.

Having run the doctors' gauntlet, Conrad moved on to the psychologists, at Wright-Patterson Air Force Base in Dayton, Ohio. For them he stared at inkblots and pictures and made up stories about them. Looking back, he came to realize it was the psychologists who did him in; he didn't take all their hocus-pocus seriously enough. When they showed him a blank card, he studied it for a moment and deadpanned, "It's upside down."

Then came endurance tests of all sorts. He was locked in a dark room, baked in a heat chamber, frozen, shaken, and whirled. When it was all over, Conrad had made it to the final cut, along with an old classmate from Pax River, Wally Schirra. Finally NASA announced that seven men had been chosen as the nation's first astronauts. Wally Schirra was one of them; Pete Conrad was not. Maybe it was just as well. During the selection process Schirra and some of the other navy pilots had gotten to talking, and they had doubts about the whole astronaut program. The stated objective of Project Mercury was to demonstrate that human beings could survive in space. The astronauts would be America's—and perhaps the world's—first space travelers. But would they really be pilots, or just passengers? And what about after the project was over? They'd go back to the navy, and who could say what effect the whole episode would have on their flying careers? Conrad, for one, wasn't sure that having flown around the earth a couple of times would do him much good. He returned to Pax and put the whole astronaut business behind him.

In the years that followed he watched from the sidelines as the Mercury astronauts soared into the limelight and then into space. If he had any lingering doubts about the program, they vanished on February 20, 1962, the day John Glenn became the first American in orbit. While Glenn circled the earth once every ninety minutes a hundred miles up, higher and faster than any jet pilot, Conrad was flying his Phantom over the Pacific. When Glenn splashed down there was an announcement over the navy radio, and Conrad, in his supersonic jet, felt a sharp pang of envy.

By that time the astronaut program was headed places Conrad never dreamed of. NASA was going to the moon. By the spring of 1962 NASA put out a call for more astronauts; this time applications were to be made on a volunteer basis. The new selection amounted to a filling of the roster for the most challenging and historic test flights of all time. However slim his chances might be, Conrad couldn't pass up the opportunity to apply. At Brooks Air Force Base in San Antonio, Conrad was pleasantly surprised to find the psychologists much less in evidence; they seemed to have been replaced by the FBI, if the number of background checks was any indication. Still, when it was all over Conrad headed back to Miramar fully expecting to be rejected a second time. The moon would be for John Glenn and Wally Schirra, but not for him.

The ready room—a large space with enough well-worn chairs for the entire squadron of thirty during a briefing—was nearly deserted. One wall was all windows, looking out on the flight line. At the opposite wall a couple of pilots lingered by a row of gray equipment lockers. The duty officer was at his desk. Everyone knew about the impending astronaut selection, and when Conrad came in they excitedly told him about the phone call from Slayton. He thought it was a joke—until he held the message in his hand. Conrad sat on the edge of the duty officer's desk and borrowed the phone to call Houston. Slayton came on the line, and the two men exchanged some brief pleasantries. Then, in the quiet, understated tone that fighter pilots often use, Slayton asked Conrad the question he would remember for the rest of his life:

"How would you like to come fly for us?"

Conrad almost fell over. He was not to tell anyone, Slayton instructed him; NASA wanted everything kept quiet until the official announcement the following week. He was to report to Houston on October 1. Conrad thanked Slayton and hung up, doing his best to hide his elation. He said nothing to the men in the ready room; he did not phone his wife. He changed into his khakis, got in his car, and started the 18-mile drive home. In his excitement he almost wrecked the car.

On that drive, it hit him for the first time that he was going to have a chance to go to the moon. He thought about the others who had been selected. He didn't know who they were, but he found himself wondering: If they really did this thing, how would it affect them? The Mercury pilots walked in the white-hot glare of celebrity even before they flew. When they splashed down they rode into a shower of ticker tape; they mingled with the Kennedys. What would it be like for the ones who came back from the moon? Conrad drove on, thinking about the incredible path his life had just taken. Back when he was at Pax River, a woman who lived across the street started calling him "Moon Man." And that was before there even was an astronaut program. Back then he was just a test pilot who loved to fly. Now he was aiming for the moon, after all. And he promised himself, *It's not going to change me.*

BOOK ONE

CHAPTER 1

"Fire in the Cockpit!"

When the moon rises beyond the Atlantic shore of Florida, full and luminous, it seems so close that you could just row out to the end of the water and touch it. In January 1967, the moon seemed to draw nearer by the day to the hard, flat beaches of Cape Kennedy. Seen from there it was no longer the governess of the tides, the lovers' beacon, the celebrated mistress of song; it was a target, a Cold War beachhead in the sky. It was NASA's moon.

Almost six years had passed since John Kennedy's challenge for a lunar landing by decade's end. The moon program had grown into an effort whose size and complexity dwarfed even the Manhattan Project. At aerospace contractors around the country, 400,000 people were hard at work on the moonships for Project Apollo. Meanwhile, Project Gemini had just come to a spectacular finale. These two-manned missions had bridged the gap between the pioneering Mercury flights and the challenge of the lunar landing. For the first time in the race to the moon, the United States appeared to have pulled ahead of the Soviet Union. With the first manned, earth-orbit Apollo flight scheduled for mid-February, all seemed on target to make Kennedy's vision a reality. But one evening late in January, that soaring optimism suddenly, terribly fell to earth.

January 27, 1967
Merritt Island Launch Area, Kennedy Space Center
Cape Kennedy, Florida

"How are we going to get to the moon if we can't talk between three buildings?"

Gus Grissom's voice was low and calm, but with an unmistakable edge of irritation. A senior astronaut, veteran of two space missions, Grissom was the commander of the first manned Apollo flight set for February 1967. On this warm January afternoon Grissom and his crew, veteran astronaut Ed White and a rookie named Roger Chaffee, were participating in a simulated countdown, the kind of routine test that preceded every mission. They were sealed inside the cone-shaped Apollo 1 command module, high atop a huge Saturn 1B booster rocket at Pad 34, one of dozens of launch complexes that lined the beach at Cape Kennedy. A few hundred yards away, inside the concrete bunker called the Saturn blockhouse, some two hundred members of the launch team heard Grissom's words. At the Capsule Communicator, or "Stony," console, a young rookie astronaut named Stuart Roosa tried in vain to answer.

"Apollo 1, this is Stony; how do you read?"

In 1967 there were so many astronauts that Grissom and Roosa hardly knew each other, but the younger man could hear the barely contained exasperation in Grissom's voice:

"I can't hear a thing you're saying. Jesus Christ . . . I said, how are we going to get to the moon if we can't talk between two or three buildings?"

No one who knew Gus Grissom took him lightly. Small and powerful, he was known as a fierce competitor. When Grissom was a young air force fighter pilot in the Korean War, the fliers would ride an old school bus from the hangar to the flight line. Only those who had been in air-to-air combat could sit down; the uninitiated pilots had to stand. Grissom stood only once. He brought the same hard-driving determination to his spaceflight career, as one of the Mercury astronauts, known as the Original 7. Even among these most elite fliers, the Seven had a status approaching royalty. Despite their rivalries they had a unique bond that came from being the first Americans to venture into the heavens. Everything the Original 7 did was energized with competition, from flying airplanes to their impromptu drag races on straight Florida roads to their adventures in the nightspots of Cocoa Beach, and Grissom was always a zealous participant. But even to other astronauts, Grissom was not an easy man to know; he was a loner among loners.

In 1967 Gus Grissom stood at the top of the active roster of astronauts, but he had started out as something of an underdog. His first spaceflight, a suborbital Mercury mission in 1961, ended in near-disaster when the hatch of his tiny spacecraft *Liberty Bell* 7 blew off prematurely after splashdown in the Atlantic. *Liberty Bell* sank; Grissom narrowly escaped drowning. He maintained it had been a malfunction, that the hatch had blown off by itself, but somehow there had been a lingering skepticism—in the press, at NASA, even among other astronauts. The doubts infuriated him. He fought to make up for that image. In 1965, after helping to design the two-man Gemini spacecraft, he commanded its successful first flight. Now, at age forty, after more than a year immersed in the development and testing of the Apollo command module, he would fly its maiden voyage in earth orbit. Commanding the first flight of any new craft was always a prize assignment, but Grissom's ambitions didn't end there. He had his sights on the most coveted mission of all, the lunar landing. More than anything, Grissom wanted to be the first man on the moon.

But on January 27, the moon seemed a long way off. For several months now, Grissom had worked to help ready his spacecraft for its mission, and he had become more and more displeased with the way things were going. When engineering problems came up, he pushed for better solutions using his experience from Mercury and Gemini, but no one seemed to be listening. It made him so mad, he confided to an interviewer, he couldn't see straight. By the time the Apollo 1 command module left the factory in Downey, California, last August, it still had dozens of separate discrepancies, some of them serious. To make matters worse, the command module simulator here at the Cape was a constant source of difficulty. Just a few days ago, to show his frustration, he'd hung a big Texas lemon on it. Today, his patience was being strained once again by a trouble-plagued test. Despite all the frustrations, Grissom was pushing to get Apollo 1 into space on schedule, not because he was reckless—Grissom hadn't lived this long by being reckless—but because problems were to be expected in any new flight program, whether a new airplane or, especially, a moonship.

Today's simulated countdown was nothing new; it wasn't considered dangerous—the Saturn booster was not fueled—or even difficult. But there was trouble almost from the time Grissom and his crew climbed into the command module cabin, around 1 P.M. First there was an unidentified odor in the breathing oxygen that reminded Grissom of sour milk; that alone had held up the test for an hour. Finally the problem was solved, and at 2:45 P.M. the pad crews installed the command module's heavy, two-piece hatch and sealed it shut. The spacecraft was pressurized with pure oxygen, just as it would be on launch day. Then came the communications trouble. By late afternoon Roosa was able to converse with the men inside the sealed spacecraft, but there were problems with the voice link to the Manned Spaceflight Operations building 5½ miles away. In the blockhouse, controllers weighed the decision of whether to abort the test. They decided to continue.

In the blockhouse, seated next to Roosa at the Stony console, Deke Slayton listened as technicians tried to fix the faulty communications. Slayton was forty-three years old. He had been a civilian for several years now, but he still carried himself with the quiet, serious demeanor he'd had as a young air force fighter pilot. Another member of the Original 7, Slayton was one of Gus Grissom's best friends, but their fortunes could not have been more different. Grissom was now a veteran of two space missions, while Slayton had been grounded since 1962 for a minor heart irregularity and was still waiting for his first chance to fly in space. Now, as chief of the Manned Spacecraft Center's Flight Crew Operations Directorate, Slayton's role included following astronauts through tests like this one.

Slayton knew only too well that Grissom wasn't happy with the way things were going. He'd had breakfast with Grissom, White, and Chaffee this morning, along with Joe Shea, the hard-driving NASA manager in charge of the command module effort, and they were all running through the litany of troubles with Grissom's spacecraft: Malfunctions in the environmental control system. Coolant leaks. Faulty wiring. Grissom bemoaned the communications trouble that plagued almost every test. "If you don't believe it," Grissom told Shea, "you ought to get in there with us."

Shea declined, largely because there was some question about whether technicians would be able to rig up an extra communications headset for him in time. But Slayton gave the matter some thought. There would be space for him to sit, in shirtsleeves, in the command module's lower equipment bay beneath the footrests. He was still weighing the idea around midday as Grissom, White, and Chaffee suited up. Slayton rode with them in the transfer van to Pad 34, and by the time they arrived he had made his decision. He would be better off in the blockhouse, he told Grissom, where he could keep a close eye on the test. Years later, he would still wonder whether or not he made the right choice.

By evening, as dusk settled onto the marshlands, technicians continued to troubleshoot the faulty communications. Searchlights came on, bathing the giant Saturn rocket in white light. Meanwhile other events in the countdown continued. Apollo 1 was on its own electrical power now, just as it would be in the final minutes before liftoff. But the communications troubles were still unsolved, and at T minus 10 minutes, the test director called a hold in the count.

At 6:31 P.M., eleven minutes into the hold, Slayton was looking over the test schedule when he heard a brief, clipped transmission from Apollo 1. It sounded like "Fire."

• • •

On the other side of the country, at the North American Aviation plant in Downey, California, veteran astronauts Tom Stafford, John Young, and Gene Cernan were sealed inside the second Apollo command module in their own spacecraft test. Besides being the most experienced space crew yet assembled—Cernan had flown once before on Gemini, while Stafford and Young each had two missions under their belts—they had a camaraderie seldom matched on a space crew. Now the three men were assigned as the backup team for the second manned Apollo mission.

This command module, like Grissom's, was a prototype called Block I. It was never built to go to the moon, but was instead designed only to fly in earth orbit, on the first Apollo flights. An improved version, called Block II, was already being developed for the lunar missions, and last month, NASA had decided that there would be only one Block I mission, Apollo 1. Stafford and his crew were here at North American, the prime contractor for the command module, to help provide engineering support for the activities at the Cape.

The work on Block I had been frustrating, in part because of the atmosphere here at Downey. North American had some very competent people, but they had never built a manned spacecraft before. Bundles of wire on the command module floor were unprotected, making them susceptible to damage. There were so many changes made within the cabin that the workmen could barely keep track of them all. Wires were constantly being rerouted, black boxes replaced. And lately, astronauts had felt a strain in their relationship with the North American engineers, who had begun to resist their suggestions.

It had been different during Gemini. That spacecraft, like Mercury, was the product of the McDonnell Aircraft Corporation in St. Louis, which had forged a harmonious relationship with NASA. The astronauts who visited the factory had no trouble making inputs into the design process. And if they had a problem, they could always take it to "Mr. Mac" himself, and he would get results. But Apollo wasn't like Gemini. The intimacy was gone now. There was no single boss who could respond to the astronauts' concerns; now there were bosses scattered throughout this massive operation.

Even within NASA, there was a disturbing lack of coordination. In Houston, some of the engineers in the Apollo program office acted as if Gemini had never existed. They had an arrogance that seemed to say, *We know this business better than you do.* Experience from Gemini—Stafford's or anyone else's—seemed to make no impression on them. They rolled their eyes at "those Apollo astronauts and their Gemini war stories." Trying to get through to them was like talking to the wall.

All of them knew of the trouble with Block I, but in the back of everyone's mind was the end-of-the-decade deadline for the lunar landing. However dissatisfied they were, Stafford and the other astronauts had been willing to put up with Block I for the first couple of flights in order to stay on schedule. Their feeling was, "Just get us airborne; we'll fly it."

But on this day, Stafford wasn't so sure. Their test, too, was in trouble: Leaking coolant lines. Short-circuits. At one point the hatch fell on Cernan's foot.

"Go to the moon?" Stafford growled, "This son of a bitch won't even make it into earth orbit."

Finally Stafford stopped the test. When he climbed out of the spacecraft, an emergency call from the Cape was waiting for him.

• • •

When Deke Slayton heard the report of fire from Apollo 1, his gaze shot to a nearby closed-circuit television monitor. It showed the picture from a camera pointed at the command module's hatch window. The window was filled with bright flame.

Suddenly there was another message from the spacecraft, this time quite clear, in a voice of contained urgency: "We've got a fire in the cockpit!"

Slayton recognized the voice as Roger Chaffee's. Chaffee was on the right side of the spacecraft, where the radio controls were. It was his job, in an emergency, to maintain contact with the blockhouse.

On the television monitor Slayton could see Ed White's arms reaching back over his head, trying to undo the bolts that held the side hatch shut. Neither Slayton nor anyone else in the blockhouse fully understood what was happening; Slayton would say later that his main concern was not fire, but smoke. But now Slayton heard another voice, clearly frantic:

"We've got a bad fire. . . . We're burning up!"

At first Slayton thought it was Pad Leader Don Babbitt, stationed next to the spacecraft, calling for help. Later, on the tapes, listeners identified the voice as Chaffee's. Seconds later—less than half a minute after the first report of fire—Slayton and the horrified controllers heard the last transmission from Apollo 1. It was a brief cry of pain.

Long seconds passed. Now the communications loop surged with activity as technicians struggled to get the hatch open. "It's too hot," you could hear them say. On the television monitor, through dense smoke, Slayton could see the pad crews approach the hatch only to be driven back by the intense heat. Roosa tried several times to reestablish contact with the crew, with no response. Several long minutes elapsed before the hatch was finally opened, and a short time after that the pad leader came on the communications loop with a terse and ominous transmission: "I'd better not describe what I see."

Physicians Fred Kelly and Alan Harter were in the blockhouse, and Slayton instructed them to go to the pad. Then he put in a call to the Manned Spacecraft Center in Houston, to set up a command center and get word to the families in case things were as bad as he feared. After several minutes the call came from the doctors, confirming what everyone had dreaded. Slayton made another call to Houston; then he and Roosa left the blockhouse and headed to Pad 34.

• • •

It was a slow Friday afternoon in the Astronaut Office at the Manned Spacecraft Center. Here, twenty-five miles from downtown Houston, built on a flat, coastal prairie that had once been owned by a Texas oilman, was a collection of modern buildings, all dark glass and white stone, set among neatly manicured lawns and artificial duck ponds. Here and there a windowless training facility or a laboratory with massive cryogenic storage tanks, hinted at the true nature of this place, but for the most part it looked more like the campus of a community college than a place where engineers, scientists, and astronauts were waging an all-out assault on the moon.

The Astronaut Office was located on the top floor of the three-story structure called Building 4. On this Friday afternoon it was nearly deserted. Most of the astronauts were out of town, chasing down some piece of the Apollo effort at contractors' plants around the country. But a young rookie named Alan Bean was here. Although he had been an astronaut since 1963—like Roger Chaffee, he was a member of the third astronaut group—Bean was still waiting for his first chance to fly in space. A few months ago, Deke Slayton had named him as the Astronaut Office representative on the manned space station project that was planned for the early 1970s. Bean didn't always hear about the details of what was going on in Apollo, but he knew Grissom, White, and Chaffee were at the Cape, getting ready for their mission.

Sometime before 6 P.M. the phone rang. It was one of the support people at the Cape. What he said sounded so strange that Bean did not at first understand:

"We've lost the crew."

The man's voice was quiet. Bean heard no anguish in it. He had to stop and think about the words. "The crew" was surely Grissom's; were the people at the Cape having trouble finding them? *Lost the crew?*

Bean answered, "Where do you think they've gone?"

The voice stumbled over more words that didn't make sense; he just didn't seem to want to tell Bean what had really happened. It took a long time for him to say it: Grissom, White, and Chaffee were dead.

Bean had barely hung up the phone when it rang again; this time it was Mike Collins, in Deke Slayton's office up in the administration building. Collins, acting as the Astronaut Office representative for the Friday afternoon staff meeting, had just heard the same news. Now there was no miscommunication, for it was one pilot talking to another and both knew what needed to be done. They agreed that Bean would coordinate astronauts and wives to go to the homes of the dead pilots. Bean called his wife, Sue, and sent her to the home of Martha Chaffee until Collins could get there. Wally Schirra's wife, Jo, and Chuck Berry, the space center's chief physician, would go to Betty Grissom's. Neil Armstrong's wife, Jan, would go to her next-door neighbor Pat White's, and Bill Anders, another member of the third group, and like Bean, a rookie, would follow.

* * *

El Lago was one of a handful of planned communities that had sprung up around the space center, scattered with ranch houses and crisscrossed with tidy streets winding through the greenery. Here was the space community's own suburb: astronauts and NASA engineers and managers all lived next door to each other. And aside from the tourists who occasionally came by looking for some sign of an astronaut as if they were on a tour bus in Beverly Hills, and the mobs of reporters who, during missions, stood watch on an astronaut's lawn as if the Fischer quintuplets had just been born inside, there was nothing remarkable about the way it felt to live here. That was precisely the point: the residents of El Lago, and of nearby Nassau Bay, and Timber Cove, and Clear Lake, clung to normalcy in the midst of the most extraordinary enterprise of the twentieth century.

It was already dark when Bill Anders arrived at the ranch house belonging to Ed and Pat White. Normally there wouldn't have been any reason for Anders to know White very well; the two had never been on a crew together, and they belonged to different astronaut groups, White a Gemini veteran from the second group, Anders a rookie from the third. Anders and his wife, Valerie, had gotten to know Ed and Pat White mostly because they lived one street apart.

Even among the astronauts, Ed White had always stood out; a strapping six-footer who had barely missed becoming an Olympic hurdler, he was known as one of the finest physical specimens in the Astronaut Office. And perhaps more than any astronaut except John Glenn, White subscribed to their all-American image. In 1965, after he became the first American to walk in space, White easily wore the mantle of a national hero. There appeared to be no limit to how far he might go. Ed and Pat seemed perfectly matched. Few women were so devoted to their husbands. Now, Bill Anders would have to tell Pat White that her husband was dead.

Bill Anders had lived with death for most of his adult life, first as an air force fighter pilot, then as an astronaut. He had been to his share of funerals. And it was at funerals that Anders had noticed something about himself that made him feel different from other people. He did not shed tears. While he felt sadness for the family, he did not grieve for the man who was killed. When he'd decided to become a fighter pilot, he'd accepted deadly risk as part of the bargain. And he knew the dead pilot would have felt the same way: "Sure, it's a shame to lose a good man—but he knew what he was getting into. He flipped a coin, and he lost."

By 1967 death had become a part of the astronaut life. Just last year Charlie Bassett and Elliot See had been killed in a plane crash, and two years before that, it was Ted Freeman who perished the same way. And none of this had made Anders hesitate to climb into the cockpit of a supersonic jet fighter, or trust his life to a pressure suit in the vacuum of a test chamber. It didn't blunt his desire to ride a moon rocket. Like all the astronauts, Anders accepted the risks of the job, and it wasn't difficult for him to do that. What was difficult—the hardest thing Anders would ever do—was to go to the house of this attractive woman in her thirties who was raising two children, bearing bad news. A short while ago Pat had picked up her daughter from a ballet lesson. When she arrived home her neighbor Jan Armstrong was waiting silently for her. She must have been surprised, then confused—after all, Ed was at the Cape, he wasn't flying tonight—and then she must have filled with dread. But it wasn't up to one astronaut wife to tell another that her darkest nightmare had come true. That task most often fell to another astronaut. Anders rang the doorbell.

• • •

When Slayton and Roosa arrived at Pad 34 ambulances waited in vain at the base of the huge launch tower, their rotating beacons flashing in the night. Entering the steel gridwork of the gantry, the two men boarded a small elevator and rode it to Adjustable Level A-8, 218 feet up, and headed across the swing arm to the small enclosure at the other end, called the White Room. Even before they arrived they were assaulted by the stench of burned electrical insulation and incinerated plastics. At last, reaching the end of the swing arm, they could see it: the command module's square hatchway, flush against the side of the White Room. Something was hanging from the open hatch; it was an arm, clad in a white spacesuit. Kelly and Harter, the two doctors, were here; they told Slayton the dead men were still inside, snared in a web of melted nylon netting that had once hung within the cabin. The doctors had been unable to remove them.

Slayton gripped the rail just above the hatchway and leaned into the cabin of Apollo 1. He could see the familiar configuration of the command module cabin: three couches, side by side; the broad center instrument panel, amid a forest of switches, knobs, and controls. Warning lights still glowed amber on the blackened panel. Much of the once spotless cabin was covered with soot. On the right side of the cabin Slayton could see Chaffee, his space-suited form motionless in his couch, still strapped in. The other two couches were empty. Slayton looked down, below the edge of the hatchway, and spotted two helmeted heads, both with clear face-plates still closed. Right below the hatch were a pair of legs, doubled up, from which the layers of space suit material had been burned off. It was impossible tell who was Grissom and who was White.

Slayton told Kelly and Harter not to do anything more until photographs could be taken. Then he and Roosa turned away from the blackened cabin of Apollo 1, carrying the smell of fire and death with them. For the rest of their lives, even after the terrible images had faded in memory, they would remember that smell.

• • •

The astronauts had always known it was only a matter of time. Gemini had had its share of close calls, none worse than when Gemini 8 began tumbling out of control: Neil Armstrong and Dave Scott had narrowly escaped with their lives. And the Gemini pilots had taken some calculated risks. For example, everyone knew that the Gemini ejection seat and parachute that served as the only means of escape in a launch emergency was effective only under very limited conditions. And there were other risks, not only on Gemini but on the comparatively primitive Mercury flights. Looking back, it was not only superb hardware and outstanding people, in space and on the ground, that had averted tragedy, it was also luck. But the lunar missions were even more complex, and among the astronauts there was an unspoken feeling that it was only a matter of time before their luck ran out.

Gus Grissom had known that. Sometime during Gemini, he

had told his wife, "If there's ever a serious accident in the program, it's probably going to be me." And a few weeks ago, at a press conference, he had said, "If we die, we want people to accept it. We're in a risky business. . . ." But he was talking about dying in space. If Grissom, White, and Chaffee had burned up in reentry, if they had perished in the fireball of an exploding booster, if their parachutes hadn't opened and they had plummeted into the ocean—any of those fates would have been easier to accept. The terrible shock of this January night, and the irony, was that they died while their spacecraft was sitting on the pad, with technicians all around them, safety just on the other side of the hatch. And yet, no one had been able to save them.

What went wrong? Even years after investigators began to sift through the wreckage of Apollo 1 piece by piece, no one could say exactly. But within weeks, the general picture became clear; the fire was a disaster waiting to happen. During the test, the command module was pressurized with pure oxygen at 16.7 pounds per square inch, slightly above sea-level atmospheric pressure. Pure oxygen can be a fire hazard even at low pressure, but at 16.7 psi. the danger grows to frightening proportions. And yet, the practice of pressurizing with oxygen on the pad had been used dozens of times during Mercury and Gemini, without mishap.

NASA had chosen pure oxygen for Apollo for the same reasons it had been used in Mercury and Gemini: it eliminated the weight and complexity that would have been required for an oxygen/nitrogen mixture. Pure oxygen was essential while in orbit, when the cabin pressure was only 5 psi. Somehow, no one had absorbed the realization that on the pad, the pressure was kept slightly above sea-level conditions. And if oxygen carried a fire risk, the command module's designers thought they had removed all possible sources of ignition from the command module cabin.

But they were wrong. Shortly after 6:31 P.M. on January 27, the review board concluded, there was a spark inside Apollo 1, probably in the vicinity of some damaged wires in the lower equipment bay at the foot of Grissom's couch. Perhaps aided by flammable fumes leaking from a nearby coolant pipe, the spark ignited some nylon netting that had been installed underneath the couches to catch dropped equipment, and the fire spread quickly. Other flammable items, including foam pads that were there to protect the interior finish during the test, fueled the blaze. Even materials normally considered flame-resistant burned as if they had been doused in kerosene. On the walls Velcro fasteners, a favorite means of securing loose gear in weightlessness, exploded in a shower of fireballs. In seconds, as the temperature soared to 2,500 degrees Fahrenheit, the command module became an incinerator, and Grissom, White, and Chaffee never had a hope of escape—because the hatch had become impossible to open.

The command module's side hatch was one of the inevitable design compromises. It was a two-piece affair, with an outer hatch and an inner hatch that opened inward, into the cabin. Some NASA engineers and astronauts, and even engineers at North American, had questioned the design, calling for a one-piece hatch that could be swung open, like the one used in Gemini. But the Apollo managers, and Joe Shea in particular, always had sound reasons for vetoing the change. Mostly it was a matter of weight: Each pound of payload cost many times its own weight in propellant to haul it off the surface of the earth and send it to the moon. The two-piece hatch was not only the most lightweight design, it was also the simplest. And to anyone concerned about air leaks during a two-week trip to the moon and back, an inward-opening hatch solved the problem: cabin pressure would keep it tightly sealed. But even under the best conditions it was very difficult to open. The inner hatch was a heavy, cumbersome metal plate secured by a set of bolts. The man in the center couch had to reach back over his head, undo the bolts using a special tool, and then lower it out of the way. Ed White had been in Apollo 1's center couch, and no astronaut surpassed him for sheer physical strength. For exercise, White and his backup, Dave Scott, used to practice opening the hatch; it was like pressing a couple of hundred pounds at the gym. But only seconds after the fire started neither Ed White nor any other human being would have been strong enough to open it. As the fire progressed the buildup of hot gases sealed the hatch shut with thousands of pounds of force. As it was, White never had a chance even to undo the bolts.

The blaze would undoubtedly have consumed the three men had it continued, but within fifteen seconds after the first report of fire, the pressure in the cabin soared to nearly twice sea-level atmospheric pressure—high enough to rupture the command module's hull. Hot gases rushed through the breach with a loud whoosh that startled the pad crews, leaving the cabin enveloped in thick smoke. But by then, the horror of being trapped in an inferno was nearly over for Grissom's crew. Seconds later their oxygen hoses burned through and carbon monoxide forced its way into their space suits. Fifteen to 30 seconds after that, the medical examiners estimated, the men lost consciousness. Within four minutes there was no hope of reviving them. Grissom, White, and Chaffee did not burn to death; they were asphyxiated.

The greatest irony was that Gus Grissom, who had almost drowned after his Mercury mission because of a hatch that opened prematurely, was claimed by a hatch that could not be opened at all.

Even the astronauts who did not see the charred spacecraft or smell its acrid odor in the Florida night were stunned by the news from the Cape that Friday evening. They were appalled to realize how much they had overlooked. They had talked about what they would do if a fire broke out while they were in space, how the flames might propagate in zero gravity. But none of them had even considered a fire on the pad. Like so many things about this disaster, it was almost beyond comprehension.

In hindsight, there was enough blame to go around. Some astronauts, in anger, singled out North American, saying its engineers, yielding to schedule pressure, had taken shortcuts. And as the investigation of the fire progressed, there were charges of mismanagement and shoddy workmanship. Nor was NASA without fault: All of this had happened under the agency's supervision.

But there was another, more forgiving view. True, North American wasn't like McDonnell. But no one had ever tried to assemble a moonship before. The Apollo command module was the most complex flying machine ever devised, an intricate package crammed full of state-of-the-art equipment. It would have been naive not to expect all kinds of things to go wrong the first time they put one together. And no one, at NASA or North American, had knowingly compromised the astronauts' safety.

And in the next weeks the astronauts came to admit what Stafford, Young, and Cernan did over a few drinks one night, that there was a hidden blessing in this disaster: the wreckage of Apollo 1 was there for the accident board to examine, not a silent tomb circling the earth or drifting in the translunar void. Although three men had died, three or perhaps six more lives had probably been saved.

On the cold, wet January day when they buried Grissom, White, and Chaffee, most of the astronauts put the tragedy of the fire behind them with an acceptance that was difficult for outsiders to understand. Ultimately, when Grissom had spoken of accepting death in a dangerous, critical undertaking, he had been speaking for all of them. Their biggest concern now was making Apollo fly. The immediate causes of the fire had to be fixed—and now there was a much needed chance to correct the long list of other inadequacies. It would take time, but they had to recover and move on. Before them was the most extraordinary goal of the twentieth century, and it came with a deadline.

CHAPTER 2

The Office

April 1967
Manned Spacecraft Center, Houston

Deke Slayton stood in the small conference room on the third floor of Building 4, facing the collection of astronauts he had called together. In the spring of 1967, NASA was moving on from the tragedy that would always be known as "the Fire." Amid lingering criticism of the agency, the accident review board had made its report to Congress. At the North American plant in California, a redesigned, fireproof Block II command module was under development, and now there was an astronaut—Gemini veteran Frank Borman—working hand in hand with new management. And here in Houston, it was time to get back to the business of making the end-of-the-decade deadline for the lunar landing. Eighteen astronauts waited for the meeting to begin.

Slayton was not one for fanfare. "The guys who are gonna fly the first lunar missions," he said, "are the guys in this room."

At first glance, there was nothing remarkable about these men. Sometimes, when they walked the halls of the space center, clad in NASA-blue flying coveralls, they were easy to spot; here, dressed mostly in sports shirts and slacks, they wore a casual anonymity. But their trim bodies hinted at lives of self-discipline, and their eyes, quick and bright, revealed an unwavering concentration born of countless adventures in the air, in a profession in which the mind simply does not wander. To the people who worked with them—engineers, simulation instructors, flight controllers—they were "just regular guys," but even they could not deny that there was something remarkable about who these men were, where they had been and, most of all, where they were going. Most of these eighteen were veterans of the two-man Gemini flights of 1965 and 1966. Those missions, as bold as they were successful, had been their training ground for the moon.

If the astronauts had been a squadron, Slayton would have been their wing commander; sometimes, in conversation with each other, they called him Father Slayton. In 1959—a lifetime ago, it seemed—there were just seven men who called themselves astronauts, and Slayton had been one of them. He was considered one of the most gifted pilots of the Original 7, but even now Slayton was still waiting for his first spaceflight. In 1962, only weeks before he was scheduled to become the second American in orbit, Slayton was abruptly grounded by the doctors. He had a condition called idiopathic paroxysmal atrial fibrillation: every so often, for unknown reasons, his heartbeat became irregular. Slayton's protests that he felt absolutely fine and that his performance was unaffected were to no avail. NASA's administrator, James Webb, ruled that the agency couldn't take the risk of sending him into space. And that was not the end of Slayton's frustration; he was banned from flying an airplane by himself.

Since then, while waging a lonely battle to get back on spaceflight status, Slayton had served as the space center's director of Flight Crew Operations, managing the affairs of every pilot at the center, including the astronauts. From his office on the ninth floor of the administration building, he oversaw just about everything the astronauts did, from training to business trips to public appearances. In the most difficult of ironies, it was Slayton who picked the crew for every mission. He would have given anything to join them, to trade his executive suite for a government-issue desk in the Astronaut Office, and they knew it.

There is no record of how the astronauts reacted to Slayton's announcement; probably they said nothing. To some, Slayton's words came as welcome reassurance; for others they were not news. For a handful—the rookies—just being in this room at all was reward enough; it meant an end to years of struggling to make it onto a space crew. But none of them, rookie or veteran, missed the significance of what Slayton said. They were finalists in an unofficial and largely unspoken competition in which the prize was the ultimate test flight, the first lunar landing. Most of them saw it the way Pete Conrad did. He knew he belonged in this room; he had known he would be here all along.

No one in the Astronaut Office was more of a natural than Pete Conrad. At the age of thirty-six, he was not only one of the most seasoned pilots in the Office, but one of the most flamboyant. He appeared to enjoy everything he did. He liked to race Formula-V cars and was given to wisecracks and practical jokes. He used foul language in the simulator. He enjoyed giving the other pilots nicknames they hated. It was Conrad, for example, who had bestowed the monicker "Shaky"—anathema for a test pilot—on his good friend Jim Lovell. For every Conrad scheme that the world knew about, there were probably three others he never pulled off. Returning to earth after Gemini 5, he planned to do a somersault across the carrier deck, just to show the world that eight days in space hadn't harmed him—but at the last moment, spotting the red carpet, the admirals, and the bands, he decided against it. Conrad was the kind of astronaut the press loved to write about.

Now Conrad had his own Apollo crew, and with characteristic enthusiasm he was ready to lead them into training as backups for an early Apollo flight in earth orbit. Eventually, he knew, his team would rotate into a prime crew assignment, and although nobody could predict what his mission might be, he had his sights on commanding a lunar landing flight—and if possible, the first one. He felt sure the other men in this room, especially the mission commanders—Wally Schirra, Jim McDivitt, Frank Borman, Tom Stafford, and Neil Armstrong—shared his ambition. But no one could predict who might make that first landing, not even Slayton, and that was part of his message on this April day. The Block II command module wouldn't be ready for its first manned flight until next year at the earliest. After that, NASA would have to clear a string of daunting technical hurdles before anyone could attempt a moon landing. Astronauts would have to check out the lunar lander and practice space rendezvous, perhaps several times, in a succession of earth-orbit missions. Then there would be test flights in lunar orbit. And then, only if everything before had gone well, the next crew would try for the landing.

But that plan could change at any time. If something went wrong on one mission, the next flight would have to make up those missed objectives. And fate could intervene without notice, just as it had with Grissom, White, and Chaffee, and a backup crew would have to step in. Slayton had put together these six crews so that no matter what happened there would be a team of competent astronauts available for any flight. As the missions progressed he would just keep rotating crews, until one of them accomplished Apollo's ultimate goal. His message this day was simple: Any crew could fly any mission. In his mind they were all equal. The real selection, he said, was made the day each of them joined the Astronaut Office. It wasn't the first time the astronauts had heard Slayton make such a statement; it would not be the last.

For some of the men in this room, like Pete Conrad, the journey to the moon had offered few obstacles so far. For the rookies, like Bill Anders, it had been laden with delay and frustration. But none of these astronauts understood precisely how they had been selected to fly these missions. And if you had asked any of them, especially the rookies, whether Slayton's statement was really true —he might as well have said that the moon doesn't go around the earth.

• • •

One morning early in 1963, six shiny Corvettes tooled through the "space suburbs" of Timber Cove and Friendswood, and converged on Interstate 45, heading for Houston. Moving onto the freeway, the six drivers encountered each other and took off. They roared past startled motorists at 100 miles an hour, passing them two at a time, one on the left, one on the right. They jockeyed for position, hunting for clear lanes, weaving through the rush-hour traffic. As quickly as they had come, the six exited the freeway and sprinted through city streets to the Farnsworth and Chambers building, where the astronauts had temporary offices. The cars pulled into the parking lot, still crackling with the heat of the race. Doors opened and out came Gus Grissom, Wally Schirra, Gordo Cooper, Deke Slayton, Jim Lovell, and Pete Conrad. The men laughed as they locked up and headed for their offices. Deke Slayton, clenching the stub of a cigar in his teeth, muttered, "Goddammit, we gotta knock this shit off. We'll get ourselves arrested."

In 1963, Pete Conrad's first full year as an astronaut, an early-morning drag race with the Original 7 was the perfect expression of the best that life had to offer. Never mind the fact that he was still a rookie waiting for his chance to fly in space; even now, he could see himself headed for the moon. Not that the moon itself held any particular fascination for Conrad. He was here because John Kennedy had directed that America was going to the moon. If Kennedy had ordered NASA to build a space station instead, he would have been just as happy to do that. But the moon was NASA's goal, and when he joined the astronaut corps the previous October, it became his. He was a member of the second astronaut group—or the New 9, as they were dubbed by *Life* magazine—and they were key players in the Apollo effort. But to the world, the only real astronauts were the Original 7.

One of Conrad's first assignments was to accompany John Glenn on visits to Apollo contractors around the country. Glenn was likable enough, and Conrad could see that as the first American to orbit the earth he had risen to a hero status no pilot had attained since Charles Lindbergh. At every airport an excited crowd would quickly surround Glenn, asking for autographs, and Glenn would smile and put down his bag and oblige them. In order to keep moving, Conrad ended up carrying Glenn's bags as well as his own. A few weeks of this were enough to shatter any illusions Conrad may have had about his status as one of the nation's astronauts.

In fact, Pete Conrad was the last person anybody would think to ask for an autograph. A wiry five feet six, his blond hair nearly gone, his features were best described as gnome-like, with mischievous blue eyes, a prominent and pointed nose, and a grin that exposed a wide gap between his two front teeth. It wasn't unusual for tourists to come through Timber Cove and the other space suburbs, and when that happened Conrad would stand out on his lawn wearing a baseball cap, scratching his balls. Getting recognized was the last thing Conrad worried about.

What mattered to every one of the Nine was flying in space as soon and as often as possible. Conrad was well aware that as a group, the Nine's credentials were even more impressive than those of the Original 7. Not only were the Nine test pilots, they had advanced degrees in aeronautics or the new field of astronautics. Some, like Frank Borman and Tom Stafford, had been instructors at the air force's test pilot school at Edwards Air Force Base. There were those, like John Young, who had set a record or two. And there was even a rocket pilot among them: Neil Armstrong, a NASA man out of Edwards, who had flown the X-15 rocket plane to the edge of space. In short, the Nine were ready and able to compete on the same field as the Seven.

But that wasn't going to be easy. Conrad had the distinct feeling that the Original 7 were not at all happy to see newcomers invading their ranks. Mercury wasn't even over yet, and already the title of astronaut was something they had to share. It came out in subtle ways—the goodies, for example. The Nine came in already knowing about the lucrative contract for the astronauts' personal stories with *Life* and Field Enterprises, which meant an extra $16,000 per year for each of the now sixteen astronauts. After years of raising a family on a military pilot's pay, Conrad was glad to hear about the money.

But there were a host of other perks the Original 7 had scored, great deals on homes, Corvettes on loan from dealers, you name it—that the Nine knew nothing about. Conrad's group had to pry information out of the Original 7. When one of the newcomers stumbled across a deal such as motel rooms for a dollar, or free tickets to Houston Oilers games, one of the Seven would say, "Oh, yeah, we forgot to tell you about that."

Of course, the cars and the other goodies didn't matter to the Nine, not compared to the real issue. The only one of the Seven Conrad knew well was his Pax River classmate, Wally Schirra. Now that he was an astronaut who had flown in space, Schirra was still the same breezy, fun loving fighter jock he'd always been, but there was a subtle difference about the way he related to the Nine. He was like a pro ball player visiting with the farm team. But Schirra never put it into words. It was Gus Grissom who came right out and said it: You're not an astronaut until you fly.

But that hardly mattered to *Life* magazine. In September 1963, *Life* ran a cover story on the Nine: "The New Astronauts: They go head over heels into training." Inside, the Nine were introduced to the American public with a fanfare that, if it didn't measure up to the Original 7's, was still good coverage. There were pictures of the Nine in training, climbing into a Gemini mockup, trying out a Gemini life raft, eating space food while floating around in a training aircraft. Then came their personal stories, with each of them talking about the role he was playing in the assault on the moon. And pictures from the home front: there was Ed White going for a bike ride with his family, Jim McDivitt out for a weekend spin with the wife and kids, Conrad and his clan seated in front of the fireplace, and so on. Even Deke Slayton had an article, with a shot of the Nine sitting around a conference table in his office, dressed in business suits, looking like the young executives of some rising corporation. Slayton told the *Life* people, "There will be plenty of flights for everybody." And he meant it.

Around this time, there was a definite change in the Original 7's attitude: they began to warm up to the Nine. With Mercury over, Gemini was on the near horizon, and it was clear that even sixteen astronauts would have their hands full planning and flying those missions. Meanwhile, Apollo was gaining momentum. NASA was already in the process of selecting more astronauts.

Of course, the Seven could afford to be friendly, because none of the Nine were going to take a mission away from any of them. That was due to an arrangement called the pecking order, a holdover from the military. The Astronaut Office was not a military outfit, but the Nine came to understand that the pecking order was a reality just the same. Here it was based not on military rank but how long you had been an astronaut; the Original 7 were permanently first in line.

But for the most part, the Nine didn't have much time to think about when they would fly in space. After ground school with the Original 7 that included jungle and desert survival training and courses on all facets of spaceflight from orbital mechanics to computers, they were swept up in the Apollo whirlwind. Each man was assigned a slice of the massive engineering effort. Conrad drew the enviable task of helping to design the controls and instrument displays for the lunar module, the spacecraft that would actually land on the moon. In the months that followed Conrad found himself standing in a plywood mockup of the lander, surrounded by painted switches and dials, imagining himself flying over a silent, cratered world. He wriggled down rope ladders, descending to a simulated moonscape. He hung suspended from a Peter Pan rig, to familiarize himself with the trampoline-bounce of walking in the moon's one-sixth gravity. Always, in the back of his mind, Conrad was certain that some day he would be doing these things for real.

Conrad knew that when some people thought of astronauts they imagined the thrill of exploration, but that wasn't what he felt. It was his *job* to learn how to walk in lunar gravity, to figure out what kind of instruments a lunar lander ought to have. He wasn't here because he had an insatiable desire to set foot on another world. For him it was not the destination but the journey, one that offered the most challenging test flying anyone could imagine. He was hungry for the chance to take control of a moonship and fly as no one had ever flown before, to take it all the way down to the surface of the moon, and when he was there he would do some good, useful work, and then he would fly it home again. The lure for Pete Conrad was not to walk on the moon, but to stand on the summit of his profession. If people didn't quite understand that, if they wondered why astronauts didn't sound like explorers when they talked, he had a simple answer. He wasn't an explorer; he was a test pilot.

As the months passed, the Nine came into their own, and at the same time the Original 7 waned. Alan Shepard, the first American in space, considered by many to be the most skilled pilot of the Seven, had been taken out of the running by an inner ear disorder and, like Deke Slayton, was grounded indefinitely. John Glenn was spending so much time making public appearances and so little time in the Astronaut Office that Wally Schirra had criticized him in a television interview for shirking his duties to the program. But Glenn was flying off on a new trajectory; there were noises about him considering a bid for the Senate in the 1964 election. Scott Carpenter's spaceflight career had also come to a premature end; unlike Glenn, he had no choice. Many at NASA felt he had botched his Mercury mission by wasting precious maneuvering fuel, misaligning his spacecraft for reentry, and firing his retrorockets late, which caused him to splash down some 250 miles off target. Though no one ever said so publicly, it seemed clear Carpenter would never be assigned to another space mission. By the end of the year, only three of the Original 7 were available to fly, Gus Grissom, Wally Schirra, and Gordo Cooper. What that meant for the Nine was that more seats would be available on the two-man Gemini flights and the three-man Apollo missions.

And so Pete Conrad found himself engaged in an unspoken but real competition for a seat on a Gemini mission. There were no stated rules, and not even any tangible measures of progress. During ground school, for example, there had been no grades; the Original 7 would never have put up with that. Besides, how would Slayton weigh a less-than-perfect showing in the classroom against golden hands in an airplane? There were some hot pilots in the second group, particularly Tom Stafford and Jim McDivitt, who had a rare combination of qualities: good stick work and good head work. They could fly with a skill and grace that made even the most complex maneuvers look like a work of art; at the same time, they had instincts that in an emergency made the difference between buying the farm, as the saying went, or living to tell about it. The others, Conrad thought, were a mixed bag; some matched him as pilots, others didn't.

Soon it was clear that the competition embraced everything they did. In the summer of 1963, Conrad and the other astronauts made trips to Johnsville, Pennsylvania, to the navy's Acceleration Laboratory, to ride the huge centrifuge, which was known simply as The Wheel. Inside the passenger gondola of The Wheel was a mockup of the Gemini cockpit, in which Conrad was put through a series of simulated reentries from earth orbit. During the exercise the doctors could monitor him on closed-circuit television. One run simulated a fast, steep reentry. Conrad felt the force of tremendous deceleration, just as he would if his spacecraft were slamming into the dense layers of atmosphere. When he hit 8 g's, Conrad could not move his arms, and still the g-forces mounted. At 12 g's, his eyeballs flattened out of focus, causing his vision to blur. Still the load kept building. When it peaked at 15 g's, Conrad felt as if his chest were being torn open. Thankfully, it didn't last long—but he came out with a punishing hangover and ruptured capillaries all over his back, as a souvenir.

Next, the centrifuge was programmed for a long, shallow reentry. This time the acceleration built up to only 5 g's—but it stayed there for five solid minutes. At 5 g's, with his chest nearly pressed against his backbone, Conrad had to force his rib cage open just to take in breath. Five minutes became an eternity. Halfway through the run, Conrad was about to stop breathing. He was sure he was going to black out. He was trying to figure out how he might manage to do that without closing his eyes so that as the centrifuge kept whirling and the doctors watched him, no one would know he'd lost consciousness. To his relief, Conrad managed to get through the run fully conscious, and thank God, because if he hadn't, Slayton would have heard about it, and it wouldn't have been good for his career.

When Conrad was back in Houston, he and the other astronauts breezily queried each other:

"Hey, how'd you do on the five-g run?"

"Oh, super! Piece of cake."

"You didn't black out, did you?"

"Me? Naaah, not at all."

And Conrad couldn't help but smile, certain that they'd probably gone through the same thing he had and that they were all lying through their teeth. And it was always that way, after a practice session in the docking simulator, or a parachute drill, or anything they did. When they were in Houston, astronauts rarely socialized together, but on those occasions when Conrad and Jim Lovell or Tom Stafford or any of the Nine gathered on a day off, and the air filled with the smell of steaks on the patio barbecue, the one thing they didn't talk about was who might get picked to fly, and when. None of the Nine was about to show a weakness like worrying about his career, not in front of another astronaut.

• • •

In 1963, while Pete Conrad and the rest of the Nine were drag racing with the Original 7 and searching for their paths into space, Bill Anders was a twenty-nine-year-old fighter pilot with one goal, and that was to become a test pilot. He hadn't grown up with dreams of flying; his childhood ambition had been to become a career naval officer like his father, who had commanded the gunboat *Panay* when it was under attack by the Japanese in 1937. He was raised to be practical and competitive; he didn't smoke or drink, and he had always sought to push the limits of whatever he was doing. There had never been any question about where he would go to college; he applied to the Naval Academy at Annapolis and was accepted. It was there that Anders gravitated toward flying. But during his first cruise on an aircraft carrier as a midshipman, it seemed as if there were an accident every other day. He saw a plane hurtle toward the carrier deck on final approach, miss the arresting wire, strike a line of parked jets, careen off the deck and plummet into the sea. Two or three other pilots died on the same cruise, all of them during takeoff or landing. Anders was willing to accept the risk of flying jets, but he wanted to take that risk in a dogfight at 30,000 feet, or evading enemy guns—not trying to land. Later, when a young air force colonel came to Annapolis to tell the cadets that his branch of the service offered a better path for advancement, Anders made up his mind. After graduation, he was off to the air force, to earn his wings.

By 1959, fresh from a tour of duty with an interceptor squadron in Iceland, Anders knew he had become a hot pilot. But he also knew he wasn't like the other squadron pilots, who lived only to fly. Flying didn't give Anders the intellectual stimulation he craved; he wanted something more. Test pilots have to be engineers as much as pilots, and that appealed to Anders. Most of all, for any hot fighter pilot, test pilot school was the next step up the ladder. Anders had already amassed the 1,500 hours of flying time in high-performance jets required by the air force's Experimental Flight Test Pilot School at Edwards Air Force Base, in the high desert of California. But when he went to Edwards to make his case, things didn't go as he'd hoped. One of the admissions people told him, "If only you had an advanced degree, you'd be a prime candidate."

"Advanced degree?"

"Yeah, we're pushing academics."

Along with his mild disappointment, Anders felt a sense of irony; he wondered what Chuck Yeager—the most famous test pilot in the world—would have thought of this new requirement. In 1947, Yeager had become the first human being to shatter the sound barrier, flying a rocket plane over the desert at Edwards. The world of test flight had changed since then, but not Yeager; he was still the prototypical rocket ace, an alloy of split-second reflexes, go-for-broke daring, and unflappable calm. There was hardly a fighter pilot in the country who didn't long to be molded in his image. But Yeager had not had so much as a college diploma.

But this wasn't Yeager's school, and Anders knew he had no choice but to accept the new direction. In fact, graduate school had been on his list of things he hoped he might accomplish. He enrolled at the Air Force Institute of Technology at Wright-Patterson Air Force Base in Dayton, where he planned to steep himself in the intricacies of astronautical engineering, then return to Edwards. It wasn't until Anders packed up his three children and his wife, Valerie (who was expecting their fourth), and moved to Dayton that he was informed that the astronautical engineering program was full. Because of his excellent math grades at the Naval Academy, the school assigned him to nuclear engineering. Maybe that wasn't so bad; the air force was planning to develop a nuclear-powered airplane, and this kind of experience would make him a natural for the project. But Anders wasn't taking any chances. Even as he labored through the two-year degree program at Wright-Pat, he went to night school at Ohio State University, taking courses in aeronautics. It was a wise decision: the nuclear airplane program was soon canceled.

And something else happened while Anders was in school. One

day in May 1961, John Kennedy announced that the United States should put a man on the moon. Valerie couldn't imagine how anyone could accomplish such a feat, and she was even more stunned when her husband said to her, "That's what I'd like to do." But it wasn't the flying that was uppermost in Bill Anders's mind, or the benefit to his career; it was the chance to be an explorer. As a boy he'd devoured tales of exploration: Magellan's circumnavigation of the globe, Lewis and Clark's ventures into the American wilderness. The *terra incognita* of the twentieth century was a quarter of a million miles away, and Kennedy was saying men would go there. If there was flying involved, so much the better, but Anders would have signed up to make the trip by barge. What he wanted most of all was the chance to walk on another world.

But the astronaut program, he knew, was beyond his reach as long as he lacked test pilot credentials, and after graduation in 1962, he was ready to try again at Edwards. By now, the air force had developed an interest in space, and the test pilot school had been replaced by the Aerospace Research Pilots School, otherwise known as ARPS, to help air force pilots gain the qualifications to be astronauts.

When Anders contacted ARPS he learned the school wasn't taking any new students for a while. Discouraged, he accepted an assignment as an engineer and instructor pilot at Kirtland Air Force Base in Albuquerque, where the air force was conducting a program of nuclear research. And he bided his time, waiting for the hiatus at ARPS to end. That happened the following year, 1963. Anders readied another application, sure that his graduate work would impress the selection committee.

But ARPS was now under the command of none other than Chuck Yeager. He told Anders, "We've changed the requirements. We're looking for flying time." Anders realized that if he'd stayed in his squadron for the last three years flying an F-101, he'd be in —some of his squadron mates were being accepted. Undaunted, Anders applied anyway. As often as he could, he took a break from his duties at Kirtland to fly to Edwards, demonstrating his skills by duplicating the flying maneuvers required for the entrance exam. Then it was back to Albuquerque, to wait for some word.

He was still waiting on a hot Wednesday early in June 1963. That day he was driving home from work in his Volkswagen microbus and listening to the news when there was an announcement that NASA was looking for new astronauts. As the announcer read the requirements, Anders checked them off in his mind: Age, thirty-five years or less. Two thousand hours flying time in high-performance jets. Advanced degree in engineering or physical science. And then—most important of all: Test pilot experience preferred, but not mandatory. Anders met every requirement on the list. He had never given serious thought to the astronaut program before now, because he knew he wouldn't get in without test-pilot credentials. But if NASA would take him, he didn't need Yeager. By the time he reached the next traffic light, he had decided to apply.

No one was more surprised than Anders when he found himself still in the running after each cut. When it came time for the interview with Deke Slayton's selection committee in Houston, Anders tried hard to anticipate what they would want. He stressed his fighter-pilot experience, of course, but he also played up his work at Kirtland on how to shield spacecraft from space radiation. NASA was concerned about the hazards from intense solar flares to moon-bound astronauts, who would be the first to venture outside the earth's protective Van Allen belts. Anders didn't think there was too much of a risk—you'd get dosed, and maybe some of your hair would fall out, but it wouldn't kill you—but he wasn't about to tell the committee that. Radiation was a real problem, he told them, and he could help them solve it. Anders returned to Albuquerque satisfied he had done his best but entirely unsure of his chances.

On October 14, three days before his thirtieth birthday, Anders got a call from Deke Slayton inviting him to fly for NASA. Not long afterward, the space school turned him down, but Anders couldn't have cared less. By then he was already getting his family ready for the move to Houston. One day, on a trip to the Pentagon to check on some paperwork, Anders ran into Yeager.

"Too bad you didn't make it," Yeager said.

"That's okay," Anders said wryly; "I got a better offer."

"A better offer?"

"Yeah, I'm going to NASA. I made it into the program."

Anders still remembered how Yeager fumed—"NASA took you, and you haven't been through the school? I'll see about that . . ." But, Yeager could bitch all he wanted, and it wouldn't make a difference. Anders had been accepted into the most elite flying fraternity in the world. Of course, "fraternity" was best used loosely in this case, as Anders would find out.

• • •

By the time fourteen new astronauts came to Houston at the end of 1963, the Astronaut Office had moved to Ellington Air Force Base, just down the road from the construction site of the new Manned Spacecraft Center. The astronauts had their offices in some rehabbed World War II barracks near the flight line, which made it easy to hop into a T-33 jet trainer and head out to a design review halfway across the country. To Pete Conrad, flying was a great escape from the barrage of paperwork—of which he'd never seen so much in his life. Fortunately, the new astronauts had arrived just in time to take up the slack, fourteen new rookies whose place on the totem pole was even lower than the Nine's. Conrad's group saw no threat from the newcomers; the first of the Fourteen probably wouldn't lift off until after the last of the Nine had splashed down. And Conrad had some good friends in the new group, including Dick Gordon, who had been his roommate on the carrier *Ranger*, and Al Bean, who had been one of his brightest students at Pax River. Conrad wished them well. And if the Fourteen came in with a nose for the goodies, the Nine weren't keeping any secrets. Neither were the Original 7. Wally Schirra even sat down with the newcomers and spelled out, clearly and in detail, how the goodies worked, the cars and the *Life* contract and the rest of it.

Before long the Fourteen were ready to take on engineering assignments. Conrad handed off his job on Apollo controls and displays to his buddy, Dick Gordon. Then he and the rest of the Nine left the new rookies to their own devices and headed off to work on Gemini. Never mind the moon, for the time being; Gemini was upon them, and every one of the Nine was itching to fly it. The first manned flight was slated for early 1965. Nine more missions were planned after that, brimming with techniques vital to the moon program like space rendezvous, dockings, and space walks. Gemini would be a test pilot's dream.

Grissom, Schirra, and Cooper were first in line for Gemini, and the Nine could only guess at how they might be jockeying for position. The two real piloting plums would be the first manned flight, Gemini 3, and the first rendezvous mission, Gemini 6. By early 1964 it was getting to be time for Slayton to fill the seats for the coveted first flight. It went without saying that Grissom, who was at the top of the rotation and had a key role in Gemini's development, would command the mission. But no member of the Original 7 would fly right seat; that was a job reserved for the rookies.

One day in April, Slayton called all the astronauts together in the briefing room at Ellington and announced that Gus Grissom was going to command Gemini 3 and that his copilot would be John Young. Backups to Grissom and Young, Slayton announced, were Wally Schirra and Tom Stafford.

Conrad didn't dwell on his disappointment. But in July, Slayton named two more crews, and they were all from the Nine. Jim McDivitt and Ed White would fly Gemini 4; their backups were Frank Borman and Conrad's buddy, Jim Lovell. Conrad was beginning to lose his patience. McDivitt, White, and Borman were air force, like Slayton. There was a lot of good-natured ribbing between the air force and navy, but it was meaningless—or was it? Was Slayton taking care of the air force?

By the end of 1964, signs of order emerged. Wally Schirra and Tom Stafford, who were still training as Grissom and Young's backups on Gemini 3, were unofficially named as the prime crew for Gemini 6, the first rendezvous mission. That was why none of the Original 7 had showed up on Gemini 4; compared to the first rendezvous, it was a much less desirable flight; there wasn't a single new objective, except staying in orbit for four days, and not much real piloting. Furthermore, the Nine realized, Slayton was setting up a rotation: You serve on a backup crew, skip two flights, and then fly as the prime crew. Presumably, as long as you didn't screw up, you could stay on the rotation indefinitely. At that point, you were "in the pipeline."

But getting into the pipeline to begin with—that was the difficult part. Conrad reassured himself that the longer he waited to fly, the better the missions got, since the last several Gemini flights would all be rendezvous flights. And Slayton had said that rendezvous experience was essential to getting a seat on a lunar mission.

In February 1965, Conrad's wait came to an end, without warning or explanation, when Slayton told him he would be assigned as Gordon Cooper's copilot on Gemini 5. It was anything but a plum assignment: a week in orbit, in a cabin no bigger than the front seat of a Volkswagen. But it was fine with Conrad. And this was just the beginning. If he didn't screw up, he'd stay on Slayton's crew rotation and keep flying. He'd have a shot at one of the rendezvous and docking missions, maybe even as mission commander. And then it would be on to the moon. As far as Conrad could see, he had it made.

• • •

Looking back on it all years later, Bill Anders would admit that he'd been unprepared for what he found when he joined the astronaut corps. He'd expected to be in pretty fast company—you don't find a collection of overachievers like the astronaut corps every day—and Anders knew that he was lucky to be here. But he never had any doubts about his own abilities, and he'd come in fully expecting that with hard work and perseverance, he would find the way into space open to him. He would never have imagined that he would still be waiting for his chance to fly more than three years later, in April 1967, as Deke Slayton and the other astronauts were sitting down to help plan the lunar missions. Looking back, he would wish he had been a little smarter about the crew selection game.

When the Fourteen reported for work at the end of 1963, the Manned Spacecraft Center was surging with activity. There were seven years left to meet John Kennedy's lunar landing deadline, and the men and women of NASA were giving their all to Gemini and Apollo without a thought for the long hours. These were the most dedicated bunch of people Anders had ever seen. And Anders fit right in; before long he was working as hard as he had in grad school. He soon came to understand, however, that the hardest part of the job wasn't the long hours, or the demands of keeping in shape, or the flying—it was simply finding a way to stand out and get selected for a space crew. From the beginning, the whole process was shrouded in total mystery. The only thing anyone knew was that Deke Slayton made the choices, and that Al Shepard, who had been appointed chief of the Astronaut Office after being grounded, was in on it too. Unlike Slayton, of whom the astronauts saw relatively little, Shepard was their day-to-day boss. Around the office they wryly referred to him as "Big Al."

Shepard wasn't like the rest of the Original 7. He had charisma to fill a room, but he lacked the good-natured bluster of Wally Schirra, or the easy-going friendliness of Gordo Cooper. He was no Father Slayton. You didn't have to be around Shepard too long to find out how well he thought of himself; he was dripping with arrogant self-confidence. Everyone knew his ready wit, his toothy grin, his appetite for a party, his penchant for obscene jokes; they also knew his icy, penetrating, blue-eyed stare. Either incarnation could surface without notice.

From what the other astronauts could tell, Shepard's influence extended well beyond the walls of the Astronaut Office. Since becoming grounded, he'd built up something of a small empire in business. He was part owner of the Baytown Bank near downtown Houston, and he had made money in hotels, shopping malls, and a variety of other ventures. He was the only astronaut who lived apart from the space suburbs, in a downtown apartment.

In the Astronaut Office, Shepard usually kept a chilly distance from his troops; a hello might be returned by a grunt, or nothing at all. And he had a way of intimidating the other pilots, especially some of the younger ones, by staring right through them and asking questions on obscure technical details, to throw them off balance. It was all very subtle; Shepard was too smart to be obvious; but many astronauts felt an inner alarm go off when he entered a room. To be sure, there were astronauts who had no fear of Shepard and some who even seemed to get along with him—though they would never use the word "friends" to describe their association. But even they knew that crossing Big Al was bad head work. Years later, one astronaut would say, "You had the feeling that if it came down to you or him, frankly, he would cut your balls off so fast you wouldn't know they were gone for a little while." More than one of the Fourteen tried to avoid him.

But there was no way around the fact that along with Slayton, Shepard had ultimate power over an astronaut's career. But what did they look for? Late in 1964, near the end of basic training, Shepard gave the Fourteen a homework assignment called a peer rating. Each man was to evaluate his colleagues' abilities and rank them in the order they ought to fly in space, leaving out himself. The peer rating gave Anders a chance to think about the competition. For example, Mike Collins and Charlie Bassett, both of whom came to NASA from coveted test-flying jobs in Fighter Operations at Edwards. And Dave Scott, who not only had been through ARPS like Collins and Bassett, but had a couple of advanced degrees in aeronautics and astronautics from MIT. And here was Dick Gordon, a Pax River grad who had been the project test pilot on the F4 Phantom and had used it to win the Bendix Trophy transcontinental air race. Gordon had been on a first-name basis with Al Shepard, Deke Slayton, and Wally Schirra before he even came to NASA. It wasn't hard to figure out that Collins, Bassett, Scott, and Gordon ended up near the top of many of the Fourteen's ratings.

Where Bill Anders fit in, he had no idea. If the measure of an astronaut was his flying skills, then Anders was ready to face off against anybody in the office. Because he'd been a T-33 instructor at Kirtland, Anders was given the job of teaching his colleagues in the Fourteen how to fly it, and he'd already encountered astronauts whose flying abilities left something to be desired. One came close to crashing until Anders grabbed the controls away from him, much to the man's chagrin. Presumably Shepard and Slayton were flying with the Fourteen enough to make their own assessments, but as far as Anders could see, flying wasn't the basis of this competition. And if not, then what was? Anders's motto was simple: Work your tail off, and someone will notice.

When it came time for the third group to divvy up pieces of

the Apollo design effort, Anders chose the environmental control system, the complex assemblage of pumps, pipes, sensors, and valves that would maintain the spacecraft's breathing oxygen, proper cabin temperature, and so on. Getting all that plumbing to work was a challenge that appealed to Anders; it fit well with his Naval Academy training. But it didn't take him long to realize that he would've been better off if he'd landed an assignment that had to do with actually flying the spacecraft. A good example was the spacecraft's hand controller, the equivalent of the control stick in a jet fighter. The hand controller was important to the pilots, and the astronaut who got it could, in the course of his efforts, circulate among the Old Heads' offices and solicit opinions. They would try out a model of the controller while wearing a space suit glove, then engage in a thoughtful discussion of the best angle at which it ought to be mounted, the "breakout forces" required to move it in a given direction, and so on. If the advice was valuable, so was the chance to be seen. The hand controller was a plum assignment.

No one asked Anders's advice on the hand controller; he was too far down on the totem pole for any of the Seven or the Nine to care what he thought. And his own assignment—the environmental control system—wasn't the best way of getting noticed. Walking into Wally Schirra's office with a section of coolant pipe just didn't produce the same effect. Most of the time, though, Schirra and the rest of the Old Heads weren't even around; in fact, the only time all the astronauts saw each other was at the Pilots Meetings that took place in the Astronaut Office conference room every Monday morning. These meetings, which were run by Shepard, gave the men a chance to coordinate their activities.

It was at the Pilots Meetings that the Old Heads would band together against a common enemy—the doctors. Astronauts and doctors were natural adversaries; every pilot knew there are only two ways you can walk out of a doctor's office, fine or grounded. The Original 7 had to put up with the doctors' alarmist predictions about spaceflight, none of which came true. They talked about having astronauts wired with heart catheters and rectal thermometers, until the men put their foot down. Now they were irritating the Old Heads by coming up with medical experiments that cluttered up the Gemini flight plans. At the Pilots Meetings, when the doctors cooked up some new plan to have the next Gemini crew wired with sensors from stem to stern, the Old Heads' rallying cry was, "Look what those bastards are trying to do to us now!" Joining the protest was probably good politics, but the truth was that Anders didn't mind the doctors or their experiments. To send people into space and not try to understand the effects was just plain stupid. But Anders kept these thoughts to himself.

Gradually Anders understood that he was different from the other pilots, and that it wasn't helping him. The most obvious difference was on his résumé: He wasn't a test pilot. There were five others like him among the Fourteen, including a wiry marine named Walt Cunningham and a tall, red-headed air force pilot named Rusty Schweickart. Like Anders, they were highly educated, but they didn't have diplomas from ARPS or Pax River. Now that they were here, the seven of them, the unspoken question in the minds of the Old Heads—Anders could sense it—was, What the hell are *they* doing here?

The answer was that the astronaut corps was evolving. NASA no longer deemed test pilot experience essential to flying in space. Personally, Anders felt the test pilot distinction was overrated. Besides, he and Cunningham and Schweickart had been selected because they had scientific expertise that made them valuable players in their own right, or so Anders had thought. In reality, it was a liability. Most of the Old Heads seemed to regard the scientists who developed experiments for flights as nuisances who thought the program revolved around them and who simply didn't understand the demands of flying in space. The fact that Cunningham had done doctorate-level research in upper atmospheric physics and Schweickart had studied astronomy at MIT didn't boost their standing in the Astronaut Office. Schweickart's assignment was to coordinate the scientific experiments, and Anders wondered if someone put him there to get him out of the way.

It was easy to see why Schweickart irritated a few of the Old Heads. Some detected an air of intellectual superiority. Not all of them appreciated his barbed wit. Schweickart didn't defer to them on technical matters; he didn't hesitate to offer his opinions, even when the Old Heads weren't interested. Rusty Schweickart was an unusual commodity in the Astronaut Office; he was a free spirit. He shared an office with Walt Cunningham, and despite some superficial similarities—Cunningham was also very bright, even more outspoken, and just as superior in his attitude—the two could not have been more different. While Cunningham was decidedly right of center, Schweickart had left-wing leanings on most of the day's social issues. It would be years before Schweickart would let his hair grow and sport a beard, prompting Pete Conrad to label him the Astronaut Office's "token hippie." But even now, under that orange crew cut, there was enough about him to break the fighter pilot mold. A few years younger than most of his colleagues, Schweickart seemed curious about everything that was going on in the world. He and his wife, Claire, hosted a literary discussion group. He showed an openness to new things, and listened in fascination to the stories of a friend who had witnessed San Francisco's drug culture. In short, there was much more to him than flying.

Anders, Cunningham, and Schweickart were odd men out. What did not make sense was that they were joined by a thin, quiet navy test pilot named Al Bean who'd been near the top of his class at Pax River and had flown in a couple of attack squadrons before coming to NASA. He'd barely missed getting selected with the second astronaut group. Bean was different from the other test pilots in the office. He didn't project the same macho image. Most astronauts went hunting or worked on cars in their spare time; Bean liked to paint. He seemed more inclined to hang out with the likes of Anders, Cunningham, and Schweickart than with the other test pilots. They became a kind of Four Musketeers, the closest thing to a pal group in the Astronaut Office. Whenever they had a free moment together, they talked about one thing more than any other: Who would be the first of their group to fly?

The suspense ended one day in the fall of 1965 when many of the Fourteen were on a geology trip in Oregon. Slayton got them all together in his motel room and announced that some of them would be moved into the Gemini flight rotation. Dave Scott would fly right seat on Gemini 8; Charlie Bassett would copilot Gemini 9. Mike Collins, meanwhile, was going to be Jim Lovell's backup on the upcoming Gemini 7 mission, and Dick Gordon would back up Scott on 8. According to Slayton's rotation, that meant Collins would fly right seat on Gemini 10, Gordon on 11. In one fell swoop, four of them were in the pipeline. No sooner had the group returned to Houston than Collins and the others soared off into the world of Gemini flight crews. They began simulator training, and got fitted for space suits, and went to mission planning meetings, while the rest of the Fourteen were left to their comparatively routine lives.

Late in the year most of the astronauts went to Boston for an introductory course at MIT on the Apollo computer and guidance system. Between lectures, Anders, Cunningham, Schweickart, and Bean took in the sights, venturing to the Concord Bridge to stand where the Shot Heard 'Round the World was fired, and then back to Cambridge for a stroll among the shops and cafés of Harvard Square. Everywhere they went, they talked not about the historic landmarks they were visiting, but about the crew selection game. And Anders had had enough.

"It doesn't work," Anders told the other three.

Cunningham responded, "What doesn't?"

"Working hard and hoping someone notices. Nobody does."

In December two more Gemini crews reached orbit. The world's first space rendezvous brought Wally Schirra and Tom Stafford in Gemini 6 face-to-face with their two bearded colleagues in Gemini 7, Frank Borman and Jim Lovell, who were midway through their own two-week mission. Deke Slayton had suggested, optimistically, that each of the Fourteen might fly a couple of space missions. But the Gemini program was already half over.

• • •

Nineteen sixty-six was America's calm before the storm. The war in Vietnam was in full swing, with 184,000 American troops in southeast Asia and more on the way, but political support for the war at home was still strong. Lyndon Johnson's Great Society was at the top of the political agenda. The struggle for civil rights was in transition; in the inner cities, hope was giving way to frustration and rage. In the Astronaut Office, Bill Anders paid little attention to these events. He would admit to feeling a little guilty about the fact that he'd made it into the astronaut corps but wasn't flying in space; he might as well be in Vietnam, doing his duty. But like the rest of the Astronaut Office, he was caught up in Gemini's spectacular progress, which was closing the gap in the race to the moon. The Soviets had not flown a manned space mission since 1965, when they had scored the first walk in space. In the meantime, Ed White had made his own space walk, and he'd topped his Russian counterpart by staying out twice as long and by using a hand-held maneuvering gun. With the completion of Borman and Lovell's fourteen-day marathon, Americans had taken a commanding lead in man-hours in space. The rendezvous and docking missions were next, something the Soviets hadn't even attempted. The flight schedule was proceeding at a manic pace, a new launch every other month, each one a totally different mission. Gemini had the flavor of the old barnstorming days of flight, transferred to the high-tech world of space. Wing walking had given way to spacewalking; aerobatics yielded to rendezvous and docking. Gemini's star was rising, and with it, the Fourteen's. Dave Scott was about to become the first of the third group to fly, on the Gemini 8 mission, and there was a member of the Fourteen on every Gemini crew from then on. None of them, however, included Anders, Cunningham, Schweickart, or Bean.

Ironically, Anders noticed, the outside world treated even rookie astronauts as if they had been anointed. He did his share of NASA PR activities—Rotary Club appearances, mayors' luncheons, and so on—and he felt a little strange; what could he talk about except that he was training to go into space—someday? But it was possible to give a speech on astronaut training and get a standing ovation. Occasionally he and Valerie took in the Houston social life, in which no party or charity ball was complete without an astronaut. But every Monday morning at the pilots' meetings, surrounded by men who had been in space, some not once but twice, Anders had no illusions.

Try as he might, Anders could not piece together an understanding of the crew selection game. The problem with Cunningham, Schweickart, and himself went beyond the fact that they weren't test pilots. Background alone wasn't the answer. Otherwise, why was Al Bean, a Pax River grad like Dick Gordon, still waiting? And look at Gene Cernan, another of the non-test pilots. Gregarious, easy-going, and well liked, Cernan fit in well with the Old Heads, and he was named as Tom Stafford's copilot on one of the later Gemini missions. Years later, Anders would decide that there weren't any selection criteria. Maybe Shepard and Slayton threw darts. Or maybe this whole thing was based on how well you fit the mold. But Anders couldn't be a chameleon. He couldn't turn himself into a hard-charging fraternity brother. And he didn't know what else he could do.

In February the Astronaut Office got a reminder that the pecking order could be overruled at any time, by death. The prime crew of Gemini 9, Elliot See and Charlie Bassett, were killed when their T-38 struck the roof of the building housing their spacecraft at the McDonnell plant in St. Louis. Backups Tom Stafford and Gene Cernan were named to replace Bassett and See. At the same time, Jim Lovell and Buzz Aldrin, who up to now had tended a dead-end assignment as backups for Gemini 10—there would be no Gemini 13—became the Gemini 9 backups, putting them in line to become the prime crew of the final mission, Gemini 12. The last Gemini seat had been taken.

It was around that time that somebody finally noticed: Anders was assigned as a Capsule Communicator on the Gemini 8 mission. Being a Capcom put him a step closer to getting on a crew. On the first day of the mission, March 3, 1966, Anders walked into the Mission Control Center just minutes after Neil Armstrong and Dave Scott had successfully docked with their unmanned Agena target. The craft was out of radio contact, somewhere over China. Jim Lovell was just ending his shift at the Capcom mike.

"You'll pick him up in about five minutes," Lovell told Anders. "Pretty boring." Lovell left the control center, leaving Anders to his Capcom debut. But that debut coincided with one of the most harrowing moments in the history of spaceflight. When Gemini 8 came back in radio range, Anders was startled to hear Armstrong report, "We've got serious problems, we're tumbling end over end. . . ."

Gemini 8 was spinning out of control, and it wasn't until many anxious minutes had passed that Armstrong stabilized the craft. By that time, flight director Gene Kranz was preparing to abort the mission. Anders radioed the countdown for retrofire to Armstrong and Scott. Half an hour later, Gemini 8 was floating in the Pacific, and almost before it had started. Anders's first real job as an astronaut was over.

Still, he must have impressed somebody, because he was soon named as Armstrong's copilot on the backup crew of Gemini 11. As glad as he was to get the assignment, he wasn't happy about working for the Gemini 11 prime crew, Pete Conrad and his old navy buddy Dick Gordon. Anders thought they were a couple of obnoxious loudmouths—until he realized that they were two of the most competent astronauts in the office. He'd never seen anyone work so hard and have such a great time doing it. After a couple of months Anders liked Conrad and Gordon so much that he would've been happy to fly in space with either of them.

Anders knew from the start that backing up Gemini 11 was a dead-end assignment, but he dug in anyway. He learned the ins and outs of flying in Gemini's right seat, including his part in a space rendezvous, until it was second nature. And in a special training airplane that could produce brief intervals of weightlessness, he sweated through hours of practice space walks, hot and nauseated in his pressure suit. There was always the chance that the prime copilot, Dick Gordon, would break his leg, and if that happened, Anders would be ready to step in.

The most amazing thing was that suddenly he mattered. People listened to what he had to say. Conrad and Gordon were hot to set a new altitude record during their flight, but they had to prove to the doctors that there wasn't an undue hazard from radiation. Anders used his nuclear expertise to help make the calculations that won the managers' okay. And in September 1966, when Conrad and Gordon soared to 850 miles, Anders listened from mission control with not only envy but a sense of accomplishment. One more Gemini flight came and went in November, and then the program was over.

By that time, the Manned Spacecraft Center was teeming with

astronauts. Nineteen new pilots had come in, joining a handful of scientist-astronauts selected the previous year and swelling the ranks of the astronaut corps to forty-six. Slayton had been wary of taking on so many pilots; even with plans for perhaps a dozen lunar landings and a space station in earth orbit, he suspected that he would end up with nineteen astronauts he couldn't use. But the NASA management had pressed him to get "manned up" for the future flights, and he'd gone along. In the meantime, those flights were in danger of being cut from NASA's budget. By the time the Nineteen arrived in Houston, they sensed that they were in for a long wait. In fact, they seemed to be ignored, as if they weren't really there. They wryly called themselves the "Original 19."

But for Bill Anders, the wait was finally over. Slayton assigned him as the lunar module pilot for an earth-orbit test of the Apollo spacecraft, with Frank Borman in command and fellow Gemini veteran Mike Collins as the command module pilot. Meanwhile, Walt Cunningham and Rusty Schweickart had made it into the pipeline too. But not all of the Four Musketeers had found deliverance. Al Bean still wasn't on a crew; in fact, he wasn't even working on Apollo any longer. He'd been transferred to the Apollo Applications Project, the earth-orbit space station planned for the 1970s. Anders could only shake his head when he thought of Bean; if he wasn't going to the moon, none of this game made much sense.

It was funny; Anders never heard any of the Old Heads talk about what it would be like to go to the moon. But sometimes, flying cross-country at 45,000 feet, he could see it through the clear canopy of his T-38, and he would think about how much he wanted to explore it. Now that he was on a crew, it was becoming a real possibility. He'd stay in he pipeline, and eventually, he'd find himself on the crew of a lunar landing mission. In the meantime, as 1967 opened, Anders shared the soaring confidence that filled the space center as Gus Grissom and his crew readied for their Apollo 1 flight.

November 9, 1967
Merritt Island Launch Complex, Kennedy Space Center

On a chilly November dawn, a crowd of spectators gazed across the still wates of a tidal basin to Pad 39-A, 3½ miles away. There, the first Saturn V moon rocket was minutes from its maiden voyage, and most of the astronaut corps was here to witness it. In the crowd, Pete Conrad gazed admiringly at the mammoth white booster as it spewed trails of vapor into the morning air.

By this time, Pete Conrad was one of the most seasoned astronauts in the office. In the sumer of 1965 he'd copiloted Gemini 5, an eight-day marathon in earth orbit. Then, on Gemini 11, he and Dick Gordon had made the first one-orbit space rendezvous. And for much of this year, Conrad had been immersed in training for Apollo. Like all the astronauts, Conrad thought it was a waste to launch a Saturn V unmanned, but NASA would never have considered putting astronauts on a new and untried rocket, especially the most powerful one ever created. And a successful launch, even an unmanned one, would clear the way for the flights to come and give Apollo badly needed momentum.

During the fall, NASA planners had firmed up the Apollo flight schedule into a series of methodical, incremental steps that are the hallmark of test flying. On Vu-graphs and blackboards around the center they were conveyed in a series of letters. Today's Saturn V launch was one of the unmanned test flights called the "A" and "B" missions. The command module would make its manned, earth-orbit debut on the "C" missions. Then, on the "D" and "E" missions, the lunar module (LM) would be added for a test of the full Apollo spacecraft in earth orbit. If they were successful, the astronauts of the "F" mission would make the first flight to the moon, for a "dress rehearsal" of the lunar landing. Then, at last, would come the first attempt to land on the moon: the "G" mission. NASA would follow these milestones, each mission bolder than the one before, closing the distance one step at a time, like a ladder to the moon.

By November 1967, Pete Conrad was beginning to think he might find himself at the top of that ladder in 1969. He and his crew were backups for the D-mission, and by Slayton's crew rotation, if everything went according to plan, they would become the prime crew of the G-mission. To Conrad, and all the astronauts, the first landing mission was considered the ultimate test flight. No assignment equaled the chance to be the crew that actually carried out Apollo's mission.

Conrad knew the lunar landing was a task of enormous complexity, the details of which had yet to be worked out. Even if everything went perfectly, would the first attempt at the actual landing be successful? It might take several tries before they succeeded. And the time left to meet Kennedy's deadline was steadily dwindling. A catastrophic failure could place the entire moon program in jeopardy.

Over the public address system, Conrad could hear the public affairs commentator calmly giving status reports as the Saturn was fully pressurized with fuel. Then the count entered its final minute.

"T minus fifty seconds and counting. We have transferred to internal power and the transfer is satisfactory. . . . T minus thirty seconds and counting. . . . T minus twenty, nineteen, eighteen . . ."

Conrad wished he were on that rocket.

". . . eleven, ten, nine, ignition sequence start. . . ." Suddenly a torrent of fire erupted at Pad 39-A. For many long seconds it continued, and still the great rocket sat motionless, as if it might never leave the earth. At zero, the Saturn slowly rose from the pad, trailing a tail of incandescent flame. It crept upward past the towering launch structure while the crowd cheered. Suddenly their cries were drowned out by an onslaught of sound. Witnesses felt the shock waves pound against their chests. The Saturn arced upward, gaining speed, and Conrad's eyes followed it into the clear morning sky, until it was gone. Nine hours later NASA had chalked up a flawless test flight of the booster and the Apollo command module, and for the first time in NASA's worst year, the moon came a little closer.

CHAPTER 3

First Around the Moon

APOLLO 8

I. The Decision

The news was bad in the summer of 1968. A nation reeling from the assassinations of Martin Luther King and Robert Kennedy now confronted more images of violence in its living rooms: the blood of young soldiers in Vietnam, the blood of demonstrators outside the Democratic National Convention in Chicago. And even within NASA's world, where these events were overshadowed by the race with the decade, there was bad news. The second unmanned test of the Saturn V moon rocket had been a near disaster. Minutes into the launch the booster began to vibrate so badly that two of the first-stage engines shut down prematurely. Later, a third engine refused to reignite in space. And if that weren't enough, there were ongoing headaches with the Apollo spacecraft. The redesigned command module was coming along well at North American, and the craft slated for Apollo 7, the command module's manned, earth-orbit debut, was already at the Cape being readied for an October launch. But the lunar module was facing one technical problem after another. From the beginning, engineers at the Grumman Corporation in Bethpage, Long Island, had struggled to keep the lander's weight from exceeding forbidden limits. And there were other woes: faulty wiring, corroded metal, and most serious of all, troubles with the LM's ascent rocket. And when the first manned lunar module was shipped to the Cape in June, quality control inspectors found 100 separate defects. At NASA, no one who heard the reports on the lander was happy with the situation. Apollo 8, the LM's first manned flight, would almost certainly be delayed beyond the end of the year, throwing the whole sequence of Apollo missions into jeopardy. The end-of-the-decade deadline for the lunar landing was slipping out of reach.

All that began to change in early August. A plan emerged, elegant in its simplicity, astounding in its boldness, that altered the course of the moon program. It was the brainchild of George Low, the quiet engineering genius who oversaw the development of the Apollo spacecraft from Houston. If Apollo 7 went well in October, Low reasoned, why keep Apollo 8 in earth orbit? Even if the LM wasn't going to be ready for its debut, the second command ship could go to the moon by itself in December. Already, during the spring, Low had quietly raised the possibility of a circumlunar flight in which the joined command module/lunar module pair would execute a figure 8 loop around the moon and then come home. His new plan was even more ambitious. Low wanted to send the command module to the moon by itself, not to fly a figure 8 loop, but to go into lunar orbit. Even without a lunar module, that would let NASA practice the elements of a basic lunar mission: navigating across the vast translunar gulf, executing the precise rocket firings to get into and out of lunar orbit, communicating across a quarter-million miles, and the critical reentry into the earth's atmosphere at hypersonic speeds. Then, by the time the LM was ready—estimates said February—Apollo would have taken a giant step forward.

But there was another reason for urgency. Reports from the Central Intelligence Agency said the Soviet Union was about to resume flying its new Soyuz spacecraft—after the first Soyuz crashed, killing its lone cosmonaut, Vladimir Komarov, in April 1967—and were on the verge of sending one around the moon. Most experts doubted the Soviets had the capability to land on the moon before the end of the decade; for one thing, they had yet to test a rocket, like the Saturn V, powerful enough to propel the necessary payload to the lunar surface. Even a lunar orbit flight was probably beyond them. But with the booster they already had, they could fire a Soyuz, with one or two cosmonauts aboard, on a trip around the moon.

From the beginning, without warning, the Soviets had upstaged the United States in space with their own spectacular firsts. In 1957 it had been the first earth satellite, Sputnik I. More than three years later it was Yuri Gagarin's one-orbit flight that stunned the world and sparked John Kennedy's decision to go to the moon. Then came the first woman in space, the first multiperson space crew, the first spacewalk. If the Soviets got to the moon first—even if they did nothing more than loop around it—the world would hardly notice the difference between that accomplishment and NASA's more difficult lunar orbit mission.

There is no way of knowing what would have happened to Low's plan if NASA Administrator James Webb had been in Washington, but the fact was he was not; he and his deputy George Mueller were in Vienna attending a conference. In their absence Associate Administrator Thomas Paine, a bright, young engineer with a penchant for the visionary, was in charge. When Paine's deputy, Apollo program director Sam Phillips, told him about Low's idea Paine immediately saw the logic in it. But it remained to convince Webb, and that might not be easy.

Webb was not an engineer. He was, however, a canny bulldog of a politician. When Kennedy said "Go to the moon" it was up to Webb to keep Congress from having second thoughts, which he did by any means of persuasion he found necessary—including a knack for knowing where congressional skeletons were hidden. Year after year, he was Apollo's champion on the Hill, where it counted most. He had persevered even as the war in Vietnam claimed more and more of Lyndon Johnson's attention and Apollo became a target of congressional opposition. If Americans reached the moon by the end of the decade, it would be due in large measure to Jim Webb. But Webb would not be at NASA to see it; he already knew his tenure would end when Johnson left office.

Webb took Paine's call at the American embassy, where there was a secure phone, and then Sam Phillips got on the line. Webb wasn't ready for what he heard. He yelled over the transatlantic phone line, "Are you out of your *mind*?" Webb reviewed what was apparent to any sane person: They hadn't even flown a manned Apollo spacecraft, and here they were with a scheme to send the second flight to the moon. And with no lunar module! All along, the LM had been thought of as a measure of safety, a lifeboat in case something happened to disable the command ship's rocket engines. Sending the command module by itself only increased the risk of what was already a risky mission. With the Fire still fresh in the memory of the public and the Congress, Webb could only imagine the effect of another space tragedy. He warned his two deputies, "You're putting the agency and the whole program at risk."

Webb was right. For all its logic, Low's plan was audacious. Many would look back on it as the boldest decision NASA ever made. Still, by the time Paine and Phillips hung up the phone to Vienna, Webb had agreed to give the idea a chance.

In Houston, they were already working on it. Low had asked Chris Kraft, director of Flight Operations, to find out whether his people could be ready to send Apollo 8 to the moon in December. They went ahead with their study in secret. When their office mates asked—"What's all this lunar stuff you're working on?"—they replied coolly, "Oh, it's just a what-if type of study . . ."

Within a week Kraft's team had an answer. By the summer of 1968, after years of intensive effort, the basics of sending a manned spacecraft to the moon were all but perfected. The biggest hurdle: finishing the computer software that mission control would need to help Apollo 8 navigate to and from the moon. Making the December launch date would be tight, but Kraft's people were confident they could do it. Meanwhile, at the Marshall Space Flight Center in Huntsville, Alabama, Wernher von Braun's rocket team reported the problems with the Saturn V were being ironed out. It remained for the Apollo spacecraft to prove itself. If all went well on Apollo 7, slated for October, there wouldn't be anything to stand in the way.

Saturday, August 10, 1968
North American Aviation, Downey, California

It wasn't unusual for Frank Borman to be working on Saturday; the past nineteen months had been among the busiest of his life. After the Fire he had plunged into the recovery effort, beginning with the accident review board at the Cape and then many long months here, at the North American plant in Downey, California. Until a few months ago Borman had all but lived here at the factory, helping to redesign the command module. Now that effort was behind him, and the Block II command module was on the assembly line, with several of the cone-shaped craft taking form in the sterile whiteness of North American's clean-room. One of those spacecraft, command module number 104—the official designation for the fourth Block II command module—was his. There were problems with 104, but that was to be expected, and Borman was here with his crew, Jim Lovell and Bill Anders, to nurse his ship along and help ready it for delivery to Cape Kennedy.

When Frank Borman walked into a room, you knew that he was in charge. Looking at him—he was a sturdy man with a square, slightly oversized head—you could still see a tough, scrappy kid. He'd been molded at West Point; at age forty he still wore his dirty-blond hair as short as a cadet's, and he still lived by the Point's simple motto: Duty, Honor, Country. The mission came first. It was always that simple for Borman. When he applied for the second astronaut group, the psychologists on the selection board said in amazement, "Nobody's that uncomplicated!" He did not deal in nuance, he did not make small talk. His hearing in one ear had been bad ever since he ruptured an eardrum early in his air force career, but there were times when he seemed not to hear very well with the other one; in meetings he'd listen to a number of views and then make a decision as if no one had spoken. When it came to his crew Borman was a model of the old-school military commander. If there was a decision to be made Borman made it; then he told his crew about it.

He hated few things more than wasting time. During meetings

if he got bored or too impatient he would get up and walk out. He made decisions quickly—so fast that if you didn't know Borman you would have thought him impulsive—and once he convinced himself he would do something a certain way it was all but impossible to dissuade him. But he was usually right. And if he seemed mostly gruff and unyielding, Borman understood people, and in addition to his talents as a pilot and engineer he was a capable manager. He was also a favorite of the NASA management. After Borman had finished his work on the Accident Board—and after appearing before Congress to say, "Let's stop the witch hunt and get on with it"—Bob Gilruth, the head of the Manned Spacecraft Center, personally asked him to oversee the recovery efforts at North American. Borman would always look back on that time, working with the engineers here in Downey, as the most productive episode of his astronaut career. Getting Block II ready took much longer than people expected, but it was a superb spacecraft, free of most of the problems of its predecessor. It had a new, one-piece hatch that could be opened in as little as three seconds. It was fireproof, and soon it would be spaceworthy. And now that Block II was on the assembly line, what Borman wanted most of all was to vindicate that spacecraft: he wanted to *fly* it.

Borman had his own command module and his own crew, but he wasn't very enthusiastic about his mission. Deke Slayton had given him command of the third manned Apollo flight, scheduled for sometime early in 1969. Eventually it would be called Apollo 9, but for now it was known by its letter designation, the E-mission. True, Borman's crew would be the first to ride the huge Saturn V booster. And during the mission they would change the high point of their orbit to a record 4,000 miles. For a few days, while they traveled that lofty ellipse, they would see the entire earth at a glance. But aside from those frills, the E-mission was a repeat of the previous flight, which belonged to Jim McDivitt. McDivitt's D-mission would be the first manned test of the entire Apollo spacecraft in earth orbit. It would include tests of the rocket engines on both the command ship and the lunar lander, and the first Apollo rendezvous maneuvers. With all those firsts, the D-mission was a test-piloting bonanza, and Borman would have gladly traded places with McDivitt. But this wasn't the first time Frank Borman was going to fly a mission he did not want.

In December 1965, after waiting three years for his chance to fly in space, Borman lifted off on what was considered the "dregs" mission of the Gemini series: Gemini 7, a two-week marathon in earth orbit. Borman and his copilot, Jim Lovell, spent fourteen days in Gemini 7's tiny cabin wired from stem to stern with medical sensors. Two weeks in a flying men's room. It wasn't an enviable assignment, but Borman took it without complaint. Dregs or no, he wanted to fly. And in his mind, he'd accepted whatever assignments might come his way on the day he joined the astronaut corps—including the E-mission.

On Saturday, August 10, Borman was knee-deep in testing command module 104 when he was called away for a phone call from Deke Slayton.

"Frank, get back to Houston right away. I need to talk to you," Slayton said.

"So talk to me now, Deke; I'm busy."

"I can't do this over the phone. Grab an airplane and get back here."

Borman was irritated at the interruption—the test was far from over—but when he reached Slayton's office and his boss asked him to close the door, he understood the urgency. Borman listened as Slayton described the CIA report on the Soviets and George Low's plan to send Apollo 8 to the moon. Borman immediately saw the logic. A successful lunar orbit mission would lift NASA out of the shadow of the Fire, restore its shaken confidence, propel it to a lunar landing. Slayton was proposing that Borman and McDivitt swap places and command modules, that Borman's crew fly in December. One thing Slayton did not say was that he had already offered the lunar mission to Jim McDivitt, who had turned it down. As for himself and his crew, Borman needed no time to weigh his decision. To him Apollo was like a war; nothing less than the nation's prestige was at stake. He wanted to see the war won. When Slayton asked him if he wanted to go to the moon, Borman said yes.

Monday, August 12
Timber Cove, Texas

Marilyn Lovell was feeling pleased with herself. She had just finished a long, hot day of hunting for bargains on vacation clothing in the August department-store sales. She could barely remember the last time the family had taken a vacation. Was it before Jim became an astronaut? But just a few weeks ago Jim had agreed to take the family to Acapulco during his week off between Christmas and New Year's. Marilyn was jubilant, and today, in the Houston swelter, she'd found some great buys. She would tell her news to Jim, who would be home tonight from a trip to California.

Marilyn and Jim had met and dated in the mid-1940s, when they were both attending high school in Milwaukee. Even then, Jim had his mind on the stars; sometimes, on clear nights, he would take her to the roof of his apartment building to give her a tour of the heavens. And during World War II, Jim had been fascinated by the reports of German V-2 rockets. Marilyn witnessed his experiments with rockets made from mailing tubes and fueled by gunpowder mixed with airplane glue. While he and a classmate were busy in the open field across the street Marilyn would sit in the apartment with Jim's mother, both of them hoping the two boys didn't blow their heads off. But this was no reckless stunt. The young experimenters took every precaution, wearing welder's goggles and gloves. And it was a good thing—one rocket exploded, sending the nose cone about 80 feet into the air. No one was injured—in fact, Jim proudly proclaimed the "flight" a qualified success.

After high school Jim went to Annapolis, where he had his sights on becoming a rocket engineer. For his senior thesis, he wrote about interplanetary rocket travel, and Marilyn, now his fiancée, typed the paper for him. When she came to the end, she read with disbelief Jim's statement that someday people might ride rockets to the moon. But in 1952 there weren't any jobs for rocket engineers to speak of, and after Annapolis, Jim found himself flying airplanes for the navy. His life had taken a different path, and to Marilyn's joy and amazement it had led to Houston and into space.

Like all the astronaut wives, Marilyn accepted the constant demands the space program placed on her marriage. Her husband's competition for flights spilled over into her own life, just as it did with all the wives. In 1965, while Jim was training for his first Gemini mission, Marilyn discovered she was pregnant with their fourth child. She kept the news to herself for weeks, because she was afraid Jim might be taken off the crew. And for the next three and a half years, as Jim plunged happily into the maw of training, going from backup crew to prime crew to backup and prime again, Marilyn was head of the household, financial manager, mother and father to their kids. During the week Jim would always call in from wherever he was, and somehow they managed to work through the day-to-day crises over the phone. But seeing him in person—that was something she enjoyed only on weekends.

Now Jim was working with Frank Borman again, and his lunar module pilot Bill Anders. But with the flight set for late next winter, the most intensive training was still ahead, and Marilyn was glad that her husband had finally made plans to take some time off at the end of the year.

When Jim came home, Marilyn proudly reported the success of the day's shopping expedition. In the midst of her excitement, she noticed Jim had a peculiar expression, and when she asked him what was wrong, he took her into the privacy of his study, where pictures from his space flights adorned dark wood paneling, and closed the door. "I hate to tell you this," he said, "but we're not going to Acapulco for Christmas."

Marilyn didn't try to hide her disappointment. "What do you mean we're not going? Where on earth do you think you're going to be if you're not going to be with the family for Christmas?" Years later, Marilyn would still get goose bumps thinking about what happened next. Jim paused for just a moment, his eyes bright, and he said, "Would you believe, the moon?"

• • •

Some people—other astronauts—had the mistaken impression that Jim Lovell flew through life on luck. Tall, relaxed, and outgoing,

Lovell got along with everybody. Frank Borman would later say that he had never worked with anyone who faced life with such consistent, good-natured optimism. But Lovell's easy-going manner masked his competitive energy and sharp mind. He'd graduated first in his test pilot class at Patuxent River, ahead of Wally Schirra and Pete Conrad. Like Conrad, he had been turned down for the Mercury selection, then made it with the second astronaut group.

Lovell had to wonder at the twist of fate that had brought him and Frank Borman together again. Lovell knew Borman well; after two weeks cooped up in Gemini 7, he probably knew him better than anyone except his wife. The two men were the same age (Lovell was eleven days younger), and although they had chosen different branches of the service (Borman the air force, Lovell the navy), they were the same equivalent rank. In Lovell's mind they had been equally qualified to command a space mission. When Slayton named him as Borman's copilot on Gemini 7, Lovell took the assignment without ever revealing his disappointment. He was a rookie, and he wanted to fly.

Two weeks in space only whet Lovell's appetite for more, but Slayton handed him a dead-end assignment as backup commander for the final Gemini mission. That changed when Elliot See and Charlie Bassett were killed; Lovell and Buzz Aldrin reached orbit for Gemini's finale. Lovell expected to move into Apollo with his own crew, but instead found himself as backup command module pilot for Frank Borman's E-mission. But in July, fate again changed Jim Lovell's plans. Borman's command module pilot, Mike Collins, was forced off the crew by a bone spur on his spine that threatened his body and his career. Collins would have corrective surgery, but even if it went perfectly he would be out of action for months. Lovell stepped in to take his place. And now, he found himself flying right-seat to Borman once more.

When Borman came back from Houston with the news that they were going to the moon, Lovell was electrified. He could still remember how, as a boy, he had devoured such science fiction classics as Jules Verne's *From the Earth to the Moon,* in which a trio of adventurers ride a craft fired out of an enormous cannon on a circumlunar voyage. It had been beyond his wildest teenage dreams that he might see human beings live out the adventure Jules Verne had imagined, let alone that he would be one of them. Even now, as an astronaut who had made the moon his goal, he found the news almost too exciting to believe. Later, under the clear canopy of a T-38 as he and Borman flew back to Houston, he had sketched a design on his knee-pad that would become the mission emblem of Apollo 8: a figure 8 with the earth in one loop and the moon in the other.

Lovell never had any second thoughts about the lunar mission. To him, it was worth the risk for the adventure alone, never mind the potential for scientific discovery. And then, to be pathfinders for those who would follow, the benefits Apollo 8 would bring to the program—who could question the logic of it? Once Lovell had a chance to get used to the idea, he wondered why that hadn't been the plan all along.

• • •

In the summer of 1968 Bill Anders's fortunes seemed to be improving all the time. He had spent much of the spring as one of a handful of astronauts who were learning how to fly the dangerous and unwieldy trainer called the Lunar Landing Research Vehicle. Neil Armstrong was flying it too, and he and Anders had a friendly competition to see who could make the better simulated lunar touchdown. No one had said, "This means you're in line for a landing mission," but you didn't get an assignment like that for no reason.

Meanwhile, he was hard at work as Frank Borman's lunar module pilot for the E-mission. As a lunar module pilot, he was developing an intimate knowledge of the lander, and the experience he'd get on the E-mission would surely make him an ideal candidate for a landing mission down the line. It looked as though Shepard and Slayton had finally recognized what he could do and were going to give him a chance to prove it. Anders figured he had an 80 percent chance of walking on the moon in the near future.

When Frank Borman came back to Downey with the extraordinary news that their mission had been changed, Anders felt disappointment underneath his excitement. He was losing his lunar module, and with it his chance to one day land on the moon. Borman tried to give him a fatherly talk, saying it wasn't so bad, that he'd still get his chance later. But Anders wasn't convinced. The good news was he was going to the moon in four months; the bad news was he would probably never walk on it.

• • •

With only four months to go until a December launch, no one had time for second-guessing, least of all Frank Borman. As mission commander he had to be intimately involved in working out the flight plan for his mission. Normally that took months, and since the circumlunar mission was an entirely new creation it should have taken even longer. But to Borman's great satisfaction—he would look back on this as the space program at its best—the basic design for Apollo 8 was hammered out in a single meeting one August afternoon in the office of Chris Kraft.

Christopher Columbus Kraft, Jr., was the engine that powered mission control. In the infancy of manned space flight Kraft had created the persona of a flight director as an actor creates a timeless role, setting the standard for all who would follow. Now in his mid-forties, the Virginia-born Kraft possessed the special brand of grace under pressure required to direct a space mission in which human lives are at stake. To his team of flight controllers, the men who manned the trenches of mission control and kept watch on every bit of telemetry from a spaceborne machine, he was stern, generous, perceptive, and inspiring, sometimes all at once. They idolized him the way an infantry division does a beloved commander. In addition to his forthrightness—you always knew where you stood with Kraft—he had the ability to sort through differences of opinion and get to what really mattered. To Frank Borman, Kraft was one of NASA's giants.

Kraft's style, honed during a score of manned and unmanned space missions, was an almost inexplicable blend of caution and boldness. Borman knew it well, for it had been Kraft who had saved Gemini 7. For Borman the last three days in earth orbit were the worst three days of his life. Thrusters went bad; fuel cells were threatening to quit; they were low on maneuvering fuel. He and Lovell were just drifting around the earth, trying to hold out long enough to fly the full fourteen days. Had it been up to him, they never would have made it, because he was ready to come down. It was Kraft's persistence and deft handling of the crises that kept Borman going to complete his mission.

On the afternoon of August 19, Borman and several of Kraft's best people, including such stalwarts as Bill Tindall—the tireless and ebullient mission planner who spearheaded the effort to work out detailed techniques for all phases of the lunar landing—gathered in Kraft's office to design the first flight to the moon.

The six-day mission was slated to begin on December 21, just after the new moon. The timing was chosen so that when Apollo 8 arrived the sun would just be rising across the Sea of Tranquillity, throwing the landscape into relief, allowing Borman's crew to reconnoiter a potential touchdown spot for the first landing.

Borman's crew would become the first to ride the three-stage Saturn V booster into space. Reaching orbit, they would leave the rocket's third stage attached. For about three hours, while they circled the earth twice, Borman, Lovell, and Anders would check the systems aboard their spacecraft, while mission control scrutinized Apollo 8 via telemetry. A failure in any of thousands of components might prompt the controllers to cancel the lunar mission and direct Apollo 8 to stay in earth orbit for a ten-day alternate mission. But if everything was in order they would send Borman's crew a historic message: Go for Translunar Injection. In that maneuver—the astronauts called it "TLI"—Borman's crew would relight the Saturn's third stage and accelerate out of earth orbit. When the engine shut down Apollo 8 would cast off the spent booster and continue on a course for the moon, 240,000 miles away. From then on, Borman, Lovell, and Anders would be in uncharted territory.

Consider the sheer scale of the voyage. Up to now, human beings had barely strayed from their home planet; the world's altitude record, set by Gemini 11 astronauts Pete Conrad and Dick Gordon, was a mere 850 miles. If the earth were a basketball, that would amount to just one inch from the surface. But in the same scale model the moon, 2,160 miles in diameter, would be a baseball 23 feet away. Getting to the moon and back would require acts of precision more demanding than any previous space flight.

To make matters more difficult, the moon is a moving target, barreling along in its orbit at a speed of 2,300 miles an hour. Apollo 8 would have to reach the moon's orbit just as the moon was arriving. Then, like a car racing a locomotive at a crossing, the spacecraft would zip in front of the moon's leading edge. After speeding behind the moon, Borman's crew would fire the spacecraft's main rocket engine and go into an orbit with a low point of 69 miles above the lunar surface—eight one-hundredths of an inch from the skin of the softball.

But that was only if everything went as planned. Suppose Borman's crew suffered a serious malfunction, an engine failure, for example, before they reached the moon? Without some other way to turn around, would they simply speed past the moon, condemned to a lonely death when their oxygen supply ran out? Fortunately, the laws of celestial mechanics—whose strictness gave space flight an extraordinary predictability—made it possible to give Borman's crew a built-in ticket home. The trajectory specialists had taken advantage of this in an elegant creation called the free return. By aiming Apollo 8 at just the right distance from the moon, they could use the lunar "gravity well" like a curve in a toboggan course, to bend its path around the moon and send it back toward earth. In theory, it was possible to fire Apollo 8 out of the starting gate so precisely that it would fly a perfect figure 8 around the moon even if Borman's crew never touched the controls. There wasn't a man in Kraft's trajectory division who wasn't so confident of this that he would have offered to go in their place.

Borman took them at their word. He wanted the trajectory people to pledge they'd keep Apollo 8 on an essentially perfect free-return path, but that was a promise they could not make. The problem, they stressed, was that it would simply not be possible to keep perfect track of Apollo 8's position in space at any given moment. There would always be some error in the measurement—but how big an error, no one could say until Apollo 8 was actually moonward bound—and without that precise knowledge, a perfect free return was impossible to guarantee.

Furthermore, there was a crucial, built-in safety measure. Unlike a rifle bullet—or, for that matter, Jules Verne's manned cannon shell—Borman's crew would be able to correct Apollo 8's path along the way with bursts from its small maneuvering thrusters. If the spacecraft drifted off course, mission control would be able to tell, and they would radio Borman's crew instructions for a so-called mid-course correction. As long as the maneuvering thrusters were functioning, there would be plenty of chances along the way to keep Apollo 8 on course.

The translunar crossing of some 234,000 miles would take 66 hours, roughly half the time it takes an ocean liner to cross the Atlantic. At last, on the morning of December 24, Apollo 8 would reach its destination. If everything checked out onboard mission control would give Borman's crew the go-ahead for Lunar Orbit Insertion (LOI). Over the lunar far side, out of radio contact with earth, they would fire Apollo 8's big Service Propulsion System (SPS) rocket engine for some four minutes—just long enough to slow to the speed necessary to go into lunar orbit, about 3,700 miles per hour. When the engine shut down there would be no more free ticket home. Apollo 8 would be a satellite of the moon, and Borman's crew would have to trust the SPS to work perfectly when the time came to return to earth.

While millions on earth would be among friends and loved ones on Christmas Eve, Borman, Lovell, and Anders would spend it circling the moon, taking pictures and observing its pockmarked surface at close range. From their orbiting platform they would scout landing sites, and take navigation sightings on lunar landmarks. Meanwhile, Kraft's trajectory people would track Apollo 8, amassing data on its orbit and keeping close watch for any changes. Already, from some of the unmanned Lunar Orbiter probes, scientists knew that the moon's gravitational pull was uneven, probably due to buried masses of relatively dense rock that they called mascons (short for "mass concentrations"). Mascons had caused small, unexpected shifts in the probes' orbits, and they would surely do the same thing to an Apollo spacecraft. Before anyone could commit a crew to a lunar landing mission, mascons had to be understood.

But here was the makings of a dilemma. Some of Kraft's people wanted Apollo 8 to circle the moon for as long as possible to accumulate the maximum amount of data on its orbit. Frank Borman, on the other hand, didn't want to spend a minute longer in lunar orbit than necessary; the longer they stayed, the greater the chance of something going wrong. In the end, with Kraft mediating, they settled on 10 orbits—a total of 20 hours circling the moon.

The first minutes of Christmas Day would bring the moment of truth for Borman's crew: the final blast from their big rocket engine, called Transearth Injection (TEI), to free them from the bonds of lunar gravity and send them toward a small and distant earth. On December 27, in the last hour of their homeward voyage, Borman's crew would cast off the spent service module and steer their command ship through one final act of precision, reentry into the earth's atmosphere at 25,000 miles per hour. The angle of approach would be critical: too shallow and the command module would bounce off the atmosphere like a stone skipping across the waters of a pond; too steep and it would be torn to pieces by the forces of deceleration. The zone of safety was a perilously narrow cone just 2 degrees wide. Think once more of the earth as a basketball; reentry was like trying to hit the edge of a thin sheet of cardboard balanced atop it.

These were the weighty matters under discussion that afternoon in Kraft's office—the boldest venture in the history of space exploration, the safety of the three men who would make the journey. And if Borman had shown some unease, he also displayed characteristic pragmatism and bluntness. Someone anxiously pointed out that the timing for a 10-orbit mission meant Apollo 8 would splash down in darkness.

"What the hell difference does it make?" asked Borman. "If the parachutes don't open, we're dead anyway, whether it's day or night." Some risks, Borman understood, weren't worth worrying about.

For now, the new mission was still kept secret, even from the other astronauts. There were still too many what-if's unanswered, the biggest being whether the command module would perform well on its maiden voyage, Apollo 7. But for Borman, Lovell, and Anders, there could be no waiting, and as the summer of 1968 neared its end they plunged into training for the first flight around the moon.

• • •

For Borman and his crew, Apollo 8 began on September 9, inside the command module simulator, a jumble of angular shapes that resembled high-tech sculpture. A carpeted stairway led to a small, square hatchway and an exact replica of the command module cabin in which every switch, readout, and control—from engine gauges to circuit breakers to hand controllers—was simulated to function like the real thing. Without ever leaving earth, Borman's crew practiced the complex task of piloting a moonship. They confronted untried maneuvers like the Lunar Orbit Insertion burn and the hypersonic reentry. Here is where Borman's crew earned their own self-confidence, practicing every phase of the mission from liftoff to splashdown until it became second nature.

Outside, seated at consoles, were the simulation instructors, young men with sharp minds and mischievous spirits who understood the command module as well as the astronauts did, perhaps even better. Much to the chagrin of any astronaut within earshot, they liked to say that they could take any reasonably bright individual off the street and, in a year's time, teach them to fly to the moon. It would have been an empty boast if not for the command module's onboard computer. Utterly primitive by today's standards, this mid-sixties vintage had only 33,000 words of memory, a fraction of any modern desktop model. But in that memory lay coded instructions for a flight to the moon, permanently written on bundles of magnetized wire encased in plastic, so that not even a total power failure would erase them. Although the computer was incapable of adding two numbers, when it came to getting to the moon and back it was as good as putting Isaac Newton himself to work on the problem. It could calculate the command module's position and path through space with the same equations of motion used by Newton to study the orbits of celestial bodies. And it was so essential to flying the command module that the astronauts thought of it as a fourth crew member. Normally it was the computer that would align the command module's gyroscopic navigation platform with the stars, fire its rocket engine with precision, and keep its antenna aimed at the earth. Thanks to the computer, the spacecraft would almost fly itself—that is, if everything worked. The premise of simulation, borne out by hundreds of hours of manned space flights, was that not everything would work.

Again and again, under the watchful eyes of the simulation instructors, Borman's crew rode the rumbling Saturn booster off the earth. Before long the instructors were throwing in simulated malfunctions—engine failures, errant trajectories, electrical trouble—and in each case the astronauts had to respond, quickly and correctly. Sometimes, for example, when the Saturn strayed dangerously off course, the only thing to do was abort. Early in the launch, Borman would simply take hold of the abort handle in his left hand and twist it, setting off an automatic chain of events to fire a rocket, called the escape tower, perched on the spacecraft's nose. It would yank the command module away from the careening booster and then depart, leaving Borman's crew to a high-speed, roller-coaster ride through the atmosphere before parachuting into the Atlantic. But if the mishap struck later, when the escape tower was already cast off and the Saturn had propelled them to the fringes of space, Borman's crew would have to cut loose from the booster and then, racing the clock, fire up the service module's big rocket engine to kick them onto the right trajectory for a safe splashdown. Emergencies like that test the mettle of a space crew, and there were plenty of them in all phases of their simulated lunar journeys: An engine malfunction during Lunar Orbit Insertion. Communications trouble on the way home. Computer failure just before reentry. Yes, the simulation instructors would readily admit: if things went wrong it would take a highly skilled and experienced test pilot to escape disaster.

But none of them could have mastered it all, not with only four months to train, and so each man had to specialize. As commander Borman had the overall mission responsibility; he also trained to steer the command module through its fiery reentry in case the computer went out. Lovell, meanwhile, concentrated on navigation; he would use the command module's sextant to make star sightings throughout the voyage to verify the craft's trajectory. In lunar orbit, he would take bearings on lunar landmarks. And if the radio went out at any time, it would be up to Lovell to get them home. And to Bill Anders fell the role of command module systems expert.

Then there was the service module, the 16-foot cylinder that would be joined to the command module's base throughout the voyage; it contained the rocket engines, propellants, oxygen, and electric-power fuel cells. All of it was Anders's responsibility, and Borman had told him in no uncertain terms, "I'm expecting you to make sure it works."

The first time Anders climbed into the simulator, sometime in 1967, he was greeted by an almost bewildering array of switches, dials, displays, and controls. Over time they grew familiar, as Anders learned to navigate the miles of wiring, the intricate array of pipes, valves, tanks, antennas, relays. Subjected to one simulated malfunction after another, he came to play the spacecraft like a virtuoso. If plumbing to the big rocket engine became blocked he rerouted the flow of propellants. When one of the power-producing fuel cells conked out he rearranged electrical connections to keep current flowing. In time, these sessions in the simulator seemed less like spaceflight than a ride in a game box. Even when he wasn't in the simulator, Anders was learning his machine, immersing himself in a world of fuel cells and evaporators, relays and accelerometers. Seeming bits of minutia that Borman simply didn't have time for became Anders's turf. To Borman, Anders was like a precocious younger brother, and it rankled him that the junior man sometimes knew more about the hardware than he did—especially when Anders realized why one of Borman's ideas wouldn't work: "Goddammit Anders, *show* me!"

In his intimacy with the machine, Anders found reassurance. He couldn't help but be impressed with the beauty of the Apollo design. Nothing exemplified this better than the SPS rocket engine, rated at 20,500 pounds of thrust, that would get Borman, Lovell, and himself into and out of lunar orbit. The beauty of the SPS was its simplicity: It was a no-frills rocket engine. There were no fuel pumps because pumps have moving parts that could break down. Instead, pressurized helium forced the propellants into the combustion chamber. It had no ignition system; none was required because the SPS burned hypergols (in this case hydrazine and nitrogen tetroxide), chemicals so reactive that they need only come in contact with each other to explode into hot gases. As long as the valves opened, that engine would *fire*. And for all its simplicity, the SPS was an extremely high performance rocket engine.

And the SPS also exemplified Apollo's other watchword, redundancy. The fuel tanks, pipes, valves, quantity sensors, everything except the combustion chamber and the huge engine nozzle itself, came in duplicate sets. If one component malfunctioned, there was a backup standing by.

But would it work? In the moment of truth would it blast them out of lunar orbit on a course for home, or would there be only silence? According to its designers, Apollo was "three nines" reliable: The odds of an astronaut's survival were .999, or only one chance in a thousand of being killed. Anders, for one, didn't believe it; the odds couldn't possibly be that good. Soon after he found out about the lunar mission he took time to ponder his chances of coming back from Apollo 8, and he made a mental tabulation of risk and reward in an effort to come to terms with what he was about to do. Frank Borman would never have conducted such an exercise.

Danger—like speed—is an experience that follows a logarithmic, ever-shallowing curve. Once you've been exposed to a lot of risk (and Anders had been, as a fighter pilot) then a *hell of a lot of risk* doesn't seem like that much more. So it wasn't just his own welfare that he considered. He thought of his wife, Valerie, and of their five young children. He had barely any life insurance; no one would sell it to an astronaut about to go to the moon. Was it irresponsible to hazard leaving them alone so prematurely? Was a seat on the first flight around the moon worth that chance?

In Anders's mind there were three factors. First, and most important, it was the greatest opportunity for adventure and exploration a man could have in the twentieth century. He would see the hidden face of the moon with his own eyes, and he would be first.

Second, as an American and as a military man he felt a sense of duty, and Apollo was his country's most important mission. As a pilot, he could not deny Apollo 8's lure: the first manned test flight of the Saturn V, the first voyage of the Apollo spacecraft to lunar distance.

Finally, Apollo 8 would put his name in the history books. Somehow—Anders couldn't say exactly why—all those things combined were worth it. If he had two chances in three of coming back—and he figured the odds were probably a good bit better than that—he was ready.

• • •

As the summer ended NASA lifted the veil of secrecy surrounding the circumlunar mission. As if to heighten the sense of urgency, the Soviets achieved a notable advance in mid-September: an unmanned spacecraft called Zond 5 flew a figure 8 loop around the moon and splashed down in the Indian Ocean, all under automatic control. This was the same type of free-return trajectory being planned for Apollo 8. Before long, many speculated, a man would follow.

Meanwhile, in Houston, the flurry of activity and anticipation surrounding Apollo 8 almost eclipsed Apollo 7, the first manned Apollo flight. And yet, Apollo 7's outcome would make or break the decision for a moon flight. Everyone knew the odds were small that a craft as complex as the command module would sail through its maiden voyage with no major problems—but that is just what happened. On October 11, Wally Schirra, Donn Eisele, and Walt Cunningham rode a Saturn 1B booster into orbit for an eleven-day shakedown cruise. Schirra's crew met every engineering objective and performed every scientific experiment in the flight plan, and then some. Only minor problems turned up in the command and service modules. The big SPS engine passed every test.

Perhaps the biggest problem of Apollo 7 was its irascible commander. Schirra's performances in Mercury and Gemini had won him admiration, but this time around he surprised observers by acting ornery and short-tempered. He caught a monster head cold early in the mission and sneezed and sniffled his way through the eleven days. Along the way, he battled with Kraft's flight controllers, canceling a scheduled television broadcast from space, then refusing to accept changes in the flight plan. At times Schirra's bluster seemed to rub off on his crew. After one test, Eisele quipped with obvious annoyance that he wanted to talk to "the man, or whoever it was, that thought up that little gem." The man was Glynn Lunney, the flight director. By the time Schirra and his crew were ready to come home, Kraft's men were so angry that one of them offered, only half-jokingly, to bring Apollo 7 down into a typhoon—and Kraft was half-ready to have him do it. Years later Schirra would pin his bad mood on some broken promises, including a decision to launch Apollo 7 under less than ideal conditions. Other astronauts felt there was no excuse; to make such outbursts with the world listening, they would say, was plain unprofessional. Cunningham would publicly surmise that the grind of training for almost a decade had simply worn the man out. But for Chris Kraft, there was another explanation for Schirra: The Fire—which had claimed his next-door neighbor Gus Grissom—had scared the daylights out of him.

Discord aside, Apollo 7 was better than anyone could have hoped for. Now NASA made it official: Apollo 8 was going to the moon. Within weeks, however, the Soviets made their next move. Eighteen months after the Komarov tragedy, they lofted Soyuz 3 with Georgi Beregovoy aboard and returned him four days later. Whether another Soyuz was being readied for a circumlunar mission, no one knew.

In Houston the space center geared up to support Apollo 8, and at the center of it all, Frank Borman tried to keep his mission from growing beyond bounds. If someone brought up an idea that wasn't strictly essential to the mission, Borman cut him off. It didn't matter how small the extra—even an improved type of in-flight meal—Borman's reaction was always swift and negative. Some of his reluctance was understandable, but to Jim Lovell, his behavior seemed entirely in character. In the last days of Gemini 7, Kraft wasn't the only one who tried to keep Borman's spirits up. When Borman talked about reentering early, Lovell encouraged him to keep going. When Borman worried they might have to come down in some remote stretch of ocean Lovell reassured him, "The navy would find us, Frank. They know what they're doing." Not that Borman's fretting affected his performance. But as far as Lovell could tell, behind that macho, take-charge exterior was a very apprehensive astronaut.

Was Borman more uncomfortable with the lunar mission than he let on? Years later, he would say that to his mind, his worry was simple: he wanted to make sure he and his crew didn't get handed a mission they couldn't perform. He'd seen Gemini commanders accept flight plans so crowded with tests and experiments that nobody could have done them all; he was determined that wasn't going to happen to him. In Borman's mind anything that wasn't essential to the mission—circle the moon ten times and come home—was inviting trouble. They were going to the moon—that was enough!

• • •

By late November, with a month to go until launch, Borman and his crew were working seven days a week, spending three or four days in the simulators at the Cape and coming back to Houston for meetings with mission planners. Now there was no time for weekend parties or dinners with friends. Sunday existed only to wade through the piles of mail on their desks. So demanding was the pace that they had less time than they wanted for exercise, and they were a somewhat tired crew by the time LBJ threw a bon voyage party for them in Washington.

As the fall of 1968 wore on, apprehension surfaced once more within NASA. Would the Soviets try to beat Apollo 8 to the moon? Due to the latitude of the Baikonur launch site in central Asia, their lunar launch window opened in early December—well before the Americans'. The weeks passed with no news from the Soviet Union. By the beginning of December, everyone at NASA realized, with great relief, that there would be no Soviet circumlunar attempt in 1968. The field was clear; now it remained to send Apollo 8 on its way.

On December 10, with only eleven days to go until launch, Borman, Lovell, and Anders flew their T-38's to the Cape for the last time. From now on, they would live there, in the Spartan crew quarters. Each man had a small bedroom, and shared a living room, a conference room with maps of the moon and the stars on the walls, and a dining room that was something like a ship's mess hall. The decor was strictly Holiday Inn, but there were compensations—not the least of which were the high-calorie meals served up by a former tugboat cook named Lew Hartzell, brimming with steak and potatoes and mile-high sandwiches. Access was highly restricted, to protect the men from last-minute illness. The sign at the entrance to the crew quarters read, "No one with a cold, or symptoms of a cold, may pass beyond this point."

In this comfortable monastery, the three men settled in for their last days on earth. Occasionally—less than they would have liked—they had a chance to go for a run or to work out in the nearby exercise room. They reviewed the flight plan, and received briefings on the readiness of their spacecraft and Saturn booster. And above all, there were the daily sessions in the simulator; at times, Apollo 8 seemed to be an exercise in switches and valves and maneuvers, not the first flight away from the earth. But on December 20, the day before launch, Borman's crew had a visitor who brought home the historic impact of what they were about to attempt.

Charles Lindbergh, one of the most enigmatic figures of the twentieth century, emerged from his retreat to visit Borman, Lovell, and Anders in the crew quarters. Forty-one years after flying solo across the Atlantic, Lindbergh appeared tall, tanned, and surprisingly fit for his sixty-six years. Accompanied by his wife, Anne, herself an accomplished pilot and author, Lindbergh arrived to have lunch with three fellow fliers about to navigate an ocean far more vast and untraveled.

To most of the astronauts Lindbergh had been a boyhood hero, and Borman was no exception. Now, in the quiet of the crew quarters, it was just one flier talking to other fliers. Gathered around the table with Lindbergh and his wife, Borman's crew and their backups shared questions, recollections, and humor. They were fascinated by his accounts of meetings with Robert Goddard, whose experiments with liquid-fueled rockets in the New Mexico desert had foretold the space age (and fired the imagination of a teenage Jim Lovell). Goddard had conceived of flights to the moon, Lindbergh said, but was daunted by the fantastic cost of the venture—he had mused, "it might cost a million dollars." With that, the room exploded in laughter.

The great flier asked Borman's crew about the navigation system that would take them to the moon. Then he told the astronauts how before his own trip, he and a friend had gone to the library, found a globe, and measured, with a piece of string, the distance from New York to Paris; from that he had figured out how much fuel he would need for the flight. Lindbergh asked how much fuel the Saturn V rocket would consume during its climb into space; one of the astronauts did a quick calculation: 20 tons per second. Lindbergh smiled. "In the first second of your flight tomorrow," he said, "you'll burn ten times more fuel than I did all the way to Paris."

• • •

Eight miles away, the Saturn V towered above Pad 39-A, looking more like a skyscraper than a rocket. Some 363 feet tall—about six stories higher than the Statue of Liberty—the Saturn was more than three times the size of the Titan missile Borman and Lovell rode in Gemini 7. It was, far and away, the most powerful thrust machine ever flown, the crowning achievement of Wernher von Braun and his team of rocket engineers at NASA's Marshall Space Flight Center in Huntsville, Alabama. Through it, the sheer difficulty of reaching the moon was made visible. It was a monument to human audacity.

For now, the Saturn stood empty. But overnight, even while Borman's crew slept, technicians would ready it for departure. By morning its enormous fuel tanks would be filled with super-cold propellants, until the rocket would contain the explosive energy of an atomic bomb. This engineering masterpiece was designed to tame that energy and liberate it in a sustained, fiery release of power. Public relations people for the contractors that built the Saturn were always coming up with new analogies to convey its incredible might. Someone estimated that the thrust from the booster's first stage engines at liftoff would equal more than twice the hydroelectric power that would be obtained if all the rivers and streams in North America were channeled through turbines. Everything about the Saturn V was grossly out of scale with the rest of the world. For example, each of the five F-1 engines that powered the first stage had an engine bell measuring 12 feet in diameter. At liftoff, those engines would deliver a combined thrust of 7.5 million pounds—about 160 million horsepower.

Three separate stages would do the work of pushing Apollo 8 off the earth and toward the moon. The first stage alone, nearly half a football field long, would burn half a million gallons of kerosene and liquid oxygen in just two and a half minutes, cutting off at a height of 40 miles, and then it would fall away. The second stage would fire for just over six minutes until its supply of liquid hydrogen and liquid oxygen was spent. By then Borman's crew would be 120 miles up, and all that would remain would be a three-minute push from the third stage to place Apollo 8 in orbit around the earth. The third stage would do its best work three hours later, when Borman's crew would execute the Translunar Injection maneuver to break the bonds of their home planet.

For now, the Saturn stood in the embrace of its steel launch tower, waiting for the launch window to the moon to open. Evening twilight revealed the moon's thin crescent glowing briefly in the west—Anders went out in the parking lot with a couple of visitors to look at it—then slipping beneath the horizon of a turning earth. At nightfall the floodlights at Pad 39-A came on, turning the Saturn into a huge glowing monument and reaching past it into the Florida night.

In the crew quarters, Frank Borman lay awake in his room, confronting his darkest fear. It wasn't blowing up on the Saturn, being stranded in lunar orbit, or being burned to a cinder in reentry. Some people had wanted him to make a tape in case he didn't come back; he'd scoffed at the idea. Sure, he had some anxiety—who wouldn't have, preparing to go to the moon for the first time? But if he thought he wasn't coming back, he wouldn't be going.

Borman was afraid of one thing: that they would be in earth orbit and some malfunction, however small, would arise, and the managers in Houston would cancel the lunar mission right there. He would be stuck with the alternate mission, ten long days in earth orbit doing nothing but keeping the spacecraft going. He hated the thought of it. And as the night dragged on, while technicians worked under the floodlights of pad 39-A, readying his booster for an early-morning launch, Borman prayed not that he would come back from the moon, but that he would have the chance to go.

II: A Hole in the Stars

Saturday, December 21, 1968

Frank Borman's sleepless night came to an end at a few minutes past 2:30 A.M. when Deke Slayton came to the door to awaken him. The night was clear, Slayton told him, and the weather at liftoff—set for 7:51 A.M.—was expected to be good. Minutes later, after undergoing a final medical exam, Borman, Lovell, and Anders sat down to their last meal on earth, the traditional astronaut's breakfast of steak and eggs. Deke Slayton was there, and Al Shepard, along with backup crewmen Neil Armstrong and Buzz Aldrin, scientist-astronaut Jack Schmitt, and the man who had envisioned this mission, George Low. Years later Anders would remember the conversation as decidedly unremarkable, his mood as matter-of-fact. And if some of the support people seemed extra careful, extra serious, even a little nervous this morning, that was in striking contrast to the three of them who were about to leave the planet.

When this brief, earthly ritual ended the three men headed for the suiting room. There technicians, wearing surgical masks as part of the preflight health quarantine, were waiting to help them into their space suits. Someday suits like these would protect astronauts on the surface of the moon, but on this flight they were merely a precaution against a loss of cabin pressure during launch. They were also fireproof, thanks to a pristine white covering of glass-fiber Beta cloth coated with Teflon. Hidden from view were layers of insulation and pressure restraints, special joints and cables to facilitate motion when the suit was pressurized, all of which made these suits seem less like garments than wearable machines. Each of the three astronauts, clad in long johns, climbed into his modern-day suit of armor. It was all familiar from tests and practice runs, but this morning the technicians had little tension breakers—a tiny stocking hanging from a paper Christmas tree for Frank Borman, a clean white handkerchief for the pocket of Jim Lovell's space suit.

Next the men donned communications hats resembling the headgear of a World War I flying ace. Oxygen hoses were mated to metallic blue and red connectors on the chest. Then came black rubber pressure gloves, joined to rotating rings at the wrists of the suit. Finally, a clear bubble helmet was lowered into place and snapped onto a metal neck ring. At that moment, each man was a self-contained universe. Aside from the occasional voice of a technician in their headsets, they heard only the sound of their own breathing; they felt cool oxygen flowing past their faces. For a time they rested, letting the pure oxygen purge their bloodstreams of nitrogen. Then, at last, it was time to go. Toting portable oxygen units they headed down a long corridor to the outside with stiff-legged strides. At the entrance to the Flight Crew Training Building they were greeted by the glare of television lights and a small crowd of well-wishers. Only a hint of applause penetrated their bubble helmets as they boarded a special transfer van for the 8-mile ride to Pad 39-A.

Brilliantly lit, the Saturn V stood naked next to its launch tower, its tanks full of cryogenic propellants, spewing plumes of vapor into the predawn darkness. The sight of it filled Borman with awe. Accompanied by a suit technician, the three men entered into the service tower, and within that complex of steel girders, massive fuel pipes, and machinery, they boarded a small elevator and ascended past the Saturn's huge first stage, then the second and third stages, to the 320-foot level. There they strode across an access arm to the small White Room. First Borman, then Anders, climbed into the spacecraft, assisted by the closeout crew. Jim Lovell, meanwhile, waited alone for several minutes outside the crowded White Room, within the metal gridwork of the access arm. From this lofty perch he gazed down at the most powerful rocket in existence, and all at once the realization came to him—*My God, they're serious!*—that the very thing they had all been talking about and practicing for four months was about to happen, that NASA was really going to seal him up in that command module and fire him off to the moon. In the far distance Lovell could see the headlights of cars making their way to viewing sites: Thousands of spectators had descended on the Cape, eager to witness the departure of the first moon voyagers. For a time, sealed within his private universe, Lovell savored this communion with awe.

Then Lovell joined his crewmates, sliding on his back into the center couch. He lay still while the closeout crew hooked up oxygen hoses and communications lines. Strapped in, fully suited, he had almost no room to move; he was literally rubbing elbows with his crewmates. To his left, Borman lay before gauges and readouts for the Saturn V; he would keep a watchful eye on the booster's performance during the ascent into space. To Lovell's right, Anders manned the controls for the spacecraft's electrical and communications systems. From the center couch, Lovell would operate the command module's onboard computer and monitor their trajectory into space.

At last it was time to close the hatch. Rookie astronaut Fred Haise, who had been inside the spacecraft checking switch positions when Borman's crew arrived, now wriggled underneath the couches and through the open hatchway, then offered his hand in farewell. The technicians swung the massive hatch closed and locked it, sealing the three astronauts inside. It was 5:34 A.M.—T minus 2 hours, 17 minutes and counting.

Inside Apollo 8 all was quiet. Within their helmets Borman's crew heard the voice of test conductor Dick Proffitt talking to them from the Launch Control Complex 3½ miles away, where hundreds of engineers monitored data from the spacecraft and booster. For most of the next hour they followed his instructions, setting switches, as part of the complex process of readying the command module for flight. Over in the right couch Anders was amazed at how calm he felt. It was just like a simulation; he was almost bored.

Around Pad 39-A the glare of floodlights yielded to a clear dawn. During a break in the switch-settings, out of the corner of his eye Anders noticed something moving; he looked over at a window in the protective heat shield, called the boost protective cover, that covered the command module. A hornet buzzed around and landed, worked for a short while, then flew off and returned. She's building a nest, Anders thought, and did she pick the wrong place to build it!

Far below, the great rocket was filling with fuel. If disaster threatened now, while the Saturn was still earthbound, the men would scurry out of their craft and into a small gondola attached to a slide wire stretching from the launch tower to a concrete bunker. And if there wasn't time for that, Borman could, with the twist of a handle, fire the escape rocket poised above the command module's nose, whisking them up and away from the disaster. But an abort from the pad involved considerable risk of injury, and that was something everyone hoped would not be necessary. Meanwhile, 1,000 yards from the pad, armored tanks stood at the ready in case the men had to be rescued. Inside Apollo 8, Borman asked Dick Proffitt, "How is the booster doing?" Proffitt assured him that all was well.

Up to now Borman, Lovell, and Anders had known, in the back of their minds, that there was the possibility that a malfunction would turn this countdown into just another practice run and they would have to climb out and try again another day. But now, as the count reached T minus 15 minutes, there seemed no doubt: they were really going. In the control center the flight surgeons monitoring telemetry from the spacecraft saw Borman's heart rate start to climb. With 7 minutes left Dick Proffitt took his final status check, and in the middle of his lengthy poll he called out, "Spacecraft" and Borman, Lovell, and Anders answered together, in full voice: "Go!"

Now the pace quickened. With just 5 minutes to go the White Room and its access arm swung away. At 3 minutes, 7 seconds, the launch pad's automatic sequencer took over, monitoring the last influx of propellants and controlling the final events before liftoff. By T minus 60 seconds all three stages were fully pressurized; 10 seconds after that the booster went on its own power. With 45 seconds to go Borman confirmed the last switch settings to ready the command module for launch. Thirty seconds, now 20. And now test conductor Proffitt began to count: "*Nine, eight, seven . . .*" For a moment the men heard faint sounds of fuel pouring through manifolds to the five huge F-1 engines.

"*. . . Ignition.*"

Suddenly the base of the Saturn spawned a cauldron of smoke and flame that gave way to a river of golden-white fire, spilling out from both sides of the launch platform. Yellow smoke billowed into the chill morning. For long seconds the behemoth strained against a set of enormous hold-down clamps, while the first-stage engines built up to full power. Up in the command module there was no noise or vibration, but somehow Borman and his crew could sense the growing power far beneath them. At T minus 3 seconds there came a distant rumbling, like thunder on the horizon, that swelled into a roar. Finally, in the midst of the heightening commotion came a sudden, mild jolt, and Borman's crew heard Proffitt cry, "*Liftoff!*"

Borman glanced at the mission clock on the instrument panel. "Liftoff," he called, his voice charged with adrenaline. "The clock is running." The Saturn ascended, seemingly wracked by spasms of uncertainty, steering nervously past the launch tower, its engines correcting and recorrecting in quick, spasmodic jerks. Up in the command module these corrections translated into sudden, jarring motions that threw the men from side to side against their harnesses. No simulation had even hinted at the violence of this ride. In the post-flight debriefing Anders would say only that he was "impressed" by the Saturn's "very positive control," but in reality, he felt as if he were helpless prey in the mouth of a giant, angry dog. After all those simulations—if the first 10 *seconds* were this different, what would the rest of the flight be like?

Long seconds passed in thunder while the rocket climbed its own length and still higher. Borman's crew barely heard Proffitt shout, "Tower clear!" The danger of collision with the launch tower past, they kept climbing. Now the rocket turned and headed onto its programmed flight path. "Roll and pitch program!" called Borman, his voice shaking with the vibrations of the ride. Meanwhile, the Cape launch center yielded command to mission control in Houston, where Mike Collins was serving as Capcom. If the booster suddenly went berserk and mission control ordered an abort, it would be Collins who would relay the command, but inside Apollo 8 the Saturn's roar was so loud that Borman's crew would not have heard him. And they could no longer hear each other; they were no longer a crew but three passengers riding in a fury of sound.

Just 40 seconds after liftoff the Saturn went supersonic and the ride smoothed out. Now there was quiet again. In Apollo 8's left seat, Borman kept a watchful eye on the trajectory readouts. If at any point the Saturn should turn angry he would be able to whisk them away from it by twisting the abort handle, setting off the escape rocket, but the beast was behaving itself beautifully. And from Houston, Mike Collins's message of reassurance came through loud and clear: "Apollo 8, you're looking good."

The Saturn tore through the atmosphere on a great bonfire column of light hundreds of feet long. Under the commands of its own gyroscopic brain the booster arced slowly over until it was almost horizontal, following the curve of the earth, picking up speed and receding into a deep blue sky. As the rocket penetrated the rarefied upper atmosphere the exhaust fanned out into a broad plume of golden flame.

In the command module Borman and his crew scanned the instruments and felt the mounting force of acceleration as the massive load of fuel in the first stage was consumed. Soon their chests begin to flatten. The g-meter registered three times the force of gravity, now four, and was still climbing. Their arms were leaden. Then, just as the g-meter hit 4½, the forces of acceleration abruptly vanished as the first stage shut down on schedule. At that moment, the men could have been sitting on a catapult. They flew forward against their straps with tremendous momentum—Anders was sure he would go right through the instrument panel—but their harnesses held them firmly. Borman felt the sudden jarring and thought warily of the stress placed on the booster.

Suddenly, right on schedule, there was a muffled bang as jets of smoke and flame heralded the departure of the now unneeded emergency escape rocket, taking the boost protective cover along with it. Daylight streamed into the cabin as the command module's windows were uncovered. For a stolen moment rookie Anders glimpsed a view available only to the space traveler, a vivid bright arc of ocean and clouds against a darkening sky.

Five minutes into the flight now; it was amazingly quiet. The g-forces had lessened. And there were more welcome words from Mike Collins: "Apollo 8, your trajectory and guidance are Go."

"Thank you, Michael," said Borman, sounding pleased.

Apollo 8 sped out of the last fringes of the atmosphere, picking up speed. Now 7 minutes. At about 8 minutes a rapid vibration set in, the same kind that had rattled the previous Saturn V to the point of malfunction. Thankfully, it did not build beyond a mild shaking, but Borman was relieved when the second stage shut down and fell earthward, its work done. At 8 minutes, 45 seconds the third stage kicked in with a mild jolt and chugged along, getting up the last bit of velocity until, 11½ minutes after liftoff, it too fell silent. Apollo 8 was in orbit.

• • •

A hundred and fifteen miles up, circling the earth at a speed of more than 17,400 miles per hour, Apollo 8 moved in that exquisite balance between gravity and momentum called orbit. Had they done nothing, Borman, Lovell, and Anders would have remained there for days, slowed only by friction with the scant upper atmosphere. The bold difference in this flight was set for just 2½ hours from now, late in their second orbit. At that time they would reignite the third-stage engine for a little over five minutes and tip the balance between gravity and momentum enough for Apollo 8 to leave earth orbit and reach the moon's gravitational sphere of influence. And that was what Borman lay awake worrying about.

"I don't want to see you looking out the window!" Borman's voice. They had a timeline to stick to; Anders knew that as well as his commander. But to be in orbit for the first time and not look outside! That was easy for Borman to say; he and Lovell had been here before. Once or twice, when Borman wasn't watching, Anders stole glimpses of the earth, a magnificent panorama of color and bright clarity that filled his window. Brilliant white plumes and swirls of cloud crisscrossed land and ocean. Entire continents swept past in minutes. Somewhere over the midnight earth—was it New Zealand?—lightning glowed in the clouds far below like flash bulbs going off under wads of cotton. And when they came over the coast of California he spotted San Diego, the scene of his childhood explorations of hills and rabbit trails. He wanted to linger here, taking in the ever-changing beauty of his home. Anders wished they weren't going to the moon—not yet.

But the time flew by. There was a small mishap when Lovell, under one of the couches to adjust a valve, accidentally inflated the life vest attached to his space suit; he would always remember the disgusted look on Borman's face. But aside from that, everything went like clockwork. What so many had doubted, including Borman, was actually happening: Apollo 8 was checking out perfectly. And at last came the word the three astronauts had been waiting for, and ironically, it came from the man originally slated for Apollo 8's center seat, Mike Collins. One of the most momentous directives ever given, it was spoken with remarkable calm and in the coded language of space flight: "Apollo 8, you are Go for TLI."

As Apollo 8 drifted through darkness over the Pacific the last minutes ticked by until the scheduled ignition of the third stage. If everything went as planned, Borman and his crew would be mere passengers while the computer did the work. With 10 seconds to go until ignition the computer gave a coded message to the astronauts, a flashing number 99. Translated, it said, "Are you sure you want to do this?" Lovell answered by pushing the button marked "PROCEED," and moments later, at Mission Elapsed Time 2 hours, 47 minutes, and 37 seconds, the third-stage rocket came to life with a long, gentle push. This time, the ride really did feel like the simulator; the men sank into their couches with barely more than the force of normal gravity.

Immediately, they sensed the rocket veering to one side as it headed out of earth orbit and onto a course for the moon. Trajectory specialists in Houston hawkeyed the moonship's path and sent word, via Mike Collins: "You're looking good here, right down the old center line." Borman kept his eye on the attitude indicator, ready to take over steering if the booster's automatic system failed. Anders monitored the pressures and temperatures in the fuel tanks. And Lovell called out their ever increasing speed from the computer readout. The numbers galloped upward: 30,000 feet per second . . . now 33,000 . . . and finally, 35,532 feet per second, some 24,226 miles per hour, the speed necessary to reach the moon on a free-return path. At that instant, 5 minutes and 18 seconds after ignition, the computer shut down the engine automatically. Apollo 8 was on its way to the moon.

From mission control Collins had good news for the departing moon voyagers: "We have a whole room full of people that say you look good." And one of those people was Chris Kraft, sitting in the back row of the control room. Kraft rarely came on the radio during a mission, and Borman was surprised to hear his exultant sendoff: "You're on your way—you're really on your way now!"

Still, inside the command module there was nothing to convey this departure to the senses, no sensation of speed whatsoever, just numbers on the computer. That changed dramatically when Borman cut loose from the spent third-stage booster, pulled away with a burst from the service module's small maneuvering thrusters, and spun Apollo 8 around. At first, the sight of the third stage itself—a hulking cylinder aglow in the unfiltered sunlight of space—caught their attention. But then, as the spacecraft turned, Borman's crew could see the place they left behind, not a landscape but a *planet*, a luminous sphere whose roundess was apparent to the eye. Apollo 8 was departing at such fantastic speed that the men could see their world receding from them almost as they watched. Already the entire globe fit neatly within the round window of the command module's side hatch.

Whatever names humans gave their earth, it deserved to be called the Blue Planet, for its dominant aspect was the vivid, deep blue of oceans. In striking contrast were the clouds, brilliant white flecks and streamers that embraced the globe, swirling along coastlines and across oceans. Where land masses peeked through, the vivid oranges and tans of the deserts were easy to spot. More elusive were the jungles and temperate zones; because their verdant hues did not easily penetrate the atmosphere, they showed up as a bluish gray with only a hint of green. And everywhere, beyond the planet's bright, curved edge, a blackness so deep as to be unimaginable.

Right now, though, it wasn't time to look at the earth; Borman was more concerned about the cast-off third stage. As the flight plan called for, Borman had pulled up within a few dozen yards of the booster, to demonstrate the maneuvers that future crews would use to extract a lunar module from its berth. But Borman, anxious to save fuel and to avoid any maneuvers that would affect their trajectory, did not want to prolong the exercise. Furthermore, he knew the booster was scheduled to blow off its excess fuel sometime in the near future—and when that happened it would be better not to be anywhere nearby. All he wanted to do was get away from it. After conferring with mission control, Borman pulsed the hand controller and fired the maneuvering thrusters to pull away.

But the third stage seemed to be following them. Already it was spewing fans of brilliant ice particles into space, reminding Borman of a huge lawn sprinkler. For the better part of an hour Borman made anxious queries to Houston on how to get away without disturbing the free-return trajectory. Lovell's attempts to realign the command module's navigation platform were to no avail; the sky was full of "false stars" from the booster, and it was impossible to find any real ones. And right now the best landmark in this dark, sunlit ocean—the earth—was out of view. When Collins in mission control outlined a small evasive maneuver, Borman replied, "Okay, as soon as we find the earth, we'll do it." In mission control Borman's words triggered brief, amazed laughter.

Finally, after more than an hour, Anders saw the world drift into his right-hand window, and after more deliberations with Houston Borman fired the maneuvering thrusters once more. Slowly, the third stage dwindled until it was just a bright star, and Apollo 8 was alone in the translunar void.

• • •

"Would you pass me the flight plan, Bill?"

Anders reached for the three-ring book floating in the air next to him. He gave the book a gentle push and it drifted across the cabin into Lovell's open hand. Apollo 8 was coasting moonward like a baseball fleeing the strike of the bat, and everything inside it—including the three astronauts—was weightless. There was a moment, back in earth orbit when he unbuckled his harness and his body hung literally in midair, suspended above his couch. For years he'd heard other astronauts talk about zero g, but there was no way to anticipate it. No simulation could have prepared him.

In the first few hours of the mission there was no time to enjoy this strange new world. But sometime after TLI, well on the way to the moon, Anders climbed out of his space suit and found a freedom unlike any he had ever experienced. Wearing only a pair of Beta-cloth coveralls over his long johns, he floated unencumbered. Suddenly the cramped cabin seemed to grow roomy. With a push of a fingertip against his couch he propelled himself slowly past the instrument panel into the open area they called the lower equipment bay, which housed storage lockers and Lovell's navigation telescopes. There he found enough room to stretch out, or to hang inverted, with his feet up by the top hatch and his head pointing at the floor. "Up" and "down" were whatever he wanted them to be. He could even float underneath the seats, among coolant pipes and storage compartments. The command module seemed to have suddenly doubled in size. And in zero g it became a wonderland. Water formed perfect shimmering, dancing spheres. Cameras twirled and tumbled with the touch of a fingertip, or lingered in midair when not in use. This was a world of action and reaction, a three-dimensional ice rink. There wasn't room for gymnastics, but in the lower equipment bay Anders had enough space to tuck his body into a ball; a nudge against the wall set him tumbling, like an acrobat magically suspended at the top of his arc. It would have been great fun but for one thing: it was making him ill. Suddenly, in the midst of his acrobatics, Anders felt a wave of nausea come over him.

For years the NASA doctors had worried about motion sickness in space, fearing that zero g would confuse the inner ear, which gives the body its sense of up and down. But no astronaut had ever returned from orbit with anything but glowing enthusiasm for weightlessness. Borman and Lovell, for example, had spent two weeks in free fall with no ill effects. But the command module had something that the phone-booth-sized Gemini didn't—room to move. The doctors feared that simply by floating around, an astronaut would push his vestibular system over the edge.

Still, Schirra's crew came back from Apollo 7 with no complaints. Maybe it came down to the individual, and there was no way of knowing who would be sick and who wouldn't. The only thing Anders knew was that he needed to be still for a while; soon he felt better. Several hours later, Anders not only wasn't sick, he was so comfortable that he felt as if he had always been weightless.

But one aspect of weightlessness was so unpleasant that even the thrill of exploration didn't make up for it. If this marvel of engineering called Apollo had one major design flaw, it was the "Waste Management System," perhaps the most euphemistic use of English ever recorded. For urine collection there was a hose with a condom-like fitting at one end which led, by way of a valve, to a vent on the side of the spacecraft. On paper, at least, it seemed like a reasonable if low-tech way to handle urinating in zero g, assuming you got over your anxiety about connecting your private parts to the vacuum of space. You roll on the condom, open the valve, and it all goes into the void where it freezes into droplets of ice that are iridescent in the sunlight. One astronaut answered the question "What's the most beautiful sight you saw in space?" with "Urine dump at sunset."

In reality, using the urine collector didn't work out so well. For one thing, it could be painful. If you opened the valve too soon, some part of the mechanism was liable to poke into the end of your penis, which tended to prevent you from urinating. And at that point, as if to confirm your worst fears, the suction began to pull you in. Now you were being jabbed and pulled at the same time, so you shut off the valve, and as the mechanism resealed itself it caught a little piece of you in it. It only took one episode like that to convince you not to let it happen again. Next time you had a strategy: start flowing a split-second before you turn on the valve. But once you began to urinate the condom popped off and out came a flurry of little golden droplets at play in the wonderland, floating around and making your misfortune everyone's misfortune. And in no time the whole device reeked; it was an affront to the senses just sitting there.

Anders got used to the urine collector, though, and he got used to mopping up afterwards. But there was no getting used to the other part of the Waste Management System. Tucked away in a storage locker was a supply of special plastic bags, each of which resembled a top hat with an adhesive coating on the brim. Each bag had a kind of finger-shaped pocket built into the side of it. When the call came you had to flypaper this thing to your rear end, and then you were supposed to reach in there with your finger—after all, nothing *falls*—and suddenly you were wishing you'd never left home. And after you had it in the bag, so to speak, you had one last, delightful task: Break open a capsule of blue germicide, seal it up in the bag, and *knead the contents* to make sure they were fully mixed. At best, the whole operation was an ordeal. In the confined space of the command module, your crewmates suffered too. One of the Apollo 7 astronauts said the smell was so bad it woke him up out of a deep sleep. When Schirra's crew came back they wrote a memo about it: "Get naked, allow an hour, have plenty of tissues handy . . ." Anders saw the memo and heard the stories, and before the mission he decided he was going to do everything in his power to avoid it. The food on Apollo 8 was specially formulated to produce as little residue as possible, but Anders wasn't taking any chances. He started his own low-residue diet a few days before launch. Six days was a long time, but he was determined. He'd go all the way to the moon and back on Lomotil, if he had to.

• • •

The hours passed in steady activity. By noon, Houston time, some five hours into the flight, all three men had doffed their bulky space suits and stowed them underneath the couches. Around 1 P.M. Lovell began taking star sightings for navigation. And there were more tasks into the afternoon—replace an air-filtration canister, look after a battery, service a power-producing fuel cell. At 6 P.M. the astronauts made the first, brief firing of the service module's SPS engine. Though it lasted only two seconds—the engine slammed Borman's crew back into their couches, then released them—it was enough to correct Apollo 8's path after Borman's earlier maneuvers to get away from the third stage. Just as important for engineers in Houston, the firing gave a crucial look at the engine's performance in space. The SPS passed its first brief test with flying colors.

By then, more than eleven hours had passed since launch, and aboard Apollo 8 it was getting to be a long day. It was time for Borman to get some sleep. The flight plan called for at least one man to be awake at all times to keep an eye on the spacecraft and maintain contact with Houston. For now, Lovell and Anders would stand watch while Borman slept. As he floated into the sleeping bag attached to the underside of his couch, Borman was more than ready for a rest, but his mind would not cooperate. It wasn't easy to just turn off the mission and fall asleep. Two hours later, still keyed up, he called down to Houston and got permission to dig out the medical kit and take a Seconal. He hated pills, but it was more important that he rest.

9 P.M., Houston time
14 hours Mission Elapsed Time
77,000 miles out

Already, a bit more than eleven hours after Translunar Injection, Apollo 8 was a third of the way to the moon. But even as Borman, Lovell, and Anders sped moonward, the earth tried to pull them back, slowing their flight. It was as if the moonship were coasting up a hill, one that became less and less steep as it went along. About two days from now, on the afternoon of December 23, Apollo 8 would reach the gentle crest of that hill, the place where the earth's gravitational influence gave way to the moon's. From then on it would begin falling toward its destination.

For now, though, there was no sense of speed—or for that matter, any normal sense of time. To Borman's crew time was told by the mission clock on the instrument panel. Their wristwatches were still set to Houston time, but all vestiges of day and night had vanished. They moved in the unrelenting glare of an unfiltered sun in a black sky. To keep the sun's heat and the frigid cold of space evenly distributed on the hull, Borman had set the spacecraft rotating slowing on its axis, making one full turn in an hour. The astronauts nicknamed this the "barbecue mode." Every once in a while, as the craft turned, the men caught sight of the earth. With each passing hour it dwindled. They couldn't see the change as they watched, but if they turned away from it and looked again later, they noticed that it was a little smaller and more distant. Presently it was about the size of a baseball held at arm's length.

At least now Anders could tell what he was looking at. It hadn't been so easy a few hours ago, when the planet still loomed big and bright. Back then, fresh out of his space suit, Anders had his first chance to savor the view, and to his embarrassment, he couldn't tell what in the world he was looking at. As a kid, he'd prided himself on being something of a geography expert. He knew every country and major city on the schoolroom globe. But the real earth wasn't like the schoolroom globe. It had *clouds,* for one thing, and the countries weren't different colors—no small detail! A large land mass peeked from beneath the clouds. Was that Africa? He could just about make out the bulge of the Sahara, the point of Cape Town. Wait a minute, he thought, that doesn't make any sense. If that's Africa, then where is South America? And what is *that* thing out in the middle of the Pacific?

It was time to go to back to basics. There was a big white patch near the edge; it had to be ice or clouds. Anders thought it looked more like ice. Winter in the northern hemisphere, and there's a big patch of ice in sunlight. That had to be Antarctica. But how could that be right when it was at the top? Then he realized: Because *we're* upside down. Anders turned himself until the white patch was at the bottom, and suddenly everything fell into place. There was the great south polar ice cap. Above it, not the Horn of Africa but the coast of Chile, and all of South America, from rain forest to coastal desert, wrapped in clouds. North America hid beneath a winter overcast, but he could spot the Florida peninsula under clear skies. In the Maritimes a cyclone's brilliant white pinwheel sprawled across the Atlantic. And in the Caribbean, the shallow waters of the Bahamas gleamed like a turquoise jewel lit from within; he would be able to spot it all the way out to the moon. He wished he could spin the planet around on its axis and see the rest of it, but it turned at its own pace, just as it always had.

Sunday, December 22
1 A.M., Houston time
18 hours Mission Elapsed Time

Borman awoke after about five hours of fitful sleep. He didn't feel well. He told Lovell and Anders he had a headache and took a couple of aspirin, then he just floated in his couch and watched the instrument panel. A few minutes went by, and the next thing Lovell and Anders knew he was retching. Anders handed him a plastic bag, and Borman went down into the lower equipment bay and threw up. Lovell flashed Anders a knowing look: Borman must be motion sick.

The episode was beginning to make Anders feel a bit sick himself, when suddenly he spotted a greenish sphere, about the size of a tennis ball, ascending slowly out of the equipment bay in a flurry of tiny bits and globules. The sight of it made him want to gag. But when it drifted closer he noticed that the blob was shimmering and pulsating in three directions at once in some kind of complex fluid vibration made possible in zero gravity. At that moment the scientist in him took over. He was about to go for a camera when suddenly the blob split in two. As if to affirm Newton's laws of motion, the twin spawns headed away from each other in exactly opposite directions, giving Anders a flash of recognition: *Conservation of momentum!* One scooted away whence it had come and the other headed right for Lovell. The man was cornered. The blob hit him on the chest and then, overcome by the forces of surface tension, spread out on his coveralls as flat as a fried egg.

By now a horrible stench had rolled out of the equipment bay. Anders left Lovell to his predicament and reached for an oxygen bottle on the wall of the cabin, meant to be used in case there was a fire. *To hell with that,* Anders thought; he slapped the mask on his face and turned it on full. Meanwhile, Borman's troubles weren't over; now he was struck with diarrhea. What a mess—Lovell and Anders had to help chase down stray bits of vomit and feces with paper towels. In a strange, detached way, Anders was reminded of hunting butterflies.

Anders floated in his sleeping bag, eyes closed, trying to relax. He was tired. Before now he would have thought that sleeping on a bed of air would have been the best imaginable, but it wasn't working out that way. Like Borman, he found it difficult to take his mind off the flight. He missed the pressure of a pillow against his head and the security of a blanket drawn up around him. The bag was clearly designed for someone as big as Lovell; Anders was bouncing around inside it like a lone pea in a pod. He steadied himself and tried to lay still. Every now and then a residual bit of vomit drifted by and he cowered. And he noticed that his body had not fully adapted to zero g. His heart, accustomed to a lifetime of fighting gravity, was suddenly too strong. As a result he heard a muffled, incessant *boom-boom-boom*—his own blood pulsing in his ears. And each time he was about to fall asleep, he was startled awake by the sensation of falling, just like the feeling he'd had in dreams on earth. His central nervous system seemed to be broadcasting an alarm, telling him what he already knew, that he was in an environment unlike anything he had ever known.

8 A.M., Houston time
1 day, 1 hour Mission Elapsed Time
120,000 miles out

When Anders awoke—he did manage a few hours of fitful sleep—he found Borman much recovered, blaming his illness on a twenty-four-hour virus. Anders suggested he reveal the incident to mission control, but Borman replied, "I'll be damned if I'm going to tell the whole world I had the flu." Anders finally convinced his commander to put a short summary on tape; the message could be sent to earth via a special telemetry channel. That way, no one would hear except the few managers who listened to the tape.

"I'll go ahead and dump this," Anders radioed Houston. He couldn't come out and say what was on the tape, but he had to find some way of getting them to listen to it soon. He suggested, "You might want to listen to it in real time, to evaluate the voice." Then there was nothing to do but wait for a response. Hours went by with no word from earth about the message. Eventually Anders found out why: it was hours before the flight controllers even had a chance to hear it. (So much for putting messages on tape, Anders thought.) Finally, at 1 day, 4 hours Mission Elapsed Time, Mike Collins called up on a special frequency:

"Apollo 8, this is Houston. We're on private loop right now, and we'd like to get some amplifying details on your medical problems. Could you go back to the beginning . . ."

"Mike, this is Frank. I'm feeling a lot better now. I think I had a case of the twenty-four-hour flu. . . ." Borman recapped the whole episode for Collins, and, to Borman's surprise, Chuck Berry came on the line to talk to him directly; that almost never happened. Unbeknownst to Borman, the episode had triggered serious talk of canceling the mission. Berry worried that Borman had a virus, and that it was only a matter of time before his crewmates caught it. But Borman told the earthbound flight surgeon that he felt much better, and that neither Lovell nor Anders had been affected. "We're all fine," he said.

Minutes later, in consultation with Berry and other managers, Apollo program director Sam Phillips decided to let the flight continue. Even if Phillips had decided otherwise, Apollo 8 was too far away for the SPS to manage a swift about-face maneuver. Borman, Lovell, and Anders were committed now: even if they had to abort their mission, they were going to go around the moon.

2:01 P.M., Houston time
1 day, 7 hours, 10 minutes Mission Elapsed Time
140,000 miles out

"Are you receiving television now?"

"Apollo 8, Houston. We just got it."

"You are getting it?"

"Okay, Apollo 8. We have a good picture."

Frank Borman had fought to keep the small television camera off Apollo 8—he wanted neither its added weight nor the demands on his time—but he had lost that battle. And just now, as he conversed with Capcom Ken Mattingly, the big screen at the front of mission control flickered to life, and there was Jim Lovell, apparently upside down, at the navigation station, making star sightings and making lunch. The picture was fuzzy, and it was in stark black and white—but to many who saw the brief telecast on the afternoon of December 22, it seemed a small miracle: a live glimpse inside a moonbound spaceship. For this first telecast from Apollo 8, Anders handled the camera while Borman narrated.

"Jim, what are you doing here? Jim is fixing dessert. He's making up a bag of chocolate pudding. You can see it come floating by." The narrow bag tumbled in the middle of the cabin. To the astronauts it was natural to see such things. But Anders had to wonder—how must it look to those who were watching? There were people who didn't believe Apollo 8 was real to begin with, that it was all a hoax perpetrated by the government. And it crossed Anders's mind that live television of three men floating inside a spaceship was as close to proof as they might get.

The irony was that apart from the weightlessness—and the view—it was pretty hard to convince *himself* that this was really happening. For one thing they didn't seem to be going anywhere. That was the paradox of this flight: faster and farther than anyone had ever gone, with no sense of motion. They might as well have been in the simulator, and yet they were more than 140,000 miles from home. They could *hear* the distance in every conversation with mission control—the pause between question and answer, while radio signals spanned enormous distance.

"This transmission is coming to you approximately halfway between the moon and the earth," Borman continued. "We have about less than forty hours to go to the moon. . . . I certainly wish we could show you the earth. Very, very beautiful." Unfortunately, that attempt failed. When they turned the camera on the brilliant blue and white planet, Houston reported only an unintelligible blob of light.

Now Borman trained the camera on Anders. "You can see that he has his toothbrush here. He's been brushing regularly." Anders twirled the toothbrush and it spun magically in the air in front of him until he snatched it back. "It looks like he plays for the Astros, the way he tries to catch those things." The Astros hadn't been doing well.

"Hey, Frank, how about a couple of words on your health for the wide world." Deke Slayton's voice. Slayton rarely came on the air, but the illness had caused a big flap in Houston and had made it into the news, and now was the chance to show the world that the crisis had passed. Inside Apollo 8, Borman smiled and waved at the camera. "We all feel fine," he said.

Only fourteen minutes after the telecast began, it was time to end it. Borman would now set the spacecraft back on its slow, thermal-control spin, and that meant the high-gain antenna could no longer track the earth. Before they signed off Lovell ducked into view. He wanted his mother to get a good look at him; today was her seventy-third birthday. Lovell looked at the camera and grinned. "Happy birthday, Mother," he said.

Monday, December 23
6 A.M., Houston time
1 day, 23 hours Mission Elapsed Time
187,000 miles out

Bill Anders was alone on watch, floating in Borman's seat on the left-hand side of the cabin. The flight was turning out to be somewhat different than he'd imagined. He had never expected to be bored. Make sure it works, Borman had told him, but everything was working just fine without him; he found himself wishing something would go wrong so he would have a chance to fix it. Not that Anders had grown complacent, far from it. He was still having trouble sleeping, in part because he couldn't stop worrying about the systems. Anders didn't want anyone else messing around with the instrument panel, and it seemed to him that Lovell was a little carefree about throwing switches. Once or twice Anders was floating in his sleeping bag, unable to sleep, and heard mission control call, "Apollo 8, we'd like you to switch to the secondary evaporators . . ." Then he saw Lovell's hand reaching for the wrong switch. Anders stopped him—*"Ah, Jim, it's the other one—"*

"I thought you were asleep," said a surprised Lovell.

Not much chance of that, Anders thought.

2 P.M., Houston time
2 days, 7 hours Mission Elapsed Time
207,000 miles out

Could Jules Verne have imagined the view from Apollo 8? The earth was so far away now that Jim Lovell could hide it behind his outstretched thumb. The feeling this evoked in the pit of his stomach was hard to convey. It was that delicious mix of exhilaration and apprehension that comes from testing yourself in dangerous conditions. (No matter how nonchalant an astronaut might act, Lovell knew, that apprehension was always there; you always wondered whether the engine would work.) And in particular, more so on this flight than either of his previous space missions, it was pure awe. Everything he had ever known was on that blue marble, and it was getting smaller by the moment. None of this came through in Lovell's voice just now, as he became the tour guide for the second telecast from Apollo 8. At last they had succeeded in showing the earth to itself. The black and white image was only a crude facsimile of the real thing, but at least it would convey some of what it meant to be a space traveler. Lovell keyed his mike.

"Houston, what you are seeing is the Western Hemisphere. At the top is the North Pole; just below the center is South America, all the way down to Cape Horn. I can see Baja California and the southwestern part of the United States . . ."

By his own admission, Lovell was addicted to spaceflight. He'd logged eighteen days on his two Gemini missions—more than any other astronaut. Now, in space for the third time, he felt as if he'd come home again. He was glad for that familiarity; it gave him the ability to relax and absorb the experience—especially the view.

Mike Collins in mission control asked, "Could you give me some ideas about the colors . . . ?"

Lovell gave it a try. "Okay. For colors, waters are all sort of a royal blue; clouds of course are bright white. The reflection off the earth appears to be much greater than the moon. The land areas are generally a sort of dark brownish to light brown in texture. . . ."

What Lovell had seen in the past two days brought him an entirely new sense of scale. On Gemini, his references were continents and oceans; now he had to think in terms of celestial bodies. The earth was a little ball off in one direction, and the moon in another, and the sun still another. And then there were the stars: unmoving, unblinking landmarks along a dark and distant shore. After two trips into space, Lovell had come to think of them as his friends. Now, on Apollo 8, the stars were his *raison d'être.*

When Apollo was first conceived it was thought that the astronauts would act not only as pilots but as onboard navigators. But the task proved so time consuming and ate up so much space in the memory of the command module's computer that planners decided to let this work be done by computers in mission control. Still there had to be a backup, in case Apollo 8 lost communications with earth. In that contingency, Lovell would use the stars to help himself and his crewmates get home. With the command module's 28-power sextant, he could measure the angle between selected stars and the earth's edge, enter the data into the computer, and let the electronic brain compute Apollo 8's location relative to the earth and the moon. In principle, it was the same as the shipboard navigation he'd used as a midshipman at Annapolis—but the setting was undeniably different. Already, he had tested his skill with practice star sightings, and the results were within a few thousandths of a degree of perfection. Later, in lunar orbit, he would take navigation sightings on craters and other landmarks over the far side. Those data, crucial to helping Kraft's trajectory people analyze Apollo 8's orbit, could only be obtained by the man onboard.

Lovell was proud of his role on the first circumlunar voyage. Of course, none of them, himself included, was immune to the stress of the mission. The difficulty in sleeping, for example. Lovell had never slept well the first night in space, and this flight was no exception. Eventually you get over it. But he also knew that of the three of them, he probably felt the least amount of pressure. Bill Anders had the rookie's burden of doing his job while adapting to a strange environment. And, like all rookies, he took his work very seriously. And Frank Borman had the heaviest burden of all. There was no doubt in Lovell's mind that command narrows one's focus, because he'd experienced that on Gemini 12. Borman's seeming uptightness—in training as well as now—was probably the commander's syndrome at work. But for Jim Lovell, who had hungered for command, there was an unexpected blessing in not getting it. He was able to enjoy the first flight around the moon more than either of his crewmates, and it was turning out to be every bit the adventure he'd hoped.

• • •

Minutes after the second telecast ended, Borman, Lovell, and Anders passed the most significant milestone since leaving earth—and yet, they were completely unaware until Houston mentioned it. Some 38,900 miles from its destination, Apollo 8 reached the top of the gravitational hill and crossed over into the lunar sphere of influence. At that moment Apollo 8 was traveling only 2,223 miles per hour, but in mission control, Kraft's flight controllers saw the craft begin to speed up. But Borman's crew felt nothing. They saw no change in the visible universe; outside there was the same, dull, starless black. To Bill Anders, the lack of tangible milestones made the voyage seem even longer. At one point, Capcom Jerry Carr asked what they could see and Anders replied, "Nothing. It's like being on the inside of a submarine."

The moon itself was nowhere to be seen. Anders had looked forward to watching it grow ever larger as they closed in until it became a huge, cratered ball in the sky, like a science fiction vision. But he had not seen the moon once on the whole trip out, not even a glimpse. Because of their angle of approach the moon was lost in the sun's glare. It was an act of faith even to convince himself that when they arrived, the moon would really be there.

Anders thought of the moon as his specialty. Before the flight, Borman was so involved with the journey that he had neither the time nor the inclination to worry about the destination. Lovell, who had his tracking tasks, had spent time familiarizing himself with lunar landmarks. But of the three, Anders had the most chance to think about and study the moon. Geologist-astronaut Jack Schmitt had spent many hours with him going over features of interest, and he'd met with other geologists about the observations and photographs he would try to obtain. Anders had his own personal flight plan for the twenty hours he would spend in lunar orbit.

Anders still remembered how, years before, he and the rest of the Fourteen had visited Kitt Peak National Observatory in Arizona, where they saw the moon's forbidding face—that place where he longed to walk—projected onto a large white table. But even then, blurred by the churning desert air, it was not fully revealed. He could only guess what it would be like to see it up close, with only a window in the way. Unfortunately, the windows on Apollo 8 were in pretty bad shape. The largest ones were clouded because of a sealing compound that had partially decomposed in the vacuum of space. Anders's side window looked as if it had been smeared with an oily rag. Only the two small, forward-looking viewports on either side of the hatch (the so-called rendezvous windows) had stayed relatively clear. Anders wondered how much he'd be able to see of the moon when they got there. He didn't have long to wait.

Tuesday, December 24
2:55 A.M., Houston time
2 days, 20 hours, 4 minutes Mission Elapsed Time

One of the paradoxes of Apollo 8 was that the three men on their way to the moon were far less able to determine their status than the flight controllers on earth. For all Borman's worries about whether Apollo 8 was staying on the free-return trajectory, there was absolutely no way for him to find out except to ask mission control. Kraft's trajectory specialists were able to detect tiny changes in Apollo 8's path by measuring the Doppler shift in its radio signals. In principle, it was the same as the change in pitch that a stationary listener hears from a passing train. Even a tiny change in the frequency of Apollo 8's signals meant something to Kraft's people. Their data was so good that when they plotted the curve you could see a little wiggle in it, due to the spacecraft's slow thermal-control spin. And they had nothing but good news. The trajectory was nearly perfect. Only two minor midcourse corrections had been necessary so far, and it looked as though Apollo 8 would get to the moon without making any more. The perfect marksman's shot that everyone had hoped for was about to happen.

Meanwhile, a constant stream of telemetry beamed from Apollo 8 to earth was picked up by the giant radio dishes of the Manned Space Flight Network, then transmitted across land lines and via satellites to Houston, where an army of flight controllers kept watch on hundreds of different components. Borman was amazed at how much they could tell about his spacecraft, more than 200,000 miles away. He had no idea, for example, whether the fuel lines in the SPS engine were as warm as they should be or frozen solid, but the systems people did, and their reports were terrific. The fuel cells were functioning even better than expected; the computer was running like clockwork; there was plenty of maneuvering fuel left. Borman had wanted a perfect spacecraft before he'd commit to the Lunar Orbit Insertion burn, and now he had it. Jerry Carr radioed the word to Apollo 8: "You're Go for LOI. You're riding the best bird we can find."

With characteristic caution, Borman had already turned the spacecraft to the precise orientation for the burn, in fact he had done it two hours ahead of time. The moment of truth, the crucial Lunar Orbit Insertion burn, would come when Borman's crew was out of radio contact, with only themselves and their machine to rely on. If the firing went as planned, this radio blackout would last 45 minutes. But if there was a malfunction and Borman decided to abort the mission, Apollo 8 would come around a good bit sooner than that. In mission control, Kraft's trajectory people would know how things had gone simply from the moment they picked up Apollo 8's telemetry.

Strapped in their couches Borman and his crew waited out the last minutes of a three-day journey. Each of the three astronauts knew they were cutting it very close to aim nearly a quarter of a million miles across space to a world 2,160 miles across, zip just ahead of its leading edge, and go into orbit just 69 miles from its surface. (The joke around the simulator was, wait till you see the 70-mile-high mountain on the far side of the moon.) Sixty-nine out of 234,000 left very little room for error. It was understandable that Borman's crew wanted something more than numbers to assess the accuracy of their path. Before the flight, the trajectory people had told them that they would not be able to see the moon as they came in. Deprived of the one seat-of-the-pants method a pilot has—eyeballing the target—they asked for something else. There was one answer, and it was Loss of Signal. "LOS," as it was called, was the moment when Apollo 8 would slip behind the moon and lose radio contact with earth. Once the craft was on its way to the moon the controllers would be able to predict the time of LOS down to the second. If it happened precisely as mission control predicted, Borman's crew would know that all the calculations were right after all.

"One minute to LOS," advised Carr. As he spoke, the mission clock read 68 hours, 57 minutes, 4 seconds.

"Ten seconds to LOS," Carr radioed. "You're Go all the way."

"Thanks a lot, troops," Anders said.

"We'll see you on the other side," added Lovell.

Borman watched the mission clock intently. At precisely 68:58:04 he and Lovell and Anders heard static in their headsets. He could hardly believe it—right to the second. He said aloud, "That was great, wasn't it? I wonder if they turned it off." Anders laughed; he could just imagine Kraft saying, "No matter what happens, turn it off."

The men were running through the checklist for the burn when suddenly the spacecraft was enveloped by darkness. Anders realized they were deep in the shadow of the moon. As his eyes adapted, he saw that the sky was full of stars, so many he could not recognize constellations. He craned toward the flat glass to look back over his shoulder, where they were headed, and he noticed a distinct arc beyond which there were no stars at all, only blackness. All at once he was hit with the eerie realization that this hole in the stars was the moon. The hair on the back of his neck stood up. *Come on, Anders,* he told himself; *you're not supposed to feel this way.*

III: "In the Beginning . . ."

Tuesday, December 24
3:53 A.M., Houston time
2 days, 21 hours, 2 minutes Mission Elapsed Time

Falling in darkness, Apollo 8 was pulled toward its rendezvous with the moon at more than 5,000 miles per hour. The spacecraft was turned so that its big SPS engine pointed forward, into the direction of flight. Borman, Lovell, and Anders would need every bit of its power, because Apollo 8 would have to slow down in a hurry, or else speed right past its goal. It would take just 4 minutes to slow Apollo 8 to about 3,700 miles an hour, slow enough to go into orbit. Inside the command module, Borman's crew set to work, running through the checklist to bring the SPS engine to life. With 10 minutes to go, rapid-fire conversation, in the jargon-rich language of spaceflight, buzzed in the command module cabin as Anders called off to Borman each item on the checklist:

"TRANSLATION CONTROL POWER, ON."

"On."

"ROTATIONAL HAND CONTROLLER NUMBER 2, ARMED."

"Armed."

"Okay. Stand by for the primary TVC check. . . ."

About 3 minutes before the scheduled ignition, Apollo 8 suddenly flew into sunlight once more. Lovell glanced through the hatch window and said, "Hey, I got the moon."

Borman asked, "Do you?"

"Right below us."

Anders looked up from his checklist at his smeared window. It looked as if streams of oil were descending slowly across the glass. *Dammit,* he thought, *whatever that stuff is, now it's running down the window!* But then his eyes refocused and he realized he was looking at *mountains.* They moved slowly past, lit by the slanting rays of the sun, trailing long black shadows. The mountains of the moon. He said quietly, "Oh, my God."

"What's wrong?" Borman said anxiously.

"*Look* at that."

"Alright, alright, come on," Borman said, "you're going to look at that for a long time."

With seconds to go, the computer gave its flashing "99" message and Lovell pushed the PROCEED button in response. Four seconds later the engine lit, pressing the men into their couches. They heard a clattering noise as a stray piece of gear fell to the cabin floor. Slowly acceleration mounted. There was no noise at all, just a gentle vibration and a smooth, steady push. Even though they were held in their couches with just a fraction of normal gravity, after three days in weightlessness it felt like 3 g's. Anders scanned the gauges: tank pressures, valve positions, fuel quantities. "Pressures are coming up nicely," he told Borman. "Everything is great."

Time seemed to slow down. Each man knew the engine must fire for the prescribed duration—no more, no less. If the engine shut down prematurely, or if it didn't deliver the proper amount of thrust, they could end up in a weird, errant orbit. If it fired even a few seconds too long, Apollo 8 would lose so much energy that it would crash into the moon. By the 2-minute mark the burn had begun to seem very long. Borman said aloud, "Jesus, four minutes?"

"Longest four minutes I ever spent," Lovell said as the engine roared silently in the vacuum.

Two more minutes passed without mishap, and then Anders counted down the last few seconds. Borman knew the computer was programmed to shut down the engine automatically but he wasn't taking any chances. At zero he pushed the shut-off button just in case.

"Shutdown," Borman said. Suddenly they were weightless once more. "Okay," Borman sighed, "go ahead."

They ran through their deactivation checklist like clockwork. Lovell queried the computer for the dimensions of their orbit. They were circling the moon in an ellipse that ranged from 69 miles, at a point above the far side, to 194 miles at the opposite point above the near side. A little over 4 hours from now, at the start of the third orbit, Borman's crew would fire one more 11-second blast to change their path into a 69-mile circle. For now, the SPS had done its work beautifully. Within a few tenths of a mile, the orbit was perfect.

"That's it," Anders said. "Dig out the flight plan."

Apollo 8 drifted above the far side of the moon while three visitors looked down at a scene of total desolation. It appeared devoid of color, apart from various shades of gray. With no atmosphere to soften the view, it was a scene of unreal clarity. If they hadn't known, the men would not have been able to tell whether the moon was sixty-nine miles away or six. Everywhere there were craters: smooth round bowls, misshapen gouges, gentle hollows, tiny BB-shot holes in the gray moon, shoulder to shoulder, one on top of another. Large craters bore on their ancient walls the scars of smaller craters. Every so often a lonely mountain rose from the bleakness, its slopes rounded and pockmarked. Everything else—every rise and fall of the landscape—was formed from the rim or the shoulder or the floor of a crater. The place looked like the deserted battlefield of the final war.

"*Whew.* Well, we answered it," said Borman with a laugh. "They're meteorites, aren't they?"

"It looks like a big beach down there," Anders said. That's what it reminded him of: beach sand darkened by the cold embers of bonfires, churned up by a big game of volleyball, but now deserted.

Lovell, meanwhile, got out the map and tried to figure out where they were. That wasn't easy. They were over a part of the moon where the sun was almost directly overhead, and the moonscape was without definition, like a bleached, rocky ocean. Finally Lovell spotted a huge crater with a dark floor, like a mountain lake. That was Tsiolkovsky crater, named for the Russian scientist who had dreamt of space flight over a century before. Lovell knew it as soon as he saw it.

Minutes later, Apollo 8 crossed over onto the lunar near side. It took several tries for Anders to make contact with earth, but once the high-gain antenna locked on to the signal it was amazing how clearly Jerry Carr's voice came through, as if he were somewhere nearby. And after Lovell passed down the essential data on the burn, Carr spoke for a curious world: "What does the ol' moon look like from sixty miles?"

"Okay, Houston," Lovell radioed. "The moon is essentially gray. No color. Looks like plaster of paris—"

"Or a beach," Anders prompted.

"—or sort of a grayish beach sand. We can see quite a bit of detail. . . ."

Frank Borman did not join his crewmates in their excited descriptions. He was more concerned with the health of his engine. In Houston the engineers were poring over strip charts of data from the burn; Borman wanted to know as soon as possible what they found out. And he wanted mission control to give him a go-ahead for each new orbit, otherwise he would prepare to leave. *Let Lovell and Anders rhapsodize about the moon,* Borman thought; *I've got to think about getting us back.*

• • •

Seen from earth on December 24, the moon was a ripening crescent. Most of the near side was in darkness, but that meant that almost all of the moon's hidden face, never before seen by human eyes, was in sunlight. Inside Apollo 8, Bill Anders manned a pair of

Hasselblad still cameras and a 16 mm movie camera, his goal to record on film as many lunar mysteries as possible. His photography plan was packed with objectives, and Anders had gotten to work minutes after Apollo 8 reached orbit. When Borman, like any tourist, asked to take a picture, Anders became the rigid one—he didn't want to take any pictures that weren't in the photo plan. Now, armed with his map and his checklist, Anders scanned the parade of craters searching for his assigned targets, and whatever else might look intriguing. The command module had never been designed as an observation platform, but it was turning out to be much worse than he'd anticipated. Only the two small rendezvous windows were reasonably clear, but the view through them was disappointingly restricted. It was sight-seeing in a Sherman tank. To make matters worse, the best maps of the far side, drawn from unmanned probe photos, weren't that accurate. Even when he *thought* he knew where he was, he couldn't find anything he recognized among the swells and hollows. And when he managed to get his bearings, it was all too easy to lose track in the scramble of setting up camera gear, changing film magazines and switching lenses. At first he hesitated to take pictures, but he decided if he was going to come home with anything he'd better just aim the camera and fire away. By the end of the third orbit, six hours into the twenty-hour lunar visit, he'd already taken many of the targets on his list, but there was still a lot left to accomplish.

The irony was that the far side of the moon was turning out to be very different from the place he had envisioned. Like most of the astronauts, he went to see the film *2001: A Space Odyssey* when it opened in the fall of 1968, and somehow, through the weeks of training, poring over the unmanned probe pictures, it was still Arthur C. Clarke's moon that stayed in his mind: a place of drama, with towering, sharp edged mountains, cliffs, and cracks. Instead he'd come nearly a quarter of a million miles to see dirty beach sand. It was a place of such unrelenting sameness—crater upon crater, hill upon battered hill—that to see it with his own eyes was almost an anticlimax. Anders realized, with some disappointment, that the moon was a less interesting world than he had imagined.

10:37 A.M.

Perhaps it is true that our most electrifying experiences are the ones that take us by surprise. Even on the first flight around the moon, in which everything was figured to the second, rehearsed in painstaking detail, an event that no one anticipated became the most moving of all. Apollo 8 was drifting over the far side for the fourth time. Borman prepared to turn the spacecraft so that Lovell would be able to sight the moon through the command module's sextant.

"Alright," Borman announced, "we're going to roll." He nudged the hand controller and the craft turned slowly until it was right side up. When the maneuver was finished, Anders glanced out the window.

"*Oh, my God.* Look at that picture over there."

"What is it?" Borman asked.

"The earth coming up. *Wow,* is that pretty." Slowly, beyond the bleached horizon, a radiant half-circle of blue and white emerged, ascending into the black sky.

"Hey, don't take that, it's not scheduled," Borman said, seizing the chance to give Anders some grief about a picture that wasn't in the photo plan.

Anders wasn't listening. He called urgently to Lovell, "Hand me that roll of color, quick, would you?" But Lovell was already joining them at the windows.

"Oh, man, that's great!"

"Hurry. Quick," Anders said. At last he slapped on the color magazine and aimed the camera with its telephoto lens.

Lovell was impatient: "You got it? Take several of them! Here, give it to me." After telling Lovell to calm down, Anders snapped the picture. "Are you *sure* we got it now?" asked Lovell urgently.

"Yeah. It'll come up again, I think," Anders said dryly.

The first witnesses to an earthrise returned to their work, each carrying the impact of the sight. For his part, Anders had been so focused on photographing, observing, and describing the moon since they arrived that it had not occurred to him to look at the earth. When it suddenly appeared, his overwhelming impression was how beautiful it was, even more so beside the barren face of the moon, and how very small.

12:30 P.M.

Apollo 8 had been in lunar orbit for more than eight hours, and Borman was in need of sleep. He left Anders in charge of the systems while Lovell attended to his landmark tracking. Here too, the computer did amazing things. Once Lovell had determined Apollo 8's position and entered it into the computer, he had only to give it the coordinates of the next target and the sextant automatically swung to the right place. It even moved to track the landmark, compensating for the spacecraft's swift motion. The results were breathtaking. He felt as if he were flying only a few miles above the surface. Peering down into craters, he spotted landslides, even a few boulders. To his surprise, he found he could see detail even in the shadows. If Anders found the moon less intriguing than he'd hoped, Jim Lovell did not share his disappointment.

Lovell's most important target lay on the near side, in the eastern region of the Sea of Tranquillity. There, mission planners had picked out a possible site for the first lunar landing. One of the main objectives of the mission was to reconnoiter East 1, as it was called, from orbit. Lovell had studied the approach that a lunar module would make before landing, and had picked out landmarks along the way. Some of them—the ones that had been discovered on the unmanned probe photos—had no names, and Lovell, following the explorer's prerogative, had named them. Now, as Apollo 8 flew over the Sea of Tranquillity, Lovell was pleased to find the familiar craters and mountains so easy to recognize. Over East 1, Lovell searched for boulders and other potential obstacles to a descending lunar module and found none. The lighting conditions were even better than he expected. It seemed a fine place for a team of astronauts to try to land. With only 69 miles between himself and the moon, Lovell wished he were making the journey.

Christmas Eve
Timber Cove, Texas

As night fell outside Houston, Marilyn Lovell left her home, got into her car, alone, and headed for St. John's Episcopal Church. Knowing she would be too busy to attend the scheduled Christmas Eve mass, she had arranged with Father Raish for a private service. She needed this, especially after the ordeal of yesterday afternoon.

Even before her husband left earth, Marilyn Lovell's life had become an emotional roller coaster. She was the only one of the three Apollo 8 wives who decided to witness the launch personally, and she had packed up her four children, from teenage Barbara to little Jeffrey, not quite three years old, and headed to Cape Kennedy, where a friend had arranged accommodations on the beach. Two nights before the launch, Jim had found time to stop by for a visit, and the two of them had driven out to see that magnificent rocket, ablaze in floodlights, that would propel him to the moon. He'd explained to her what the launch would look like, that the Saturn would veer off to one side as it lifted off, to avoid hitting the launch tower. She thanked God that she had known what to expect; otherwise it would have scared her to death. He'd seemed so confident that night, so excited, so ready to do what he had been trained to do. She wished she could have faced the prospect of his moon trip as well as he did.

From the time she and the kids arrived back in Houston, it seemed the house was full of people. Marilyn was glad of that, glad for all the activity. It was one of the best customs of the space community; in times of greatest stress, the astronaut wives really came together and helped each other. There was always someone coming by with food, or offering to watch the kids or run an errand.

The squawk boxes—small speakers NASA had installed to pipe in the conversations between Apollo 8 and mission control—were a constant presence. She had kept one ear open, listening for her husband's voice, thrilled to hear him describe what he was seeing. Best of all were the television transmissions. It had been wonderful to see Jim on Sunday, smiling, wishing his mother a happy birthday. But her children had their own ways of absorbing the experience

—or not absorbing it. Yesterday afternoon, when she got the children together for the second telecast, thirteen-year-old Jay was outside; she was barely able to get him into the house once the transmission began. When he did sit down, he was decidedly moody. He complained that he couldn't recognize his father's voice. He asked, "How fast are they going now?" The answer—that Apollo 8 had slowed to only a few thousand miles per hour—disappointed him.

Whatever her children might think, Marilyn was spellbound to see pictures of the cloudswept earth, and to hear Jim talking about it. On the screen, the TV networks superimposed Apollo 8's distance from home. Already it was 200,000 miles, hard enough for her to fathom, and it kept increasing as she watched.

No sooner had the TV show ended than her children dispersed, and for the first time since the mission began Marilyn was alone, with nothing to distract her from her darkest fears. On the squawk box, she heard Jerry Carr saying that Apollo 8 had crossed into the moon's gravitational influence. Marilyn wasn't particularly attuned to the technical details of her husband's work, but she understood one crucial thing: it was only a matter of hours until the men would fire their rocket engine and go into orbit, and that engine had to work perfectly, along with the entire, complex machine her husband was flying. Twenty hours later that engine would have to be perfect one more time, or she would never see Jim again. She had known this all along, of course, but she had pushed the thought out of her consciousness. She said nothing about it to Jim. And she certainly wasn't going to discuss it with any of the other wives. There were certain unwritten rules here, carryovers from the test pilot business, and one of them was that you just don't talk about things like that.

And so Marilyn had succeeded in banishing her terror—until yesterday afternoon. Suddenly overwhelmed, with no one there to see, she broke down. Several minutes later the doorbell rang; the daughter of a friend had come by with some food. Seeing that Marilyn had been crying, the girl told her mother about it, and by evening Marilyn's house was full of people once more. Her neighbors had come to spend the night with her, to share her anxiety, and her hope, for what was about to happen.

Marilyn barely slept. During the night she came out of her bedroom to find the living room floor strewn with bodies, the sleeping forms of her dear friends. At three-thirty in the morning, everyone waited by the squawk box in silence as Apollo 8 headed for lunar orbit. She heard her husband say, "See you on the other side," and then, after an agony of waiting, his voice came through the static once more to announce success.

All day today, Marilyn had tried to keep up with what was happening. She knew Jim was making landmark sightings, and she had a special interest in them. One night at the Cape before the launch, when she and Jim were alone, he had given her a present: a large black-and-white closeup photo of the moon from one of the unmanned probes.

"What's this?" she asked.

"I just wanted you to see where I'm going to name a mountain for you." Near the upper-right-hand corner of the photo, Jim pointed to a triangular-shaped mountain sticking up from the dark plains of the Sea of Tranquillity. Jim said it was one of the most important landmarks leading up to the landing site. He was going to call it Mount Marilyn. The name wouldn't be official, not unless it was approved by the International Astronomical Union, but that didn't bother her. As far as she was concerned, nobody could take away such a splendid gift. But on this Christmas Eve, after the trauma of the past three days, Marilyn needed something more.

When Marilyn arrived at St. John's, the organist was practicing, and the church was filled with the sounds of Christmas hymns. Candles glowed everywhere; Father Raish had arranged to have them lit for her. At the altar she celebrated a private communion; she left feeling renewed. And on the way home, she looked up and to her amazement, there was the crescent moon. She could barely comprehend it: Jim was *there*.

3 days, 8 hours, 55 minutes Mission Elapsed Time

"Well," Lovell yawned, "did you guys ever think that one Christmas Eve you'd be orbiting the moon?"

"Just hope we're not doing it on New Year's," said Anders with fighter-pilot gallows humor.

If Lovell got the joke, he didn't show it. "Hey, hey, don't talk like that, Bill," he said quietly. "Think positive."

Borman was awake now. It was getting to be a long day. They had been in orbit for nearly fourteen hours and they still had six hours to go before they would leave the moon. He could tell that his crew was tired. Lovell was hard at work on his landmark tracking. There was weariness in his voice whenever he spoke. And he was making mistakes. Several times he punched the wrong commands into the computer, triggering warning tones and startling Borman and Anders. Anders was probably tired too, racing around to keep up with his photo plan. Borman knew how tired *he* had felt a few hours ago, before he got some rest. *The flight plan is just too full,* Borman thought. They still had a TV show to do during the ninth revolution. Then the Transearth Injection burn. Compared to that burn, Lovell's navigation was secondary, and so were Anders's photographs. The most important thing was getting home. Borman knew what he had to do.

On the radio, they heard Mike Collins in mission control asking about some of Lovell's landmark sightings.

"Apollo 8, Houston. We'd like to clarify whether you intend to scrub control points one, two, and three. . . ."

Borman keyed his mike. "We're scrubbing everything. I'll stay up and keep the spacecraft vertical, and take some automatic pictures, but I want Jim and Bill to get some rest."

Anders couldn't believe what he was hearing. The last thing he wanted to do was waste time sleeping in lunar orbit. He still had stereo pictures to take, dim-light photography, and filter work, and there were the targets north-of-track he had to finish up. He didn't feel tired. Was Borman serious?

Borman looked at the overcrowded flight plan. "Unbelievable, the details those guys put in here," he said to Anders. "A very good try, but completely unrealistic. I should have warned you."

"I'm willing to try it," Anders said gamely.

"No," Borman said. "You try it, and then we'll make another mistake."

Lovell started to speak but Borman cut him off—"I want you to get your ass in bed! Right now!"

"I can do another rev," Lovell said.

"No, get to bed. Hurry up. I'm not kidding you, go to bed."

Anders thought of all the unexposed film. He hid his frustration and asked his commander, "What do you want me to do?"

"Go to bed. We'll get that thing going when we get to daylight," Borman said, indicating the camera. "Then you guys go sack out for two hours."

So that was it. Like it or not, Borman had the authority to send them to bed, in the interest of keeping his crew alert in a dangerous situation. Just now, Mike Collins radioed, "We agree with all your flight plan changes. And have a beautiful back side; we'll see you next time around." No one in mission control was going to argue with Borman's decision.

The radio fell silent once again as Apollo 8 coasted out of contact with earth and into total darkness. Lovell had gone to his sleep station, but Anders remained in his couch, tending the cameras, hoping he might hold Borman off long enough to take some more pictures when they came into sunlight again.

"We're doing fine," Borman said quietly. "Why don't you go to bed?" Anders was about as close to arguing as he'd ever been. But in a spacecraft almost a quarter of a million miles from home, an argument with his commander would have been tantamount to mutiny. Still, he tried to hang on.

"This is a closed issue," Borman said. Anders asked about the movie camera. "I'll just click it on when the time comes," Borman said. "You should see your eyes. Get to bed. Don't worry about the exposure business, goddammit, Anders, get to bed. *Right now.*" Anders had no choice.

"You want me to take some pictures?" Borman offered. "Okay, I'll take care of it all." As Anders headed for his sleeping bag Borman was saying, "A quick snooze, and you guys will feel a hell of a lot better."

An explosion of light came again over the far side, and the sun cast long shadows behind gray mountains and inside countless holes. In his sleeping bag, Anders craned his head to look past his couch through the tiny rendezvous window. Now he did feel tired; he could hardly keep his eyes open. Ironically, this was the best view of the moon he'd had on the whole flight. And something on the stark ground caught his eye, a feature that stood out from the pulverized sameness. He was all but certain he was looking at a region of old lava flows. This was what he'd been looking for, some sign of volcanic activity in the highlands. He could hear the Hasselblad clicking away on automatic in Borman's window; he hoped they were getting this. Even if they were, Anders was aware that he was bringing home something more important. From lunar orbit, the earth looked no bigger than the end of his thumb, and yet, on a cosmic distance scale a quarter of a million miles was nothing at all. He knew that if he were to go a hundred times farther out—so far into the lonely dark that the earth would shrink to a point of light—he would barely have left home. He couldn't help but think that the cosmos would continue to turn as it always had if suddenly there were no earth. But how little that mattered when it appeared, blue and radiant, rising beyond the lifeless moon. In that moment he saw a thing of inexplicable fragility; later he would liken it to a precious Christmas tree ornament. And if the earth was only a mote of dust in the galaxy, that blue planet was everything to him and the creatures living on it. On his way into a fitful sleep, Anders began to realize: *We came all this way to explore the moon, and the most important thing is that we discovered the earth.*

• • •

While Anders and Lovell slept, Borman floated in the commander's left-hand seat. Spaceflight was quiet. There was none of the constant vibration of an airplane, the steady whine of jet engines. Whenever he fired a maneuvering thruster, there was a thump of solenoids opening the valves, but otherwise, as long as they kept the cabin fans turned off, there was only quiet.

Borman knew Anders was upset about the unplanned sleep period. Borman liked Anders, but he understood that Anders just didn't have the experience to always see the big picture; Borman had been flying ten years longer. The real role of the commander on these missions wasn't to fly the spacecraft; there was precious little of that. It was to make the crucial decisions. And if ever there were a crucial need, it was to keep his crew sharp for the Transearth Injection burn.

But there was something else, before the burn, that Borman had come to realize was extremely important. The flight plan called for two TV transmissions from lunar orbit; the second was set for the ninth rev, a couple of hours from now. The Public Affairs people had told him, "There will be more people watching those shows than have ever listened to a single human being in all of history. Say something appropriate." And with the help of a friend in Washington, Borman found something. He had it reproduced on fireproof paper and placed in the back of the flight plan; after that he didn't give it a minute's thought.

Before the flight, he'd barely thought about the spiritual impact of going to the moon. But now that he was here, he couldn't deny it. To see the moon so desolate, looking like the earth must have looked before life—or how it would look after nuclear war—was more sobering than he could have anticipated. But what moved him most was his own planet: the only color in the universe. To see the earth rising beyond the moon on Christmas Eve was all the confirmation of a Creator that Borman needed. Now that he was here, he was glad to have that TV camera: he wanted to share his new perspective with humanity. What the three of them were about to do was perfect. It was time to get Lovell and Anders up for the telecast.

8:11 P.M., Houston time, Christmas Eve
3 days, 13 hours, 40 minutes Mission Elapsed Time

"Here it comes!" Borman nearly shouted when he saw his world ascend once more from behind the desolate moonscape. On that earth, in town squares and living rooms, pubs and offices, half a billion people were tuning in for a broadcast from the three men circling the moon.

"This is Apollo 8, coming to you live from the moon," Borman began. "Bill Anders, Jim Lovell, and myself have spent the day before Christmas up here, doing experiments, taking pictures, and firing our spacecraft engines to maneuver around. What we'll do now is follow the trail that we've been following all day . . ."

He'd gone over the plan with Lovell and Anders before they started transmitting. First, each man would say what impressed him most; then, just before Apollo 8 flew into darkness, they would give a joint reading of the message.

"The moon is a different thing to each one of us," Borman told his huge audience. ". . . I know that my own impression is that it's a vast, lonely, forbidding type of existence or expanse of nothing. . . ." His words came out with an awed, sober cadence. "And it certainly would not appear to be a very inviting place to live or work."

When Lovell's turn came, he spoke eloquently of two worlds, the lonely one he was orbiting and the "grand oasis" he had left behind. Anders described the moon's spectacular appearance near lunar sunrise and sunset, where long shadows made the landscape look jagged and forbidding. For about twenty minutes, the three men took their viewers on a tour of the landmarks passing below them. Over the Sea of Crises they flew, and then the Sea of Fertility, and then the Marsh of Sleep, and then, at last, the Sea of Tranquillity. In the distance, they could see the place of long shadows; soon they would be crossing into night. Anders gave the introduction.

"We are now approaching lunar sunrise. And for all the people back on earth, the crew of Apollo 8 has a message we would like to send to you." Anders held the flight plan in front of him and began to read:

> *In the beginning, God created the heaven and the earth; and the earth was without form and void, and darkness was upon the face of the deep; and the spirit of God moved upon the face of the waters.*
>
> *And God said, "Let there be light," and there was light.*
> *And God saw the light, that it was good.*
> *And God divided the light from the darkness.*

Now it was Lovell's turn:

And God called the light Day, and the darkness He called Night.

And the evening and the morning were the first day. . . .

As Lovell read, Anders thought, "We're trying to say something fundamental. This isn't just another space mission; it's a new beginning, for all of us."

. . . And God called the firmament Heaven.

And the evening and the morning were the second day.

The shadows lengthened on the Sea of Tranquillity as Borman closed the reading:

. . . And God called the dry land Earth, and the gathering together of the waters called He Seas.

And God saw that it was good.

"And from the crew of Apollo 8, we close with, Good night, Good luck, a Merry Christmas, and God bless all of you, all of you on the *good earth.*"

El Lago, Texas

The night was crisp and clear in Houston. When the verses from Genesis came down a crescent moon shone high overhead, and after the telecast ended more than one witness went outside to look at it. In El Lago, Susan Borman had a house full of friends and relatives. Susan was a model of composure—in other words, she seemed no different now that her husband was circling the moon than she was at any other time.

The other wives could not look at Susan Borman without feeling some amazement. Always, she was an impeccable military wife: white gloves for formal occasions, hair always perfect, always neat and well dressed. Her clothing alone was an accomplishment on an astronaut's salary. Valerie Anders, raising four children, had to scrimp on her wardrobe, and when her mother saw her on television she would ask Valerie later, "Why are you always wearing the same dress?" The answer was that it was the only nice one she had.

This Christmas Eve, Susan had carried herself with characteristic poise. She'd been up all the previous night, listening to the conversations between Houston and the moon on the squawk boxes. In the morning she'd gone to church, pausing outside to speak with reporters, and then back home by midday.

After the evening telecast, Valerie Anders left her house, also full of people, to be with Susan during the critical Transearth Injection burn. Now the two women and several visitors waited by the squawk box in the Borman kitchen. They heard the voice of Ken Mattingly, the young, serious astronaut serving as Capcom, read up a long list of numbers and technical shorthand. It was almost midnight when Mattingly advised, "Three minutes to LOS" —Loss of Signal as Apollo 8 flew behind the moon for what everyone hoped would be the last time.

"All systems are Go, Apollo 8," Mattingly said. Susan heard her husband say, simply, "Thank you." Then the squawk box fell silent. And now came the worst moment of her long, private agony.

Her ordeal had begun the previous August, as soon as she learned of Frank's moon mission; from that moment she'd been sure he wasn't going to come back. She was furious with the NASA managers for sending him. There was nothing new about this terror; it had been with her off and on ever since Frank started flying. But it didn't take long for her to learn what was required of her: Keep smiling; keep your fears to yourself. In the squadron, she and Frank had seen some of the men wash out because of wives who couldn't keep their worries to themselves; after a while it affected a man's concentration. She knew it was wrong to complain, but when they were at Edwards, and Frank was assigned to fly zoom tests in that silver beast they called the F-104, in a pressure suit she knew was outdated, she'd pleaded with him not to go. He'd said to her, in exasperation, "*Look*—There is something you've got to get through your head. There's more to this life than just living." And she had only begun to understand it then, that nothing mattered to him more than carrying out his mission.

Over the years, she embraced her role, and even picked up some of her husband's fighter-pilot bravado. When someone in his squadron got killed, she told herself, "that would never have happened to Frank." She did the same thing when they came to NASA—until the Fire. When Ed White died—this magnificent physical specimen—Susan realized that no one, not even Frank, could have gotten out of that burning command module alive. And she knew that as long as he stayed an astronaut he was waiting his turn to die. That was when Susan began to escape into alcohol. No one knew; she drank in private, and was careful never to appear intoxicated around her family. It wasn't hard to hide her drinking from Frank; he spent most of 1967 at North American. She told herself her own problems were minor compared to the stress he was under, and when he came home for an overnight visit she made sure there were no troubles to greet him. She blamed her unhappiness on herself. And when Frank accepted the circumlunar mission, she acted as if nothing were wrong.

But it was one thing to pull off the charade for her husband; it was another to manage it under the media microscope. On launch day, Susan's facade finally cracked. Not with the reporters who were camped on the lawn; she did fine with them. It happened when the cameras invaded her house. Producer David Wolper, who was making a documentary on Apollo 8, had asked NASA for permission to film Susan and the children as they watched the launch and the Translunar Injection burn. Dreading this, Susan had voiced her reluctance, but Frank told her, "I'm sorry, but NASA wants us to do it. It's for the good of the program, and that's the way it's going to be." Wolper arrived at dawn with a back lot's worth of camera equipment. They put microphones in the kitchen cabinets to catch bits of candid conversations. And they left with footage of a very anxious Susan Borman.

However worried she might have looked on camera, that was nothing compared to the anguish inside her, and when Chris Kraft paid a visit a couple of nights later, she made no attempt to cover up. "If you think the Fire was bad, wait until these guys get stranded in lunar orbit!" It would take days for the condemned men to die, circling until their oxygen ran out. She could just imagine what the press would do with that story. She could imagine the NASA man who would come to tell her that Frank was dead, and she could picture the big memorial service they would hold. With a strange kind of logic, she decided that no government official would write her husband's eulogy; that was her job. And she told Kraft that too.

But that was her only outcry. Her husband was commanding the first circumlunar voyage, and if she was expected to play the smiling, confident leader of his support crew, then that is what she would do. Twenty hours ago, just before Apollo 8 slipped behind the moon for the first time, she had sent him a special message, via Jerry Carr in mission control: "The custard is in the oven at 350."

"No comprendo," she heard Frank say, and then a moment later, "Roger." He understood; it was an old line from their Edwards days. Frank used to say to her, "You worry about the custard and I'll worry about the flying." She wanted him to know she was playing along. But now, with only silence from the squawk box and nothing to do but wait, it was all she could do to keep her composure as the time neared when Apollo 8 was to reestablish radio contact. She could not have known that in mission control Kraft was enduring his own agony of waiting, and that even the engineers who knew the SPS engine like the back of their hands were sweating out this silence like nothing before.

IV: "It's All Over but the Shouting"

Wednesday, December 25
Manned Spacecraft Center, Houston

The clocks in mission control crept toward a Mission Elapsed Time of 89 hours, 28 minutes, and 39 seconds, and the tension was palpable. If the Transearth Injection burn went as planned, Apollo 8 would reemerge at that time, 19 minutes past midnight on Christmas Day. If the engine didn't fire, contact would come as much as 8 minutes later.

The mission clock read 89:28:39. Seconds passed in silence. Suddenly a cheer went up from the flight controllers: Telemetry from Apollo 8 began to register on their screens. It took a few more minutes for earthbound antennas to lock onto the signal, and finally, they heard Jim Lovell's voice:

"Houston, Apollo 8. Over."

"Hello, Apollo 8," Mattingly replied. "Loud and clear."

"Please be informed there is a Santa Claus."

"That's affirmative," Mattingly responded gratefully. "You are the best ones to know."

• • •

Once the mission is done, go home. Not only was that Frank Borman's attitude, it was most astronauts'. Last August, in Chris Kraft's office, Borman had asked the trajectory people to get him home in two days, but the speed required would have trimmed the margins of accuracy on the trajectory dangerously close, as well as subjecting the command module to undesirable stresses during its reentry into the earth's atmosphere. Borman settled for 2½ days.

The voyage began with a spectacular view of the entire moon as Apollo 8 climbed away from it like a jet on afterburner. Then there was nothing to do but tend the systems, listen to news reports from mission control—headlined by messages of congratulations that were pouring in from around the world—and catch up on sleep. Around midafternoon, the men gave a televised tour of their home away from home. While Anders demonstrated how to prepare a freeze-dried meal in space, Borman told Mike Collins in Houston, "I hope you all had better Christmas dinners today than us." But Borman spoke too soon. When the TV show was over, they discovered a surprise waiting for them in Apollo 8's food locker, wrapped in foil and tied with red and green ribbons: real turkey with stuffing and cranberry sauce. This was a so-called wetpack meal developed by the military, one of the innovations Borman had fought to keep off the flight. It was also by far the best meal of the voyage. And there was another surprise, courtesy of Father Slayton: three tiny bottles of brandy. Borman was annoyed. "Put it back," he told his crew. He wasn't about to risk someone in the public raising a ruckus; if they made a single mistake on the rest of the flight the brandy would get the blame. The bottles went unopened (Lovell would say later that neither he nor Anders had any intention of opening them).

But there were other packages, and these were meant to be opened: Christmas gifts to the three men from their wives. Susan Borman had sent cuff links made from a pair of St. Christopher's medals that had gone through World War I with the late husband of a dear friend. From Marilyn Lovell there were cuff links and a man-in-the-moon tie tack, and from Valerie Anders, a gold "8" tie tack, replete with moonstone.

For Bill Anders, the trip home was a long, quiet, and boring fall. At one point, Mike Collins mentioned that his son Michael had asked who was driving up there. Anders replied, "I think Isaac Newton is doing most of the driving right now." This should have been a welcome chance to catch up on all the sleep he'd missed on the way out, but Anders, at least, wasn't having much luck there. Somehow it worked out that when he was trying to sleep Borman and Lovell were awake, and they got into small talk. Because of Borman's bad ear, they yelled a lot—"YOU THINK THE OILERS HAVE A CHANCE?" Anders was only thankful that in zero g he could survive on so little sleep, because he wasn't getting much.

At the other end of this fall lay the high-speed reentry into the atmosphere. When Anders was four years old the circus came to the small California town of Vallejo, where his family was visiting, and his grandfather took him to see it. There was an enormous tent, and a man climbed up a ladder—to his young eyes it looked about eight stories high—and dove off into what seemed a tiny tub of water. The boy talked about it for weeks. Anders hadn't thought of that circus dive in many years, but it came back to him when he looked across the lunar distance at the earth: *That's what that guy did in Vallejo!* He couldn't help but think, "I sure hope we hit that thing."

Flying the reentry was a task reserved for the command module computer. If it worked, Borman would just sit there and monitor. If it broke down, he would have to take over and fly it, and he'd worked hard to help create the techniques to do that. They'd probably be off target for splashdown—nobody could fly it as well as the computer—but they'd be alive. The tough part was going to be flying through the periods of high g's. It was hard enough in the training runs in the centrifuge, when you could barely lift your arm, but after six days of weightlessness it would be even tougher. Flying the reentry was one piloting job Borman would just as soon not have.

Even a perfect reentry would subject the command module to extreme stress. In Gemini, the ride down from earth orbit was long and slow, but Apollo 8 would be coming in at 25,000 miles per hour, and the forces of heat and deceleration would be far greater. Temperatures around the command module would soar to 5,000 degrees Centigrade, and their lives would depend on the heat shield on the craft's blunt end. It was made from a substance called phenolic epoxy resin whose protection came not from resisting the intense heat—no one had found an alloy that could do that and still be light enough to use on a spacecraft—but from giving in to it. Just as the boiling water in a kettle absorbs the heat of the stove and keeps the pot from overheating, the heat shield would become white hot, then char and melt away, taking with it the awesome heat of reentry. And when the fiery plunge through the atmosphere was finished, there would still be one more critical event, the blossoming of three 80-foot parachutes to lower Apollo 8 to the waters of the Pacific.

For Frank Borman, that would mark the end of his astronaut career. He had decided beforehand that Apollo 8 would be his last mission. Slayton had all but offered him the first landing, but Borman turned it down. He appreciated Slayton's confidence, he told him, but he doubted he could get his crew ready for a landing mission in time. He knew they would be disappointed—Anders in particular. But Borman had decided it was time to move on. If Apollo was a war, then a crucial battle was almost won; let someone else have the final victory.

In Borman's mind, the truth of it—which would have come as a great surprise to Anders and Lovell—was that Apollo 8 had been less difficult than he'd expected, far less stressful than Gemini 7. Apollo 8 was turning out to be a wonderful finale to his test flight career. There was only one thing Borman wanted now. After the ordeal of Gemini 7, nothing compared with the high he felt standing weak-legged on the carrier deck, his mission accomplished. And now, as Apollo 8 sped homeward, that was what Borman was looking forward to most of all.

Friday, December 27
9:31 A.M., Houston time
6 days, 2 hours, 40 minutes Mission Elapsed Time

Borman: Look who's coming there, would you?
Anders: Yeah.
Borman: Just like they promised.
Lovell: What?
Borman: The moon.

With just six minutes to go until reentry, the brief appearance of an old friend, rising beyond the dark curve of the earth's night side—at exactly the moment the trajectory specialists had predicted—was welcome reassurance that Apollo 8 was aimed right for the middle of the corridor. Just minutes earlier Borman had flipped a switch to cast off the now unneeded service module. Unprotected, it would meet its end as a shower of meteors over the darkened Pacific. Inside the command module, the conversation sounded like a movie script:

"Well, men, we're getting close!" Borman said.

"There's no turning back now," Anders said.

"Old mother earth has us," Lovell said. He was right; though they could not sense it yet, the men were returning to the earth as they had left, at fantastic speed.

"It's getting hazy out there," Anders said. "Does that mean anything? Every time you fire a thruster."

Now Borman gave the spacecraft to the computer. From here on, the autopilot would fly them in. He glanced out the window. "God, it *is* hazy out there, isn't it?"

"That's sunrise," Anders said.

"Yeah, that might be sunrise," Lovell added quickly. But suddenly they all knew it wasn't sunrise at all, but something far more strange: the glow of ionized gas. The command module was slamming into the outermost fringes of atmosphere so fast that atoms were being stripped of their electrons, creating a glowing plasma. Borman and Lovell had seen a similar glow on their Gemini reentries, but never this bright. And it was only beginning. "God damn," Borman said, "this is going to be a real ride. Hang on."

Still they were weightless in their couches, but soon, they knew, the command module would begin to decelerate. A light on the instrument panel would come on when the g-forces measured 0.05; at that moment, the real reentry would begin. From then on things would happen fast and furious.

"Got it! O-five g," said Borman. "Hang on!"

"And they're building up!" called Lovell. G-forces mounted as the command module slammed into the denser layers of air.

"Call out the g's," Anders reminded Lovell.

"We're one g." After six days in weightlessness 1 g felt like 3. Seconds later the men were pressed into their couches with tremendous force. Lovell groaned with the sudden deceleration.

"Five!" Lovell's voice was thick with the strain of five times his normal weight.

"*Six!*" An elephant was sitting on their chests. But the worst was over. Within moments the g-forces began to slack off. The command module had slowed to orbital speed now. Borman, Lovell, and Anders were captives of the earth once more.

A cold white light flooded the windows, as bright as daylight. It was like flying inside a neon tube. Borman had never seen anything so weird. He glanced over at Lovell and Anders, who were bathed in this unearthly glow.

Outside, Anders could see flaming objects, undoubtedly tiny pieces of the burning heat shield, flying past his window. Every so often what looked to be a fist-size chunk shot by and Anders thought, "Jesus, we can't take too many of those . . ." He kept waiting to feel heat building up at his back; it never came. While the head shield charred and melted, inside the cabin the temperature hardly rose a degree. Far below, to those who were fortunate enough to see it, the command module was a glowing meteor in the night sky.

The computer, meanwhile, was doing a miraculous job. Under

its control the command module turned and soared along a precisely crafted roller-coaster ride. Apollo 8 dipped down into the denser air, then ascended briefly for a respite from the heat and the g-loads, then dipped and climbed once more. At last, it headed down for the final descent, falling through the predawn darkness like a stone. Borman said, "It's almost all over but the shouting."

Now they were back in radio contact with Houston. At about 100,000 feet the altimeter sprang to life. The command module plunged toward the Pacific at a thousand feet a second. All that was left was the parachutes.

The command module cleared 30,000 feet and there was a loud crack as the parachute cover flew off. Then another crack.

"There go the drogues," Borman said, announcing the release of three small, stabilizing parachutes. Now there was a loud whoosh of air as a vent opened to let cabin pressure equalize with the outside.

"Should be approaching ten K," Anders called. "Stand by for the mains in one second."

They heard a crack.

"You see it?" asked Lovell.

"Can't see a thing," Borman said, peering up into the night. But the altimeter had slowed its readings; the chutes had to be okay. Now a loud mechanical groan filled the cabin as the excess fuel was dumped overboard and the command module thrusters spat flame; in their pale light Borman and Anders glimpsed three great canopies of red and white—beautiful, perfect parachutes.

Now their headsets filled with the chatter of recovery helicopters. "*Apollo 8, Air Boss 1, you have been reported on radar as southwest of the ship at twenty-five miles . . . Welcome home, gentlemen, we'll have you aboard in no time.*"

As Borman, Lovell, and Anders went through the checklist for splashdown—

"CABIN PRESSURE RELIEF VALVES, CLOSED!"

"Got it!"

"DIRECT O2, OPEN!"

"Open!"

—they were all but drowned out by the radio traffic.

"This is Recovery 2, I see the chutes, I see the light, level with me at precisely four thousand feet . . ."

Anders was yelling at the top of his lungs—"Floodlights to post-landing!"—for Borman and Lovell to hear him.

"This is Yorktown. *Affirmative, we do have capsule in sight—"*

"Turn him down," Anders said. "Christ, we can't get anything done."

"Alright," Borman asked, "anything else we missed?"

"Negative," said Anders. "Stand by to release the mains." That was a bit of teamwork that Borman and Anders had worked out in advance: when they hit the water, Anders would push the right circuit breaker, and Borman would flip a switch to cast off the chutes. Otherwise the wind might drag them through the water until they tipped over.

"Brace yourselves," Anders called.

"Well, wait," Borman said, "we've got two thousand feet yet."

"I don't know if we have or not," Lovell said. "They were reporting us lower."

"Oh, they were?"

Moments later the command module hit the water with a tremendous jolt. Water poured in through an open vent, so much that for an instant Anders thought the spacecraft's hull had cracked with the impact. The water poured all over Borman, who was poised to release the chutes, and before he could do anything the wind dragged the command module upside down.

Now it was up to three balloons in the spacecraft's nose to inflate and set them upright. Meanwhile Borman, Lovell, and Anders hung in their straps, upside down, in a dark, warm, suddenly quiet spacecraft, tossed about by a rough sea. Ten-foot swells set them pitching and rolling while helicopters circled overhead, waiting for the first light of dawn before dispatching swimmers.

Anders wryly took stock of the situation: Here they were, in the middle of the Pacific, hanging upside down in the dark, like bats, with bits of trash from the command module floor raining down past them. Not quite what came to mind for returning space conquerors.

Borman had never had a strong stomach for sea voyages, and it wasn't long before he was sick. This time Lovell and Anders showed him no mercy, saying, "What do you expect from a West Point ground-pounder?"

At last they were upright again, and the swimmers had arrived to secure a flotation collar around the command module.

From one of the circling helicopters came a voice: "Apollo 8, is the moon made of Limburger cheese?"

"No," radioed Anders, "it's made of *American* cheese."

The words of the returned moon voyagers belied the condition inside Apollo 8. Borman, Lovell, and Anders were positively grungy. For one thing they needed a shave, Borman less so than his crew, but he'd taken so much grief about his scrawny beard after two weeks on Gemini 7 that this time he'd arranged for an electric razor to be waiting for him in the recovery helicopter. After six days of living in a flying toilet they didn't even notice how bad it smelled inside Apollo 8. But when that first swimmer opened the hatch, he reeled backwards as if he'd been kicked in the head. Inside, the three men noticed a strange smell too: fresh sea air.

Life rafts waited outside the hatch, and minutes later the chopper lowered a Billy Pugh net and lofted each man, one at a time. As Borman ascended he glanced down at spacecraft 103, his ship, bobbing in the Pacific, and felt gratitude. And a few minutes later he stood smiling with crew on the deck of the U.S.S. *Yorktown* while hundreds of sailors cheered, waving American flags in the morning light.

It was the middle of the night when Borman, Lovell, and Anders stepped off the plane at Ellington Air Force Base, clean, rested, dressed in caps and blue coveralls. Susan was there with her sons to greet Frank and take him home. They arrived in El Lago to a celebration of candles and welcome home signs. Her husband had flown around the moon—would anything ever be the same? Well, yes; her husband would. As they went into the house Frank spotted the dog's dinner sitting in a corner, uneaten. He turned to his sons and said, "Why is that still sitting there? You know the rules." Not that she was surprised. But she had to laugh.

• • •

In the last days of 1968, there was a single image—pure, awesome, even holy—to counter a year's worth of violence. It was a photograph of the earth, rising beyond the battered and lifeless face of the moon. Apollo 8 was more than a successful space mission; it was a bright moment for a nation experiencing its first pangs of self-doubt. Even as Vietnam threatened to become a war America could not win, here was an American triumph. Not long after Borman, Lovell, and Anders were back in Houston, Borman got a telegram from someone he had never met. It said, "You saved 1968."

And for NASA, 1969 held great promise. The way was clear now for them to go the rest of the way, to meet Kennedy's challenge, to land on the moon. In his Christmas cards, Alan Bean, backup lunar module pilot for Apollo 9, wrote, "It's going to be a wonderful year for all of us."

CHAPTER 4

"Before This Decade Is Out"

I: The Parlay

The first moon voyagers were back on earth, and the impact of their voyage was considerable. In one masterstroke Apollo 8 had given the United States a clear lead in the space race. Publicly, at least, the Soviets were now talking about missions in earth orbit without even mentioning manned lunar flights, as if that had been their plan all along. At NASA Headquarters, no one was ready to write off the competition; intelligence reports said the Soviets were still working on a "super booster" for their own moon landing program. But in Houston, everyone felt the surging confidence that took hold of the Manned Spacecraft Center. For the first time, John Kennedy's end-of-the-decade lunar landing deadline seemed within reach. Indeed, if everything went according to plan, NASA would make it with months to spare. The architects of Apollo were talking about a landing for Apollo 11, slated for a July liftoff—earlier than anyone, including the oddsmakers in the Astronaut Office, would have dared guess.

To land on the moon—after only four manned Apollo missions? To many the plan seemed wildly optimistic, considering the tremendous challenges that lay ahead. The lunar module, which had passed an unmanned test in earth orbit, had never been flown with a man aboard. The space suit designed for the first moonwalks had never been tested in the vacuum of space. No Apollo crew had attempted the intricate and crucial rendezvous between the lander and the command module, which would have to be practiced not only in earth orbit but around the moon. Remarkably, NASA was planning to soar over those hurdles with just two missions: Apollo 9, an earth-orbit flight crammed to the hilt with tests, and Apollo 10, a full-up "dress rehearsal" of the landing mission in lunar orbit.

The lunar module's manned debut was the primary goal for the earth-orbit Apollo 9 mission, commanded by Gemini veteran Jim McDivitt and set for a late-winter launch. McDivitt's team—veteran Dave Scott and a red-headed, irreverent rookie named Rusty Schweickart—would practice the critical rendezvous maneuvers and give the entire Apollo spacecraft a thorough workout. Schweickart would perform Apollo's first space walk, to test the space suit and life-support backpack. By any measure, Apollo 9 would be an extraordinarily ambitious mission. In May, if nothing went wrong on Apollo 9—and that was a big if—Tom Stafford's all-veteran Apollo 10 crew would head for lunar orbit for the dress rehearsal. While John Young circled alone in the command module, Stafford and his lunar module pilot, Gene Cernan, would steer their LM within 50,000 feet of the moon and in so doing demonstrate every maneuver for the landing mission, save the final descent to the surface. Then they would rejoin Young and head home.

Both of these missions were staggering in their complexity. By comparison, Apollo 8 had almost been simple. Both would have to be nearly perfect for Apollo 11 to attempt a landing in July. Attempt was really the right word; there was no assurance that the first mission assigned to land would succeed. Nevertheless, as 1969 opened there was more than a little speculation, both inside and outside the Astronaut Office, on who would get the chance to try. Of course, anyone who went through the arithmetic of Slayton's crew rotation had a good idea of who it would be. The crew that backed up Apollo 8 would skip 9 and 10 and fly the next mission, Apollo 11. Pete Conrad had figured that out some time ago, and he didn't like the answer.

Since early 1967, Conrad and his crew had been the backups for Jim McDivitt's team, training for the first manned LM flight. Conrad and his lunar module pilot, Al Bean, had racked up hundreds of hours in the Lunar Module Simulator, helping the engineers wring out the bugs in its computer software. Conrad hoped that with that kind of experience under his belt, if the timing worked out just right, he might one day find himself in command of the first lunar landing. Conrad would have readily admitted it: he wanted the first landing as much as anyone in the Astronaut Office. And the numbers were on his side; he was backing up Apollo 8, and that meant he'd fly Apollo 11, a feasible, if optimistic bet for the landing. But Conrad's fortunes changed when NASA decided to send Apollo 8 around the moon. In August 1968, when Frank Borman's crew swapped places with Jim McDivitt's, so did their backup teams, and Conrad's place in line went to Neil Armstrong.

Now it looked as though Armstrong's team, a crew with hardly any lunar module experience, was about to be assigned to the landing. That's what bothered Conrad most. After all, he and Bean knew more about flying the LM than any other astronauts. One day Conrad took his misgivings to Slayton, who said what he always said: Any crew can fly any mission. And besides, Conrad knew, Slayton wasn't about to pull him off backup duty for Apollo 9, where he was sorely needed. Of course, there was still a chance that his own mission, Apollo 12, might turn out to be the first landing, and around the space center in January 1969 more than a few bets were on Conrad. What Conrad did not know, because Slayton told next to no one, was that Slayton had almost thrown his any-crew-can-fly-any-mission credo out the window.

For years the media had tried to break the code of Slayton's crew selections, theorizing that he must be picking the best astronaut for each mission, carefully matching crewmen for some combination of skills and temperaments. Slayton also knew that some NASA managers wanted him to hand-pick crews. And he had always resisted. He knew anything other than an orderly crew rotation would make a shambles of Astronaut Office morale. But what about the first lunar landing—was that important enough to put the very best astronaut in command? NASA's upper echelon had always considered Jim McDivitt and Frank Borman as prime candidates for that mission, and in truth, both men were highly regarded by Slayton. Late in 1968, Slayton reasoned that as the only veteran lunar crew, Borman's team would have an edge that might make the difference between success and failure. And on the chance that Borman did not succeed, Slayton was ready to put McDivitt's crew right onto Apollo 12 instead of Pete Conrad's. Slayton knew he'd face a firestorm of protest from the pilots waiting in line—no astronaut had ever been ushered directly from one prime slot to another—but he was willing to put up with that. But Slayton's plan fell by the wayside in the fall of 1968 when Borman turned down the offer.

Many years later, the myth would endure that somehow NASA had hand-picked Neil Armstrong to command the first lunar landing mission. In one version it was because he was the best test pilot among the astronauts; in another, because he was a civilian and NASA was eager to avoid any connotations of militarism on the first landing. Both theories were wrong. In truth, the crew for the ultimate test flight was chosen not by design, but by chance, just as Slayton had always said it would be.

On Monday, January 6, 1969, Slayton summoned Armstrong to his office and told him that he was planning to assign his crew to Apollo 11. Buzz Aldrin would be his lunar module pilot, and Mike Collins, who had fully recovered from his surgery, would be his command module pilot. It wouldn't be official for a few days, but assuming it was approved, they would be in line for the first lunar landing.

Neil Armstrong had gone to work that Monday morning with only a suspicion, not an assurance, of what might come his way. Looking back, he would say that he was very pleased, but not wildly elated, as he left Slayton's office. The sobering reality was that a host of untried procedures would have to be defined, evaluated, and finally boiled down into neatly scripted checklists before he and his crew could even begin their most intensive training. There were just seven months for him, Collins, and Aldrin to master the complexities of the most demanding space mission ever attempted. Even more significant in Armstrong's mind on this January day was the lunar module, and the tremendous challenges facing Apollos 9 and 10. The odds that both missions would be flawless, as NASA's plan required, were small indeed. Armstrong was not at all sure that when July came, landing on the moon would really be his mission.

• • •

"It was a day, the first of many, I'll bet, of walking on eggs, or normalcy tinged with hysteria." Joan Aldrin didn't usually write entries like that in her diary, but that is how she described life on the day after she learned her husband was on the crew of Apollo 11. In 1954, on one of their first dates, Buzz had informed her that people would land on the moon sometime in this century. And she had thought, How ridiculous.

When she met Buzz—the nickname came from his baby sister, whose efforts to pronounce "brother" came out "buzzer"—Joan was working on an acting career. She did a little television work—a walk-on part on "Playhouse 90," a few lines on "Climax." Years later, in Houston, she would help open the Clear Creek Country Theater, but late in 1954 she put all that on hold to marry this young blond fighter pilot, and to go with him to Germany. And aside from wild statements about moon trips, Joan thought she had married a fighter pilot like the others in his air force squadron. She learned otherwise when he traded the flight line for the classroom in 1959, as a graduate student at MIT. At first Buzz had intended to stay only long enough to earn a master's, then apply to the test pilot school at Edwards. But the academic life appealed to them both, and soon he was working on his doctorate. He was also talking about the astronaut program. Buzz knew NASA required test pilot credentials, but he was betting that would change. In the meantime, he was hard at work on his thesis on piloting techniques for space rendezvous. He chose it knowing that rendezvous would become a critical part of NASA's moon program, when astronauts returning from the lunar surface would have to link up with their command ship.

At MIT, Joan discovered a side of Buzz she hadn't seen before: his extraordinary intelligence. He would pace the floor and talk excitedly about his thesis work. She wanted to support him in any way she could, and that meant she listened, but she didn't comprehend very much of what she heard. There were many people who didn't understand Buzz. His mind was on things that were far removed from day-to-day existence. Like many very smart, intense people, he wasn't too comfortable socially. He understood the intricacies of orbital rendezvous, but he didn't know how to make simple, light conversation. But he was always steady, always calm, always logical. He was her anchor, and she was, quite simply, in awe of him.

Not until years later, when Buzz finally got his chance to fly in space, did he show a crack in his armor, and even then, she didn't immediately recognize it. Buzz came home from Gemini 12 elated at the flight's success, but that was followed by a brief and minor depression. She had never seen him like that before. When she thought about it, she realized that he was going through the same letdown she experienced whenever a show closed; she didn't know when she would get to do another one.

Now Buzz was going to be in another show, and it was the big one. When he told her, she put up a calm and happy front, but whenever she thought about the lunar landing she felt sick. This was something new for Joan. She hadn't been afraid when Buzz was getting ready for Gemini 12, despite the fact that he was going to spend a total of five hours walking in space, more than any previous astronaut. Space walks had been done before, she told herself, and Buzz would just do his better. But this time the old confidence wasn't working. No one had ever landed on the moon.

March 7, 1969
Manned Spacecraft Center

In the quiet of mission control, Buzz Aldrin stood behind the Capcom's console and listened to the voices from space. He felt at home here; in some ways he preferred the company of the flight controllers to the other astronauts'. He had never shared the fighter-jock bravado of his colleagues, but among Kraft's men he felt the bond of common academic interests. With them, Aldrin had helped to work out the techniques for the events now unfolding 155 miles up. The climactic moment of Apollo 9 was under way.

"Okay, Dave, I can see you. Boy, are you bright . . . Okay, I'm at nine hundred fifty feet, ten feet per second."

To the rest of the world, the ten-day earth-orbit mission of Apollo 9 must have seemed decidedly mundane, coming as it did on the heels of the first flight around the moon. But few astronauts would have been surprised to learn that Jim McDivitt had turned down the circumlunar mission to stick with this one. They wouldn't have said so publicly, but many saw Apollo 8 as little more than a ride—no real flying involved. But Apollo 9 was a test pilot's feast. In truth it was far more difficult, more ambitious, and in some ways more dangerous than Apollo 8. For the engineers, the demands of getting two manned spacecraft ready for flight were headache enough. The simulators kept breaking down. There were days, during training, when McDivitt would go home and tell his wife it couldn't be done, it would never all come together in time. Miraculously, it did, and McDivitt's crew was launched on March 3. They figured that if they accomplished half of what was in the flight plan, they'd call the mission a success.

By March 7, the mission's fifth day, they'd already come close to doing it all. Besides a complete shakedown of their command module *Gumdrop*, they'd activated the docked lunar module *Spider* and test-fired its descent engine. Schweickart had suffered a debilitating bout with motion sickness that almost canceled his space walk, but he recovered in time to go outside, wearing the lunar space suit and backpack, for thirty-eight minutes. And today came the climax. While Scott remained in *Gumdrop*, McDivitt and Schweickart climbed into *Spider* and undocked. For six hours the two men flew a craft with no heat shield, venturing up to 111 miles from the safety of their command ship. Then they fired their engine and headed back to Scott, beginning the long and intricate dance called space rendezvous. As Buzz Aldrin knew only too well, an astronaut flying a rendezvous couldn't use the instincts he'd honed in airplanes. This was a completely different way of flying.

Rendezvous was a domain ruled by the arcane statutes of orbital mechanics. A spacecraft in orbit is like a ball bearing whizzing around inside a deep, curved funnel. A ball thrown into the funnel will "orbit" at a height and speed that depends on the amount of energy it has. A ball with a lot of energy will circle at the upper end of the funnel; one with less energy will circle lower down. If the ball is down toward the neck of the funnel, it will circle faster than one up near the mouth, even though it has less energy. If the balls are spacecraft around the earth, the funnel becomes the invisible "well" of gravity which all orbiting objects must fight in order to stay in orbit. The closer the spacecraft is to earth, the stronger the force of gravity, and the faster it must go to balance that pull. The farther away it is, the weaker gravity's pull, and the slower the spacecraft travels.

From a pilot's standpoint, the analogy makes clear the most important rule of orbital mechanics: Height and speed are inextricably linked. To slow down, the spacecraft must be kicked into a higher orbit (by adding energy with a burst from a rocket). Conversely, speeding up requires dropping into a lower orbit (by using the rocket as a brake). A pair of astronauts who start out behind their target must lower their orbit until they catch up, but not for too long, or they will overtake it. If they start out ahead, they must raise their orbit, go slow for a while, and then descend in time to meet their target. Every burst of speed, every bit of braking changes their height and therefore their speed. Catching the moving target—and staying there once they've arrived—becomes a feat of great complexity.

At MIT, Aldrin had foreseen the importance of rendezvous to NASA's moon program. He knew that special onboard computers were being developed for Gemini and Apollo to help astronauts fly a rendezvous. But what if the computer broke down? With the help of specialists in MIT's Charles Draper Laboratory, Aldrin worked out the techniques the pilot could use to take over and fly the final stages of a rendezvous by hand. At NASA, Aldrin had been instrumental in designing the rendezvous schemes for Gemini and Apollo. He would look back on those months, working with mission planners, as one of the most demanding and rewarding experiences of his life.

Now, listening to the words from Apollo 9, he knew that McDivitt and Schweickart were on their final approach to the command module. Inside *Gumdrop,* Dave Scott could see the returning lunar module:

"You're upside down again!"

"I was just thinking, one of us isn't right side up," McDivitt answered.

"Boy, you've got contraptions hanging out all over."

"That's show biz. Okay, I have us at about three hundred seventy feet. . . ."

Around the space center, Buzz Aldrin—the first astronaut with a Ph.D. after his name—was known as an intense, brilliant theoretician whose interest in rendezvous was nothing short of fanatical. The other astronauts called him "Dr. Rendezvous," and Aldrin lived up to the name. For anyone who would listen, he offered a discourse on his ideas brimming with orbital-mechanics jargon, whether his listener could keep up with him or not. Once, when Aldrin's wife was away, Walt Cunningham invited him to dinner, then was called out of town unexpectedly. Somehow Aldrin never got word of the change. At the appointed hour he showed up for dinner, where he enthusiastically lectured Cunningham's wife, Lo, on out-of-plane errors and line-of-sight velocities until the small hours of the morning.

It was hard to think of anything besides rendezvous that had been the subject of more meetings and discussions within and outside the Astronaut Office. Knowing the ins and outs of rendezvous was almost synonymous with being an accomplished astronaut. No astronaut knew more about it than Aldrin; and yet, to his regret, he had always felt like something of an outsider in the office. Aldrin was sure it was because of his résumé—not only was he not a test pilot, but he wasn't a member of the navy clique that he felt dominated the astronaut corps.

Other astronauts would later say it had more to do with his personal style. To them his self-confidence had a way of seeming like arrogance. He could be direct to the point of bluntness. Aldrin saw no reason to conceal his pride at his academic achievements, but the Old Heads rolled their eyes when he showed up wearing a tie clasp fashioned from a pair of air force wings, with his Phi Beta Kappa key dangling from it. And if the other pilots knew less about the theory of space rendezvous than he did, then they also understood that an astronaut didn't have to be a theoretician in order to fly.

If Aldrin drove the Old Heads up the wall, he was also right more often than they would have cared to admit. But he sometimes fell prey to that well-known syndrome of technical organizations known as Not Invented Here. When the astronauts were concerned about lighting conditions during the rendezvous' final stages, Aldrin suggested that the lunar module fly upside down as it approached the command module. In zero g, of course, it would make no difference, and it would keep the sun's glare out of the lander's windows. The Old Heads scoffed at the idea—*Fly upside down?* But eventually when it was time to make a decision, they went ahead and quietly adopted the plan, as if that had clearly been the way to go all along.

When Aldrin finally did get a chance to fly in Gemini 12, he got a chance to put his rendezvous expertise to good use. He and Jim Lovell had just reached orbit and were chasing their Agena target rocket when their rendezvous radar malfunctioned, which in turn prevented the men from obtaining crucial data from their computer. While Lovell did the flying, Aldrin got out a set of rendezvous charts—which his own efforts at MIT and NASA had helped to create—and computed the maneuvers himself, and the pair completed their rendezvous successfully. And after a string of aborted space walks on earlier flights, Aldrin made the most successful walks of the Gemini program. Chris Kraft himself would later say Aldrin had turned in a superb performance.

"Dave, I just can't see it. Let me get in a little closer."

As *Spider* maneuvered into position for docking, Scott radioed instructions to guide McDivitt, who was blinded by the sun's glare: "Just keep coming easy like that. . . . You ought to go forward and to the right a little. . . . There you go. . . ."

At last the two ships met. A buzzer sounded as a set of docking latches on the two craft snapped shut. A relieved McDivitt said, "I haven't heard a song like that in a long time!" Apollo 9 had fulfilled all its major objectives. At that moment, Aldrin knew that Apollo 10 would also succeed, and that he and Armstrong would attempt to land on the moon. On March 24, NASA made it official. For Aldrin and his crewmates, training now went into high gear.

Friday, April 18, 1969
Building 9, Manned Spacecraft Center, Houston

Inside the cavernous expanse of Building 9, a lone figure clad in a bulky white space suit and backpack stood at the center of a small crowd of technicians. He stooped slightly under the burden of the equipment necessary to keep him alive in the vacuum of space. After the technicians had attended to the necessary adjustments, he strode stiffly to a silvery mockup of a lunar module, and stood in one of its bowl-shaped footpads, ready to begin the strange spectacle of a practice moonwalk.

"All set, Neil, if you read me."

"Yeah, I read you," said Armstrong, answering the technician who was acting as Capcom.

"Okay, proceed."

Armstrong held on to the ladder that was attached to the mockup's front landing leg, and lifted his left foot over the footpad and onto a bed of sand. As he had practiced many times over the past few weeks, he tested his weight on the sand as if it were the moon, and described what he saw.

"Okay. Checking the bearing strength, and we're leaving a one-quarter-inch footprint. And we have a poorly sorted sand-and-gravel aggregate which does not stick to the boot. Range of the groundmass is from one centimeter down to below the resolution of the eye." Now he stepped off the footpad and took a few steps. "My balance seems to be very much like earth simulations." It was an optimistic thing for Armstrong to say, weighed down by 200 pounds of gear, but he had reason to expect that the real thing would be easier. In the past two months he had glimpsed what it might be like to walk in the moon's one-sixth gravity, inside the KC-135, a converted air force cargo plane with a long, padded cabin. By flying a carefully planned parabolic trajectory, the pilot of the KC-135 could create about half a minute's worth of simulated lunar gravity, giving Armstrong time to practice scooping up rocks, or handling gear, or simply get accustomed to the surreal lightness that came with every step. With a few dozen parabolas per flight, a session on the KC-135 was enough to let Armstrong taste the freedom that the moon would offer him once he was there. But the bulk of his work was here, in Building 9, where a moonwalk was as hard as a day on a construction site.

Like a polar bear lumbering about in his pen, Armstrong went about his work while a small crowd of technicians and training specialists looked on. As he moved, he struggled not only against the suit's weight but its stiffness. Pressurized at 3.5 pounds per square inch, the suit was a rigid balloon in which every movement required effort. The gloves were clumsy, and it was especially difficult for Armstrong to manipulate a camera or grasp a geologic hammer. Simply opening or closing his hand was like squeezing a tennis ball. As he reached into a pocket on his space suit thigh and pulled out a collapsible long-handled scoop, technicians with headphones could hear his controlled but labored breathing. "We're beginning the contingency sample," Armstrong said. "I have the collector . . ."

Standing at the sidelines, a fully suited Buzz Aldrin watched his commander at work and waited for the proper moment to join him. It had been decided in the summer of 1968 that the first landing mission would feature a single moonwalk; by February 1969 its duration had been decided at about 2 hours and 40 minutes. For months now, the people in the space center's Crew Systems division had been working with him and Armstrong to make every minute on the lunar surface as productive as possible. The two men would collect rock and dust samples, lay out a few simple scientific instruments, and take pictures. All of it had to be rehearsed to the letter, until the tasks became second nature. Never in the history of exploration had 2 hours and 40 minutes been so carefully planned.

"Okay," Armstrong said, "we're ready for the second man to come down now. . . ." Aldrin lumbered onto the sandy training area and began his work. Some of the NASA doctors, Aldrin knew, were predicting that he and Armstrong would have trouble adjusting to lunar gravity, that they would wear themselves out working in pressurized suits. Aldrin was sure they were wrong. With nearly five and a half hours of space walking under his belt, he'd proved that an astronaut could do useful work outside a spacecraft, in the three-dimensional ice rink called zero g. Aldrin was certain that walking on the moon, with the luxury of a gravity field, would be even easier. It was the part that would come afterward that worried him.

Media attention had always been the single biggest adjustment for the astronauts. Most of the Original 7 had been woefully unprepared for the barrage of attention that greeted them the day they were introduced to the press. The spotlight had dimmed somewhat since then, but not much, and most of the astronauts had learned to deal with the constant demands of their celebrity. There were even some, like Gene Cernan, who seemed to enjoy talking to the press about their experiences. But it wasn't so easy for Aldrin. Publicity was a double-edged sword. Whatever enjoyment he got from the recognition, it was far outweighed by his discomfort. He hated the question that awaited him on his return from Gemini 12: What did it feel like in space? He could go on at length about the technical aspects of the experience—his mobility during the space walk, or the computations he made during the rendezvous—but don't ask him to plumb his own feelings. It was the worst kind of bind. Sitting under the TV lights, he couldn't just shrug off the question with a pat response; he was too much the perfectionist for that. And so he was caught, struggling to find an honest answer, when the only one he could think of was, "I don't know."

Aldrin's unease was so great that given a choice, he would have preferred to be on the second lunar landing, or the third, instead of the first. For one thing, he would be able to put his scientific talents to greater use on a later mission, when the emphasis would be more on exploration. Even more important, there would be less attention. No one, save Aldrin's wife, knew that when he learned he would be on the lunar landing crew, he had briefly considered bowing out. He'd rejected the idea just as quickly. No astronaut had ever resigned from a mission. If he asked to leave this landing crew, he might never get onto another one.

Years later, Aldrin would say his mixed feelings were thrown into sharper focus by one issue in particular. In fact, it was the first question asked at the first press conference after the crew announcement. "Which one of you gentlemen will be the first man to step onto the lunar surface?" Deke Slayton had been sitting up on the stage with Armstrong, Aldrin, and Collins, and he told the press that the question hadn't been decided. Armstrong added, "It's not based on individual desire," and said it would be worked out as the training progressed, according to how best to accomplish the mission objectives. Of course, they all knew very well that there were very few astronauts who *didn't* want to be first to walk on the moon.

At one time, there had been a preliminary version of the checklist for the moonwalk, and on it the words "LMP EGRESS" came before "CDR EGRESS"—in other words, the lunar module pilot would go outside ahead of his commander. Aldrin could see sound reasons for having it that way. Armstrong had his hands full with the landing, and with the mission commander's responsibility for the flight's overall success. Instead of adding more to that workload, didn't it make sense for Aldrin to take the lead when it came to the moonwalk? It had been the same way in Gemini: The copilot always made the space walk while the commander stayed inside and flew the spacecraft. And there was the matter of physical conditioning. The moonwalk was going to be physically demanding, and Armstrong was anything but an exercise fanatic. More than one astronaut remembers that Aldrin paid a visit, checklist in hand, explaining the operational merits of having the lunar module pilot get out first.

There was nothing extraordinary about an astronaut campaigning for a pet cause, but this was no ordinary issue, and the Apollo crew commanders who heard about Aldrin's idea had strong reactions. For them, it was simple: The commander gets out first. Of Aldrin, they wondered, "Who is he kidding?" The Gemini precedent didn't apply, because a lunar module sitting on the moon wouldn't be in flight—it would be *in port.* And as any naval officer knows, the protocol on such matters is clear: When the ship comes to port the skipper is always first down the gangplank.

But in the end, there was something more decisive, outside the whims of human emotions. It was the lunar module's front hatch. The lunar module cabin had about as much space as a large broom closet. For one astronaut to go outside, the other man would have to hold the hatch open and stand back in his corner. The departing astronaut would then get down on his stomach and wriggle through an opening only 32 inches square. Getting out of the LM wasn't like going down a gangplank; it was more like being born. Years earlier, with the first landing still a long way off, Grumman engineers had designed the hatch so that it opened from left to right—that is, toward the lunar module pilot's side of the cabin. The only way Aldrin could get out ahead of Armstrong would be if the men first changed places. That would have been possible for two men in street clothes, but not encased in pressurized suits and massive backpacks. One day Armstrong and Aldrin tried it, fully suited, in a lunar module mockup, and they damaged the cabin. Deke Slayton was there; he saw the damage, and he told himself the situation would have to change.

Mission planners had quietly come to the same conclusion in February, but well into the spring, Armstrong and Aldrin had not learned of any decision. But Aldrin heard rumors that Armstrong would be named to go out first, because he was a civilian and because NASA wanted to keep the first steps on another world free of any militarism. The implied slight against the air force, and the thought that the matter would be decided on political grounds, instead of on operational considerations, angered Aldrin. But he knew there wasn't any point in maneuvering. It was best to be direct about it; he would take the issue to Armstrong. Their time on the Apollo 8 backup crew had left them with a sort of friendship—not exactly drinking buddies, but hardly strangers. Still, as he entered Armstrong's office he wasn't sure how his commander would take what he was about to say.

"Neil, we've got to come to some kind of decision on this," Aldrin said, somewhat self-consciously. "I'm sure you know how I may feel about it . . ." The words were out. There was an awkward silence; Armstrong looked uncomfortable—was he embarrassed at Aldrin's directness? At last he spoke:

"I'm aware of the historical significance of the decision, Buzz, and I'm not about to rule myself out."

Years later, however, Armstrong would have no memory of

their conversation. He had been aware that Aldrin had a certain focus on the issue, he would recall, and he felt sorry, because he didn't care one way or the other. In fact, as Aldrin himself had suspected, Armstrong had nothing to do with the decision, and he was never asked for his opinion until after it was made. Slayton came by to say the plan was for him to go out first, and did he agree? In Armstrong's mind the new plan was not only more practical but far safer. He told Slayton, "Yes, that's the way to do it."

In Slayton's mind, the decision was clear, hatch or no hatch. He recommended, and the managers agreed, that Armstrong, as Apollo 11's commander, and as a senior astronaut, should be the first man to set foot on the moon—assuming, of course, that he and Aldrin were able to get there. By mid-May, that rested with the success or failure of Apollo 10, which was about to leave the earth.

II: "We Is Down Among 'Em!"

Monday, May 19, 1969
1:49 P.M., Houston time
Aboard Apollo 10, 128,000 miles from earth

"If people want to know what kind of men go to the moon, there's a good look at one right there." Having prepared his earthbound audience, Gene Cernan pointed the television camera at his commander, and a gum-chewing Tom Stafford, his balding head hidden under his communications hat, smiled back. Cernan said, "Could you believe it?"

"Some people still don't," drawled Stafford.

On earth, where Stafford's grinning countenance filled television screens across the world, Capcom Bruce McCandless played straight man: "I'm surprised you all have not set this to music."

Cernan answered, "Oh, you want music? Well, we'll give you some music. . . ." On the monitors, the image blurred and shifted, then came clear again, and there was the earth; this time, thanks to Apollo 10's new TV camera, it was in full color. "Here it comes," Cernan said, as one of his crewmates activated a small portable tape recorder. "This is just so that you guys don't get too excited about the TV and forget what your job is down there." And then, the sounds of Frank Sinatra and the Nelson Riddle Orchestra came floating down from halfway across the translunar gulf:

Fly me to the moon, let me play among the stars
Let me see what spring is like on Jupiter and Mars
In other words, hold my hand
In other words, darling kiss me. . . .

The second voyage to the moon was underway, complete with background music.

There was a time, in the first part of 1968, when Tom Stafford thought he might have a chance of making the first landing. And more recently, there had been others at NASA, most notably George Mueller, head of the agency's Office of Manned Space Flight, who had pushed for a landing on Apollo 10. From the beginning, Mueller tried to quicken the pace of the moon program. It had been Mueller who insisted on "all-up" testing for the Saturn V, that is, launching a completely assembled booster instead of testing one stage at a time, as the more conservative engineers had wanted. Mueller's impatience had probably saved months in the race with the end of the decade. On the eve of Apollo 9's spectacular success, Mueller, like others at NASA, was asking, why in the world send the entire Apollo spacecraft to the moon—with all the risks involved—and not try to land?

One reason was that Stafford's lunar module, built before Grumman enacted a super weight-saving program, was too heavy to land. There was some talk of letting Stafford use Armstrong's LM, the first one built that was light enough for a landing, and postponing Apollo 10 a month to allow the switch. But some, particularly Chris Kraft, raised strong objections. There were too many unknowns, he said. His trajectory people didn't understand the moon's lumpy gravitational field well enough yet to predict what mascons would do to the paths of the orbiting command and lunar modules. Would they pull the lander off course for its descent to the surface? And while the astronauts explored the moon, would mascons pull the command module off course for the rendezvous ahead? NASA needed more navigation data from Apollo 10 before it could commit the next crew to a landing. Furthermore, Kraft's flight controllers needed experience in communicating with two separate spacecraft at lunar distance. Sam Phillips, the tough, exacting air force general who served as Apollo program director, listened to all sides of the argument and decided that the dress rehearsal was not only desirable, but crucial.

Tom Stafford agreed. He'd wanted the first landing as much as anyone else, but he wasn't about to campaign for a mission he knew was beyond accomplishing. Now was the time to find the hidden unknowns and solve them, so that Apollo 11 would be able to concentrate on the landing itself. And Apollo 10 wasn't simply a repeat of Apollo 9 in a different place; new procedures were required for a rendezvous in lunar orbit. Stafford and Cernan would take their lander and descend to 50,000 feet above the lunar surface, where they would make a critical test of the landing radar. Then, from this close vantage, they would scout Apollo 11's proposed landing site in the Sea of Tranquillity before rejoining Young in the command module. By any measure, the dress rehearsal was a grueling mission; it seemed to Stafford's crew that they had more to do on Apollo 10 than all the others combined.

With five Gemini flights among them, Stafford and his crew were the most experienced team yet sent into space. They had a camaraderie that came not only from having trained together for more than two years, but from a long personal history with one another. Stafford, a veteran of the first space rendezvous in 1965, went on to command Gemini 9 with Cernan as his copilot. He had come to know Cernan as a quick study with a sharp mind. Cernan also liked to party, and he liked the attention that came with being an astronaut; once or twice Stafford had to pry him away from some reporter and back to work. It was Cernan, as lunar module pilot, who had the major responsibility for the lander's systems. Stafford and John Young went a bit further back; they had been shipmates aboard the battleship *Missouri* twenty years earlier. Besides being a good pilot and a hell of an engineer, Young had a wonderful dry wit—a real down-home, country-boy type of delivery. And he was an expert on the command module systems; Stafford felt lucky to have him on the crew.

Now, a day into their journey to the moon, Stafford, Young, and Cernan were following the trail blazed by Apollo 8. None of them believed for a moment that the second lunar voyage might somehow be routine, and their suspicion was confirmed as soon as their Saturn V thundered off from Pad 39-B. Having heard the reports from Borman's and McDivitt's crews, Stafford and his crewmates thought they knew what to expect, but the ride was even more violent than they'd imagined. And the launch was only a prelude to the real heart-stopper. Three minutes into the Translunar Injection burn, the cabin began to vibrate with a strange buzzing sensation that the men both heard and felt. Something was wrong with the booster.

"Okay," Stafford radioed to Houston, "we're getting a little bit of high-frequency vibrations in the cabin. Nothing to worry about." Those words were for the benefit of mission control. In reality, Stafford was sure the mission was over, and so were his crewmates. Cernan, in the right-hand seat, was already thinking ahead to the abort maneuver they'd have to perform to get back to earth. The vibrations worsened until Stafford could barely read the instruments. His heartbeat quickened as he tried to puzzle out what might be happening. The only thing he could think of was a form of aerodynamic shaking that pilots call flutter. If it's bad enough, flutter can rattle an airplane to pieces. But how could flutter occur in a vacuum? Stafford held the abort handle in his left hand; with a twist he could shut down the booster and end the mission. But he told himself, "No way. We've come this far—if she blows, then she blows." The three men held their breath as the nearly six-minute burn continued. Stafford anxiously eyed the computer's velocity readout, which was climbing toward the 35,000-plus feet per second needed to send Apollo 10 to the moon. He said silently, "C'mon, baby . . ." At last the booster shut down, on time, and Apollo 10 was right on course. Later, engineers would blame the vibrations—which were well within the spacecraft's design tolerance—on some of the booster's pressure-relief valves, and would remedy the situation for future missions. Apollo 10's first heart-stopper turned out to be false alarm, but Stafford and his crew understood: they could take nothing for granted.

Wednesday, May 21
9:49 P.M., Houston time
3 days, 11 hours Mission Elapsed Time

It had ridden into space nestled within the third stage of the Saturn V, hidden from view by a set of protective panels, waiting to be extracted from its berth. That had happened shortly after Translunar Injection, when John Young separated the command module *Charlie Brown* from the booster and pulled away. At the same time, the protective panels departed from third stage, exposing the lunar module *Snoopy*, its metallic skin gleaming in the sunlight. Young then activated a special docking probe in *Charlie Brown*'s nose, and slowly steered the 32-ton command ship back toward the lander. The two ships met in a scene Freud would have loved, as Young guided *Charlie Brown*'s docking probe into a conical port, called the drogue, in *Snoopy*'s roof. Tiny "capture latches" on the tip of the probe fit inside a hole at the drogue's center. Young retracted the probe, drawing the two ships together, until twelve spring-loaded "docking latches" snapped shut, firmly mating the two craft.

During the three-day voyage, *Snoopy* had waited, dormant, docked to *Charlie Brown*'s nose. Now, on Wednesday evening, Young opened the command module's forward hatch and removed the docking mechanism, opening up a short tunnel between the two craft and clearing the way for Gene Cernan to enter the lander. When he did, pushing off the command module's floor and floating through the tunnel, it was like entering another world. Opening *Snoopy*'s hatch, he found himself staring at the floor of the tiny cabin, as if he were hanging by his toes. He tucked his body into a spin until he was upright in the small, gray space. The lander was a strange machine, even from the inside, but Cernan had come to know it well. In front of him was a square instrument panel, packed with switches, gauges, and displays. At waist height were two sets of hand controllers, one for each man. There were two small triangular windows, one on either side of the main panel, and a smaller, rectangular rendezvous window in the ceiling on the commander's side. More panels, covered with circuit breakers, lined the side walls. Bundles of wiring and all kinds of plumbing were visible along the floor, exposed because covering them would have added too much weight. They gave the cabin the look of a boiler room. Behind him, atop a small ledge, the can-shaped cover for the lander's ascent engine protruded into the cabin. There were no seats; none were needed. The lunar module was the first true spacecraft: it flew only in the void of space.

On the outside, the lunar module was as alien as the world it would land on. With no need for aerodynamic sleekness, it was an unlikely amalgam of angular shapes and boxes, rocket engines and antennae, doors and landing gear. No spacecraft before it had so completely merged form and function, and the result was a definite shift in appearance toward the organic. In the clean room at the Cape, where Cernan and Stafford had last seen it, the lander was a metal chrysalis, landing gear folded against its sides in preparation for the launch shroud. Now, in space, it was a huge robotic insect, with two triangular windows for eyes and a square hatchway for a mouth, antennae jutting at all angles, and four foil-clad landing legs. Even its name seemed to say it was alive; the astronauts called it "the lem."

The reason the LM looked so strange was simple: it had to be as lightweight as possible. The Grumman designers achieved the biggest weight savings from splitting the craft into two pieces. A boxy descent stage held the rocket used for the landing, as well as the landing gear and supplies needed for the moonwalk; it would be left behind on the moon. Perched atop the box was the angular ascent stage, which contained everything else—the crew cabin with its controls, supplies, breathing oxygen, and electronics, as well as a separate rocket for the ascent from the lunar surface back to orbit. Before long, however, the LM was hopelessly overweight, and the Grumman people were obsessed with making it lighter. In short order, the seats called for in early designs were scrapped, since the astronauts would not need them in weightlessness or the moon's one-sixth g. Then all traces of aerodynamic curves were shaved off.

By 1965, with 95 percent of the LM's design finalized, the weight-trimmers were still looking for ounces.

In the end, the lander became a strange mix of strength and fragility. The skin of the descent stage was only a Mylar wrapping stretched over a frame. In the ascent stage, the walls of the crew cabin were thinned down until they were nothing more than a taut aluminum balloon, in some places only five-thousands of an inch thick. Once, a workman accidentally dropped a screwdriver inside the cabin and it went through the floor. Now, in space, it seemed deceptively flimsy. When the cabin was pressurized the front hatch bulged outward. *That* had scared John Young, who was in the command module wearing nothing but a pair of long johns; he was up there muttering, "I didn't know I was volunteering to go on this damn thing in my underwear . . ." No wonder Jim McDivitt called his LM "the tissue-paper spacecraft." But it was sturdy enough—even more than enough—for the moon.

For about two hours, Cernan was to check *Snoopy*'s systems in preparation for the next day's dress rehearsal. But when he arrived, he found himself floating in a snowstorm of white fiberglass. It had somehow blown out of an insulation blanket on the tunnel wall, and had found its way into the LM via a pressurization valve. Before he could do anything more, Cernan had to vacuum the fiberglass bits out of the air. Stafford arrived to lend a hand and burst out laughing. There was Cernan with bits of white stuck to his hair, his eyelashes, his mouth. All Stafford could think of was chicken-plucking time at the poultry house.

There followed a couple of hours of concentrated work. Cernan was tunnel-visioned into his checklists, and when there was a lull, he stopped and consciously tried to take in what was happening. On his first space flight, the Gemini 9 mission three years earlier, Cernan's moment of realization had come during a two-hour space walk. He was floating next to the spacecraft, with no sense of speed as he whizzed around the earth at 17,500 miles per hour, about to make an unsuccessful attempt to test a rocket-powered backpack, and he took a moment to look around at the planet passing beneath him, at the craft that was his sanctuary in the void, at the blinding sun, and the dark of space.

The same thing happened now, inside *Snoopy*, as Cernan looked up from his checklist. The two triangular windows were still covered by translucent yellow shades, but he could see the moon's surface through them, drifting slowly and silently past, eerie because of the yellow tint. He wasn't quite sure it was real. And he said to himself, "Do you realize where you really are?"

Thursday, May 22
1:35 P.M., Houston time
4 days, 2 hours, 46 minutes Mission Elapsed Time

Charlie Brown and *Snoopy* flew in formation through the unfiltered lunar sunlight. For about half an hour, the two craft had drifted above the craters while the three men prepared for what lay ahead. An hour from now, Stafford and Cernan would fire *Snoopy*'s descent engine and enter a lopsided ellipse called the descent orbit, with a low point of only 50,000 feet. On a real landing, at that 9-mile altitude, the lander's engine would reignite for the final descent to the surface. But on this mission, Stafford and Cernan would simply coast along, then soar up to 215 miles, then swoop down once more. On the low passes, the pilots would reconnoiter Apollo Landing Site 2, a small patch of moonscape near the southwestern edge of the Sea of Tranquillity. Then Stafford and Cernan would cast off their descent stage and fire the ascent rocket to rejoin Young.

Now, with a burst from *Charlie Brown*'s maneuvering thrusters, Young pulled away. Inside *Snoopy*, Stafford and Cernan saw the command module shrink into the distance.

"Have a good time while we're gone, babe," Cernan called.

"Yeah," Stafford said. "Don't get lonesome out there, John."

Cernan added wryly, "And don't accept any TEI updates."

"Don't you worry," said Young.

It was the kind of banter that often comes to fighter pilots when their necks are out. Stafford and Cernan knew only too well that they were flying a craft that could not take them back to earth, that if something went wrong down there at 50,000 feet, Young would have to come get them. Ultimately the lunar landing was a walk across a long, high wire, and they were about to take the penultimate step.

2:35 P.M.

Now came the moment of truth. It was time for Stafford and Cernan to fire *Snoopy*'s descent engine to enter the descent orbit. It would take about half a minute to shift the low point of *Snoopy*'s orbit from 69 miles to roughly 50,000 feet. The timing was critical, for if the burn went even two seconds too long, the lunar module would crash into the front side of the moon. Meanwhile, in *Charlie Brown*, John Young listened in and tracked *Snoopy* with the command module's sextant, ready to come to the rescue if Stafford and Cernan had a malfunction that prevented them from getting into the proper orbit.

At the proper moment *Snoopy*'s computer flashed "99" and Cernan pushed the PROCEED button. The engine fired soundlessly in the vacuum.

"We're burning, John," called Cernan. Inside *Charlie Brown*, Young could see the glow from *Snoopy*'s engine. Half a minute later, the burn was over, and *Snoopy*'s computer showed they were right on target.

Almost immediately Stafford and Cernan sensed their diminishing altitude. Within 20 minutes they were dramatically lower. The moon's curved edge flattened out. As *Snoopy* flew onward, mountains appeared on the horizon; as Stafford and Cernan advanced they could see that the mountains were actually the rims of huge craters. Enormous boulders dotted the land—some looked as big as five-story buildings; after the flight Stafford would find out they were ten times that size. There were cliffs that must have been four or five thousand feet high. The mountains of the lunar highlands looked close enough to touch.

Suddenly the men were stunned by the sight of their first earthrise. Now they were in contact with Capcom Charlie Duke, and Cernan's voice soared. "Houston, this is *Snoopy*! We is Go and we is down among 'em, Charlie!" Down they flew, heading for Landing Site 2. Shadows lengthened as they coasted into the realm of lunar morning. Now they were over Tranquillity's plains, smooth, like wet clay; to Stafford, they seemed almost like the desert near Edwards. Later, he would report that Site 2 looked smoother than he expected, at least in the center of the area. But he would warn that if Armstrong and Aldrin were off-target, and especially a few miles downrange, the terrain would be rougher, and they had better have enough fuel to maneuever around and look for a safe spot.

Stafford and Cernan had flown routinely at 50,000 feet, but this was unlike any airplane ride. No one had ever flown this fast at such low altitude. Following the laws of orbital motion, *Snoopy* had traded its altitude for speed, until it was now nearing 3,700 miles per hour, more than five times the speed of sound. Stafford and Cernan were barnstorming the moon.

5:33 P.M.

Flying upside down and backward, Stafford and Cernan prepared to rejoin the command module by firing the lander's ascent engine. First, they would cast off *Snoopy*'s descent stage by firing a set of pyrotechnic bolts. "That mother may give us a kick," Cernan warned. "You ready?" Suddenly, Cernan saw the moon's horizon spin wildly; the lander was tumbling out of control. Still on hot-mike, his words heard by the listening earth, Cernan blurted, "Son of a bitch . . . What the hell happened?"

Stafford quickly punched the button to get rid of the heavy descent stage, to give him more authority with *Snoopy*'s small maneuvering thrusters. For eight long seconds he struggled to regain control, and then *Snoopy* was still. After the flight, the incident would be traced to a combination of minor and easily correctable failures. Ten minutes later, in darkness, the ascent engine started Stafford and Cernan on their journey back to John Young. Thirty-one hours after that, early in the morning of May 24, the second team of moon voyagers headed back to earth.

In Houston, Neil Armstrong and his crew were into their final months of preparation. When Tom Stafford's crew came back from the moon, only the landing remained. Everything else was now part of the collective spaceflight experience—because when one team of astronauts and flight controllers learned something on a mission, it was as if all of them had done it. The achievements of one mission were there for the next to build on, and the mistakes would not be repeated—or so everyone hoped. The way was now clear for Apollo 11 to attempt the ultimate test flight.

III: Down to the Wire

He was alone, high above the flat Texas landscape, with only the sound of the air rushing past his canopy. From the ground, you might have seen him, wafting on the thermal currents, circling in wide arcs. Whenever he got the chance, Neil Armstrong took time out from the pressure of his life as an astronaut to go soaring. He loved this kind of flying—unpowered, pure, mentally demanding—as much as he did any other; in short, there wasn't anything about flying that didn't interest him. But it was here, apart from the world, harnessing the wind, that Armstrong found his greatest relaxation. Almost from the moment his name was announced as the commander of Apollo 11, the press tried to know him, and for the most part he gave them shy smiles and brief, almost cryptic answers to their questions about his thoughts and feelings. By his own admission, he much preferred to talk about ideas than people. What emerged from their typewriters became a kind of legend: the small-town Ohio boyhood, the love affair with airplanes that began earlier than he could remember, the way he devoured books as a schoolboy, how he built airplane models and tested them in a homemade wind tunnel, how he got his pilot's license before he learned to drive. They wrote of his achievements in the Korean War, where he flew seventy-eight combat missions and was awarded a couple of air medals. They chronicled his career as a test pilot and an astronaut. But like his sailplane, Armstrong was essentially beyond their reach. There had always been that quality about him, even in the summer of 1955, when he arrived at Edwards Air Force Base, where a small group of fliers was exploring the unknowns of high-speed flight for the National Advisory Committee on Aeronautics. Edwards was a place of blast-furnace heat, howling winds, and utter desolation, but it was heaven on earth for pilots. Dawn came still and clear, spilling over distant mountain ranges onto a smooth, hard expanse of clay that seemed as vast as the cloudless blue canopy above. Armstrong loved the dawn at Edwards and reveled in its short-lived serenity—short-lived, because on any given morning the stillness was shattered by the sonic booms of pilots unleashing the most exotic flying machines in existence. Somehow, the primitive beauty of Edwards was the perfect backdrop to the unfolding of the future. And it was the perfect place for Neil Armstrong.

NACA's High Speed Flight Station was perched on the edge of Rogers Dry Lake. The NACA fliers epitomized a new breed of test pilot, engineers as much as aviators, that had emerged since the Second World War. Unlike the air force pilots at Edwards, who seemed to treat a new airplane as if it were a romantic conquest, losing interest as soon as they'd used it to set a record, the NACA pilots delved into lengthy, often tedious analyses that were the heart and soul of test flight. The combination of meticulous research and all kinds of flying opportunities—the mundane as well as the exotic—was what made Edwards a place to be cherished. Armstrong would look back on his years at Edwards as the most fascinating period of his life.

There were two things about this quiet young man, whose clear, boyish face made him look even younger than his twenty-five years, that the other NACA pilots found remarkable. Not his flying; he was a skilled aviator, with an impressive grasp of aerodynamics, but there were better stick-and-rudder men at Edwards. No, the first thing they noticed was his intellect; everything he did, even casual speech, seemed to be the result of a great deal of thought. The second was his remoteness. Armstrong often kept people at arm's length. He rarely engaged in idle conversation and steadfastly guarded his privacy. He and his wife, Jan, hadn't been there very long before they left the community that clung to the base and moved into a cabin in the Juniper Hills, among the Joshua trees, without electricity or running water. Was such a man knowable? One friend would say years later that at his core, Armstrong was Scottish, with the moral code of a Highlander: rock-solid integrity and a relentless memory of any injustice committed against him or his kin. In time, the NACA pilots realized that Armstrong wasn't aloof; he was shy. Once they got past his great reserve, they found warmth. Once he became a friend, he was a good friend. If he could be reticent, then he could also become so involved in conversation while driving that his passengers nervously eyed the road. Under the serious layers lurked a tart and understated wit. If he was a consummate loner, then it was also true that a postflight party in full swing usually saw Armstrong at the piano, pounding out a bit of ragtime; he might be the last to leave. But even his friends could only guess at what Armstrong was thinking, what really drove him. As fellow NACA veteran Milt Thompson said, years later, "I knew him, but I didn't know him."

At Edwards, Armstrong saw the blue desert sky and envisioned the trackless void beyond. In the mid-fifties space travel was something most people were careful to avoid talking about in serious tones. But it wasn't far-out to the people at Edwards, and Armstrong soon sensed that the road to space began on the high desert. The air force and NACA were sitting down to design a sleek, winged rocket plane called the X-15, a descendant of the rocket-powered X-1 that Chuck Yeager had used to break the sound barrier here a decade earlier. Carried aloft under the wing of a B-52 bomber and then released, the X-15 would zoom to the fringe of the atmosphere, then glide to an unpowered, or "dead-stick," landing on the lakebed. Armstrong was among the handful of pilots slated to fly it.

In the meantime, there was Sputnik, which set off a chain of events that moved spaceflight toward reality faster than Armstrong could have anticipated. By 1959 NACA had become the National Aeronautics and Space Administration, and Project Mercury was in the headlines. Some—though not all—of the Edwards men snickered at the idea of "astronauts" who would be essentially passengers from start to finish. The NASA pilots knew that *they*, not the Mercury astronauts, were on the cutting edge, and that they would fly into space at the controls of the X-15's more advanced successor. Soon that craft too was on the drawing board: Dyna-Soar, a manned glider designed to be lofted by a Titan booster and then land on a runway. Armstrong was named NASA's prime pilot on Dyna-Soar. He had no interest in becoming a Mercury astronaut.

By 1962, in its third year of operations, the X-15 was streaking through the desert sky at unheard-of velocities—more than five times the speed of sound—and record altitudes. That April, on his sixth X-15 flight, Armstrong piloted the sleek, black arrow to 207,000 feet, high enough to see a bright and gently curving horizon beneath a black sky. For a moment he lingered there at the top of his arc, suspended between aviation and space flight, before returning to earth.

The view from 207,000 feet only whetted Armstrong's appetite; he felt he had glimpsed his own future. But that future, he realized, was not in the X-15, or in Dyna-Soar, which now seemed doomed never to leave the drawing board. Already, John Glenn had demonstrated that astronauts were more than passengers by taking over manual control of his Mercury spacecraft when the automatic system malfunctioned. Armstrong realized he'd been hasty in rejecting the astronaut program. It wasn't easy to think about leaving Edwards, but NASA was headed for the moon. The week he flew the X-15 to the heights, NASA put out the call for the second group of astronauts. One day, Armstrong and a fellow NASA pilot named Bill Dana were mulling over their respective plans. Dana said he thought there was a much better future flying the Supersonic Transport that everyone was talking about. Armstrong told him, "You can do whatever you want to about that, but space is the frontier, and that's where I intend to go."

June 1969
Flight Crew Training Building, Kennedy Space Center

Neil Armstrong and Buzz Aldrin stood side by side in the Lunar Module Simulator. A soft green light filled the enclosure, emanating from dozens of electroluminescent dials and readouts. "Okay, Neil," said a voice in their headsets, "we'll give it to you at pitchover minus 30 seconds." It was the voice of one of the simulation instructors who sat at a console nearby. On the instrument panel, gauges jumped to life and registered engine temperature, chamber pressure, fuel and oxidizer readings, thrust, and the other vital signs of a lunar module in flight. A tape meter scrolled down ever diminishing readings of altitude; another showed their speed of descent. Two "8-ball" indicators showed the LM's orientation in space. Armstrong and Aldrin scanned each gauge, just as they would in the moments before the lander began its final descent to the surface. Suddenly a coded message flashed on the LM's computer display. "P sixty-four," Aldrin announced, and suddenly, through the two small triangular windows, a bright moonscape swung upward into view. Once again, as they had done many times over the past few months, Armstrong and Aldrin were about to confront Apollo's greatest unknown.

"Space is the frontier," Neil Armstrong had said seven years earlier, and with the lunar landing he had found a piloting task at the frontier of spaceflight. Nothing like it had ever been attempted. The liftoff from the moon, on which his and Aldrin's lives would depend, would be a rocket launch that was basically like many others; only the launch site would be extraordinary. But a manned rocket *landing*—that was something for which there was no prior experience. The airless moon offered no alternatives.

The trip from 50,000 feet down to the surface would be, in essence, a long, controlled fall from orbit, a trajectory governed by the same exacting laws of motion as orbit itself. Called the Powered Descent, it would begin with the LM flying horizontally over the moon at a speed of just under 3,800 miles per hour, and would end with a touchdown as gentle as a leap from a bar stool. There were two keys to the Powered Descent. One was an engineering marvel: the world's first rocket engine with a throttle. With it the LM could descend at a whole range of speeds; if necessary it could even hover in the sky like a helicopter. The second was the LM's onboard computer. More than half the lander's weight was fuel for the descent rocket, but the fuel budget was so tight that there would have been little chance of making it to the surface without computer control. The computer wizards at MIT had programmed it to govern the entire descent. It would compute the precise amount of thrust needed from the descent engine at any given moment. It would keep track of the LM's distance to the designated landing site. It would share control of the spacecraft with the pilot, or, if asked, give up control altogether.

In their simulations of the Powered Descent, Armstrong and Aldrin began where Tom Stafford and Gene Cernan left off, 9 miles above the moon. With the lander tipped back horizontal, they lit the descent engine and throttled up to full power, using its thrust to brake their orbital speed. Obeying Kepler's laws, the lander fell moonward as it slowed. In this early phase the craft was oriented with its windows facing the moon. This allowed Armstrong to observe landmarks passing underneath and use them to judge whether he and Aldrin were on the proper trajectory. Then, some 4½ minutes into the descent, the craft rotated until it was windows-up, allowing the LM's landing radar to bounce echoes off the surface, to determine altitude. At about 6 minutes, the computer throttled the engine down to 55 percent thrust, and the long brake continued. Throughout this period, if everything went well, Armstrong and Aldrin did little more than watch the gauges.

Finally, 8½ minutes after engine ignition, the final phase of the Powered Descent began. Some 7,500 feet above the moon the computer pitched the lander forward until it was nearly upright—this was called the pitchover maneuver—so that its engine would not only brake the craft's forward speed but keep it from falling too fast. From then on the lander followed a slanting path to the surface like a car braking on a long, straight mountain road, with the computer carefully riding the throttle like a motorist's foot on the brake. With only 4 minutes from pitchover to touchdown, things happened very fast. In that sense, landing on the moon was similar to Armstrong's X-15 flights. Some X-15 pilots talked of a kind of "fast time" when the clock seemed to race. There wasn't time to read checklists; they had to have the entire flight plan memorized. Pilots sometimes talk about being "ahead of the airplane," a state of mind that enables you to think one step ahead of your craft, the better to cope with emergencies. But that was nearly impossible flying a rocket plane at five times the speed of sound—and it wouldn't be much easier in a lunar module, descending to the moon.

For Armstrong, pitchover was the moment when the piloting part of the landing began. With the cratered plains of the Sea of Tranquillity spread before him, he scanned the advancing moonscape, looking for familiar landmarks he had memorized from photos. And by using a special grid called the Landing Point Designator, he could see for the first time where the computer was aiming to land. The LPD grid was marked off in degrees on the LM's double-pane window, and was used like a gunsight. By sighting along the grid at the proper angle, which was supplied by the computer, Armstrong was able to spot the landing point. If he didn't like what he saw, he could tell the computer to change its aim by nudging the LM's attitude controller with his right hand: A nudge to the left or right shifted the aim point accordingly; tilting the controller backward or forward moved the landing spot up- or downrange.

Meanwhile, Aldrin interrogated the computer, asking it for the LM's height and speed and comparing the results with the data from the radar. Then he relayed this information to Armstrong. Aldrin knew very well that his title of "lunar module pilot" was a misnomer, because all the flying during the landing—and there might not be much of it—belonged to Armstrong.

In theory, Armstrong knew, he could let the computer fly the LM all the way to touchdown, without touching the controls. In reality he wasn't about to do that. It wasn't just that as a self-respecting test pilot he wanted to be at the controls at the moment of landing; it was because the computer was blind. It might bring him and Aldrin down on an absolutely perfect trajectory, right into a crater or a boulder field. Armstrong planned to take over from the computer when he was about 500 feet above the moon.

Even then, Armstrong would let the computer do most of the work. Flying the LM entirely by hand was so difficult that he wanted to avoid it at all costs, and unless the computer went out, he wouldn't have to. Instead, he would maneuver the LM by tilting it slightly to one side, and let the computer continue to ride the throttle. Armstrong knew there was always the possibility that something would go wrong—from communications failure to a malfunction of the descent rocket—and force him to abort the landing. In that case, he could press the button marked ABORT STAGE, setting off a dramatic chain of events. In an instant, pyrotechnic bolts would sever the connections between the descent and ascent stages. At the same moment the ascent rocket would blast to life, boosting Armstrong and Aldrin back toward orbit. But an abort, especially one at low altitude, carried its own risks. If the ascent engine didn't light, there would be no time to find out why and do something about it; Armstrong and Aldrin would crash in a matter of seconds. Even if it did fire, the ascent stage might not separate cleanly from the descent stage. And even if there were no malfunctions, there was the problem of finding Collins, with the timing for the rendezvous now completely disrupted; Armstrong and Aldrin would face a long and complicated journey back to the command module. And it would have been several hours since Aldrin had aligned the LM's guidance system with the stars. An abort would force the men to fly with a guidance system that could be significantly in error. And so the last thing Armstrong wanted to do was abort—not only because it would mean failure, but because it could be even more risky than the landing itself.

As the LM got very close to the moon, Armstrong and Aldrin would enter the most hazardous portion of the descent. Somewhere in the last 200 feet, they would be too low to abort successfully if the descent engine quit, for the LM would be going too fast for the ascent engine to arrest the lander's plunge and start the ascent stage upward again. The astronauts, borrowing a term from helicopter pilots, called this part of the descent the "dead man's curve."

Even if nothing went wrong, Armstrong would have his hands full. Already, he would have had to find a safe landing spot, free of large craters and boulders. By the time the LM was 100 feet up, Armstrong would have to arrest nearly all of the craft's forward motion and begin a slow, careful vertical descent.

The computer would still control the throttle, but Armstrong would be able to make small adjustments by using a special toggle switch. By clicking the switch up or down he could increase or decrease his rate of descent by one foot per second, repeating this as many times as necessary to get the change he wanted. It seemed to Armstrong to be a strange way to control a craft, and he was skeptical that the switch would prove effective in the actual landing. But he would have to wait and see until he was 100 feet above the moon.

In these final moments Armstrong's gaze would be directed almost entirely at the moon, and he would rely on Aldrin's steady reports on altitude, horizontal speed, and descent rate. The first part of the LM to touch the moon would be three long metal probes attached to the footpads; at that moment, a blue light labeled LUNAR CONTACT would glow on the instrument panel. Aldrin would be watching for the contact light, ready to call it out. At that moment, Armstrong would shut down the engine, and the LM would fall the remaining three feet to the surface.

Perhaps the toughest thing about landing on the moon was that there was no such thing as a second chance. Even in carrier landings—and Armstrong had flown plenty of those in Korea—there was usually the chance to circle and try again if something went wrong the first time. But in the lunar landing he and Aldrin would have to stay on top of any problems that might come up, even as the computer brought them closer to the moon. No surprise that the simulator instructors knew Armstrong and Aldrin as serious students. Aldrin, who was becoming a virtuoso with the LM's computer, was a bundle of self-confidence, ready to debate with instructors and experts on any technical matter.

Armstrong was all business. As the man who Mike Collins would call "far and away the most experienced test pilot among the astronauts," Armstrong had seen his share of close calls, and more than once had faced malfunctions he had never seen in simulations, including one on his Gemini 8 mission that nearly cost him his life. It was March 5, 1966, and Armstrong had just accomplished the first space docking, linking up with an Agena target rocket 185 miles above the earth. He and his copilot Dave Scott were monitoring the instruments on the joined spacecraft when suddenly they began to tumble. The men were sure that the problem was a stuck thruster on the Agena. If they let it continue, Armstrong knew, the tumbling might break the two craft apart. He undocked, and the Gemini began to gyrate faster and faster, spinning a full turn every second, as the earth and sun alternately flashed by the windows. Armstrong's vision blurred as he searched the instrument panel and he said to Scott, in a voice tinged with wryness, "I gotta cage my eyeballs." Armstrong brought all his energy to bear on the problem, trying methodically to isolate the stuck thruster, but he could not.

If the spin worsened, he and Scott were in danger of losing consciousness; then, it would be only a matter of time before the Gemini broke apart. Armstrong was forced to turn off the main thrusters and switch on a backup system, reserved only for reentry, an action which was tantamount to aborting the flight. Half a minute later he had halted the spin—and lost the mission.

Other than disappointment, Armstrong carried no special psychic burden from that incident. He had no score to settle. He had been in the test flight business too long not to view Gemini 8 as just another page in the history of his profession, another encounter with the unexpected. Now, training to land on the moon, it was the unknowns that Armstrong worried about most.

For the first few weeks of training, the instructors went easy on Armstrong and Aldrin. In one practice landing after another, the two pilots worked out the teamwork they would need to handle the descent. By the time Stafford's crew came back from the moon, Armstrong and Aldrin were landing confidently and smoothly. And then, the instructors decided, the honeymoon was over, and they were nothing short of diabolical in their efforts to push their students to the limit. Now there was a Gemini 8 every day, and Armstrong was unflappable. The instructors would freeze the pointing mechanism for the descent rocket, so that the nozzle could no longer be aimed from side to side, and Armstrong would compensate by tilting the entire lunar module when he wanted to change direction. They would make a maneuvering rocket stick in the on position when he was in his final, hovering descent, and Armstrong would cant the LM against the unwanted thrust as if he were leaning an airplane into a crosswind.

Every so often, in their efforts to reach new levels of ruthlessness, the instructors threw in a set of emergencies so complex, so difficult that no one could have handled them. With other astronauts this usually brought words of protest from the cockpit: "What the hell are you giving me that for? That's *negative training.*" But no matter what they did to Armstrong the instructors never heard an angry word from the simulator. Instead they'd find out only in the discussion afterward, sitting around the console, when Armstrong's clear, pale face would turn red, and his usually restrained voice would crack, and the instructors would know they had pushed the "war" too far this time. Minutes later Armstrong would be back at the controls, taking another try, and he never ended a session without making it down successfully.

By the end of May, the simulator at the Cape was linked via radio to the Mission Control Center in Houston. These so-called integrated simulations were designed to test not only the astronauts but the flight controllers. Integrated simulations were considered essential, and for good reason. Spaceflight is not a solo venture by daring pilots; it is a *partnership* between the astronauts and mission control, and nothing demonstrated this more than landing on the moon. For the dozens of flight controllers who worked for Chris Kraft, the first lunar landing was a coveted goal just as it was for the astronauts.

The flight director who would be in charge during the landing was Eugene F. Kranz, a former air force fighter pilot who had flown F-86's in Korea. With rugged features and a light blond crew cut, Kranz had the look of a drill sergeant. During Gemini, one of his flight controllers nicknamed him General Savage. Inside, Kranz was as sentimental as they come. Like most people at NASA he was an unabashed patriot who could get misty when he heard "The Star Spangled Banner." At the start of the work day he would go into his office and listen to John Philip Sousa marches, to get his blood flowing. He was a devout Catholic and a man of strong beliefs, and at his core he believed in the exploration of space. The space program wasn't just a part of his life; it *was* his life.

At age thirty-five Kranz was already a veteran flight director. He'd begun his tour in mission control in 1960, as Chris Kraft's assistant flight director during Mercury, then as a flight director himself in Gemini; then he had moved on to Apollo. He'd been in the flight director's chair for more emergencies than he cared to think about, including Armstrong and Scott's brush with disaster in Gemini 8. Every nerve was tuned to the process of making decisions under pressure. During the landing Kranz would sit in the third row of the Mission Operations Control Room presiding over a team of flight controllers who would keep track of every system onboard the descending lunar module and monitor its trajectory down to the moon. Each controller would in turn rely on his own team of "back room" experts, down the hall from the MOCR.

Kranz knew his team of flight controllers the way a battle commander knows his troops. These young men, like all the people who worked for Chris Kraft, were exceedingly bright. It was extraordinary what these people could do, together, during a mission. Kranz had seen it happen time and again: in the compressed moments of a launch or some other critical phase, they became one giant conglomerate brain, twenty minds wired together in parallel, each focused on some small piece of the whole event. In such situations, they could solve almost any problem that came their way, given twenty seconds to work on it. In twenty seconds a controller could look at the problem, talk to someone in his back room, think, talk to someone else, come back to the first person, and make a decision. And all the while, he would be monitoring the events around him, listening not only to the conversations on the flight director's loop, but the air-to-ground, and perhaps one or two other loops. The amount of information processed by one controller was staggering. And the entire team was trained for that kind of split-computer mentality. With that kind of brain power at work, twenty seconds could be a long time.

Kranz and his team started simulating lunar landings as soon as Apollo 10 ended. At first, they worked on their own, using computers to take the place of simulators, and instructors instead of astronauts. After that they moved on to the integrated simulations. Sometimes Armstrong and Aldrin were in the simulator at the Cape; at other times it might be the backup crew, or the Apollo 12 team. In any case, the simulations were so realistic that it was impossible to tell the difference from a real flight. People who sat in on them experienced gut-wrenching tension.

The simulation instructors in Houston waged war on Kranz and his team relentlessly as their counterparts at the Cape did with Armstrong and Aldrin. Their goal was to find the open seams in their decision making, to trap them right at the instant when two options overlapped, so that if they made the wrong choice, or picked the right one but didn't implement it in time, there would be no recovery. Kranz was the point man. If an emergency came up, he would have to decide whether to continue or order an abort. And that was where things went bad, in late May.

It was difficult for the controllers to master the complexity of the Powered Descent. It was especially tough for them to judge when the astronauts had entered the dead man's curve. Then there was the added wrinkle of the lunar distance. It took 1.3 seconds for a message from the spacecraft to reach earth, and another 1.3 seconds for mission control's reply to reach the astronauts. That delay was also part of the simulations. But the most difficult part was getting the feel of how to handle malfunctions during the Powered Descent. At first the controllers were so conservative they'd abort at the first sign of trouble. Afterward, the simulation instructors would point out to Kranz's men that they could have kept going, if only they had done such and such. . . . And the next time, the team would be so daring that the problems would overtake them, and the LM would crash into the moon. By the first part of June, there had been so many crashes that Kranz wondered if he and his team would ever get it right.

Kranz wasn't alone. During integrated simulations every manager in Houston, Chris Kraft included, was listening to the squawk box in his office as if it were the best radio show on the air. After the first crash, the black telephone behind the flight director's console rang, and it was Kraft. Despite his concern, he almost joked about it—"I see you let it get away from you." The next time the black phone rang, perhaps a week later, Kraft wasn't joking anymore: "What's the matter with you guys?" Kranz wasn't sure he knew. Maybe the answer was that no one had ever tried anything like this before.

Kranz had the distinct impression that the astronauts were getting frustrated too. It was hard enough just getting the complex simulator at the Cape—not to mention the roomfuls of computers in Houston—to work long enough to get in all the integrated simulations required in the training syllabus. What the crew didn't need was a team of controllers who couldn't respond correctly to an emergency. Kranz was beginning to doubt his own abilities.

One day late in June, Armstrong and Aldrin stood side by side in the lunar module simulator, descending to a simulated moon. The pitchover maneuver came right on time. Aldrin scanned the gauges, and called out data on altitude and speed. Armstrong surveyed the moon for a landing spot. Suddenly, it all went bad. The attitude indicator began to tumble; a thruster had stuck on. Aldrin looked out and saw the moonscape tilting crazily, looming ever larger. He was sure that if they continued the descent they would be killed. But Armstrong did not abort. Aldrin hesitated to tell his commander what to do, but he felt like a helpless passenger in an Indy race car. At last he said, through clenched teeth, "*Neil—hit abort.*" Finally Houston gave the order: "Apollo 11, we recommend you abort"—but it was too late. The TV image of the moon froze, and the gauges ceased changing, and Aldrin knew he and Armstrong were scattered across the moon in a thousand pieces. He looked over at Armstrong, who was still absorbed in the problem.

That evening, Aldrin and Collins were having a nightcap in the living room of the crew quarters. While Collins sipped a beer, Aldrin dipped into a bottle of scotch and began to let off steam about the morning's simulated disaster. Everyone had been listening, he told Collins, and Armstrong knew that. And yet he hadn't aborted; in fact he'd been indecisive. And this was going to be recorded as a *crew failure* . . . As the scotch bottle emptied, Aldrin's voice grew louder and more strident. Suddenly a door opened and there was Armstrong, in his pajamas, a frosty look on his face. "You guys are making too much damn noise out here," he said. Aldrin realized he had heard every word. Collins excused himself and went to bed, leaving his crewmates to their discussion.

Aldrin talked it out with his commander. Armstrong said he didn't think anyone would see the incident as a black mark on the crew. By the next day, all seemed forgotten, but Aldrin could not deny that a certain coolness had set in between himself and Armstrong since the whole business about who would be first man out of the LM, and this had not improved things any.

Armstrong, for his part, wasn't concerned about having crashed in an integrated simulation. In truth, he'd thought the exercise had been a good test of the mission control, and was anxious to see how they would handle it. And they hadn't. As for being indecisive—he was testing his own limits and abilities in the one place where he could safely afford to. After all, you couldn't kill yourself in the simulator.

• • •

Meanwhile, throughout the spring of 1969, Mike Collins was waging his own battle for mastery of the command module simulator. If Armstrong and Aldrin had the landing to worry about, then Collins had just about everything else, for the command module pilot's job now included, in addition to navigation, being in charge of the burns into and out of lunar orbit, as well as the reentry into the earth's atmosphere. Collins too fell prey to a team of ruthless simulator instructors, and when he made a mistake it meant being stranded in lunar orbit, burning up in the earth's atmosphere, or plummeting into the ocean with parachutes still packed away. By June, Collins was steeling himself for the final weeks of battle. He would master his command module and his mission, and that was all there was to it.

Collins had joined the lunar landing crew after the most difficult episode of his career. He recovered from surgery in time to help send his former crewmates to the moon. As Borman, Lovell, and Anders spoke from lunar orbit Collins listened from mission control and felt a special connection to events, and a special envy. When Apollo 8 splashed down Collins stood amid cheers, waving flags, and cigar smoke, and was overcome by emotion. Leaving mission control, he shed his tears in private. Now, on the lunar landing mission, Collins would have the role of staying behind in lunar orbit while his crewmates made history. Fully aware of how people expected him to feel, he told interviewers with complete honesty that he felt no frustration. He was going 99.9 percent of the way, he said, and that was fine with him.

But in joining the crew of Apollo 11, Collins had taken his place among two men who were in many ways his opposite. He was as easygoing as Aldrin was serious, as accessible as Armstrong was remote. Alone among the three, he professed no love of machines. He barely tolerated the computers that had taken over spaceflight. The instructors would hear his voice from inside the simulator, "All I do is punch buttons!" Collins's tastes ran toward fine wines and good books. He dabbled in oil painting and cultivated roses in his Houston garden. To reporters faced with Armstrong's inscrutability, Aldrin's technical relentlessness, Collins was a breath of fresh air. He fielded their queries with good humor; his face seemed to say that yes, these are interesting questions. When asked, he did his best to explain his two crewmates. He spoke of Armstrong's "almost towering intelligence" and his sly wit, and said with a smile, "He has his own barriers erected." In truth, Collins admired Armstrong and Aldrin for their brilliance and technical skills, and felt fortunate to be flying with them. But he wished he knew them better. He lamented the fact that the three of them seemed to communicate only about technical information. Even now, after months of training together, he felt he barely knew them. He liked Armstrong but didn't know how to close the distance his commander kept. Aldrin, as Collins would later write, "is more approachable; in fact, for reasons I cannot fully explain, it is *me* that seems to be trying to keep *him* at arm's length. I have the feeling that he would probe me for weaknesses, and that makes me uncomfortable."

But Collins understood that close friendship was not required to fly a good mission. What was required, in his case, was knowing how to rescue Armstrong and Aldrin if they got into trouble during the landing or the ascent from the lunar surface. There were eighteen different variations of emergency rendezvous, and Collins had to learn how to fly them all. But even if he never needed his rendezvous "cookbook," Collins would have to take care of docking the command ship with the lander, then removing the docking mechanism from the connecting tunnel. For the latter task, there was a horrendously long and opaque checklist loaded with terms like "capture latch release handle lock." If he couldn't get the probe and drogue out of the tunnel, Collins would later write, "I was supposed to get out the tool kit and dismantle it. Me, who couldn't repair the latch on my screen door. I hated that probe, and was half convinced it hated me and was going to prove it in lunar orbit by wedging itself intractably in the tunnel. . . ." If that happened, Collins knew, Armstrong and Aldrin would have to make a risky space walk to return to the command module.

In Collins's mind, the eighteen rendezvous scenarios and the probe and drogue swirled around twin responsibilities: He must get his crewmates back to earth safely, and he must not do anything to screw up the mission. During the spring of 1969 Collins felt the eyes of the world upon him as he never had before, and his labors were tinged with anxiety. He also knew that Apollo 11 was putting a great strain on his wife, and that was something he wanted to stop as soon as possible. One day when they were flying a T-38 to the Cape together, Deke Slayton offered to put Collins back in the rotation after Apollo 11. Collins knew he'd probably command one of the later lunar landings, but he declined Slayton's offer. He had already decided that, assuming it was successful, this flight would be his last.

• • •

As the first weeks of June passed, Gene Kranz and his team had turned the corner. They had nearly mastered the challenges of the Powered Descent. They had firmed up the mission rules—critical guidelines for how to handle emergencies—and were becoming increasingly confident. But concern was mounting elsewhere, especially at NASA Headquarters, where managers worried that the astronauts were tiring and would not be ready in time for a July launch. And there were moments when Armstrong had his doubts; it just seemed that there was too much to do before the scheduled launch date of July 16. Should NASA postpone the mission until the next launch window in August? There wasn't much time to decide, because the Saturn V was on the pad, and technicians were getting ready to load propellants onto the lunar module and command ship. The hypergolic fuels used were too reactive to sit in the spacecrafts' plumbing for more than a few weeks without risking corrosion. If NASA wanted to postpone the launch, they would have to decide soon.

Deke Slayton came by the simulators at the Cape one day to

find out how the astronauts saw things. Collins told Slayton he was ready, but he knew the burden rested on his crewmates. Aldrin did not want the training to stretch on indefinitely, feeling that they might be only marginally better by August. It came down to Armstrong. He told Slayton it would be tight, but they would be ready. On June 12, after holding a ninety-minute teleconference with Slayton, Wernher von Braun, Chris Kraft, and other Apollo managers, Apollo program director Sam Phillips made it official. Apollo 11 would leave earth on July 16, as planned.

Monday, June 16
Ellington Air Force Base, Houston

A machine that looked like a strange, skeletal prehistoric bird ascended into the muggy air at Ellington Air Force Base, whining and belching puffs of vapor like a possessed steam calliope. The machine turned and hovered in midair for an instant, then flew slowly over the concrete apron. This was the moment that made NASA managers chew their fingernails. Neil Armstrong was piloting the Lunar Landing Training Vehicle, one of the most unforgiving flying machines ever built. For Armstrong, it was also one of the most essential: it let him train to land on the moon.

The LLTV looked as if it had been put together at an aerospace garage sale. A crisscross of metal struts with no skin formed the body of the craft, leaving fuel tanks, engines, and plumbing fully exposed. Four legs stuck out like a bedstead. The key to the trainer was that it had two independent means of propulsion. A jet engine supported five-sixths of its airborne weight. Armstrong didn't concern himself with the jet, but maneuvered the LLTV using a pair of rocket engines, powered by hydrogen peroxide gas, that simulated the lunar module's descent engine. Small jets mounted around the LLTV's metal framework mimicked the LM's attitude-control thrusters.

In May 1968, Armstrong had ejected from an earlier version of the trainer when it ran out of attitude-control fuel and became unstable. The accident board traced the incident to a design flaw. No sooner had that problem been solved than NASA test pilot Joe Algranti bailed out of the new LLTV in December; this time it was an aerodynamic problem. Bob Gilruth was pressing Slayton not to let any more astronauts fly it, but Slayton resisted, saying he would sure as hell rather risk losing one at Ellington than above the surface of the moon. And Armstrong agreed, knowing that only the LLTV could prepare him for the last few hundred feet of the Powered Descent.

Throughout the past few days Armstrong had piloted the unwieldy craft. At an altitude of a few hundred feet, he activated the lunar simulation mode, and the LLTV flew just as if it were in a vacuum and in the moon's pull. Almost every instinct Armstrong had developed from a lifetime of flying airplanes was wrong now. The LLTV was balanced on the thrust of its rocket engine like a dinner plate on a magician's broomstick. To start moving in one direction, Armstrong had to tilt the LLTV slightly, letting the rocket engine push him where he wanted to go. Because of the one-sixth g, he had to tilt the craft six times as far as he would have in normal gravity. He had to be careful not to tilt too far, or would risk falling out of the sky. At the same time, with the engine's thrust pointed off to one side, he had to ride the throttle in order to keep from losing altitude. And once he started moving he would keep going, just as if he were flying in a vacuum; to arrest his motion he had to tilt the LLTV in the opposite direction. Simply changing his landing point a few dozen feet became an act of supreme coordination.

The thing that most surprised Armstrong when he first flew the LLTV was how sluggish it was. Each maneuver had to be anticipated well in advance, because it took a surprisingly long time to get going and just as long to stop. To make matters even more difficult, the LLTV carried enough fuel for only about 6½ minutes of flight; each simulated landing lasted 3 or 4 minutes at most. It was precisely because the LLTV was so unforgiving that it was invaluable to Armstrong. It conditioned him to the "fast time" clock once more, for each time Armstrong brought the LLTV down he knew that his fuel supply was dwindling, that each maneuver skirted the edge of the machine's stability, and that he had only one chance to make it. In a setting no more exotic than Ellington Air Force

Base, he had to play it for keeps, just as he would in the airless sky of the moon.

The LLTV drifted slowly over the runway, no faster than a car pulling into a garage. To those watching from the flight line it was an eerie sight, as if it were controlled by some force originating beyond the steamy midday air. Finally the machine started straight down, slowly, like a dragonfly deciding whether to alight. At last the spindly craft came gently to rest. Ground technicians informed Armstrong that he had only 90 seconds' worth of fuel left. "Understand," he answered, and lifted off on another run. By day's end he had completed eight flights in the machine. He was confident now that when the time came for the real thing, he would be ready.

• • •

Ironically, Armstrong and Aldrin had little time to study the world they would visit. Armstrong was clearly interested in the science of his mission, and the geologists had long considered him a promising student. For Apollo 11, he made time in the overcrowded training schedule for a field trip to the mountains of westernmost Texas. The press found out in advance and almost turned it into a circus. They followed the astronauts' cars with a caravan of their own. *Time* magazine hired a helicopter that roared overhead and made it nearly impossible for Armstrong and Aldrin to hear what the geologists were saying. Aside from that one outing, they depended on briefings by geologists from the Manned Spacecraft Center and from the U.S. Geological Survey in Flagstaff, Arizona. Armstrong also asked scientist-astronaut Jack Schmitt to act as their liaison with the scientific community.

As Armstrong made a point of saying at a preflight press conference, the surface of the moon was not an unknown place in the strictest sense, thanks to the unmanned Lunar Orbiter and Surveyor probes. But even with the naked eye, it's easy to see two different types of terrain on the lunar near side: the dark splotches called *maria* (the Latin word for "seas") and the bright highlands, which scientists call the *terrae*. A good pair of binoculars reveal that the *maria* are relatively smooth compared to the heavily pockmarked highlands, and by 1969 most geologists agreed that if Armstrong and Aldrin managed to touch down on Mare Tranquillitatis—the Sea of Tranquillity—they would find an ancient lava plain made of basalts like those found in Hawaii and other volcanic locales. And judging by the pictures sent back from the Surveyor landers, they would also find what was apparently a ubiquitous layer of fine-grained debris, probably formed by the continuous rain of meteorites over the eons.

But when it came down to it, the probes could only radio back numbers. They could not completely dispel the aura of mystery around the moon, even in the minds of some scientists, and there were some dire predictions. Cornell astronomer Thomas Gold insisted that the moon was covered by a layer of fluffy powder dozens of feet thick. He warned that the LM would sink out of sight as soon as it touched down; he told NASA that Armstrong and Aldrin should drop brightly colored weights as they descended and watch to see that they remained in view, or else abort the landing. None of his colleagues—not even pictures from the Surveyors, resting unharmed on the surface—could dissuade him.

Other forecasts were harder to refute. One theory emerged that charged particles emanating from the sun had so altered the rocks over eons that they would burst into flames as soon as they were exposed to oxygen inside the lunar module; no one would know the answer until it happened. And no amount of speculation could answer the question of whether there was life on the moon. No geologist thought there was. It offered an almost unimaginably hostile environment, devoid of air and water, exposed to the vacuum of space, bathed in deadly solar and cosmic radiation, pelted incessantly by micrometeorites. One geologist pointed out that if he wanted to build a sterilization machine, he would construct something like the surface of the moon. And yet, the implications if even one microbe survived the trip back to earth on a space suit or under a fingernail were so unsettling that NASA could not ignore the possibility, especially when voiced by several U.S. government agencies and other countries. NASA constructed an $8 million Lunar Receiving Laboratory at the space center to store the lunar samples in total biological isolation and house the astronauts in the last two weeks of a twenty-one-day quarantine, while officials nervously waited to see if they came down with any symptoms. Quarantine was something Armstrong and his crew would just as soon have done without, but they had no choice.

The weak link in the plan was that no one had figured a way to get the astronauts from the command module to the LRL without exposing them briefly to the environment. A special quarantine trailer would house them for the trip from the Pacific back to Texas, but that left getting from the command module to the trailer. That's where the Biological Isolation Garments came in. After splashdown, the astronauts would open the command module hatch just long enough to let the recovery swimmers toss in three BIG's; after zipping them on the men would climb into the waiting life rafts, scrub each other with disinfectant, and ascend to the helicopter in the recovery net. From then on they would be in quarantine. There would be no speeches on the carrier deck; after a short walk from the helicopter to the quarantine trailer they would close the door behind themselves and view the world through a pane of glass—there and in the LRL—until the quarantine was ended. In the first days of July 1969, that moment seemed all too distant.

Saturday, July 5, 1969
Manned Spacecraft Center, Houston

With just eleven days to go until launch, Armstrong, Aldrin, and Collins spent the July 4 weekend at home, a last visit with their families before heading to the Cape. There had been precious little time off these past seven months; even their time at home was often claimed. A specialist would come by to give an informal briefing; evenings were spent holed up in the study with flight plans and training documents. Most of the time, though, they weren't even home. *Life*'s July 4 issue, with the cover story, "Off to the Moon," included the customary report from the home front—Armstrong fishing with his sons, baking homemade pizza, and playing piano duets with his wife, Jan; Collins trimming roses in the backyard;

Aldrin taking his kids to AstroWorld—but the truth was, those things would never have happened if an outing with family hadn't been a PR requirement.

Since January, Slayton had tried to keep the press at a distance, simply because Armstrong's crew had so much training to pack in. Even so, the men had made time for this or that reporter to come by the house and ask questions about their lives and their mission. Finally, Slayton gave in and agreed to a last press conference, and Armstrong's crew spent most of Saturday, July 5, talking to the media. At this point, the men were well into their twenty-one-day pre-mission medical quarantine, and so on this summer afternoon they strolled onto the stage wearing hospital masks and did not remove them until they had taken their places inside a plastic-enclosed booth. A few reporters grinned back at them from behind their own masks. One asked whether any precautions had been taken to prevent the men from catching germs from their own families. Collins answered, "My wife and children have signed a statement that they have no germs. . . . Seriously, there are no special precautions being taken." But the journalists directed few questions to Collins; they were much more interested in his crewmates, and especially, his commander. For seven months now, Armstrong had been telling interviewers that he wished the press would convey that Apollo 11 was a massive group effort, that it was a mistake to focus on him, but he had not been successful. At the press conference one reporter suggested to him that, as the first man to set foot on the moon, he would be so famous that his personal life would cease to exist. He added, "Do you have any thoughts on this prospect?"

"I suppose," Armstrong said, smiling shyly, speaking in characteristically measured words, "if there is any recognizable disadvantage to being in the position I'm in then that's it. I think that's a fair trade."

On Monday, July 7, Armstrong, Aldrin, and Collins headed back to the Cape for the last time. For the next nine days their world was a high-tech monastery, equally divided between the simulators and the crew quarters. Richard Nixon had planned to visit them here on the night before launch to have dinner, but canceled after Chuck

Berry publicly fretted that he might infect the astronauts. Privately, Armstrong and his crew fumed about the gaffe; the president was no more likely to harbor germs than the dozens of people they worked with every day.

For Mike Collins, the incident was a momentary distraction from the "awesome sense of responsibility" for the mission he was about to fly. That pressure was very much on the mind of NASA administrator Tom Paine when he dined in the crew quarters on Thursday, July 10. Over dinner, Paine made an extraordinary promise to Armstrong's crew. Don't take any unnecessary risks to accomplish the mission, he told them. If anything should go wrong, don't hesitate to abort. He would see to it that they would not have to get back in line for another flight; they would be assigned to the very next mission to try again.

For Collins, Paine's promise took some of the pressure off, but not for Neil Armstrong. He was only too aware that the nation's prestige was riding on this mission. It was impossible *not* to be aware, in the fishbowl they'd been living in since January. And although he felt ready, and sensed that Collins and Aldrin did too, he knew also that the landing would test the entire Apollo system —the hardware, the mission control teams, and themselves—to the limit. Even on the morning of July 16, 1969, as Armstrong led his crew out into the TV lights and onto the transfer van, with an old comb and a package of Life Savers in the pocket of his space suit, he had little doubt they would make it safely back to earth. But the landing, in his mind, was still a fifty-fifty proposition.

CHAPTER 5

The First Lunar Landing

APOLLO 11

I: The *Eagle* Has Landed

Saturday, July 19, 1969
5:30 A.M., Houston time
2 days, 21 hours Mission Elapsed Time
Aboard Apollo 11, 177,000 miles from earth

Buzz Aldrin opened his eyes and floated in the darkness, collecting his thoughts, remembering where he was. All was still inside the command module *Columbia*; only the hum of the cabin fans broke the total silence of the void. Every so often strange flashes appeared. Aldrin did not know what they were, but they seemed to be something actually entering the cabin, perhaps a vagabond cosmic particle decaying in the command module's atmosphere. He had seen the flashes last night too; now he made a mental note to mention them to Armstrong and Collins. They were still asleep, down in their sleeping bags. Aldrin had spent the night up in his couch on the right-hand side with a lightweight headset taped to his ear in case Houston tried to call during the night. Nobody kept watches on a spaceflight anymore; ever since Apollo 8 they had done away with that in the interest of getting better sleep. On this mission especially, sleep was important.

They had agreed before they left the earth, he and Armstrong and Collins, to take it easy on the way out to the moon, knowing that it would be a mistake to arrive in lunar orbit tired the way

Borman, Lovell, and Anders had. It came down to a state of mind: they would convince themselves, as they coasted out to the moon, that the mission had not really begun, and that it would not begin until the moment when Armstrong and Aldrin floated into the lunar module *Eagle*, undocked from *Columbia*, and started the final descent to the moon.

Aldrin was up early—the end of the rest period wasn't for another two hours—but he knew they would be reaching the moon in about seven hours, and he wanted to know whether they would be making a last midcourse correction before they got there. He keyed his mike.

"Houston, Apollo 11."

Aldrin heard only static; he guessed that one of the tracking stations was about to go out of contact, carried by the earth's rotation, and that the Manned Space Flight Network was in the process of switching to another big dish. He lifted the shade from *Columbia*'s right-hand window and looked for earth, but outside there was only dull, starless black.

It would have been a tall order, pretending the mission hadn't started yet, if Armstrong, Aldrin, and Collins hadn't flown before. They knew what it was to work in a strange and hostile environment. None of them had any trouble with motion sickness on the way out. On the contrary—Aldrin thought floating inside the command module was even more fun than space walking, for it offered all the freedom of movement with none of the limitations of a pressurized space suit. With *Eagle* attached to the command module's nose Apollo 11 was like a small, two-room space station. The two joined craft coasted moonward, twirling slowly in the sun's glare.

Aldrin hadn't gone inside *Eagle* until the day before, when he and Armstrong made a scheduled inspection of its cabin. He'd gone in first, making the topsy-turvy passage "up" from the command module and "down" into the lander; he would later call it the strangest sensation of the entire voyage. It was Aldrin's first time inside *Eagle* in two weeks, and it felt good to be back. Armstrong arrived—hanging from the ceiling, by Aldrin's reckoning—with the television camera. Before the flight they had barely enough time to learn how to turn the camera on, never mind learn how to point it accurately or stage anything like a professional-looking show. It was "challenging," to say the least, when mission control came on the air ten minutes before a TV show started and announced matter-of-factly that 200 million people were standing by for the broadcast; they're *all* watching; now—what are you going to show us?

But this one went well. For most of it Aldrin was on camera, giving his audience a look at various items of equipment—a movie camera they would use to film the Powered Descent, a flight plan, one of the protective visors he and Armstrong would wear during the moonwalk. From what Charlie Duke in mission control said, the TV show was an unqualified success; much of the world picked it up live, and the picture was so clear you could read labels on the instrument panel.

But such diversions aside, Aldrin hadn't felt quite on top of things since they left earth. There was that feeling of being slightly "behind the airplane" that he'd had as a squadron pilot whenever he flew a new jet for the first time. Even though everything was going smoothly, even though he was doing things he had done literally hundreds of times in the simulator, it was the *reality* of it all—seeing the earth beyond the windows, floating free, knowing that the world was listening and watching almost everything they did—that put him just slightly off balance. And there was always another "performance" ahead, whether it was a midcourse correction burn or a television show.

"Houston, Apollo 11." This time there was an answer. It was Ron Evans, the Capcom who was working the "night shift." With the three of them asleep Evans's shift got pretty dull, and he sounded glad to make contact. He told Aldrin the trajectory was so good that there wasn't going to be a midcourse, and he could go back to sleep for another couple of hours. So Aldrin settled in and tried, without success, to sleep again. After a few minutes he was aware of some activity beneath the couches; Armstrong and Collins were awake. They shed their lightweight sleeping bags, emerged from behind the seats, and the three men began the business of their fourth day in space. Fuel cells needed to be purged. There was breakfast to fix. And there was an unspoken but undeniable tension in *Columbia*'s cabin. Each of them could feel it: the ruse was getting harder and harder to pull off. They were about to put their lives and their abilities on the line.

Armstrong, Aldrin, and Collins had not seen the moon on the way out, but according to the flight plan they were supposed to take some pictures of it a few hours before braking into lunar orbit. As they finished breakfast, a sudden darkness came around them, and for the first time in the flight the sky was full of stars, too many to count, each with a steady, gemlike brilliance. They had flown into the lunar shadow. Through the windows of the slowly turning spacecraft they looked out at the place where the sun had once been, and there was the moon: a huge, magnificent sphere bathed in the eerie blue light of earthshine, each crater rendered in ghostly detail, all except for a third of the globe, which was a crescent of blackness. As their eyes adapted to the darkness they saw that the entire moon was set against a gigantic ellipse of pearly white light, the glowing gases of the sun's outer atmosphere, which stretched beyond the moon into the blackness. Somehow in these strange, cosmic illuminations the moon looked decidedly three-dimensional, bulging out at them as if to present itself in welcome, or, perhaps, warning.

7:25 P.M., Houston time
3 days, 10 hours, 53 minutes Mission Elapsed Time

Even before Apollo 11 left earth, Neil Armstrong knew the approach to his landing site as well as he had known the desert towns along the approach to Edwards. He'd spent some of the quiet hours of the trip out to the moon studying the photographs radioed from the unmanned probes and carried back aboard Apollo 10. Now that he was in lunar orbit, and he could look down on the Sea of Tranquillity with his own eyes, *Eagle*'s landing approach was even easier to recognize than Stafford and Cernan's pictures had suggested. The very moment of Apollo 11's departure from earth had been timed so that tomorrow, during the landing, the sun would be 10 degrees above the horizon, throwing the *mare* into relief, so that he would be able to spot rough ground as he flew *Eagle* to a landing.

Landing Site 2 was an ellipse measuring 11½ miles by 3 miles, or about as long as Manhattan island and half again as wide. There were many places that were more exciting, geologically speaking, but they were also much riskier or harder to get to. On later landings there would be time to explore; for now, the geologists would be happy with any place Armstrong and Aldrin managed to visit. And so Site 2, chosen for its blandness, was a completely unremarkable stretch of *mare,* close to the lunar equator and thus easiest to reach. Only Stafford and Cernan had seen it close up, but that was from 9 miles. What would it look like from 500 feet? Would it offer a safe place to land a lunar module? Armstrong was cautiously optimistic. Soon enough, he would find out for certain.

Late in the day, while Aldrin went into *Eagle* to make systems checks, Armstrong lingered at *Columbia*'s window, hoping to glimpse the landing site itself. Below, a crater called Moltke, only a few miles wide, glowed in the light of lunar dawn. Around it, the *mare* was cut by flat, narrow valleys like desert roads. Just to the north, almost enveloped by night, lay the place where he and Aldrin would try to land. Collins asked, "Can you see the landing site?"

Armstrong peered down among the long shadows. "I'm not sure," he said. Suddenly he and Collins heard Aldrin's voice in their headsets.

"I can see it," Aldrin said from inside *Eagle,* contained excitement in his voice; "I got the whole landing site here." After a moment Armstrong and Collins could see it too, barely emerged from the shadows. From this height, Landing Site 2 was tiny, but it was possible to make out some details—and it didn't look encouraging. In this illumination, with even the tiniest features thrown into jagged relief, it looked so forbidding as to be untenable. It was difficult to accept that while they slept that night, the rising sun would tame this unsettling scene. Right now it looked like the last place anyone would want to land.

8:32 P.M.

"Amazing how quickly you adapt," Collins said brightly at dinner. "Why, it doesn't seem weird at all to me to look out there and see the moon going by, you know?" But behind his calm words, Collins harbored unspoken concern for his crewmates. He could not tell whether they felt anxiety, for they seemed entirely relaxed. Of the three of them, he was least comfortable with the risks they were undertaking, most conscious of the fallibility of complex machines. He had come to see the flight as a long and fragile daisy chain of events, and was only too aware that at any time the chain could break. Now he felt something like an anxious parent with two children about to go away on a long trip. He offered to take the watch for the night, encouraging Armstrong and Aldrin to sleep underneath their couches. "You guys ought to get a good night's sleep, going into that damn LM." As his crewmates readied their supplies for the next morning Collins said flatly, "I thought today went pretty well. If tomorrow and the next day are like today, we'll be safe."

Sunday, July 20
12:18 P.M., Houston time
4 days, 3 hours, 46 minutes Mission Elapsed Time

"You cats take it easy on the lunar surface. If I hear you huffing and puffing I'm going to start bitching at you." Collins's gentle admonition came as *Columbia* and *Eagle* coasted through another sun-drenched orbital noon, with undocking only minutes away. When the time came, Collins pushed the button to release the LM and called out, "Okay, there you go! Beautiful!"

Pushed by the spring action of the docking mechanism, the two craft drifted apart. Armstrong carefully steadied *Eagle* with a few quick bursts of thruster fire. For several minutes *Eagle* and *Columbia* drifted in formation, orbiter and lander, robot spider face to face with command ship. Collins fired off pictures as the lander turned a slow pirouette before him.

"I think you've got a fine-looking flying machine, there, *Eagle,*" Collins offered, "despite the fact you're upside down."

"Somebody's upside down," Armstrong said.

"You guys take care," Collins radioed.

Armstrong replied simply, "See you later."

Collins gave a short burst of the maneuvering thrusters and *Columbia* pulled away until *Eagle* was merely a point of light. Collins had given some thought to the chances of accomplishing this mission, and he had independently arrived at the same fifty-fifty odds as his commander. For the next few hours, he had work to do, tracking *Eagle*'s flight over the craters and relaying this information to Houston. But when it came time for the Powered Descent, he would simply listen, like everyone else.

2:35 P.M., Houston time
Mission Operations Control Room, Manned Spacecraft Center

It was quiet in mission control, the way it almost always was during a mission. At the front of the room, a giant screen showed a green-colored map of the moon, with a tiny, moving command module and LM; it showed that *Columbia* and *Eagle* were flying over the far side, out of radio contact. Already, the controllers had passed up the data to *Eagle* for the Descent Orbit Insertion burn and the Powered Descent. Soon the spacecraft would reappear, and if all had gone according to plan, *Eagle* would be on its way down to 50,000 feet.

Gene Kranz and his team of flight controllers—they were called the White Team—sat at their consoles, wearing lightweight headsets, anticipating the Acquisition of Signal. For the moment their television screens, normally full of up-to-the-minute data, were static. Behind the controllers, in a glassed-in gallery that looked out over the control room, NASA managers, current and former astronauts, and other VIP's crowded together to witness the drama to come. Just looking at them drove home the fact that this was not a simulation.

In the third row of the MOCR, Gene Kranz sat at the flight director's console wearing a brand new white vest. When he was on shift he always wore a white vest, made for him by his wife, Marta. This one was white brocade with silver thread; it was a special vest for a special occasion. To Kranz it seemed as though months had passed since the trials of June, but in the last weeks they had finally mastered the Powered Descent. By launch time, Kranz knew he had a winning team. From the time he had awakened this morning, Kranz was sure that Armstrong and Aldrin would land on the moon today. He had gone to church and then to his office, where he had played his Sousa tapes. A few minutes after eight that morning, he took the flight director's chair in mission control to begin the shift for the descent phase. He felt unwavering confidence in the spacecraft, the astronauts, and most of all, in his men. In his mind, it was crucial that he exude that confidence.

Kranz looked down from his console to the first row of the MOCR, the place the flight controllers called "the Trench." There was Bob Carlton, who watched the LM descent engine like a hawk; his call sign on the loop was "Control." Jerry Bostick was "FIDO" (for Flight Dynamics Officer); he would monitor the tracking data as the LM descended. And Steve Bales, the expert on the LM's guidance system, who would be called "Guidance." At twenty-six, Bales was no younger than many of his fellow controllers, but he seemed especially boyish; it was his enthusiasm, his slightly unkempt looks. It was Bales's job to keep tabs on the LM's computer, landing radar, and its trajectory down to the moon.

Off to Kranz's right, in the MOCR's second row, a half-dozen other men rounded out the team, keeping track of the other systems aboard the LM. Each controller, plugged into his own "back-room" specialists, funneled data and advice to Kranz in the third row. And directly in front of him, there was astronaut Charlie Duke, the young South Carolinian whose drawl had answered the excited reports from a barnstorming Tom Stafford and Gene Cernan. Charlie Duke knew more about the lunar module than any astronaut who wasn't already on an Apollo crew; that expertise had prompted Neil Armstrong to ask him to serve as Capcom for the landing.

A short time ago Kranz had ordered Security to lock the doors to the MOCR. Then he'd switched to a separate communications loop, one that the VIP's and the pool reporter couldn't hear, and talked to his men. "Okay, gang, we've had a good training period," he began. "And today, we're really going to do it, we're going to land on the moon. This is the final exam. . . ." He reviewed the key milestones of the Powered Descent, and made sure his controllers were clear on their own ground rules for feeding him information over the loop. Then Kranz gave a pep talk worthy of a battlefield commander. "This is the best team I've ever worked with," he said. "I have ultimate confidence in you people. . . . What we're about to do now, it's just like we do it in training. And after we finish the sonofagun, we're gonna go out and have a beer and say, 'Dammit, we really did something.' "

2:46 P.M.

Telemetry readouts streamed into the control center as *Columbia* came out from behind the moon. Seconds later, Kranz heard Mike Collins tell Charlie Duke, "Listen, babe, everything's going just swimmingly." Almost two minutes passed before *Eagle* reappeared, and almost immediately communications with the LM became spotty. The signal would drop out and then return. Kranz sweated out the interruptions, knowing that only minutes from now he would have to give Armstrong and Aldrin a go ahead for the Powered Descent, and when that time came his controllers would have to have good data. How much data was a judgment call that only Kranz could make. If everything was going well, they could get by with only a few seconds' worth. But even if things were bad, Kranz was willing to stretch rules, because he knew the risks entailed in an abort. Several astronauts were seated next to Duke, listening in; one of them was Pete Conrad. He suggested to Duke that *Eagle* yaw slightly to one side to improve the signal strength. That worked.

3:06 P.M.

Inside *Eagle*, 50,000 feet above the moon, Armstrong and Aldrin stood side by side, anchored by harnesses to the LM floor. Four-day beards darkened their faces. Within their bubble helmets, they scanned the instrument panel as the time for Powered Descent approached. Their mouths were dry from the pure oxygen flowing through their space suits. Already they had pressurized *Eagle*'s fuel tanks, called up the proper computer program, and checked their trajectory by sighting on the sun with the LM's navigation telescope. With forty seconds to go, Armstrong made sure that Aldrin had turned on the movie camera to record the descent. Armstrong set the switch to arm the descent engine. Then, seconds after Aldrin pushed the PROCEED button, the two men spoke at once: "*Ignition.*"

"Just about on time," Armstrong said. The descent engine came to life so gently that Armstrong and Aldrin heard and felt nothing. Only the gauges told Armstrong it was firing. Less than half a minute later it roared to full thrust and the cabin filled with a soundless, high-frequency vibration.

Once more communications dropped out. "They've lost you," Collins radioed to his crewmates. A moment later, after Aldrin switched to a different antenna, communications returned.

Now Neil Armstrong turned his attention to the moon. *Eagle* was facedown, and through his small, triangular window Armstrong could see landmarks he recognized. Each checkpoint was appearing 2 full seconds ahead of schedule, and since *Eagle* was going nearly a mile a second, that meant they would come down about 2 miles beyond their aim point. Armstrong keyed his mike.

"Our position checks downrange show us to be a little long," he radioed to earth. Now *Eagle*'s engine throttled down, exactly on schedule, and Armstrong realized the computer was not aware of the error.

The long brake continued, and now, 46,000 feet above the moon, it was time for Armstrong to turn *Eagle* over on its back so that its landing radar would point at the moon. When he did, he and Aldrin found themselves looking out at the earth, afloat in blackness.

At 40,000 feet the landing radar came to life, blurting information on speed and altitude to *Eagle*'s computer. From these data the computer continually revised its trajectory calculations, and *Eagle* shuddered with corrective bursts from the maneuvering jets. Armstrong was surprised to hear them fire so often, much more often than in the simulator. No smooth ride at this stage; they were lurching their way down to the moon.

Aldrin, meanwhile, began a running dialogue with the computer, checking its height calculations against the data from the radar. As expected, the two disagreed by several thousand feet. Aldrin knew the radar echoes were more reliable, and he planned to tell the computer to accept those data, but first he wanted mission control to take a look. To do that, he keyed in a command to tell the computer to display the difference, or delta-H, as it was called. Suddenly, the men heard the high-pitched buzz of the Master Alarm in their ears. On the computer display the PROG light glowed amber.

"Program alarm," Armstrong radioed. It was the crispness of his words, rather than the tone of his voice, that conveyed urgency.

Quickly, Aldrin queried the computer for the alarm code, and "1202" flashed on the display. Aldrin did not know just what 1202 meant—and this was not the time to dig out the data book to find out—but it had something to do with the computer being overloaded with too many things to do. He had never seen this kind of alarm in a descent simulation; now he wished it would just go away.

In mission control Gene Kranz felt as if he were in one of those bad simulations back in the dark days of early June. He had almost been glad when the problems began as soon as *Eagle* came around the moon, beginning with the spotty communications. Kranz welcomed a little adversity at the starting gun; it would get his men thinking instead of waiting tensely. A few malfunctions made the whole thing seem more like a simulation and less like history. But the problems snowballed. Even as the Powered Descent began, critical data from the spacecraft kept cutting out, then returning for brief moments—just barely long enough for Kranz to let the landing continue.

In the midst of it all Kranz had done what he called "going around the horn," polling his men. He knew their voices so well that he could pick them out in the web of simultaneous conversations filling his headset. As each man spoke Kranz listened for signs of strain; he heard solidity. And none more so than young Steve Bales, the LM computer expert, who came back with such unbridled enthusiasm—"*Go!*"—that Kranz almost burst out laughing.

Now, with the LM's computer threatening to abort the landing, Bales was his most important man. Back in the first week of July the same kind of computer alarms had come up during a simulation with the Apollo 12 crew. Kranz had ordered an abort, and afterward the simulation instructors had really let him have it. If he had only been familiar with the alarms, they told him, he could have kept going. For the rest of that day and that night Kranz sat down with Bales and some of the computer experts from MIT, studying each type of alarm and what to do if it came up. So when Kranz heard Neil Armstrong call "twelve-oh-two," he knew it was serious. Whether they could continue or not was up to Steve Bales.

But the complexities of the LM's computer were too much for one person. Bales wasn't certain what the 1202 alarm meant. He put the question to one of his back-room experts, Jack Garman. Garman knew that for some reason—no one knew why—*Eagle*'s computer was saying, "I have too many things to do in my computation cycle, so I'm going to give up and start at the top of the list." As Kranz waited for an answer from Bales, he heard Armstrong tensely radio, "Give us a reading on the 1202 program alarm." Before Kranz could speak, Bales responded, "We're—we're Go on that, Flight." Garman had told Bales that as long as the alarm was intermittent, not continuous, everything would be okay. But if the alarm returned and didn't go away, the computer could give up working altogether, and that would mean an almost certain abort.

Inside *Eagle*, Armstrong and Aldrin heard Charlie Duke's urgent but assured words, "*We're Go on that alarm.*" Once more, Aldrin queried the computer for the delta-H; once more an alarm rang in their headsets. Again a message from Steve Bales came to them, via Charlie Duke: mission control would keep tabs on the delta-H, alleviating some of the computer's workload. Above the moon, and in Houston, everyone hoped the fix would work.

Halfway through the Powered Descent, right on schedule, *Eagle*'s computer throttled the engine back to half its maximum power. Suddenly Armstrong and Aldrin felt themselves grow lighter. The long brake was over now. It was time for *Eagle* to pitch over from its faceup position and begin the final descent. Armstrong and Aldrin waited intently for the computer to execute the maneuver. Just as planned, 7,500 feet up, *Eagle*'s maneuvering thrusters fired to pitch the craft forward. In *Eagle*'s windows the flat horizon swung upward into view, and Armstrong looked out at the cratered plains of the Sea of Tranquillity, bright in morning sunlight.

Armstrong checked the altitude and speed: 5,000 feet up, 100 feet per second, just as expected. For a moment he took control of his craft, pulsed the maneuvering thrusters, then gave the ship back to the computer, satisfied that *Eagle* would respond when it was time for him to take over. Now his eyes went back to the gauges. Three thousand feet up now, descending at 70 feet per second, about 48 miles an hour. *Eagle* was right on the planned trajectory. He heard Duke say, "You are Go for landing."

This was the time for Armstrong to watch for his landmarks and look for a good place to set down. Before he could do so, Aldrin announced, "Program alarm. Twelve-oh-one." In Houston, Steve Bales was ready. Before Kranz could finish asking him what 1201 meant, Bales shot back, "Same type; we're Go."

Inside *Eagle,* Armstrong's eyes went back to the moonscape. He gazed past the grid on his window and said to Aldrin, "Give me an LPD."

Aldrin queried the computer and told Armstrong, "Forty-seven degrees."

"Forty-seven," Armstrong repeated. He sighted along the window grid, past the 47 degree mark. He could see the target, still more than a mile in the distance, but advancing rapidly. It looked promising. "That's not a bad-looking area," he said blandly. Suddenly the alarm was back. Aldrin had no sooner cleared the alarm than it sounded again. Again, Armstrong's attention was diverted by the threat of an abort, while *Eagle* flew onward. When the alarms quieted down the moon was only 1,000 feet below him, and he did not like what he saw. A crater as big as a football field was just ahead, surrounded by a field of boulders, some as big as Volkswagens. The computer was blindly taking them there, down into the middle of the boulder field. And for an instant, Armstrong weighed the matter. He was all but certain those boulders would prove to be pieces of lunar bedrock. If he could find a safe place to land just short of the boulders, he and Aldrin would no doubt find some prizes for the geologists. But *Eagle* was going too fast; there were just too many rocks. It was time for him to take over. He switched to ATTITUDE HOLD and pitched the lander forward until it was almost level, letting the descent rocket brake its fall without slowing its horizontal flight. Only 350 feet up now, *Eagle* skimmed over the boulders and headed toward safer ground. Armstrong planned to set down on the first clear place he could find.

Buzz Aldrin did not know about the crater or the boulders, and he heard nothing from Armstrong, who was too busy flying to get out more than a few clipped words every now and then. And Aldrin was too busy to look out the window. His eyes went back and forth between the gauges and the computer readout, his hands went to the computer's keyboard to extract the critical information. And his voice, heard on hot-mike by Armstrong and the listening world, was a steady stream of data: "Three hundred and fifty feet; down at four. Three hundred thirty, six and a half down. You're pegged on horizontal velocity." His voice was almost electronic.

In mission control, Kranz and his controllers heard only Aldrin's stream of numbers. Telemetry told them that Armstrong had assumed semimanual control of the lunar module. No one knew about the football-field-sized crater; they knew only that *Eagle* was no longer following the nominal landing profile, that it had slowed its descent and was still moving at a good clip over the moon. Gene Kranz knew then that the partnership had all but dissolved, that the "center of gravity of the decision-making process" was no longer some point midway between himself and the moon. It was Neil Armstrong. Charlie Duke knew it too, and he said over the loop to Kranz, "I think we'd better be quiet." There was nothing to do but listen to Aldrin's voice and hope that the fuel held out.

Armstrong flew onward, sharing control of *Eagle* with the computer. As he cleared the big crater he was careful to pitch *Eagle* back again to avoid building up too much speed. Below, he could see a string of boulders; he banked slightly to the left to get away from them. The response was sluggish and familiar. The lunar module was a much better flying machine than he had expected, easier than any simulation. The little toggle switch to control their rate of descent —the one he had been skeptical of back on earth—was working well. Now Armstrong pushed it several times to slow their fall.

The moon rushed up at him, new terrain advancing quickly over the horizon. He heard Aldrin: "Three hundred feet, down three and a half, forty-seven forward." His heart pounded. Still there was no clear place. The clock seemed to race. He had to buy time to search for a spot. Armstrong slowed *Eagle*'s descent rate.

"How's the fuel?" Armstrong's voice was quiet, even relaxed.

"Eight percent," Aldrin answered. That was less than they'd had in the simulations. Now, at last, Armstrong saw what looked like a patch of smooth ground, just ahead. "Okay," Armstrong said, "looks like a good area here."

Aldrin stole a moment to glance out. On the bright ground 250 feet below was a dark silhouette clearly recognizable as a lunar module, bristling with antennae and landing gear and ringed with a halo of sunlight. Then he went back to the gauges, his voice more insistent as he fed Armstrong data. "Two hundred twenty feet. Thirteen forward, eleven forward. Coming down nicely."

Now Armstrong saw that the place he had selected was no good. "I'm going right over a crater," he said, his words lost in Aldrin's numbers. "I gotta get farther over here." Again he nudged the hand controller forward, leveling the craft, using the last bit of forward speed to clear the crater. And just beyond it he saw where he was going to land. It was a smooth, level place about 200 feet square, bounded on one side by a few large craters and on the other by a line of boulders. He knew they were getting very low on fuel. As if to emphasize this the DESCENT QTY light now glowed on the instrument panel. Ninety seconds of fuel left, and 20 seconds of that had to be saved for an abort. But Armstrong had his landing place. And only 100 feet separated *Eagle* from the moon.

It was crucial to bring *Eagle* straight down, with no horizontal motion; otherwise there was the risk that the touchdown might break off a landing leg. Armstrong trained his vision on a place just beyond the landing point, which he would use as a reference to judge *Eagle*'s height and motion all the way to touchdown. Now he noticed that everything was wrapped in a transparent haze. The blast of the rocket was disturbing the dust of the moon. As *Eagle* descended the haze became a sheet of rushing streaks that flew away from him in all directions, obscuring the surface. They confused his perception of motion like a fast-moving ground fog blowing across a runway. But he could see rocks on the surface, sticking up through the blur like islands, and he fixed his gaze on them.

Armstrong clicked the rate-of-descent toggle until *Eagle* descended no faster than an elevator. He focused on the rocks and on Aldrin's numbers: "Sixty feet, down two and a half, two forward, two forward." He heard Charlie Duke's voice: "Sixty seconds." A minute's worth of fuel left until he and Aldrin would be forced to abort.

In mission control, stomachs tightened. No one knew about the big crater and Armstrong's efforts to avoid it. They knew only that in almost every simulation Armstrong had landed by this point. And everyone, from the controllers riveted to their displays to the VIP's who watched in agonized silence, knew that every second brought Armstrong and Aldrin closer to their abort limit. Even now, it was impossible to know how it would end.

Fifty feet above the moon. Now thirty. *Eagle* was drifting slowly backward and Armstrong did not know why, but he knew he must not land while he could not see where he was going. He pulsed the hand controller, struggling to arrest the unwanted motion. He was displeased with himself, sure that he was not flying *Eagle* smoothly. He wished he could buy more time, but he was too low on fuel to slow the descent any further. Twenty feet to go. He'd stopped the backward drift but still wrestled with a sideways motion that had crept in. They were flying the dead man's curve now, too low to abort if the engine quit, but in the back of his mind Armstrong knew that if that happened they'd be okay, they would just fall onto the moon. Dust blew furiously. Once more, words of caution came from earth: "Thirty seconds." Then Buzz Aldrin said, "Contact light."

Armstrong had planned to shut the engine down at this moment: the engineers had warned him that if the rocket got too close to the surface, the back-pressure from its own exhaust might blow it up. But he was so absorbed in flying that he forgot about that. With the engine still firing, *Eagle* settled onto the moon so gently that neither man sensed the contact. Quickly Armstrong hit the ENGINE STOP button and said, "Shutdown."

Now there was a blur of activity as Armstrong and Aldrin set switches and keyed numbers into the computer, while Aldrin rattled off the postlanding checklist:

"DESCENT ENGINE COMMAND OVERRIDE, OFF—ENGINE ARM, OFF—413 is in."

Then there was a moment of quiet, and the two men turned to one another in the tiny cabin. Their eyes met, their bearded faces grinned at each other inside bubble helmets, and their gloved hands clasped. Armstrong keyed his mike. "Houston, Tranquillity Base here. The *Eagle* has landed."

The answer from earth was like a sigh of relief. "Roger, Tranquillity, we copy you on the ground," radioed Charlie Duke. "You got a bunch of guys about to turn blue. We're breathing again. Thanks a lot."

They had done it. For an instant, Armstrong and Aldrin savored relief and elation that the greatest challenge was behind them. For Armstrong it was more than a personal high; hundreds of thousands of people had worked for the better part of a decade to share this triumph. And for himself, the landing had been everything a pilot could ask for. It had been a close call, but that just sweetened the victory. There was no way to know exactly how much fuel remained when they touched down—the gauges just weren't that accurate—but it was something like 20 seconds' worth left before the abort limit. Of course, Armstrong knew, 20 seconds is a long time.

II: Magnificent Desolation

Tranquillity Base
3:17 P.M., Houston time
4 days, 6 hours, 45 minutes Mission Elapsed Time

Seconds after *Eagle*'s rocket engine shut down the dust particles departed on long, flat trajectories, and the stillness of a billion years returned to the Sea of Tranquillity. Inside *Eagle*, Armstrong and Aldrin put their elation on hold; there was work to do. Now that they were on the moon the most important thing was to get ready to leave it, in case of an emergency. That meant checking *Eagle*'s systems, entering abort data into the computer, and aligning the guidance platform to the weak lunar gravity. That last item was Aldrin's task, and while he attended to it, meticulously sighting on stars with the navigation telescope and conversing with the computer, Armstrong had a chance to survey the place where they had landed.

Eagle had come to rest on a broad, level plain, pockmarked with craters a few dozen feet to a fraction of an inch across, and scattered with rocks and boulders. In the distance Armstrong could see ridges that might have been twenty or thirty feet high, but it was hard to tell: there were no buildings, trees, or any other features normally used to judge size and distance. The lack of atmosphere gave an unreal clarity to the view, better than the clearest day on earth. Hills and boulders at the horizon were as sharp as the rocks next to *Eagle*'s footpads. Beyond that bright edge, as empty as the margins of a fifteenth-century map, was the blackness of space.

Most amazing to Armstrong was the strange play of light and color. Directly ahead, to the west, the light of the rising sun was brilliantly reflected by a landscape of light tan. This gave way on either side to a dimmer, grayer tan, and when he craned to look off to the side, where the ground was crisscrossed with long morning shadows, he saw an ashen gray.

It was not a hostile scene. Somehow it did not look like a place where an unprotected man would perish in seconds; on the contrary, it seemed inviting, as if he and Aldrin might descend to the craters in beach clothes and get a suntan. As he looked out, Armstrong wondered where he and Aldrin had landed. With the distraction of the computer alarms he'd missed all his checkpoints on the way down. Now he searched the horizon for some feature he might be able to identify, but found none. With a wry smile he radioed Houston, "The guys who said we wouldn't know where we were are the winners today." He knew there must be a small army of controllers and geologists working on it. And he hoped that sometime in the next 21½ hours Mike Collins would be able to spot them with *Columbia*'s navigation sextant.

Wherever they were, Armstrong was ready to explore. To the north, a line of boulders broke the rolling smoothness; they looked to be within easy walking distance. Armstrong suspected they had been ejected from some large crater nearby, perhaps that giant crater he had avoided during the descent. They were probably pieces of lunar bedrock; the geologists would undoubtedly want samples. He might even be able to run back to the crater and take some pictures.

Armstrong turned his gaze to the LM's small overhead window, and he searched the velvet blackness for stars, but saw only the blue earth. "It's big, and bright, and beautiful," he told Charlie Duke. Then, after running through the countdown for a simulated lunar liftoff, he and Aldrin turned off most of *Eagle*'s systems. The lander would be dormant until it was time to leave the moon; that would come early tomorrow afternoon, if all went well.

Nassau Bay, Texas

Outside the Aldrin house, where it had begun to rain into the muggy afternoon air, a group of reporters assembled on the front lawn. Already they had spoken to Jan Armstrong and Pat Collins about this day's incredible events; now they wanted Joan Aldrin's reaction. For some time, they waited for her to emerge.

Inside, Joan was savoring relief. Buzz had always said that the moment of truth would be the liftoff from the moon, but in her mind it was the landing. When the Powered Descent began Joan was in front of the television, surrounded by family and friends, her emotions under control. She listened intently to the squawk box, but it was hard to hear what Buzz was saying, and when she did it was so technical that she did not often understand. Fortunately, Jerry Carr and Rusty Schweickart were there to explain what was happening. Then *Eagle* was very low, and she could barely stand the tension. Rusty told her that the fuel was now down to a matter of seconds, and that Neil and Buzz had not yet found a place to land. The words overwhelmed her. She stood in the silence of that crowded room, holding onto a doorway, her eyes brimming with tears, and listened to her husband's voice. Finally, in the midst of the numbers, she heard Buzz say, "Okay, engine stop." Those words she understood: they had made it. She found the embrace of Buzz's uncle, Bob Moon, and then headed for her bedroom.

By the time the reporters and the TV crews gathered, Joan was on a high. After a time, she went out to them, sheltered from the rain by an umbrella held by a NASA public affairs man. All the times that she had faced the press and dutifully concealed her terror were behind her. She radiated life, and she used this moment as if it had been made for her: here was a stage to play on. One of the reporters asked, "What were you doing when they landed?"

"Well," said Joan in a stage whisper, "I was holding onto the wall. I was praying." The reporters did not pick up on her performance. As they asked the same questions they had already asked the other two wives—What are your plans for the moonwalk? Will you let the children stay up to watch?—she wondered, *My God, what is wrong with them?* They seemed drained of energy. She would rouse them.

"Listen!" cried Joan, her eyes wide. "Aren't you all excited?" For a moment she was silent, suddenly in command of her small audience, and then, she let loose her jubilation: "*They did it! They did it!*"

5:11 P.M.

It was hard to believe that two men could land on the moon and go to sleep before setting foot on it, but that was what the conservatively minded flight plan called for: In case Armstrong and Aldrin had to make an emergency liftoff and rendezvous, they would need to be rested. Before the mission, Armstrong had approved the early, four-hour sleep period knowing full well that, barring any problems, he and Aldrin would almost surely reject it on the moon. He wasn't about to say anything to the press—if for some unforeseen reason he and Aldrin ended up sticking to the original plan, they'd write,

"Astronauts step on moon, four hours behind schedule." But now, there was no reason to wait. *Eagle* was in perfect order. One-sixth g felt entirely natural; in fact, they liked it better than either normal gravity or weightlessness—it had much of the buoyancy of zero g without the disorienting lack of up and down. And so Armstrong and Aldrin agreed: They would go out early. Armstrong called Houston to suggest that the moonwalk begin at about eight o'clock in the evening, Houston time, some five hours ahead of schedule. Almost immediately Charlie Duke came back with a go-ahead.

"You guys are getting prime TV time," Duke said.

5:57 P.M.

In the history of exploration, one of the few moments that could compare with this one came on May 29, 1953, when Edmund Hillary and Tenzing Norgay became the first humans to stand atop the windswept summit of Mount Everest. In their brief minutes atop that ultimate peak, Hillary's actions were those of the conqueror; he aimed his camera down each ice-crusted ridge at the lands below and snapped the pictures that would prove to the world they had made it. The Nepalese sherpa Tenzing, meanwhile, hollowed out a place in the snow and filled it with offerings to his God. For him the climb was not a conquest but a pilgrimage.

Now it was Buzz Aldrin who enacted a spiritual observance in a strange and distant place. In the weeks before launch, he had searched for some gesture that would be worthy of the moment, and he had decided to celebrate Communion. Deke Slayton had warned him against broadcasting any religious observance over the air; NASA was still coping with a controversy stirred by the Genesis reading on Apollo 8. Aldrin's Communion would have to be a secret one.

Now that it was clear that he and Armstrong were on the moon to stay for a while, Aldrin took advantage of a quiet moment. He opened the stowage pouch that contained his personal mementos and removed a plastic bag containing a small flask of wine, a chalice and some wafers, and set them on the little fold-down table just beneath the keyboard for the abort guidance computer. He keyed his mike. "This is the LM pilot speaking. I'd like to take this opportunity to ask every person listening in, whoever and wherever they may be, to pause for a moment and contemplate the events of the past few hours, and to give thanks in his or her own way." Armstrong looked on, an expression of faint disdain on his face (as if to say, "What's he up to now?") while Aldrin went on with his ceremony. Released in the gentle gravity the wine poured slowly and curled gracefully against the side of the cup. Aldrin read silently from a small card on which he had printed words from the book of John:

I am the wine and you are the branches
Whoever remains in me and I in him will bear much fruit;
For you can do nothing without me.

In Nassau Bay, Joan Aldrin—who had settled into a warm and wonderful state, playing old Duke Ellington records with one ear tuned to the squawk box—marveled when her husband asked for a moment of silence. Though she did not know he was taking Communion, she saw it as happy evidence of a hidden dimension. All this time she had suspected Buzz was so caught up in the technical side of his mission that he had missed its significance—but now she realized she had been wrong.

If Aldrin's Communion marked a very personal observance, then Neil Armstrong had his own ceremony to think about. Almost from the moment the world learned that he would be the first human being to set foot on the moon he had been asked what he would say as he crossed that threshold. His mail had been full of suggestions, including passages from the Bible, verses of Shakespeare, and countless others. Everyone from the press to the simulator instructors brought it up. Not even by leaving earth could he escape; Collins and Aldrin asked about it on the way to the moon.

If it hadn't been for the fact that everyone made such a big thing of it, Armstrong wouldn't have focused on the matter at all. The landing was the flight's greatest achievement, and in Armstrong's mind, it amounted to the first human contact with the moon.

But to a public estranged from the technology of this journey, a landing was less meaningful than a footstep. If it was natural for them to want historic words for historic occasions, they were nevertheless asking them of a man who does not deal liberally in words. But now, on the moon, Armstrong knew he could delay no longer. As he thought about the first step he would take from *Eagle*'s footpad he pondered the inherent paradox—a small step, yet a significant one—and he knew what he would say.

7:21 P.M.

The moon cast its light into *Eagle*'s cabin as Armstrong and Aldrin began the most critical and the most tiring part of the entire moonwalk: suiting up. They were already behind schedule simply because there had been things to do that had never been part of the practice runs—such as stowing the trash from dinner—and they took longer than the men expected. But this was no time to rush, and Armstrong and Aldrin worked with the care of skydivers packing their chutes, following the checklist to the letter. First they pulled on the lunar overshoes, whose rubber soles had coarse treads designed to give sure footing on alien soil. Next they strapped on the Portable Life Support System—the massive backpack that had been the bane of their existence during training. On the moon each one weighed just over twenty pounds; Armstrong and Aldrin had no trouble hefting them with one hand. Still, they felt the *mass* of the packs—that was undiminished—and despite their efforts to avoid bumping into control panels and each other in the cramped cabin, that happened more than once.

Oxygen hoses were next; their metal fittings locked into receptacles on the front of their suits. Then came hoses for their water-cooled underwear. Water from the backpacks would circulate through a network of tiny tubes woven into the undergarment. This method of cooling a man inside a space suit was so effective that there was almost no way for him to become overheated, the way Gemini spacewalkers had; tests had showed that he would tire himself out before that happened. Armstrong and Aldrin carefully locked each hose in place, and then locked the *locks* in yet another level of security. Both men were on hot-mike now, and the radio transmissions that came down from the moon sounded like strange, high-tech poetry: "Locks are checked, blue locks are checked. Lock-locks, red locks, purge locks . . ."

Onto each man's clear bubble helmet went a special outer helmet equipped with a gold-plated visor to reflect the sun's unfiltered glare. On their chests, they wore small control units for their radios and to display readings on the backpack. Each methodical step brought them closer to setting foot on the bright ground beyond *Eagle*'s windows.

"All set for the gloves," Aldrin said, and each man pulled on a space age version of a knight's gauntlet, with coverings of woven steel-fiber, and rubber fingertips that afforded some measure of dexterity. With the flick of a switch, each man started the pumps and fans in his backpack and heard the familiar, reassuring hum of machinery that would keep him alive, and felt the whoosh of oxygen past his face. Their ears registered increasing pressure as the suits inflated to 3.5 pounds per square inch. Now Armstrong and Aldrin were self-contained, mobile spacecraft.

9:28 P.M.

All that remained was to vent *Eagle*'s oxygen into space, but even that took longer than expected. Aldrin opened the valve and the men watched the pressure reading creep downward. After three minutes it was four-tenths of a pound; a minute later, two-tenths.

"Let me see if it will open now," Aldrin said, reaching for the hatch handle. It stayed firmly shut. The pressure read one-tenth of a pound and holding. Neither man wanted to tug on the thin metal door for fear of damaging it. Finally, Aldrin peeled back one corner to break the seal; that did it.

"The hatch is coming open," Armstrong radioed, excitement creeping into his voice. As it did so, the last wisps of *Eagle*'s atmosphere rushed outward in a flurry of ice particles, and the two men stood in the vacuum of space.

While Aldrin held the hatch open, Armstrong sank to his knees and carefully moved his suited bulk through the opening. He moved onto a large platform called the porch, with large handrails on either side, that bridged the hatchway with the ladder. When his boots met the top rung, he grasped the handrails and raised himself upright. After five days of floating within the confines of a spacecraft, the change in visual scope was profound. The sensation of height, absent in deep space or in orbit, returned to him. Before him, the shadowed, foil-clad bulk of his lunar module; beyond, a pristine wilderness.

He could not descend yet; for one thing, the world was waiting to see the event. Armstrong pulled a D-ring on *Eagle*'s side and an equipment stowage tray lowered like a drawbridge. On it a small TV camera began transmitting to earth, where Cliff Charlesworth and his team of controllers listened and waited. Moments passed, and then Armstrong heard Capcom Bruce McCandless radio, "We're getting a picture on the TV!" On the big screen in mission control a strange, almost abstract black-and-white image flickered into existence. The front leg of the lunar module slanted across a tableau of black sky and bright ground, and at the top, the shadowy form of Neil Armstrong descended, one rung at a time, toward the moon.

When Armstrong reached the bottom rung he paused. The legs were designed to compress with the force of landing, bringing the ladder closer to the surface in the process. But *Eagle* had touched down too gently for that to happen; Armstrong was still more than three feet up. For a moment he dangled his foot in space, then launched himself into a slow-motion fall, landing on both feet inside the foil-covered footpad. Before he went any further, he wanted to be sure he could get back up. He sprang upward and almost missed the bottom rung, but at last managed to steady himself. Satisfied, he descended once more.

Standing in deep shadow, Armstrong looked down at the soil just beyond the footpad, and as he had done many times in training, he described what he saw for the benefit of the scientists on earth. "The surface appears to be very, very fine grained as you get close to it; it's almost like a powder . . ." In simulations his voice had been decidedly matter-of-fact; now it was laced with excited curiosity.

Grasping the ladder with an upraised glove, Armstrong turned to his left and leaned outward. "Okay," he said, "I'm going to step off the LM now." Silently, carefully, he raised his left boot over the lip of the footpad and lowered it to the dust. Immediately he tested his weight, bouncing in the gentle gravity, and when he felt firm ground, he was still, one foot on the last vestige of earthly things, the other on the moon. He spoke: "That's one small step for man"—now a pause—"one giant leap for mankind."* Again he tested his weight and was reassured to find that his boot penetrated only a fraction of an inch. Still holding on, he stretched out his toe and dragged it backward several times, furrowing the soft ground. Dust clung like soot to the light-blue sole of his boot. Having made this first, tentative exploration, Armstrong lowered his right foot and stepped sideways, both hands resting on the big horizontal strut of *Eagle*'s landing gear. And at last, after bouncing up and down a few more times, he let go of *Eagle* and stood on the moon.

Armstrong moved away from the lander with the halting steps of a man learning to walk again. He moved with a shuffling, stiff-legged gate; it was difficult to bend at the knee and movement came mostly from the ankles and the toes. But he felt buoyant, something between walking and floating. Heavy and light were redefined: his space-suited body, 348 pounds on earth, now weighed only 58 pounds. It was almost familiar—the simulations were that good—and it was even easier to move around than he had expected.

He knew that the first order of business was to collect a small bag of soil, called the contingency sample, that would serve as the scientists' hedge against an aborted moonwalk. But he would do that in sunlight, and he wanted to take care of getting a Hasselblad down

* The quote would be forever footnoted. Armstrong would report after the flight that he had intended to say, ". . . one small step for *a* man," but the indefinite article is definitely missing from the transmission. In 1971, asked by writer Robert Sherrod whether the "a" had been lost in transmission or simply forgotten, Armstrong, clearly savoring the ambiguity, replied, "We'll never know." To this writer, the cadence of Armstrong's words on tape suggests that the "a" was forgotten, not lost.

to the surface while his eyes were still adapted to the darkness. This he accomplished with some effort as he and Aldrin operated a special conveyor line. Then, with the camera mounted on the control unit on his chest, still standing in the lander's shadow, Armstrong snapped the first pictures taken on the surface of another world.

But pictures were not supposed to be his first priority, and after a minute McCandless reminded him about the sample. Aldrin added his own reminder. "Right," Armstrong said quickly, finishing his panorama, then he reached into a pocket on his thigh and pulled out a collapsible handle with a detachable bag at one end. He moved into sunlight for the first time, the glare penetrating his mirrored visor like a thousand-watt spotlight. Turning away, Armstrong began to dig into the surface, and what he found surprised him. Everywhere there was the same soft powder, and yet here and there he met resistance. He managed to scoop up enough dust to fill the bag, and even managed to snare a couple of small rocks; the geologists, he told himself, would get their money's worth.

"That looks beautiful from here, Neil," Aldrin said. He was talking about the sample, but Armstrong responded as if he had meant the moon. "It has a stark beauty all its own," he said, excitement finally invading his voice. It *was* beautiful. It had the serenity of the high desert of Edwards, only here was the ultimate desert, complete in its stillness, and in its starkness. When he turned, he saw the same peculiar transformation from bright tan to ashen gray he'd seen from the LM windows. And when he held the contingency sample in his hand, the mystery of the moon's color deepened: The soil in the bag was almost black, like powdered graphite.

Holding the now unneeded collector handle, Armstrong considered throwing it like a javelin, but thought better of it, and instead gave it an underhand toss. It sailed away on a long, lazy trajectory, spinning in slow motion in the sunlight and traveling an impossibly long distance before landing in the dust.

"I didn't know you could throw so far, Neil," joked Aldrin. The man on the surface of the moon answered with a delighted laugh in his voice, "You can really throw things a long way up here!" Armstrong was elated, and for good reason. The moon offered him firm ground and good footing. Working in one-sixth gravity, in contrast to the grueling training sessions, was easy. Barring a major problem with equipment or with *Eagle,* the first moonwalk was bound to be as successful as anyone had hoped.

10:10 P.M.

"Are you ready for me to come out?" Not a trace of eagerness sounded in Buzz Aldrin's voice, even though he had been watching Armstrong walk on the moon for fourteen minutes. Now it was his turn. While Armstrong radioed guidance, he emerged onto the porch. He offered a mild joke about making sure not to lock the hatch on his way out, which got a laugh from his commander. Then, ever methodical, Aldrin made his own descent, describing his progress to earth. But when he was standing in the footpad, looking out at the moon, his powers of description momentarily left him. He saw disorder, and yet there was a precision, he would say later, the precision of rock and dust. There must be some combination of words that would describe it, but Aldrin could only utter, "Beautiful view!"

Armstrong agreed, "Isn't that something? Magnificent sight out here." Hearing this, Aldrin suddenly had the words he was looking for. With quiet wonder in his voice, he said, "Magnificent desolation." Holding the ladder with both hands, Aldrin swung both feet out of the footpad and onto the moon.

The checklist called for Aldrin to check his balance and stability, and that he did, twirling and leaping like a dancer in slow motion, feeling the strange inertia of his backpack. To compensate for the mass of the pack, he had to lean forward at a seemingly impossible angle; on earth, he would have fallen over. But the pull of this small world was so mild that he could not easily tell when he was standing exactly upright. Looking into the distance, Aldrin scanned the plains of the Sea of Tranquillity. The land curved gently but noticeably away from him, all the way out to the horizon, which was only a mile and a half away. He could actually *see* that he and Armstrong were standing on a sphere.

Aldrin's eyes went to his feet, where a fascinating display of motion took place every time he took a step. Each footfall launched a spray of particles that sailed outward in perfect arcs, unhindered by an atmosphere, all coming to rest roughly the same distance away. Intrigued, he kicked his foot like a child on a playground, sending streams of dust flying gracefully into space. He looked at his own footprints and marveled at their sharpness, as if he had placed his foot in talcum powder. And always, he radioed his observations to earth.

And so, two men at the edge of human experience went about their work, their faces hidden by mirrors, their voices so unrevealing that most of the time only people who knew them well could hear the excitement in them. They talked about the mechanical behavior of the soil and the appearance of the rocks, and it was all very technical, all under control. The first men on the moon were not about to indulge in excited exclamations or elaborate statements of wonder—not just because the first lunar landing was so laden with history, but because it was not in their nature.

10:40 P.M.

By international agreement no nation could claim the moon, even one that managed to go there. That was reflected in the plaque on *Eagle*'s front leg, bearing the inscription, "We came in peace for all mankind." But it was the United States that had accomplished the feat, and NASA had decided that the Stars and Stripes would be raised during the moonwalk. Already, Armstrong had mounted the TV camera on a stand about sixty feet from *Eagle*, where it would broadcast the rest of the activities. Now he and Aldrin unfurled an American flag, stiffened with wire so that it would fly on an airless world, and struggled to plant it in the dust. As hard as they tried they could push the flagpole only six or eight inches into the ground. For a moment it seemed the flag would fall over in front of the worldwide audience, but at last the men managed to steady it; then they backed away.

Posing for Armstrong's camera, Aldrin looked at the banner and felt a swell of patriotic pride and humility come over him. He thought of the thousands of people who had helped get it to the moon, and the millions who must be watching him and Armstrong at this moment. He had an almost mystical sense of the unity of humankind, so strong that he felt as if he and Armstrong were not alone. Aldrin marveled at the paradox: No one had ever been farther from earth, and yet no one had ever been the object of more attention.

On the other end of that paradox, an estimated 600 million people, a fifth of the world's population, were indeed watching and listening, the largest audience for any single event in history. Across the United States it was a hot July evening, and in department stores keeping summer hours, and at "moonwalk parties," and in bars suddenly visited with an unaccustomed silence, a fantastic, high-tech stage play was unfolding on every working television set. It was a scene of utter stillness, except for two figures who bounded and leaped like snowmen brought to life, with *Eagle*'s spidery form as a backdrop. The picture seemed ghostly, as if it had lost some of its substance crossing the quarter-million-mile distance to earth. In all, the images from the moon were like a window on a dream.

In Nassau Bay, Joan Aldrin watched in quiet amazement. Earlier, when Buzz first appeared, she kicked her feet and blew kisses at the screen. She laughed and felt ready to cry at the same time. As she watched Neil and Buzz move jerkily about, she thought of a silent movie, or an old cartoon. It couldn't be real, she thought. And yet, here it was: men were walking on the moon, and one of them was her husband.

But walking, strictly speaking, was not the right word for what Buzz Aldrin was doing now as he took center stage to "evaluate the various paces that a person can use traveling on the lunar surface." Aldrin took off on a slow-motion jog, heading for the TV camera. Each step launched him into space, his body suddenly a projectile on a ballistic arc, suspended in mid stride, until he landed in a spray of powder. Time slowed; he was at the top of the arc waiting to come down. The mass of the backpack required him to anticipate changes in direction well in advance—just as Armstrong had done, flying the LM—and as he ran he kept his eyes out four or five steps ahead, watching for rocks or craters. Now he bounded across the moonscape on two feet. "So-called Kangaroo Hop does work, but it seems that your forward mobility is not quite as good as it is—as it is in the conventional—more conventional one foot after another." As he ran he looked like a science-fiction version of Eadweard Muybridge's turn-of-the-century movies of the human figure in motion. Aldrin fully expected that when he was back on earth the engineers would use the videotape to make careful measurements of his motions (much as Muybridge had done so long ago) to aid future moonwalkers. Instead they would be content simply to hear him tell about it.

Aldrin was in the middle of his experiments in locomotion when he heard McCandless say, "Neil and Buzz, the president of the United States is in his office now and would like to say a few words to you." Armstrong responded formally, "That would be an honor."

Aldrin suddenly felt his heart pound with anticipation. He was taken by surprise; later he would learn that Armstrong had known the president might call, but had not mentioned it. The two men faced the TV camera and stood still; moments later they heard Richard Nixon's voice:

"Hello, Neil and Buzz, I'm talking to you by telephone from the Oval Room at the White House. And this certainly has to be the most historic telephone call ever made from the White House. . . ." Throughout the moonwalk, Aldrin had the slightly discomforting sense of being a part of something bigger than himself. He noticed a kind of detachment from the event, as if he were watching it unfold before him, somehow beyond his control. And it seemed especially so in these moments, standing before the flag, listening to the President. As he listened he wondered what he might say in response; he decided he would not say anything.

"For one priceless moment, in the whole history of man, all the people on this earth are truly one. One in their pride in what you have done. And one in our prayers that you will return safely to earth."

There was a silence, and then Armstrong responded, "Thank you, Mr. President. It's a great honor and privilege for us to be here, representing not only the United States but men of peace of all nations . . . men with a vision for the future. . . ." To some listeners, Armstrong's voice seemed thick with emotion, as if he were on the verge of tears. Years later, Armstrong would say wryly that in answering the president with a few hundred million people listening he was probably concentrating on trying to say something that made sense.

"Thank you very much," Nixon said, "and all of us look forward to seeing you on the *Hornet* on Thursday."

"I look forward to that very much, sir," Aldrin said. The two men raised their gloved hands in salute, then turned away from the camera and went back to work.

10:55 P.M.

If Armstrong and Aldrin had climbed back into *Eagle* at that moment and blasted off to rejoin Collins, their mission would have been accomplished. The flag was up, and there was in Armstrong's pocket a small bag of the moon. He and Aldrin had already demonstrated that future explorers would be able to work in this alien environment. Anything from now on was frosting on the cake. On earth, this was the point at which many moonwalk parties started to break up. But in Houston a team of now quite frustrated geologists watched like children looking at a toy store on closed-circuit television. Armstrong would not let them down. He'd been preparing to collect samples when the president called; now, while Aldrin set about inspecting and photographing *Eagle*, Armstrong grabbed a long-handled aluminum scoop and began prospecting.

There wasn't much time. Armstrong was allowed only about ten minutes to gather enough rocks and soil to fill one of two aluminum sample containers, or "rock boxes." The geologists called this the "bulk sample" and it was intended to be a fairly quick grab. Later, he and Aldrin were to spend time carefully collecting and photographing the so-called documented sample. But Armstrong wasn't at all sure what the rest of the moonwalk would bring, and in case he didn't get a chance for the documented sample he wanted to select as varied a collection now as he could.

Aside from the stiffness of his suit, which fought almost every movement, there was one-sixth g to contend with: no matter how careful he was, when he lifted the scoop from the ground half the contents went sailing away like pieces of Styrofoam. Simply getting the sample over to the LM was a real challenge, but Armstrong persevered, and after several minutes the box was full. Now the task was to seal the box, which would preserve the samples in a lunar vacuum for passage to earth. But that proved to be a struggle, and when he was finished, the entire bulk sample operation had taken longer than planned.

Roughly an hour was left in the moonwalk, and there was still the work of setting up the two scientific experiments. Armstrong realized there would not be nearly enough time for all the exploring he wanted to do. Already he realized the moon was far more interesting than he'd expected. As he accompanied Aldrin on an inspection of *Eagle* his attention was constantly drawn to another interesting feature. Some of the small craters had at their centers bits of something shiny, with a beautiful metallic luster. He had no idea what they were, but they looked just like blebs of molten solder on a workshop table. He wished he still had the scoop in his hand. Here and there he saw what looked like transparent crystals lying in the dust; the biggest was the size of a walnut. He would have to come back for these things later, during the documented sample, if there was time.

11:39 P.M.

The Sea of Tranquillity was more rugged than Armstrong had expected—all bumps and hollows—and not an ideal place to set out a pair of scientific instruments. But about 50 feet from *Eagle* he and Aldrin managed to find a fairly level spot to deploy a solar-powered seismometer to detect moonquakes and an array of prisms that would serve as a reflector for a laser beam from earth, to help scientists measure the precise distance from the earth to the moon. But now, as he and Aldrin finished with the experiments, Bruce McCandless had good news: Mission control was offering a fifteen-minute extension. It wasn't much, Armstrong knew, but it would surely help. He'd already abandoned thoughts of inspecting the boulders to the north; like everything else on the moon, they were farther away than he'd thought. So was the giant crater he'd avoided during the descent; he'd fully expected to see its boulder-strewn rim to the east, but it was over the horizon. But there was the smaller crater he'd flown over just before touching down; it was definitely reachable. According to the timeline, he and Aldrin were to start the documented sample now, but Armstrong figured a quick reconnaissance of the crater would be more valuable than the one or two rocks they could pick up in the same amount of time. In any case, he'd already covered his bets with the bulk sample. Without a word to Houston, while Aldrin made his way back to *Eagle*, Armstrong took off running.

Long, loping strides carried Armstrong into the sun's glare to the edge of a pit that looked to be 80 feet across and 15 or 20 feet deep. *Eagle* was nearly 200 feet away, looking like a scale model. Armstrong wished he could climb down to the crater floor and pick up a piece of lunar bedrock, but he knew he mustn't try; if he got into trouble Aldrin's helping hands were a long way off. He clicked off a series of pictures, hoping to document on film what he had no time to investigate or even describe; then he headed back to the LM. Armstrong had been gone for only about three minutes, but it was the only real exploring he would have a chance to do.

The rest of the moonwalk passed in a rush of activity. There was no time for both men to collaborate on a documented sample. Instead, mission control put Aldrin to work hammering a metal tube into the ground to obtain a core sample, a task that proved even more difficult than planting the flag. Meanwhile, Armstrong scurried about with a pair of long-handled tongs, in search of rocks that would best represent this locale of the Sea of Tranquillity. He wished he had time to collect some of those mysterious, shiny blebs he had seen earlier, or one of the clear crystals he had spotted in the soil; he couldn't find any of those things now. Even as he worked, Bruce McCandless was telling him to press on; time was short. Time, Armstrong would later note, was a strange commodity on the moon. While their mission proceeded with an accuracy of minutes or seconds, he and Aldrin were on a world where a day lasts a month, where time seems to crawl. Looking at this landscape of craters, rocks, and dust he had the feeling that he was seeing a snapshot of a world in steady-state, that if he had been here a hundred thousand years ago or if he returned a million years from now he would see basically the same scene. But after two hours and thirty-one minutes, he had barely come to know the place. And yet, the knowledge that would emerge made these two and a half hours precious beyond measure.

Monday, July 21
12:12 A.M., Houston time

With a loud and welcome noise oxygen rushed into *Eagle*'s cabin. If the dire predictions about lunar dust catching fire when exposed to oxygen were true, Armstrong wryly mused, then his whole suit was going to burst into flames, because he was covered with grime. Neither man was surprised when nothing happened. But when they took off their helmets, they immediately noticed a pungent odor that reminded Armstrong of wet ashes in a fireplace and to Aldrin smelled just like spent gunpowder; it was the smell of moon dust.

Armstrong and Aldrin took pictures of each other's smiling, bearded faces, and of Tranquillity Base, now looking very much like an expedition site. Beyond the flag, standing in its frozen wave, was the television camera on its stand, and still farther away were the two scientific experiments, everything as still as a ghost town. The ground near the LM was covered with their footprints, each with its sharply chisled pattern of treads. There was in them something akin to immortality: those prints would remain fresh for perhaps a million years, subject only to the constant rain of micrometeorites from space.

After eating a late dinner they added to their expedition's legacy, opening the hatch once more to toss out the backpacks and a bag of other unneeded gear. When Armstrong learned from mission control that the seismometer had picked up the jolt of the backpacks hitting the surface he teased, "You can't get away with anything anymore, can you?" After fielding questions on the geology of the area, Armstrong and Aldrin prepared for a rest. Even though they had been up since 5:30 A.M., Houston time, and it was closing in on 3:30 in the morning, they were still keyed up, and Armstrong doubted they'd actually sleep. He also knew that a very full, very critical day lay ahead: the second half of John Kennedy's challenge had yet to be fulfilled.

III: "Before This Decade Is Out"

In lunar orbit

Every two hours, *Columbia* circled the moon in silent passage with its lone occupant, Mike Collins. The command module was close quarters for three men, but with Armstrong and Aldrin gone it was almost roomy. He'd folded up the center couch and stowed it underneath the left-hand seat, so that now there was a clear aisle from the side hatch to the lower equipment bay. The extra room would be needed in case there was a problem with the docking mechanism, forcing Armstrong and Aldrin to make an emergency space walk from the LM to the command module. For now, it gave Collins unaccustomed freedom. He scurried from his couch to the lower equipment bay and back again, checking systems, making navigation sightings, and attending to a host of housekeeping chores. Command module number 107, the spacecraft he had nursed through its checkout at Downey, California, and piloted across nearly a quarter of a million miles, was purring along without a single malfunction. Collins felt so confident that when one of the command module's cooling circuits grew too cold he chose not to follow mission control's advice to go through a lengthy malfunction procedure. Instead, in his solitude over the far side, some instinct prompted him to see if the machine might cure itself. He checked his switch settings and waited—and the coolant temperature promptly rose to normal. *Columbia* would give Collins no worry in his 22 hours alone. His anxieties were focused elsewhere, with the two men on the surface of the moon.

When his crewmates began the Powered Descent, Collins was listening in, his "cookbook" of rendezvous scenarios at the ready in case they had to abort. In the final minutes he heard them grapple with computer alarms and wondered how serious they might be. As *Eagle* flew onward Collins was, like everyone else, a spellbound listener. When it was over he heard Charlie Duke tell Armstrong and Aldrin there were smiling faces in mission control and all over the world, and Collins radioed, "Don't forget one in the command module."

But no one, not Armstrong and Aldrin nor anyone in mission control knew just where *Eagle* was. The location would be a helpful, though not essential, piece of information for his computer to have during tomorrow's rendezvous. It fell to Collins to try to find the LM on the surface, using the command module's 28-power sextant. This task was a little like looking down on Manhattan from a height of 69 miles, trying to spot a single Greyhound bus with a pair of binoculars—and all the while, moving at 3,700 miles per hour. It would have been pointless to try if not for the command module's computer, which could aim the sextant precisely at any feature under *Columbia*'s path and keep it fixed on the target, compensating for the command module's swift motion. Two hours after *Eagle* touched down, Houston had radioed up a set of coordinates and Collins was at the eyepiece. When he arrived over the landing site the sextant whirred into position, and Collins searched frantically for a glint of light. *Columbia*'s speed allowed him only about two minutes to search any given area within the long ellipse of Landing Site 2, and the sextant's field of view was so narrow that he could scan only one square mile at a time. Two frantic minutes later, Collins had come up empty. Each time he went around from the far side, mission control had a new set of coordinates for him to try, but on his map one guess was as much as 10 grid-squares away from the last. It didn't take Collins long to realize that no one had a handle on the problem. His search continued fruitlessly for the rest of his 22 solo hours.

If Collins could not see his crewmates on the surface, he could still hear them, via a special moon-earth-moon relay link set up by mission control. For some reason it went off sometime after the landing, leaving him feeling distinctly left out. As Armstrong and Aldrin prepared to go outside Collins asked once again to listen in, and mission control restored the relay. He'd hoped to be listening when Armstrong set foot on the moon, to finally hear his long-awaited words. But the timing didn't work out; Armstrong was just wriggling through *Eagle*'s hatch when *Columbia* slipped behind the far side; by the time he reappeared Armstrong and Aldrin were putting up the flag with 600 million people as witnesses. If, as he suspected, the TV commentators were describing him as a lonely man, they were wrong. He was savoring something as unique as a moonwalk: the experience of the solo moon voyager.

Collins moved through a continual succession of sun-drenched lunar day, soft earthlight, and unyielding blackness. For 48 minutes out of each orbit, from Loss of Signal to Acquisition of Signal, he knew a solitude unprecedented in human history. Before the flight he had been asked more times than he could count whether the thought of being alone in lunar orbit worried him. No, he had answered, I *like* being by myself. To a fighter pilot it was the essence of flying: alone in your craft, in control of your craft. It was nothing less than the purest form of freedom. There were lonelier places than the far side of the moon, and Collins had seen them; he'd taken an F-86 over the Greenland icecap in the dead of winter, hundreds of miles from rescue in the event of an emergency, and felt more anxiety than he did right now. His minutes over the far side were his quiet time, a respite from the constant chatter of mission control on the radio. He was anything but lonely.

Collins would later write of his far-side passages, "I am alone now, truly alone, and absolutely isolated from any known life. I am it. If a count were taken, the score would be three billion plus two over on the other side of the moon, and one plus God knows what on this side. I feel this powerfully—not as fear or loneliness—but as awareness, anticipation, satisfaction, confidence, almost exultation. I like the feeling." As if to capture it, during a quiet period, Collins took the movie camera, held it out at the end of his reach, and turned it on his own bearded face for a few moments, like a man sailing around the world alone.

Into the small hours of Monday morning, Mike Collins circled and worked and waited, waited for tomorrow's moment of truth, when the real success of this mission—getting Armstrong and Aldrin back—would hang in the balance. As his crewmates settled in for the night in *Eagle,* Collins was finishing up his own very long day, covering the windows, turning out the cabin lights, and thinking of his days as an altar boy in the National Cathedral in Washington, D.C., when he used to snuff out the altar candles after a service. Before the flight he'd thought he might have some misgivings about going to sleep if there were a problem onboard. But *Columbia* was working like a marvel, and he drifted easily into weightless slumber.

Monday, July 21
12:52 P.M., Houston time
Tranquillity Base

Armstrong and Aldrin stood side by side at *Eagle*'s controls, helmets and gloves locked in place. The first launch from another world was two minutes away. Armstrong had spent the night perched on the ascent engine cover, but had not slept at all. Aldrin, who had curled up on the floor, had managed only a few hours of fitful dozing. The problem was that the LM was no bedroom. Moonlight flooded the cabin through the translucent window shades; the instrument panel was aglow with luminescent switches and dials. And it was cold. The men hadn't anticipated that with the shades in place and all the systems turned off there would be no source of heat in the cabin. By the time they realized what was happening it was too late; there was no way to fix it. The oxygen flowing into their space suits only made them colder; they lay in their suits shivering. Hoping the cabin oxygen might be warmer, they took off their helmets; that only let in the high-pitched whine of the LM's coolant pumps. By the time Ron Evans gave the wakeup call they had given up trying to sleep. They were still too keyed up to be tired, and their thoughts centered on one thing: getting off the moon.

The ascent engine, hidden under the can-shaped cover behind them, had only 3,500 pounds of thrust, but that was enough to propel the ascent stage from the lunar surface into orbit. It was another of Apollo's engineering marvels, for it was even simpler in design than the Service Propulsion System engine. Like the SPS it burned hypergolics that ignite on contact, eliminating the need for an ignition system. Once the valves opened, fuel would flow into the combustion chamber, and the engine would fire. That would have to happen two minutes from now.

Before the flight Neil Armstrong had worried about those valves, and he'd suggested to the engineers that they consider replacing the electrical actuating system with a mechanical one that he or Aldrin could trigger by hand if the normal method failed. The engineers considered and rejected the idea; they had high confidence in the electrical system. And Armstrong knew there were several redundant ways to fire the engine; if necessary they could bypass the computer. One of those would work; there was no other way to think about it.

To Aldrin the thought of being stranded on the moon forever simply didn't exist. To conjure that dark thought would have been to go against the whole philosophy behind the mission: Everything had been stacked to ensure their survival. And now, as he and Armstrong followed the checklist through the final minutes of their launch countdown, Aldrin assumed that at zero the lunar stillness would yield to the power of a rocket come to life.

But inside *Columbia,* Mike Collins could not be so confident. Lunar orbit seemed remarkably safe compared to the spot Armstrong and Aldrin were in. Perched motionless on the surface with a single rocket engine to get them off, they belonged to the moon. The engine must work, and it must work long enough for *Eagle* to reach some kind of orbit. Collins was prepared to rescue them if they couldn't make it all the way up to 69 miles. He could drop down to 50,000 feet, but not too much lower than that; some of the lunar mountains were 20,000 or 30,000 feet high. And now, as he waited for liftoff, Collins could no longer push aside his darkest fears. "My secret terror for the last six months," he would later write, "has been leaving them on the moon and returning to earth alone; now I am within minutes of finding out the truth of the matter. If they fail to rise from the surface, or crash back into it, I am not going to commit suicide; I am coming home, forthwith, but I will be a marked man for life and I know it. Almost better not to have the option I enjoy." As Armstrong and Aldrin made their final preparations for leaving the moon, Collins listened and sweated out the most anxious moments of his career.

12:54 P.M.

With 45 seconds to go, Armstrong reminded Aldrin of the last actions they would take on the surface of the moon: "At five seconds I'm going to get ABORT STAGE and ENGINE ARM. And you're going to hit PROCEED."

"Right," Aldrin said.

"And, *that's all,*" Armstrong added wryly. If everything worked, he and Aldrin would just be along for the ride from the moment of liftoff until they reached orbit. Now Aldrin began the final countdown: "Nine, eight, seven, six, five, ABORT STAGE; ENGINE ARM, ASCENT; PROCEED—"

He pushed the button. For a fraction of a second there was stillness, and then, suddenly, there was a muffled bang of pyrotechnic bolts, and then a smooth, steady push, like a high-speed elevator, as *Eagle* ascended from the moon.

"We're off," Aldrin exulted. "Look at that stuff go all over the place." Outside, a spray of gold foil and debris from the descent stage flew away in all directions. The flag toppled to the dust. And the Sea of Tranquillity fell away as *Eagle,* ascending in an unreal quiet, headed for lunar orbit.

3:54 P.M.

For the first time in the flight Mike Collins let himself believe they were really going to pull it off. He had spent the past three and a half hours laboriously punching data into his computer, ready to take over if necessary, but his "cookbook" of emergency rendezvous procedures had gone unneeded. And there, in the eyepiece of his sextant, Collins could see a small black dot: *Eagle,* climbing up from the craters so steadily that it seemed to be riding up to him on rails. It was the happiest sight of the whole mission.

Collins floated back to his couch. Through the rendezvous window he could see *Eagle* slowly closing in, its thrusters spitting flame as Armstrong braked for the final approach. Even as he steered *Columbia* into position for the docking Collins raced from one window to the other, taking Hasselblad pictures and movies. And as the last steps of the dance were played out, Collins suddenly called out, "I got the earth coming up behind you—it's fantastic!" Collins captured the sight on film—*Eagle*, the moon, and the tiny blue and white world. He would always remember the moment: all of humanity captured in a single photograph, minus only himself, the photographer.

• • •

"Get ready for those million-dollar boxes," Armstrong yelled up the tunnel to Collins. As he handled the two weightless containers, snugly zipped into white cloth bags, he could feel the mass of the rocks inside them, and he was careful not to move too quickly as he passed them through to Collins. When they were safely stowed in *Columbia* he passed up a small white pouch and told Collins, "If you want to have a look at what the moon looks like, you can open that up and look. Don't open the bag, though." Collins unzipped the pouch and saw a small Teflon bag filled with black soot. Armstrong laughed, "You'd never have guessed, huh?"

"What was that bag?" Collins asked.

"The contingency sample," Armstrong said.

"Any rocks?"

"Yes, there's some rocks in it too. You can feel 'em, but you can't see 'em. They're covered with that—graphite."

And there was plenty of that "graphite" on their space suits; Armstrong thought they looked like chimney sweeps. Before he and Aldrin could rejoin Collins they tried to vacuum it off, not just to be tidy, but as part of the procedures to prevent "moon germs" from reaching earth. With no real vacuum cleaner they had to use a brush attached to one of the LM's air hoses; as they had suspected it turned out to be a vain attempt. Lunar grime had worked its way into the fabric; the suits would never be clean again.

When the cleanup was done and all the unneeded gear was piled in the tiny cabin, Armstrong and Aldrin exited to *Columbia* and closed the hatch. *Eagle* was now dead weight. Collins flipped a switch and the ascent stage drifted away. For his part, Collins was glad to get rid of the craft that had been nothing but a worry to him for six days, but in Armstrong and Aldrin he noticed a quiet sadness. Without a heat shield, there was no way to bring *Eagle* home; no museum would ever put it on display. It would linger in lunar orbit while mission control monitored each component's final hours of life. Long after it had become a dead ship *Eagle* would spiral downward until it crashed, blasting a modest new crater in the dust.

Tuesday, July 22

Just after midnight, Houston time, the SPS engine roared to life, and three minutes later Armstrong, Aldrin, and Collins were on their way home. The burn was as flawless as the one that had put them in lunar orbit two days before, and in the middle of shutting down systems and reading trajectory data off the computer, Mike Collins spoke for all of them: "Beautiful burn, SPS, I love you, you are a jewel! *Whoosh!*"

Thursday, July 24
1:16 P.M., Houston time
Aboard the carrier *Hornet*

Armstrong, Aldrin, and Collins stepped out of the helicopter onto the lower deck of the carrier *Hornet* looking like men from another world. Outfitted from head to toe in gray-colored Biological Isolation Garments, they peered through face masks clouded with perspiration and waved to a crowd of sailors and visiting dignitaries whom they saw only dimly. Despite rubbery legs unaccustomed to earth's gravity they made their way quickly to the open door of the silvery quarantine trailer. A NASA doctor followed them and closed the thick, windowed door behind them.

Three and a half days later they arrived in Houston sealed within the trailer as if they themselves were lunar samples. For the next two weeks they lived within the Lunar Receiving Laboratory, recounting all aspects of the flight in minute detail. Aldrin described the strange flashes he had seen on the way to and from the moon; no one had an explanation, but after a time Aldrin noticed that Armstrong seemed annoyed whenever the subject came up.

As the days passed the men joked about being jailed, and at the close of one debriefing session they called out to the engineers on the other side of the glass, "You know where to find us! We're not going anywhere!" In the off hours there were movies, like *Goodbye, Columbus,* and card games; Collins beat Armstrong repeatedly at gin rummy. They had company, including doctors, a NASA public affairs officer, and some unexpected arrivals—a few scientists who were accidentally exposed to lunar samples. And though the LRL wasn't a bad place—it had a bar, and an exercise room—time dragged. At the end of one debriefing, when asked, "Any other comments?" Collins said quietly, "I want out."

That would come on a hot August night, when the men would be released into a world changed, for at least a time, by what they had done. Armstrong hoped that the first lunar landing would inspire people to believe that seemingly impossible problems could be solved. As for its impact on their own lives, neither Armstrong nor his crewmates could guess what lay ahead. Until now, they hadn't had time to think about it. But there would be months on the banquet circuit, including a world tour; then each man would find his way into a new life.

For now, sitting in the LRL, Buzz Aldrin had time to ponder the significance of what he and his crewmates had been a part of. Back on the *Hornet,* they had watched videotapes of the news coverage of Apollo 11. There was Walter Cronkite, exulting at the lunar touchdown. Then awed crowds gathered around TV sets, witnessing the first footsteps on another world. For the first time, Aldrin sensed the emotional impact of the first lunar landing. For a man attuned to irony, here was something worth pondering: While the three of them were a quarter of a million miles away, much of humanity had been spellbound by a midsummer miracle. What a moment that must have been. Aldrin turned to Armstrong and said, "Neil, we missed the whole thing."

BOOK TWO

CHAPTER 6

Sailors on the Ocean of Storms

APOLLO 12

I: The Education of Alan Bean

When the men of Apollo 11 got out of quarantine, Richard Nixon had them and a few hundred guests out to Los Angeles to celebrate. By all accounts it was a hell of a party. Sometime well into the evening, one of the other astronauts in attendance—by now more than a little drunk—raised his glass. "Here's to the Apollo program," he said heartily. "It's all over." In a sense, he was right. John Kennedy's challenge to "land a man on the moon and return him safely to the earth"—the goal that had steered NASA for eight years—had been met. But was Apollo's mission over? Was the lunar landing simply an engineering demonstration, like the first flight across the Atlantic? There were those who thought so, even some within NASA. Kennedy had said nothing about a second lunar landing, or a third. You wouldn't ask Lindbergh to fly the Atlantic again, they said; why go back to the moon?

But in the summer of 1969 Apollo was only part of a much bigger question; the future of the space program as a whole was undecided. When Richard Nixon took office the agency had tried to gain early support for a manned space station in earth orbit. But the president deferred the matter by creating a Space Task Group to formulate recommendations for NASA's future. By July, NASA administrator Tom Paine was working with the STG on a plan that picked up where John Kennedy had left off. Paine, like his NASA colleagues, believed that there was something implicit in Kennedy's challenge beyond its words, that it was a call for the United States to become a spacefaring nation. Apollo had given the country the technology to go to other worlds; it had only to exploit that capability.

At NASA Headquarters, George Mueller and other planners had put together a far-reaching plan that Paine made even more ambitious in adapting it for the STG. The task group's timetable called for a twelve-man space station and a reusable space shuttle as early as 1975, depending on funding. By 1980 the station would have grown into a fifty-man space base; five years later there would be a hundred men in orbit. Meanwhile, there would be a base in lunar orbit by 1976, with a base on the lunar surface two years later. Then, as early as 1981, the first manned expedition to Mars would depart from earth orbit.

The plan was extraordinary—but it was not new. Almost twenty years earlier the same basic scenario had been mapped out in the pages of *Collier's* magazine by Wernher von Braun and other "space experts." At the time, von Braun was criticized for trying to sell the public a science-fiction vision of the future. In the summer of 1969, Paine was trying to turn von Braun's vision into reality. Like everyone at NASA, Paine hoped that the spectacular success of Apollo 11 would create a groundswell of support in the White House and in Congress to propel the space program onward and upward. Fortunately, Vice President Agnew, the chairman of the STG, was extremely enthusiastic, especially about the missions to Mars. The group's report would be ready for presentation to the White House in September. But already, there were signs it would not be well received. Since 1965, the year Apollo funding reached its peak, NASA's budget had steadily declined. Nixon staffers had told the agency that this trend would continue, and had indicated that even the first building block in the STG plan, the earth-orbit space station, would be a tough sell.

Meanwhile, Apollo moved on. Jim Webb had needed all his persuasive abilities to convince Congress and the Bureau of the Budget to pay for enough Saturn V's to fly missions through Apollo 20. He had done so on the premise that no one knew how many flights would be necessary to meet Kennedy's challenge. Now that the first lunar landing had been accomplished sooner than anyone expected, there was enough hardware to fly nine more lunar landings. Paine intended to make good on Webb's foresight. The handful of doubters aside, NASA had no intention of abandoning the moon. And if anyone wondered what would come from going back to the moon, they had only to be inside the windowless Lunar Receiving Laboratory at the Manned Spacecraft Center, on the evening of July 25, 1969.

In the LRL, five geologists dressed in white, hospital-style clothes and caps, stood around a vacuum chamber. A big, powerfully built technician reached into a pair of space-suit arms attached to one side of the chamber to open a silvery container about the size and shape of a large tackle box. Inside, preserved in a lunar vacuum, were pieces of the moon. For the better part of a decade, geologists had labored to extend the disciplines of their science to an alien world by remote observation. What they had managed to learn about the moon from their telescopes and then, the unmanned probes, testified to the power of human intelligence. But they had always longed for the moment when they could probe the moon with their state-of-the-art laboratory instruments—or simply with their own eyes.

After long moments of effort, the box was open. The technician removed a mesh covering, and a strip of foil that was part of a scientific experiment, and set them aside. Then he stepped away from the chamber and let the scientists look. With television carrying the event live, Harvard geologist Clifford Frondell peered into the chamber and blurted, "Holy shit! It looks like a bunch of burnt potatoes!" The rocks were so covered with charcoal-colored dust that the geologists couldn't tell anything about them. At this moment the curiosity that gripped them was not scientific, but human: these were pieces of the moon.

Two nights later the first sample to be cleaned was raised up inside the chamber while Frondell and the other scientists watched. Instantly they recognized it as a piece of basalt. Familiar minerals glinted under the chamber lights. In the days and months to come, the scientists would try to coax secrets of lunar history from the rocks and dust of the Sea of Tranquillity. And this was just the beginning. The moon had been transformed from a light in the sky to a world ripe for exploration, and the geologists had already picked out candidate landing sites for the landings to come, an ambitious and spectacular roster of missions. One team of astronauts would set down at the edge of a huge, winding canyon called Schröter's Valley, where perhaps a billion years of lunar history might be exposed, layer-cake-style, in its walls. Another would visit Marius Hills, a field of low domelike bumps which the geologists hoped might be small, ancient volcanoes. The grand finale, for Apollo 20, might be a descent into the yawning amphitheater of Copernicus crater. These missions would be scientific feasts, with three-day stays on the surface and lunar roving vehicles or perhaps one-man lunar flyers. Their goal, as ambitious as any in science, was to answer the most basic questions about earth's nearest neighbor: How did the moon evolve? Was it really the cold, geologically dead world it seemed to be? Where did the moon come from? Solving these mysteries could open doors to even grander ones, for on that pockmarked, lifeless mass might be preserved the earliest history of the solar system, long erased on earth. The scientists' fondest hope was that the moon would tell human beings how their own planet came to be.

By November 1969, three astronauts were ready to open the door to these explorations with the flight of Apollo 12. For one of them, the moon represented the end of a long, private journey of waiting and perseverance.

Friday, November 14, 1969
11:07 A.M., Eastern time
Pad 39-A, Kennedy Space Center, Florida

With 15 minutes to go, reality grabbed hold of Alan Bean. Sealed in his space suit, lying on his back inside the command module *Yankee Clipper,* he felt his heart suddenly pound with anticipation. The spacecraft was on internal power now, and in the right-hand seat Bean had just put the fuel cells on the line; now he scanned the gauges for the electrical system. Over in the left-hand couch,

Pete Conrad was talking to test conductor Skip Chauvin, nearing the end of the pre-launch checklist. Dick Gordon, in the middle couch, was making last-minute switch settings. It was very quiet; everything was just the way it had always been in the simulator. But this was no simulation. Silently, Bean told himself they were really going.

Already today, the crew of the second lunar landing had faced the threat of postponement because of the weather. Conrad, Gordon, and Bean had arrived at the pad under overcast skies. While they ran through the checklist a hard November rain lashed at the spacecraft atop its Saturn V booster. Rivulets of water found their way underneath the boost protective cover and danced across the command module windows. But the weather was erratic; the skies would seem to clear for a time and then gloom over again. Three and a half miles away, in the Launch Control Center, launch director Walter Kapryan deliberated and occasionally polled Houston for an opinion. Finally a report from an air force weather plane tipped the scales: ceilings acceptable, winds within limits, no lightning for 19 miles. They would go.

It was strange how unreal it had all seemed, up to now. For eight months Bean had been training to go to the moon and talking about it, and yet it was hard for him to believe it was really going to happen. Even this morning, as he sat down to a launch-day breakfast of steak and eggs with Conrad and Gordon, suited up, then rode out to the pad, it all felt like a practice run. But now, as cryogenic propellants flooded into the Saturn's tanks, Bean knew that the moment of fire and noise was almost upon him, and with it, the end of six long years of waiting. He had spent more time as a rookie than any astronaut in his group. Just why, Bean could only guess, and only in retrospect. But none of it mattered now.

Five minutes to go. "Pete, you guys have a good trip," radioed Skip Chauvin.

"Yes sir," said Conrad calmly, confidently. "Sure appreciate everything."

"Hold off the weather for five more, will you?" added Dick Gordon.

For Bean, every day of the past eight months had been an

adventure, and the best part of it was training with Conrad and Gordon. The three of them had a bond that went deeper than their mission. They shared a history that began at naval air stations in Florida and California, and the test pilot school at Patuxent River, Maryland. A seat on a lunar mission was as much as any rookie could ask for, but going with Conrad and Gordon was almost too good to be true.

"One minute," Chauvin radioed. Pete Conrad put out his gloved hand and for a moment the three men clasped hands.

Chauvin called, "Thirty seconds, Twelve."

"Roger," said Conrad.

At 14 seconds, Chauvin began to count down. Seconds later the three men heard the distant rumbling of the Saturn's engines igniting. As Chauvin kept counting down, the spacecraft was engulfed by vibration. There was some noise, but not too much—they could still hear Chauvin's voice, counting down to zero, even as the vibrations increased. Somewhere in the midst of that commotion, Apollo 12 left the earth.

"Liftoff," called Conrad, "the clock is running." By now the whole spacecraft was shaking, but not as bad, Conrad thought, as he had expected. He scanned the instruments as Dick Gordon called out the time—"Three seconds . . . six seconds . . ." It seemed to take forever for the Saturn to rise past the launch tower, and when it did, Conrad radioed cheerfully to Houston, "That's a lovely liftoff!"

In the center seat, Dick Gordon glanced through a tiny window in the boost protective cover. "Everything's looking great," he told Conrad. "Sky's gettin' lighter."

Conrad kept a hawkeye on the 8-ball, the artificial-horizon indicator that showed the spacecraft's orientation. It registered the Saturn's slow roll as the beast steered onto the proper heading. Everything was right on schedule. It was like a perfect simulation. "Roll's complete," he announced.

Suddenly, out of the corner of his eye, Conrad was aware of a bright flash of light outside. At the same instant, a long burst of static filled his ears. He felt the spacecraft tremble. He spoke rapidly over the intercom, his voice flat: "What the hell was that?"

Only Conrad had seen the flash, but suddenly all three men heard the sound of the Master Alarm ringing in their headsets. Conrad glanced over to the center panel and was startled to see almost every light that had anything to do with the electrical system glowing brightly. He could hardly believe his eyes. Simulation upon simulation, and he'd never seen this many warning lights at once. Over the intercom, he started to read them aloud:

"AC BUS 1 light, all the fuel cells—I just lost the platform!" Conrad saw the 8-ball tumble aimlessly. More warning lights came on, confirming that the command module's navigation platform was out of commission. Conrad called Houston. "Okay, we just lost the platform, gang. I don't know what happened here; we had everything in the world drop out." Only a slight strain could be heard in his voice.

In the right seat, Bean was mystified. He had seen so many electrical crises in the simulator that he could, just by looking at the pattern of warning lights, recognize any given malfunction almost immediately. But he'd never seen so many lights before. The thought flashed through his mind that something had severed the electrical connection between the command module and service module. Had the emergency detection system sensed some problem with the booster, triggering the escape tower and whisking them away? No, he'd have felt a tremendous jolt. And yet, something was seriously wrong with the electrical system. Or was it? The meters showed that the spacecraft was still drawing electrical power, although at a lower voltage than normal. Could he even trust the gauges? "There's nothing I can tell is wrong, Pete."

Conrad keyed his mike and read the list of warning lights to Houston. He hoped mission control would be able to tell more than he could.

In mission control, a young flight director named Gerry Griffin heard Pete Conrad describe, in one breath, the longest list of malfunctions he had ever heard. Griffin couldn't believe this was happening, not on his first mission as a flight director. He was certain he'd have to abort the flight. His voice calm, Griffin called on John Aaron, the bright, twenty-four-year-old flight controller in charge of the electrical system, and heard only silence. Griffin called again: "What do you see?" Aaron's problem was that aside from a bewildering maze of warning lights on his console, there wasn't anything *to* see. On his screen, telemetry from the spacecraft had been replaced by an undecipherable pattern of numbers. But Aaron had encountered the same problem during a practice run the year before, and he knew how to fix it. He could visualize the command module instrument panel, and the switch, labeled Signal Condition Equipment, that the astronauts had to set to recover the data. He said quickly and confidently, "Flight, try S-C-E to Aux."

This control was so obscure that Gerry Griffin had no idea what it was; neither did Capcom Jerry Carr as he radioed the request to Apollo 12. And neither did Pete Conrad. It was Al Bean who knew where to find the switch, and moments later, Aaron had his telemetry back. For some unknown reason, the spacecraft's fuel cells had been knocked off-line. Unless they could be reconnected, the command module would have only its batteries for power, and they were reserved for reentry. At the flight director's console, Gerry Griffin weighed the possibility of an abort.

Still the Saturn sped onward. Conrad glanced up and saw bright sunlight. They were above the clouds; they were heading in the right direction. Whatever had happened to them, it hadn't touched the booster or its guidance system. They'd make it into orbit, but Conrad was afraid he'd wind up with a dead spacecraft when they got there.

The g-meter had passed 3 and was still climbing. Now the intercom buzzed with excited voices, thick with the weight of acceleration: "Try the buses—get the buses back on the line. . . . I've lost the event timer. . . . Two minutes, EDS AUTO. . . ."

Bean heard Carr say, "Apollo 12, try and reset your fuel cells now." Bean was reluctant to do that without knowing what had gone wrong. Conrad said, "Wait for staging." They were coming up on that chaotic moment when the first stage would drop off and the second stage engines would ignite. Conrad and Gordon told Bean, "Hang on!"

The first stage fell away and the three men were slammed against their harnesses, just as they had expected. Seconds later, Bean revived the stunned fuel cells. Everything was back on line, apparently none the worse for wear. As the second stage did its work, Conrad offered a theory: "I'm not sure we didn't get hit by lightning."

Conrad was right. As it lifted off, the Saturn had trailed a column of flame and ionized gases which stretched all the way to the ground. Tearing through the rain clouds it became the world's longest lightning rod. Thirty-six seconds after liftoff, a bolt of electricity discharged right through Apollo 12 and onto the launch tower 6,000 feet below. The command module had shut itself off in response to the tremendous electrical surge. A second strike at 52 seconds, unnoticed by Conrad and his crew, had wiped out the navigation platform. Both lightning bolts were recorded by an automatic movie camera near the launch pad, as analysts would later discover.

Now, as Bean reconnected the fuel cells, the warning lights blinked out one by one. The platform was still out, but they would deal with that once they were in orbit. "Twelve, Houston," called Carr. "You're right smack-dab on the trajectory." The second stage continued its long, smooth push, and inside *Yankee Clipper* tension evaporated. Pete Conrad let out a giddy, high-pitched giggle, like a schoolboy who had just sneaked out of class without getting caught. Then he laughed, and Gordon and Bean laughed with him—

"Was that ever a sim they gave us!"

"There were so many lights up there I couldn't read them all!"

—and they laughed all the way into orbit.

• • •

To Pete Conrad, the launch of Apollo 12 exemplified one of the most important differences between flying airplanes and flying in space. Conrad could remember narrowly avoiding a midair collision with another plane when he was at Pax River; after the miss his heart pounded for several minutes. But during the lightning strike it was different. There was no terror, nor would there be at any other time in Conrad's four spaceflights, because things just didn't happen fast enough for that. And it proved one of spaceflight's emerging maxims: If you don't know what to do, don't do anything. Conrad never gave the abort handle a moment's thought.

But now, in orbit, Conrad worried that the lightning had damaged something, and that Houston was going to call off the mission. They'd ride around the earth a couple of times, and then they'd have to go home. The first order of business was to check out *Yankee Clipper*'s systems, and as soon as Apollo 12 reached orbit Conrad put his crew to work. Dick Gordon went down in the lower equipment bay, sweating through an effort to realign *Yankee Clipper*'s navigation platform with the stars. He kept talking to Bean, saying, "I don't see anything, Al," and for a while he wondered—what happened to the *stars*? Finally he realized he hadn't given his eyes a chance to adapt to darkness. Bean, consulting the star chart, told him he ought to be seeing the constellation Orion, and sure enough, there it was in the telescope, with brilliant Sirius nearby. Gordon made the alignment, with little time to spare.

During breaks in the work, Conrad and Gordon gave Bean a guided tour of earth orbit. "Here comes sunrise, Al," they said, or, "Hey, there's an island down there," and then in darkness over Africa, "Look down there; those are campfires," and he looked up from his checklist to witness the most amazing sights of his life. But for Bean, as well as his veteran crewmates, the lightning strike had been just as memorable.

"That'll give them something to write about tonight," said Conrad, thinking of the press. "I'll bet your wife, my wife, and Al's wife fainted dead away."

Gordon said, "I bet they did when you started calling out about eighteen lights."

"Every time I close my eyes," Conrad said, "all I see are those lights." For Conrad, Gordon, and Bean the launch of Apollo 12 had already become a sea story that they would tell again and again, for the rest of their lives.

To their relief, everything on the command module (and the lander, which was undergoing scrutiny in Houston via telemetry) seemed to be checking out perfectly. At 2 hours, 28 minutes Jerry

Carr radioed the message they'd been waiting for: "Apollo 12, the good word is you're Go for TLI."

"Whoop-de-do!" crowed Conrad. "We're ready! We didn't expect anything else."

What mission control did not tell Conrad was that they feared the lightning had damaged the pyrotechnic system used to deploy the command module's parachutes. Chris Kraft in Houston had conferred with other NASA managers at the Cape, and they decided to continue the mission. The rationale was simple: Conrad and his crew would be just as dead if the parachutes didn't work now as they would after coming back from the moon, 10 days from now.

Minutes later Conrad, Gordon, and Bean were strapped in, waiting for the third stage to relight. When it did, all three men felt it rumbling away, speeding them out of earth orbit. Conrad said, "Al Bean, you're on your way to the moon."

"Yeah," Bean replied. "Y'all can come along if you like."

• • •

There was a picture on the wall of Dick Gordon's study in Nassau Bay, a signed photograph of a young, flight-suited Pete Conrad—with a bit more hair—posing next to a Phantom jet. Conrad was about to leave Miramar Naval Air Station in San Diego to report to his new job as an astronaut. By the time that photograph was taken Conrad and Gordon had become best friends. They had shared the snap of the catapult on the deck of the carrier *Ranger*, where they had roomed together and flown missions with their squadron. Their friendship was cemented in weeks at sea and in the nightspots of San Diego. And in the summer of 1962, they had shared the hope of making it into the second astronaut selection. Gordon didn't find out that Pete had succeeded until the public announcement in the middle of September. Soon after, Gordon got a letter from Robert Gilruth, the director of the Manned Spacecraft Center, saying that he was sorry, but although his qualifications had been excellent . . .

Gordon's feelings could not have been more divided: joyful that

Pete had made it; heartbroken that he wasn't going with him. Maybe Pete was just being his optimistic self, or maybe he had a premonition of how things would someday turn out. But when he gave Gordon that portrait, he wrote on it, "To Dick: Until we serve together again."

It was no exaggeration to say that the high point of Dick Gordon's life were the three days he and Pete Conrad spent in orbit on Gemini 11. Even before launch, they had earned a reputation as the cockiest, not to mention the most fun-loving, team of astronauts ever to fly: Conrad, short, balding, and wisecracking, and Gordon, not much taller, but more formidable, with the rugged face and build of a boxer. Gordon's friends knew him as a man of strong emotions. He was such a ladies' man that Conrad—much to Gordon's dislike—took to calling him "Animal."

Unlike many astronauts, Dick Gordon had not had boyhood dreams of being a pilot. Growing up in Seattle, his family hard hit by the Great Depression, he had dreamed of the priesthood. In college he majored in chemistry but considered pursuing a professional baseball career. Finally he settled on dentistry. But soon after graduation his life was interrupted by the Korean war, and it was only then, in the navy, that Gordon found his calling. Gravitating toward the excitement of aviation, he was hooked after his first training flight. After the war, flying with a squadron that had the reputation of being the cockiest in the navy, Gordon won top honors for his precision in the maneuvers used to deliver nuclear weapons. Then it was on to the test pilot school at Pax River, where his friendship with Pete Conrad began. Gemini 11 solidified the bond between the two men, on duty and off. They thought alike; they flew alike. In the simulator or in a T-38, they anticipated each other's moves as if they were communicating by telepathy. Everything about them said, "I'm the best—see if you can get one up on me."

By the time Conrad and Gordon lifted off in November 1966, they already had enough stories to last a long time—and there were more waiting for them in earth orbit. An hour and a half after launch Conrad and Gordon shared the triumph of the first one-orbit rendezvous and docked with an Agena target rocket. A day later, Gordon climbed outside Gemini 11, trailing a thirty-foot umbilical and faced the most dangerous experience of his life. His assignment was to attach a Dacron tether from the Agena to the nose of the Gemini in preparation for a later experiment. But his body kept floating off the Gemini, and with no means of anchoring himself he worked so hard that he overloaded the cooling system in his suit. By the time he had attached the tether his heart rate had reached 180 beats per minute. His breathing was heavy; his eyes stung with perspiration. Inside Gemini 11, Conrad was gripped with concern for his partner. He knew that if Gordon became incapacitated there would be no way to pull the man, in his pressurized suit, back into the tiny cabin; Conrad would have no choice but to cut him loose and go back without him. Unwilling to order his copilot back inside, he waited, hoping Gordon would make that decision on his own. Conrad would remember those moments as the scariest he had ever experienced in a spaceflight, and that included the lightning strike on Apollo 12. But the flight also had its funny moments. On the third day Gordon took a second, much more leisurely space walk; this time he merely stood up in the hatch and shot astronomical photographs. During a lull, as Gemini 11 drifted in daylight over the Atlantic, Conrad fell asleep, his arms sticking out in front of him in his pressurized space suit. He awoke with a start and said to Gordon, "Hey, Dick, would you believe I fell asleep?" To which Gordon answered, "Huh? What?" He'd been asleep too, standing in the void, whizzing around the world at 17,000 miles an hour.

When Pete Conrad moved on to Apollo, there was no question that Gordon would come with him. Under other circumstances, the two men might have landed on the moon together, but Deke Slayton had a rule that command module pilots, like commanders, had to be veterans of space rendezvous. But the lunar module pilot's seat could be filled by a rookie, and Pete Conrad knew just who that rookie should be.

• • •

From the very beginning, Alan Bean saw the world differently from other test pilots. Long before he ever flew, as a boy in Fort Worth, Texas, he was captivated by the beauty of the airplane in motion. In the summer of 1956, the twenty-four-year-old Bean, thin and bright-eyed, with the well-toned body of a college gymnast, arrived at Jacksonville Naval Air Station. He was the youngest and newest member of attack squadron VA-44. In a group photo taken sometime afterward, two things about Bean stand out: his ear-to-ear grin and his tie, which is almost the only one that's perfectly straight. Even as a squadron pilot, Bean's sense of aesthetics was very much intact. He longed to fly the best looking planes, like the sleek F-8U fighter. Unlike the other pilots, who spent their free time fixing up old cars, Bean took night classes in oil painting.

Bean had always been a loner, but he felt close to the fliers in VA-44. It was a place where everyone had a nickname: there was Dancing Bear and Rockie Pie, Spanky, Dinky, Tiger Jack, and Von Du Quick. Sometimes they called Bean "Sarsaparilla" because he never touched a drink; sometimes they just called him "Beano." Once, on liberty in Paris, the pilots went drinking and decided to put Sarsaparilla to good use. While they drank inside, Bean did gymnastics outside. Before long he drew a crowd. His squadron mates then came out, one by one, to look for a date.

When it came to flying, Bean wasn't a "natural" like some of his squadron mates who seemed born to be great aviators. But whatever Bean lacked in innate talent he made up in effort. Through sheer determination he mastered the subtleties of precision bombing from a speeding jet and turned himself into one of the best weapons delivery pilots in the squadron. After a two-year tour at Pax River, Bean felt he'd reached the pinnacle of his profession. And while he shared the other fliers' drive to be the best, he was always aware that his outlook differed from theirs, even after he left Pax River for another attack squadron in Cecil Field, Florida. In 1962, during the Cuban Missile Crisis, there was talk of sending Bean and his colleagues into combat. The other pilots wanted it so bad they could taste it—after all, this was what they had been trained for. Bean would have gone in a minute, to defend the interests of his country, but he was hardly excited about it. There was no love of combat in his soul, only a love of flying.

By that time, Bean had his sights on a new goal: to become an astronaut. At the suggestion of one of his former instructors from Pax River, he applied for the second selection in 1962. He made it into the final group of thirty-five pilots, but no further; the instructor, whose name was Pete Conrad, went all the way. The following year Bean applied again; this time he made it.

At NASA, Bean was surprised at how different it felt to be in the Astronaut Office than it had in the squadron or at Pax River. Camaraderie was overshadowed by competition for flights. And for some of the rookies, Al Shepard was an ominous presence. To Bean, Shepard seemed like a tiger shark swimming around in a tank full of fish. He didn't go after the big fish; he liked to take a bite out of a minnow every now and then. For quite a while Bean felt like a minnow. And once again Bean felt he saw things differently from the other pilots. The Old Heads' animosity for the doctors, for example, made no sense to him. So what if they didn't like Chuck Berry; did that mean they shouldn't cooperate with him? Another day it might be some PR requirement—the television people would want to film astronauts in training, and the Old Heads would grumble. Bean thought, "What's the problem? We won't even know they're there." And he didn't hesitate to say so in front of Shepard and Slayton. Nor did Bean share the Old Heads' view of the non-test pilots in the office. He gravitated to Anders, Cunningham, and Schweickart precisely because they *were* different. They thought about things other than flying and work. They examined their lives. Rusty Schweickart taught him to understand points of view very different than the ones he had learned in the military. He marveled at Bill Anders's intellect, not only about science but about the ways of the world. And he admired outspoken Walt Cunningham for his directness. For Bean, being with the three of them was an education.

But Bean wasn't learning the right lessons about being an astronaut. He followed Bill Anders's motto (Work hard and someone will notice), and when he was named as backup commander on one of the last Gemini missions, he thought his day had finally come. Never mind that Gemini would be over before he had a chance to fly; he was sure Slayton and Shepard had recognized what he could do and that they were grooming him for Apollo. But in the fall of 1966, he got his assignment: astronaut representative on the Apollo

Applications Project, the space station planned for earth orbit in the 1970s. Bean's heart sank as he saw his chances for a lunar mission evaporate. While he was flying a desk, Anders, Cunningham, and Schweickart were getting ready to fly in space, and maybe even to walk on the moon.

• • •

In the fall of 1966, when Deke Slayton asked Pete Conrad to pick a lunar module pilot, he requested his old student from Pax River, Alan Bean. Slayton told Conrad that Bean was unavailable; he was assigned to Apollo Applications. Why Bean was put on AAP, Conrad had no idea, but in his own mind it was a bad deal. He couldn't help but wonder whether somebody had put Bean there to get him out of the way. It wouldn't have been the first time Bean had gotten in trouble, and Conrad had been around for that too. In 1960, Bean arrived at Pax River and became one of Conrad's students. Conrad had met the younger man four years earlier in Jacksonville, where they flew in sister squadrons. At Pax, Bean was quieter and more serious than many of the other pilots Conrad taught. He wasn't the best aviator in his class, but he was near the top, and definitely one of the brighter students. The thing Conrad noticed most about Bean was his persistence. Once he latched onto a problem he was absolutely tenacious about finding a solution, and he wouldn't let go until somebody listened. Bean also tended to speak his mind, a little too much for his superiors. When it came time to give the graduates their assignments, Conrad found out that the brass were planning to send Bean to Electronics Test, the boondocks of test flight. Conrad went to the school commander and argued Bean's case for better duty, and his intervention won Bean a place in Service Test, where pilots wrung out new airplanes before clearing them for squadron duty. For a pilot of Bean's caliber it was a far better place to be.

But in 1966 there was nothing Conrad could do to rescue his former student. Slayton wasn't letting him have Bean, and that was that. So Conrad chose another astronaut he knew, a big, friendly bear of a marine named C. C. Williams, who had also been one of his students at Pax. Williams stood just over six feet tall. During the third astronaut selection he made it under the height limit—just barely—by spending the night before the physical jumping up and down to compress his spine. Williams brought a gentle, self-effacing presence to the astronaut corps. He would set his jaw and utter with mock seriousness, "I'm a marine. I'm a trained killer. . . ." As the first bachelor in the Astronaut Office he was the envy of his colleagues. This distinction ended when he married a former water-skiing performer from Florida's Cypress Gardens. Williams fit in well with Conrad and Gordon, and the three of them leaped into training as the backup crew for the first manned lunar module flight.

• • •

Meanwhile, Alan Bean toiled in the backwaters of Apollo Applications, oblivious to Conrad's efforts. In the navy, he had learned to meet adversity with a positive attitude. His commander in Service Test at Pax River would always remember Bean for taking an unpleasant job and doing it better than any junior officer he'd ever seen. And it turned out that when NASA put out a call for new astronauts, the man was on the navy's astronaut selection committee. He made sure Bean was on the list. Bean made up his mind that he would do the best job on AAP that anyone had ever seen.

But as time passed, Bean began to see that doing a great job wasn't enough if the right people didn't know about it. Slayton and Shepard wouldn't notice him unless he learned to promote himself. And he'd have to make it look unintentional; that would take some work. Some of the Fourteen were masters at it; Bean would study their technique. And he would learn when to keep his mouth shut. Playing politics was abhorrent to him, but he had no choice.

But these realizations had come a few years too late. No one knows how long he would have stayed there in what Pete Conrad called Tomorrowland, if fate had not intervened. On October 5, 1967, C. C. Williams was flying a T-38 from the Cape to Mobile, Alabama, to see his father, who was dying of cancer. Near Tallahassee the airplane suddenly went into an uncontrollable aileron roll. The jet lost all its lift and became a ballistic projectile, accelerating earthward at a horrendous rate. Williams followed the procedures for emergency ejection, but the T-38 was going so fast that he was already far too low for his parachute to save him.

Not long afterward, Bean was at Ellington and ran into Pete Conrad on the flight line. Since coming to NASA, Bean hadn't really had much contact with Conrad. They were in different astronaut groups, and aside from the pilots' meetings, there had been only occasional, brief encounters. Now his old teacher said, "Have you got a minute, Al?" Conrad had told Slayton he wanted Bean as his new lunar module pilot; Slayton had agreed. Bean couldn't believe what he was hearing. Deliverance had come.

• • •

They called him Beano, and he fit right in. His quiet seriousness complemented Conrad and Gordon's bravado. He didn't race cars, as they did. When it came to training, Conrad and Gordon had a certain restlessness about them—Let's get on with it, that was their style—but Bean delved into the writing of checklists as if it were his only profession. He sweated the details Conrad and Gordon hated. Even his eating habits were single-minded; he ate spaghetti almost every night of the week. They gave him all the razzing due him as the rookie on the crew—but they respected him, and each other.

They soon became a fixture at the Cape, where they were backing up Jim McDivitt's crew for the first LM mission. Now that he was on a crew, Bean realized that his whole outlook had changed. The every-man-for-himself undercurrent of the Astronaut Office no longer applied to the three of them. They were in this together, and they were going to fly the best mission ever flown. Together, they had more camaraderie than just about any Apollo crew.

That didn't mean the competition was over; to the contrary, it gave the men a chance to fuel the friendly inter-service rivalry that permeated the astronaut corps. They were the first all-navy Apollo crew, and Conrad was always glad for a chance to get some mileage out of it. McDivitt's crew was all air force, and Conrad would say things like, "Well, Jim, the navy's always glad to help out the air force . . ."

Then there was that fateful day in 1968 when, in the plan to send Apollo 8 to the moon, McDivitt's crew swapped places in line with Frank Borman's. Bean still remembered how Pete and Dick reacted. It was a bad deal, they said. Until then, it looked as if they might have a shot at the first lunar landing, but this was going to screw all that up. And Bean was just so happy to be on a flight—any flight—that it never occurred to him to think about being first or second to land on the moon.

By the fall of 1969, now training for the second lunar landing, Conrad, Gordon, and Bean had been at the Cape so long they seemed like permanent residents. Conrad had gotten to know a Cocoa Beach car dealer named Jim Rathmann, a former Indianapolis 500 winner who made fast friends with the astronauts from the Original 7 on down. Thanks to Rathmann, General Motors gave the astronauts great deals on sports cars, and sometime during the training for Apollo 9, Conrad had worked a deal for three matching gold Corvettes, and had Rathmann's shop customize them with futuristic black trim and a small "CDR" for mission commander Conrad, "CMP" for command module pilot Gordon, and "LMP" for lunar module pilot Bean. The security guards got used to seeing the three Corvettes tool out the gate of Patrick Air Force Base in the early evening and head down route A-1A for Cocoa Beach. They were like a bunch of squadron buddies. They shared each other's lives as well as work. In the morning, before a simulator run, the conversations sounded like the ready room at the squadron—"Let me tell you what happened last night!" You wouldn't find the backup crew, Dave Scott, Al Worden, and Jim Irwin, telling jokes about adventures from the night before when they came in for work. And it was that way because Conrad set the tone. Conrad was the glue.

In the simulator, Conrad was spectacular. The instructors would throw everything in the book at him, and there wasn't anyone who could solve problems faster, or react quicker. If Conrad had one weakness, it was his language. When decorum was required, Conrad carried himself with all the poise of a Princeton-educated navy officer, but most of the time he raised colorful to an art form. In the simulator, he whistled and hummed and cracked his chewing gum so loudly that the instructors winced under their headsets. And when the malfunctions got thick, he swore like a sailor. The instructors smiled and shook their heads—"What's this guy gonna do during the *flight?*"—and in the next minute they'd break up laughing, because they'd hear Bean's quiet voice in the background: "Yep, that's astronaut talk. A-OK. I gotta learn that."

The instructors' worries were nothing compared to those of the NASA Public Affairs people. When it looked as if Conrad might have the first landing in his pocket, they chuckled at the thought of this little wisecracking, balding guy with the gap-toothed grin as the first man on the moon. Now, faced with the prospect of Pete Conrad on the airwaves, they were decidedly uneasy. They had barely recovered from the criticism that followed Gene Cernan's profanity during Apollo 10. (Never mind that Cernan had been in a life-or-death crisis a quarter of a million miles from earth; the letters came anyway.) And they weren't the only ones who were worried. One day during training Bean said to his commander, "You know, you can't talk like that on the flight," and Conrad replied that it was people like himself, who swore a lot on the ground, who never slip up over the air during a flight. Furthermore, he told Bean, people like you, who hardly ever swear, are the ones who do. That was the end of it, but Conrad had to laugh at the thought that Bean would worry about that until the end of the mission.

• • •

Pete Conrad had been disappointed not to fly the first lunar landing, and there were plenty of NASA people, at the Cape and in Houston, who were surprised when he didn't get it. He was in mission control when Neil Armstrong and Buzz Aldrin landed on the moon. And he was there when everyone, from the NASA managers in the control center to the geologists in a back room to Mike Collins in lunar orbit, were trying to figure out *where* Armstrong and Aldrin had landed. Sam Phillips turned to Bill Tindall and said, "Next time, I want a pinpoint landing."

Phillips's request was understandable; there was no point in having the geologists painstakingly choose a landing site when there was no certainty of being able to get there. But Tindall thought it was impossible. After all, *Eagle* had landed four miles away from its aim point. The trajectory people had identified several sources for the error, and had figured out how to prevent them from recurring, but one unknown persisted, namely the moon's lumpy gravity field. Even now, no one knew exactly what mascons would do to the lunar module's path, so predicting its precise trajectory ahead of time was impossible. But when Tindall convened the trajectory experts, a young mathematician named Emil Schiesser made a breakthrough. The key was the Doppler effect, the apparent shift in frequency of light waves or sound waves emitted from a moving object as detected by a stationary observer. You can experience the Doppler effect standing next to a highway: the horn of a passing car seems to rise in pitch as it speeds toward you, then fall as it moves away. The same phenomenon changes the apparent frequency of radio waves received from a spacecraft moving toward or away from the tracking stations on earth. Tiny Doppler shifts were already being used by controllers to analyze the trajectories of Apollo spacecraft during their lunar voyages.

Radio signals from a LM in lunar orbit, Schiesser pointed out, have a predictable pattern of Doppler shift. The effect is strongest when the lander is flying over the edge of the moon as seen from earth, and weakest when it is over the geographic center of the near side. If planners could predict the pattern of Doppler shifts, they could compare that information with the actual shifts they detected. The differences would in turn reveal whether the lunar module was off course, and by how much.

Schiesser's idea was brilliantly elegant, but it left the problem of how to give that information to the astronauts in a form that would be easy to feed into the LM's guidance computer. Soon that answer too emerged in Tindall's meetings. The solution was to fool the computer into thinking the landing point itself had moved, rather than the lander. That change required entering only a single number.

At the same time, mission planners deliberated on where to send Apollo 12. They could simply have picked out a specific crater, but planning coordinator Jack Sevier had a better idea. The unmanned Surveyor 3 probe was perched on the vast lava plain called the Ocean of Storms, where it had landed in April 1967. The geologists had already identified the Ocean of Storms as one of their candidate targets for the second landing; they suspected its rocks would be younger than, and perhaps chemically different from, those in the Sea of Tranquillity. Now Surveyor 3 became the target for the first pinpoint lunar landing. It was a bold decision to commit the system to such a visible measure of success or failure; if Conrad and Bean missed, everyone would know it. But that was exactly the point: This goal would drive NASA to achieve what had first seemed impossible.

Conrad and Bean were to spend 31½ hours on the moon, some 10 hours longer than Armstrong and Aldrin. During that time they would take two moonwalks, each lasting about 3½ hours. Most of the first excursion would be devoted to setting out an autonomous scientific station called the Apollo Lunar Surface Experiment Package, or ALSEP. The second walk would be an extended geologic traverse, culminating with a visit to Surveyor 3. Conrad and Bean would cut off pieces of the probe so the engineers could see what had happened to it during its thirty-one months on the lunar surface.

Conrad was thrilled with his mission. If he couldn't make the first landing, this was the next best thing. He and Bean weren't simply going to get down in one piece, grab a few rocks, and take off. They had the first lunar surface operations plan, a timeline packed with good, useful work for the scientists. And if they really did manage to land next to the Surveyor, they would open the way for the pinpoint landings the geologists wanted.

As the months of training wore on, the simulator instructors at the Cape considered Conrad, Gordon, and Bean one of the sharpest, most competent teams that had ever trained for space. Conrad joked that they should solve NASA's crew-selection worries by volunteering to fly every mission from then on, rotating seats on each flight: next time it would be Gordon-Bean-Conrad, then Bean-Conrad-Gordon. . . .

By the day before launch, Conrad felt ready. The mission he had worked toward for seven long years was finally upon him. That night, he and Gordon and Bean were joined for dinner in the crew quarters by Tom Paine. Later, in the parking lot, Paine made a promise to Conrad: If some problem came up and he didn't get to land, Paine said, he would put Conrad and his crew on the very next mission—so they shouldn't do anything rash. Conrad thanked Paine for his kind offer and said good-bye. He was halfway up the stairs when he realized Paine had made the same promise to Neil Armstrong.

II: Shore Leave

Monday, November 17
11:45 P.M., Houston time
3 days, 13 hours, 23 minutes Mission Elapsed Time

"Twenty-four hours from now, Beano, we're on our way down, pal. That's when I get nervous! Find that little muthuh! And I'll land it right side up!" Pete Conrad was anxious, and he wasn't trying to hide it. But right now, as he and his crew settled in for their first day in lunar orbit, the atmosphere inside *Yankee Clipper* belied his tension. In the background, Frank Sinatra crooned "The Girl from Ipanema" on the tape recorder. The conversation was decidedly relaxed, unlike the sparse exchanges of Armstrong and his crew. It was the banter of three friends on vacation together in the wilderness.

"If they made up a Hollywood movie, just like this, you wouldn't believe it," said Bean.

"What do you mean?" asked Conrad.

"Listening to this music on the back side of the moon."

Gordon countered, "You got something against music?"

"Nobody would buy it," Bean said. "This is cornball—you gotta be *hard* out there."

"Say, the biggest thing I missed on Gemini 5 was not having any music . . ." said Conrad. What he would have given for that tape player then, cooped up in that tiny cabin for 8 days! For Apollo 12 he'd brought along some of his favorite tunes, mostly country-and-western stuff (like Bob Wills and the Texas Playboys' version of

"San Antonio Rose") that sent Gordon and Bean into hiding. Bean's tape was Top-40, which they all liked well enough—especially the bubble-gum hit called "Sugar Sugar." When it came on during the trip out from earth, the three of them would hold onto the struts in the command module and bounce weightlessly to the beat, dancing their way to the moon. Conrad would say it emphatically in the debriefing: people have got to have some entertainment on these flights; you can't just look out the window.

Conrad was happy with the way things were going. Since the lightning strike, they hadn't had a single problem worth mentioning. That was the best way to start a flight—get all the trouble out of the way early. And next to Gemini, the command module was like a hotel. There was warm water for rehydrating the food and the coffee; there was toothpaste, and shaving. By now, Conrad was looking forward to getting cleaned up. "Hey, I'm gonna take a bath tonight," he announced. "Take a bath, shave, get all cleaned up, good night's sleep—" He turned to Bean. "You got anything else to do tomorrow? All right, that's what we'll do then. We'll go for a little lunar landing, how's that? Unless you got something better in mind—a little surfing at the beach, or something."

"Hell, yes," Bean said.

Gordon joined in. "How about the back-side sand? Go play in the sand on the back side."

"*Yeah,*" Conrad laughed. "Let's take the LM down and land on the back side. Wouldn't that shake 'em up?"

But behind the banter, Conrad was ever mindful of what was coming up. No doubt about it, the stakes had changed with this mission. Just getting the LM down in one piece wasn't good enough. The real test was finding Surveyor 3.

According to the scientists, the Surveyor was sitting on the slopes of a worn, old crater 656 feet across. Immediately around it were several other craters, a bit smaller and sharper. It was this clump of craters that Pete Conrad would have to locate among thousands of others when his lunar module pitched over for its descent to the moon. Seen from the east, the view Conrad would have during his approach, the craters looked like the outline of a snowman, with the Surveyor sitting in the pit of its fat belly; he simply named it the Surveyor crater. He named the other craters appropriately: Head, Left Foot, Right Foot. On the photographs, there appeared to be a fairly smooth area on the near-right side of the Surveyor crater, and that was where Conrad would try to land. If that place—which became known as Pete's Parking Lot—turned out to be unsuitable, or if Conrad couldn't get to it, he must try to land close enough to the probe for him and Bean to reach it on foot without difficulty.

Before the flight, Chris Kraft had told Conrad not to stress the Surveyor when he talked to the press; otherwise if he landed off target the press would say the mission had failed, even if he and Bean accomplished all the other objectives. But to Conrad, the bottom line was that if he and Bean didn't find Surveyor 3, all their planning would be for naught. The geologic traverse was designed around the various craters of the Snowman. And the future exploration of the moon depended on the pinpoint landing capability. It was up to mission control to put his lunar module on target, and it was up to Conrad to land.

Tuesday, November 18
4:00 P.M., Houston time
4 days, 5 hours, 38 minutes Mission Elapsed Time

"I'm about as jumpy as I can be this morning." For Alan Bean, the anticipation was over; now it was time for him to do the thing he had been training to do all these months. Stick to the checklist. Be steady. Don't throw the wrong switch. Bean couldn't deny it; this was the big day.

"Oh, you noticed?" Conrad said wryly, as the two men ate breakfast over the lunar far side. Neil Armstrong had never said a word about anxiety, but Conrad wasn't Armstrong. "I just hope we find the old Snowman! Then I hope we find a place to land! Then I hope I can set it down all right!" Between bites of Canadian bacon, Conrad confessed, "It's driving me buggy. I just don't know what I'm gonna see when I pitch over. You know"—Conrad laughed at the thought of his own helplessness—"I'm either gonna say, *Aaaaaaa! There it is!* or I'm gonna say, *Freeze it, I don't recognize nothin'!*"

"If you don't recognize a thing," Bean offered, "just tell me. I'll look out my side, and you look at the computer for a few seconds, and let me see if I see anything out there."

Below them, the sun cast long shadows over a battlefield of craters. The near side of the moon didn't impress Pete Conrad, but the far side, with its enormous bumps and hollows—that did amaze him. He looked down at what seemed to be a string of small volcanoes. "That's fantastic," he exclaimed to Bean. Then he thought for a moment about where they were, and he said quietly, "How'd we ever get here anyway?" And he and Bean laughed.

• • •

It had always amazed Alan Bean that in all the time they had spent training together, Dick Gordon had never once showed the disappointment he had to feel every time he heard Bean introduced as the guy who was going to the moon with Pete Conrad. There was never a trace of sarcasm from Gordon, never an ironic remark. He had done everything to make Bean feel welcome on the crew. Bean knew that if the situation had been reversed he could not have handled it as well. Now, Gordon was suspended at the end of the docking tunnel between *Yankee Clipper* and *Intrepid,* looking down at Conrad and Bean. It was a moment for good-byes, and yet none were said.

"I guess I've gotta close the hatch now," Gordon said. And he looked at his two friends for a moment, and then sealed the tunnel. Bean wondered whether he would ever see Gordon again.

Minutes later Gordon flipped the switch to release them, and for a few minutes *Yankee Clipper* and *Intrepid* flew in formation while Gordon took pictures. Then, as Conrad and Bean watched, Gordon pulled away.

Gordon kept the spindly, four-legged craft in sight through the 28-power sextant. When Conrad and Bean fired their descent engine to drop out of lunar orbit Gordon was looking straight up *Intrepid's* engine bell. The spacecraft carrying his friends seemed to become a glowing ball, like the view into a jet engine climbing on afterburner. The glow lasted just under half a minute, then died.

Wednesday, November 19
12:42 A.M., Houston time
4 days, 14 hours, 20 minutes Mission Elapsed Time

If the lunar module belonged to any astronaut, it was Pete Conrad's. He knew everything about the LM cabin because he'd helped to design it. Back in 1963 he'd been in on some of the major changes—taking out the seats and letting the pilots fly standing up; replacing the round, forward hatch with a bigger, square one that could accommodate an astronaut wearing a bulky backpack; adding a small rendezvous window above the commander's head (*that* had been a battle; they'd tried to tell him, the pilot, that he didn't need one), and the list went on. As *Intrepid*'s descent engine ignited 50,000 feet above the moon, Conrad suddenly flashed back to five years earlier, when he was visiting the Grumman plant on Long Island, and standing in a mockup of the LM, then known as "the bug." The instrument panels were nothing but drawings pasted onto plywood, but Conrad had stood there and imagined himself descending to the surface of the moon. Now it was happening. *Intrepid*'s engine came silently to life, right on schedule. Bean welcomed the acceleration; after five days of weightlessness it felt good to be *standing* again. Conrad scanned the gauges and *Intrepid* continued its long ride down.

Ever since the pinpoint landing became his mission last summer, Conrad had wondered whether Kraft's people really could pull it off. But they were confident—so confident, in fact, that Conrad sometimes had trouble taking them seriously. One day Conrad was talking to trajectory specialist Dave Reed about his landing point. Reed asked, "Where do you want me to put you?" Conrad doubted it would make much difference what he told Reed—after all, Armstrong and Aldrin had landed four *miles* off target—but Reed was talking like a travel agent making an airline reservation. Conrad went along and picked a spot, with due consideration to sun angle, traverse distance, and the like. Then after a few simulated landings, he changed his mind and went back to Reed for a new spot a few hundred feet to one side. Without batting an eye, Reed set about figuring out the necessary changes to the software, and Conrad couldn't believe the man was serious. He blurted, "You can't hit it anyhow! Target me for the center of the Surveyor crater." Reed answered, "You got it, babe." But Conrad would believe it when he saw it.

12:51 A.M.

At 8,000 feet, with pitchover almost upon them, Conrad leaned forward in his space suit, straining to glimpse the Snowman. He told Bean, "I think I see my crater. . . . I'm not sure. . . ." Then, right on schedule at 7,000 feet, *Intrepid* pitched forward for the final descent. Suddenly Conrad beheld a ghostly, black-and-white panorama, seemingly too stark to be real—ten thousand shadows inside ten thousand craters. Conrad felt a stab of unease: he couldn't find a single feature he recognized. But when he sighted along the 40 degree mark on his window, suddenly—"Hey, *there it is.* There it is! Son of a gun, *right down the middle of the road!*" Far in the distance, Conrad could see the Snowman, tiny, almost lost in the sea of craters, *Intrepid* was headed for the very center of the Surveyor crater. Even as Bean tried to read him LPD angles, Conrad was too excited to listen—"Look out there! I can't believe it! Fantastic!" But the targeting was almost too good. Conrad wasn't about to land in the middle of the Surveyor crater. As the computer flew *Intrepid* down, Conrad told it to shift the aim point short and to the right of the crater. He would give Pete's Parking Lot a try after all. He heard Jerry Carr's voice: "*Intrepid,* Houston. Go for landing."

A thousand feet above the moon, Bean stole a glance out his window and saw a bright field of craters, with the Surveyor crater as big as life, directly ahead. They were coming in at great speed. Up to now, the landing had seemed like a simulation, but this was almost more than he could stand to look at. It was amazing, even frightening. Quickly, he turned his gaze back to the instruments. His voice was calm as he told Conrad, "Looks good out there, babe; looks good."

But Pete's Parking Lot didn't look very good to Conrad; it looked more like Pete's battlefield. Again he shifted the landing point, this time further downrange.

"You're at five hundred and thirty feet, Pete," called Bean. "You're all right!" But they were still coming in like a bullet. Conrad wanted time to slow down and look for a landing spot, and he could afford to; they were fat on fuel. As *Intrepid* descended through 400 feet, Conrad took over. He pitched the craft back to kill their forward speed. *Intrepid* slowed, but not as soon as Conrad wanted. He saw the Surveyor crater drift past and said, "Gosh, I flew by it." Conrad looked out to his left, just beyond the crater's northwest rim, and saw where he wanted to land: a smooth area between Surveyor crater and Head crater. He banked *Intrepid* hard to the left. The craft responded with the same sluggishness as the LLTV at Ellington. Bean, who had never flown the landing trainer, was surprised to see the 8-ball in front of him tilt so sharply. He wondered what Conrad was doing. He said in a calm Texas drawl, "Hey, you're really maneuverin' around."

"Yep," Conrad answered, too busy to say anything more.

"Come on down, Pete," Bean coached, his eye on the rate-of-descent gauge. Bean knew that Conrad must not descend too slowly while they were this high up; it would cost too much fuel. "Two hundred feet, coming down at three," Bean said. "Need to come on down."

"Okay," Conrad answered. In the simulator, he had always waited until they were down to 100 feet or so before arresting the last bit of forward speed and starting the final, vertical descent. But some instinct now prompted him to level off early. It was the best decision he made, because within moments, his view of the surface began to blur. *Intrepid* was kicking up a tremendous amount of dust, far more than Armstrong and Aldrin's movies had suggested. The dust shot away from him in bright streaks, rushing to the horizon.

"Ninety-six feet, coming down at six. Slow down the descent rate," Bean called quickly, urgently.

The dust blanket thickened. Conrad could see nothing but streaks, with a few rocks sticking up here and there. He had no idea whether there were any craters directly underneath him, but he would have to take his chances.

"Lookin' real good," said Bean, his eyes glued to the gauges. "Fifty feet, comin' down. Watch for the dust." Bean didn't see the dust storm already raging outside his window.

Still *Intrepid* crept downward. Conrad's eyes flicked back and forth between the window and the instruments. It was absolutely the worst way to have to fly, with his attention split, but he had no choice. The gauge that was supposed to display lateral motion seemed to be broken, forcing him to look outside to make sure he wasn't drifting sideways or, even worse, backwards.

"Thirty feet, coming down at two," Bean said. "Plenty of gas, babe, plenty of gas. . . ."

"Thirty seconds," warned Jerry Carr in mission control. Bean answered, "He's got it made!"

Conrad was now flying entirely by instruments. He had planned to wait until touchdown to shut off the engine, but suddenly, he saw the blue glow of the Contact Light and instinctively his hand went to the ENGINE STOP button, and *Intrepid* fell the last few feet to the dust with a firm thump. Conrad had logged a hundred seconds of stick time—the only real flying he would get on Apollo 12—but they had required every ounce of experience from twenty years of piloting. He had landed on target; he was sure of that. And just now his friend and former student was slapping him on the back, saying, "Good landing, Pete! Out-*stand*-ing, man!"

5:38 A.M., Houston time
4 days, 19 hours, 16 minutes Mission Elapsed Time

"You're headed right square out the hatch. Wait, wait, wait—Come forward a little. There you are." With Bean giving him guidance, a fully suited Pete Conrad crawled out of *Intrepid*'s front hatch and onto the porch. Caution dictated that he move slowly in the lunar vacuum, but Conrad was impatient. From the moment of touchdown he couldn't wait to get outside; he had to get down to the surface and find out once and for all whether the Surveyor was really there. Looking out the window it was impossible to tell where they were. Like Armstrong and Aldrin, he and Bean couldn't gauge distances or sizes, and that made it all but impossible to figure out what craters were in front of them. But when Conrad finally stood on the top rung of the ladder, he could actually see part of the Surveyor crater—in fact, he realized with a mixture of relief and alarm, he had landed *Intrepid* not ten yards from its rim. He headed down the ladder while a color TV camera broadcast the scene to earth. Conrad knew people wouldn't remember the third man to walk on the moon; there was no need to make up something momentous to say. But he did have a quote; in fact, he had a bet to win.

The bet had its origins in the heat of a Houston summer afternoon when Conrad and his wife, Jane, were by the pool, entertaining Italian journalist Oriana Fallaci. Conrad had known Fallaci since 1964, when she came to Houston to write a book on the astronauts. Fallaci never had any trouble speaking her mind, especially when it came to bureaucracy. And that afternoon she was convinced that NASA's bureaucrats had told Neil Armstrong what to say when he stepped on the moon. Conrad tried to convince her otherwise, but she was certain of it. Conrad persisted; he couldn't swear that Armstrong had written "One small step for a man, one giant leap for mankind," but he was sure that whoever had penned it, Armstrong had chosen what he would say. "Look," he told his guest, "I'll prove it to you. I'll make up my first words on the moon right here and now."

"Impossible," Fallaci pronounced in her thick accent. "They'll never let you get away with it."

"They won't have anything to say about it, Oriana. They won't know about it until I'm on the moon." Conrad had a good idea: Since he was nearly the shortest guy in the Astronaut Office, why not say . . .

When Fallaci responded, "You'll never do it," Conrad answered, "How about five hundred bucks?" They shook on it.

Conrad reached the last rung of the ladder, held on with both hands, and jumped. He fell to the footpad with a gentle bump.

"Whoopie! Man, that may have been a small one for Neil, but it's a long one for me."

As Neil Armstrong had done, Conrad held on with his right hand and placed his left boot on the moon. He swirled his foot in the dust. "Ooh, is that soft and queasy. . . . I don't sink in too far. . . ." Still holding on, he planted both feet on the surface, then let go. His first steps brought him into the blinding glare of the sun. He turned away from it, moving in small, floating steps across the shadowed ground. Leaning forward at an impossible angle to compensate for his backpack, he felt as if he might fall over at any moment.

"Well, I can walk—pretty well—ah—Al, but I've got to watch where I'm going." Conrad kept walking until he could look past *Intrepid*'s foil-covered bulk to the east, across a vast bowl twice the size of a football stadium, and he could hardly believe what he saw. On the crater's shadowed far wall sat a tiny, spindly white shape: Surveyor 3. He let out a high-pitched cackle. "Does that look neat! It can't be any further than six hundred feet from here. How about that?"

In Houston, Capcom Ed Gibson radioed congratulations, but his words were drowned out by the sound of applause. The trajectory people had done it. Conrad was elated.

But the trip to the Surveyor would wait until tomorrow. Right now, Conrad had work to do. And after a few minutes, everyone listening in Houston and around the world heard an unaccustomed sound from the lunar surface: the sound of Pete Conrad humming. "*Dum de dum-dum-dum . . .*" Nothing in particular, just tuneless, mindless humming. "*Dum diddee dum-dum-dum . . .*"

Inside *Intrepid*, Bean didn't notice the humming at first. He'd heard his commander do the same thing in the simulator, and flying a T-38. Even for Pete, though, this was more than usual. Every few minutes the tune that had no tune crept onto the earth-moon airwaves. In the press room at the Manned Spacecraft Center, the reporters covering the flight half-jokingly wondered if Conrad was on an oxygen high. But to more accustomed ears, this was simply the sound of Pete Conrad in his no-sweat mode. None of his worries had been realized. And now Conrad saw with delight that it was even easier to do his work in one-sixth g than he'd expected. If everything kept going this well, the mission—not to mention some lunar rocks and a few pieces of Surveyor 3—was practically in the bag.

6:13 A.M.

For half an hour, Alan Bean had watched as his commander danced over the craters, 20 feet below. Now he would follow. As Conrad gave directions, Bean inched backward through the hatch and descended the ladder. When he reached the footpad Conrad was waiting with his camera: "Okay, turn around and give me a big smile. Atta boy. You look great. Welcome aboard." For the rest of his life, people would ask Bean what he was thinking at this moment, and he would tell them that his thoughts were on one thing: his checklist. On his left cuff, he wore a checklist that looked like a small spiral notebook; Conrad had one too. On those pages, almost every minute of their time on the moon was accounted for. Before the flight it had been Bean, with his love for detail, who had delved into the task of helping to write that checklist. But before he set about his assigned tasks, there was something he wanted to do. Reaching into a storage pocket on his thigh, he pulled out his silver astronaut pin, the one he had worn on his lapel during his six years as a rookie. When he returned to earth, he would receive the gold pin of a space veteran. Now, with a few halting steps, Bean stood at the edge of the Surveyor crater; he flung the pin into it.

According to the checklist, Bean had 5 minutes to gain his balance and learn to walk. He was amazed at his new buoyancy: "You can jump up in the air—"

"Hustle, boy, hustle! We've got a lot of work to do." Conrad's voice. Conrad had one rule: Stay on the timeline. It was true that in their two moonwalks the two men would spend more than 7 hours outside, about three times the amount Armstrong and Aldrin had, but all of that time was packed with objectives, and Conrad did not want to fall behind. Right now, he was setting up an umbrella-shaped S-band antenna to improve communications with earth. He called to Bean, "How about doing some useful work, like getting that TV camera going."

"Okay," Bean answered brightly, "good idea. Let's get that TV out and show everybody." That didn't go as planned; while Bean was carrying the camera on its tripod away from the LM, he accidentally pointed it too close to the sun for a few long seconds. By the time Bean realized the mistake it was too late. The moonscape on televisions in Houston and across the world had changed into a jumble of black-and-white shapes. Nothing Bean tried, even rapping on the camera with his geology hammer, seemed to work. History's second moonwalk had become a radio show—but with Pete Conrad on the moon, who needed television? When he wasn't talking to Bean, he was talking to himself, or he was humming, or laughing. In the midst of some task he would suddenly cackle, and no one knew why—except the backup crew, Dave Scott and Jim Irwin, who had arranged for a few . . . *additions* to the cuff checklist. They had adorned the pages with cartoons of Conrad and Bean as Snoopy astronauts, just like the comic-strip beagle's space-faring persona. But what really made Conrad laugh were the *Playboy* pinups, reduced to about three inches square, annotated, of course, with proper geologic terminology: "Don't forget: Describe the protuberances. . . ."

The funny thing was, there *were* a couple of mounds poking up from the undulating plain. The geologists had asked him to keep an eye out for unusual features, and here they were. They were a few hundred feet from the LM; Conrad made a mental note to get over and take a look at them, perhaps after the ALSEP work.

7:14 A.M.

"We're making our move, Houston." As Conrad announced their departure, Bean lifted what looked like a strange, square set of barbells and began walking west. The barbells were actually two pallets loaded with the ALSEP's scientific instruments. There was a seismometer to measure moonquakes, a magnetometer to look for a lunar magnetic field, another sensor to sniff out the moon's incredibly tenuous atmosphere, and others to search for ions in the moon's vicinity and analyze high-energy subatomic particles emanating from the sun. There was also a central transmitting station to relay the ALSEP's data to earth. Together, these experiments comprised the first full-fledged scientific station to be set up on another world.

Bean planned to carry the ALSEP several hundred feet from the LM, far enough so that the instruments would not be affected in any way by the dust kicked up when he and Conrad blasted off tomorrow. As he walked, Bean could feel his heart pounding with the effort to grip the bar with his stiff, pressurized gloves and hold the packages out in front of him. "I'm going to set it down and rest," he said. Meanwhile, Conrad ran out ahead of him, scouting a good area to lay out the experiments.

In their 2½ hours on the Sea of Tranquillity, Neil Armstrong and Buzz Aldrin had never ventured more than a couple of hundred feet from their lander. Now, 1 hour and 48 minutes into this first moonwalk, Conrad and Bean were extending the reach of lunar exploration more than twice over. Bean walked until he came to a level place about 500 feet from the LM.

Bean couldn't count the number of times he and Conrad had practiced setting up the ALSEP, on a simulated lunar surface behind the Flight Crew Training Building. To an unaccustomed eye the experiments looked like strange, delicate creations. In truth each was an engineering marvel, built to withstand lunar heat and cold and to perform on a minimum of electrical power—and be small and lightweight to boot. Bean knew each of them well. There was the seismometer, which looked like a silver paint can atop a round, silver drop cloth. And the magnetometer, with its three, gold-foil-tipped arms reaching into the vacuum. The small ion-detection experiment, with its ridiculously short legs joined by a spider web of wires. And the squat little solar wind spectrometer with its odd facets and bubbles. Each of them would make incredibly sensitive measurements to probe the moon and the void around it.

Laying out the ALSEP wasn't the kind of work people expected an astronaut to be doing on the moon; it was more like arranging garden furniture than like exploration: Undo bolts. Set each experiment on the ground. Tamp the dirt and make the instrument level. Make sure each one is pointed in the proper direction with respect to the sun, using a shadow indicator, and that each is the proper distance from the others. And—keep them from getting dirty. This was the job, and as he worked, Bean felt pride of accomplishment.

There was Conrad, a few yards away, working on the Central Station. His white suit glowed in the sunlight, except from the knees down, where it looked as if it had been dragged through a coal bin. Bean knew that Conrad was not immune to awe; you had only to be in the same spacecraft with him orbiting the moon. He'd talked a blue streak about the craters and mountains and lava formations and how amazing it all was. The important thing was, it never slowed him down. He was always spring-loaded, ready for the next event. Around him, the Ocean of Storms stretched in all directions, an undulating plain that rose and fell with craters of every size. Surprisingly, if Bean didn't think about it, the moon didn't really seem like an otherworldly place. But if he looked high into the black sky, he saw the earth, and the sight of it was enough to bring home the electrifying realization of where he was. And he told himself, *You don't have time now; you can think about this after the flight.* But just now, things were going well, and when he began to deploy the magnetometer Bean allowed himself a moment of fun. Removing a set of Styrofoam packing blocks, he took one of them and gave it a sidearm fling, and it sailed into the distance for an impossibly long time, tumbling in slow motion against the black sky, before landing. He took a second block and called out to Conrad, who was busy working.

"Pete? Pete!"

"What?

"Watch this." Bean made an underhand toss. The white shape ascended on a long, lazy arc and hit the dust. "Boing!"

"Stop playing," Conrad laughed, "and get to work. Come on, maybe they'll extend us until four and a half hours. I feel like I could stay out here all day." It was true; they could have handled a moonwalk twice as long as the 3½ hours allotted, and that's what Conrad would say in the debriefing when he was back on earth.

It took Conrad and Bean over an hour to set up the ALSEP, but when they were done it resembled an odd, five-pointed star with the central station in the middle and the experiments radiating from it on bright orange ribbons of cable. They looked just like the ones in training—except for one problem: It was almost impossible to walk past an experiment without spraying dirt on it. In no time the clean white surfaces were sprinkled with black powder. There wasn't any point in trying to wipe it off; that would only have ground it in.

"I remember how they took care of this white paint," Bean said. "You had to have gloves to touch it. Remember?"

Conrad laughed. "Yeah, they got kind of a problem here."

Well, Bean decided, there wasn't any point in worrying. Dirty or not, the experiments would just have to work. Finally the ALSEP was laid out and ready, and Conrad activated the central station. At that moment, in a back room down the hall from mission control, scientists gathered excitedly around a set of chart recorders. On the tracing for the seismometer, they saw little squiggles and bumps—not from moonquakes but from Conrad's and Bean's footfalls as they headed away from the ALSEP to gather samples.

In lunar orbit, aboard *Yankee Clipper*

Every two hours a tiny, unblinking star appeared in the eastern sky above *Intrepid*, ascended to the zenith, then vanished in the west. It was bright enough for Pete Conrad to see with the naked eye from the LM cabin. The star was *Yankee Clipper*, carrying Dick Gordon on a 38-hour solo voyage around the moon. Before the mission, Gordon had wondered whether he would think of the earth and feel his awesome separation from humanity. Instead, he savored his aloneness—and mostly he was too busy to think about anything besides the flight plan. But when there was a lull, his thoughts turned to his friends on the Ocean of Storms. Gordon envied Conrad and Bean more than he could say. Mike Collins had said he was perfectly happy going 99.9 percent of the way, but that wasn't how Gordon felt. Walking on the moon, not orbiting it, was the name of the game; it was as simple as that, and it had been his goal from the day he joined the astronaut corps. Gordon rationalized his position, telling himself that down the line, on one of the later missions, he would get his chance. But when he watched Conrad and Bean climb into that lunar module and pull away, he wished he were going with them. His crewmates knew it; Conrad had said he wished the LM held three people so they could all land together. When Conrad steered the lander to a touchdown Gordon was listening, and as the dust settled on the Ocean of Storms he radioed congratulations and told his friends, "Have a ball."

On the next orbit, when he flew over the landing site, he was ready to try to spot *Intrepid* on the surface. Unlike Mike Collins, Gordon knew exactly where to look, and he knew the navigation was so good that all he had to do was tell the computer to point the sextant right at the Snowman. At the proper moment the optics whirred into position and there, just off the rim of the Surveyor crater, he saw a point of light with a needlelike shadow, and he knew he had found them. "I have *Intrepid*," he announced to Ed Gibson, and as he got closer, he could almost convince himself he could see details, four landing legs sticking out from the descent stage, though he knew the sextant wasn't powerful enough for that. He was in the middle of reporting *Intrepid*'s position when he spotted another point of light nearby: "I see Surveyor! I see Surveyor!" Gordon was jumping up and down, he would say later—if it's possible to do that in zero g. If he could not make the first pinpoint lunar landing, then he had been the one to discover that it had been a success. He told Gibson, "That's almost as good as being there."

When Conrad and Bean took their moonwalk, Gordon was there vicariously. Thanks to mission control's relay he could listen to his friends as they bounced along. At one point, as the two moonwalkers wrestled with a balky piece of gear, Gordon couldn't help but laugh at what he was hearing. Later, he would tell his crewmates, "You were raising hell about some stupid device, and I was laughing my ass off."

In the moonwalk's last hour, Gordon prepared for his most important task of the day, a burn of *Yankee Clipper*'s SPS engine to adjust his orbit. If you could stand on a fixed platform in space, you would see that the moon slowly turns as it orbits, at exactly the rate needed to keep the near side pointed at earth. You would also notice that the orbit of *Yankee Clipper* stayed in a fixed orientation, with the result that each new circuit of the moon brought Dick Gordon slightly west of his previous position. If he did nothing, then over the 31½ hours that Conrad and Bean were on the moon *Yankee Clipper*'s orbit would shift far enough from the landing site to ruin the prospects for tomorrow's rendezvous. This 14-second firing would nudge *Yankee Clipper* into the proper orbit.

By now Gordon was very tired. Of course, test pilots are used to performing while fatigued, but he did *not* want to make a mistake with the whole world watching. But now, it was just him inside *Yankee Clipper*, surrounded by a forest of switches and dials, doing a task that had been performed by three men on every previous lunar mission. As he prepared for the burn he could hear Conrad and Bean going on excitedly about mounds and rocks, and the chatter was so incessant that it drowned out mission control, just when Ed Gibson was trying to read up crucial data on the burn. "If you're going to talk to me you're going to have to cut out the relay," Gordon told Houston. "It's impossible with those guys yacking." Gibson finally had to tell Conrad and Bean to be quiet for a few minutes. For safety's sake Gordon read his checklist over the air as he performed it, so that the flight controllers could check his work. Finally the engine lit and he was slammed into his couch for 14 long seconds. When it was over a relieved Gordon began to prepare for sleep. Without his crewmates to help out, everything took longer than usual. It was another two hours before he finished dinner and the lengthy cleanup for the night. He thought, Now's the time to think up all sorts of fancy prose to tell people what it's like to be the lonesome man up here. In truth, he was so tired he was glad just to go to sleep.

III: In the Belly of the Snowman

On the Ocean of Storms
5 days, 5 hours, 25 minutes Mission Elapsed Time

Wide awake, Alan Bean lay in his space suit on a Beta-cloth hammock that was strung to the sides of *Intrepid*'s cabin. Immediately above him, lying fore-aft, Pete Conrad was asleep in his own hammock. Bean looked at his watch. It was 3:45 in the afternoon, Houston time; he and Conrad were halfway into a planned nine-hour sleep period. But sleep did not come easily to Bean. For one thing, his space suit was uncomfortable, even without helmet and gloves. He would have preferred to take it off, but that wasn't a good idea, not with all this dust; there was too much risk of clogging a zipper or a wrist ring. But the suit wasn't all that kept Bean awake.

Bean had always felt that he thought more about the risks of the job than other astronauts. Still, Bean knew, you can never be sure what goes on in another man's mind, and it wasn't the kind of thing anybody ever talked about. It wasn't that he was anxious about anything in particular; it was more a heightened awareness of his surroundings. Bean heard the whine of *Intrepid*'s cooling pumps. A few hours ago, he and Conrad had just fallen asleep when one of the pumps changed pitch, and both of them awoke with a start, then went back to sleep. He looked up at Conrad's hammock. Even in a space suit Conrad's body barely weighed enough to sag the cloth; it lay almost flat in the lunar gravity. Before the flight Conrad had told him, "Don't worry, if anything goes wrong, it'll be something you've never seen before." The lightning strike proved him right. But that had not been a harrowing experience for Bean—although, he thought wryly, it would have been if he'd understood what was really happening. Pete had understood, and he'd been cool enough for all three of them. Of course, when it came to the landing on the moon, Conrad had done his share of worrying—that was only natural—and when he did, Bean was able to be reassuring, because he knew that if anybody could pull it off, Conrad could. Bean didn't worry much about the piloting end of

Apollo 12; in his mind a mechanical problem was the thing to watch out for. All through the mission, even with the exhilaration of the ride, and the wonder of such incredible sights, Bean heard the background noise of his own awareness, reminding him that one of the rules of this game is that all machines eventually fail. For Bean, being on the moon was laced with that awareness.

Bean thought about the TV camera. He wondered why it didn't work. He had no way of knowing that the full-strength lunar sunlight had burned the light-sensitive coating right off the vidicon, that it was beyond hope. The camera should have been ready in time for them to train with it at the Cape, but all they'd had was a block of wood. All he knew was that somehow, he had managed to screw it up, but there was nothing he could do about it now.

Bean thought about the Surveyor. Over dinner, he and Conrad had talked about how the shadowed crater wall looked so steep they might not be able to walk along it safely. They would have to wait and see. For now, Bean knew, he had to get some rest, or his performance would suffer. If you're not doing something productive, Bean told himself, you should be sleeping. Finally, he drifted into a light slumber.

Meanwhile, in the top bunk, Pete Conrad soon awoke—not because his mind was too active, but because he was in pain. The right leg of his space suit had been misadjusted before the flight and was slightly too short; now, as he lay in his hammock, his suit bore down on his shoulder like a vice. He called down to Bean and told him they would have to do something about it before the moonwalk. They got up, took down the hammocks, and then, while Conrad sat on the ascent engine cover, Bean set to work. The leg of the suit was adjusted by a set of cords laced around the calf like sutures. Bean had to undo each cord, which was tightly knotted (because nobody wanted it to come undone), let it out a little, and retie it. The whole process took about an hour. Then Conrad called Houston, two hours ahead of schedule, and the two men began the new day.

Wednesday, November 19
10:10 P.M., Houston time
5 days, 11 hours, 38 minutes Mission Elapsed Time

Morning lasts a week on the moon. For seven days, the sun slowly climbs in the black lunar sky, shrinking the shadows of rocks and craters. By lunar noon, temperatures climb to 225 degrees Fahrenheit. It takes another week for the sun to descend and then vanish below the western horizon. During the frigid two-week lunar night, temperatures plummet to 243 degrees below zero. Then the cycle begins anew.

When Conrad and Bean ended their first moonwalk it was about 6:30 A.M. local moon time. Nearly thirteen hours later, they stepped outside for the second moonwalk. But in lunar time only half an hour had elapsed. The sun had climbed only a few degrees, and yet everything looked slightly different. It almost seemed to Bean that a new landscape had taken shape while they slept. He kept noticing rocks that he thought he'd missed the day before; soon he realized they were the same rocks under different lighting. The colors of the surface—the spectrum of grays and tans and browns that changed as he looked in different directions—seemed a little more vivid. But the real change was in the Surveyor crater. No longer did its walls seem steep and forbidding; it had been transformed into a gentle bowl. Getting to the probe looked to be easy. Along the way, Conrad and Bean would test their skills as lunar field geologists. For years, and especially in the last few months, they'd been schooled in such esoterica as mineral identification, the characteristics of impact craters, the proper techniques for collecting samples. They had honed their skills on the lava flows in Hawaii and the desert of west Texas. When they left the earth they did so not only as pilots but as surrogates for the geologists who would have given their right arms to be in their place.

To save precious minutes on the surface, the geologists had sat down ahead of time and planned four possible traverses for Conrad and Bean to follow. About two weeks before the flight Conrad decided he wanted to carry maps, and the geologists had quickly drawn them up on a set of photographs. There wasn't time to get them on the official manifest, so Conrad arranged to have them stowed with his personal items. But that wasn't the only unofficial item Conrad had wanted to bring: a long-time collector of hats, he'd arranged for the crew systems people to make a giant, blue-and-white baseball cap that would fit over the top of his space helmet. He was going to put it on and wait for everybody on earth to notice as he bounded past the TV camera. Unfortunately, nobody could figure out a way to get it into the LM in secret. Never mind; Conrad had managed to smuggle something else in the pocket of his space suit, and when the time came, he and Bean would have their own little caper on the lunar surface.

It turned out that *Intrepid* lay right on traverse number 4, which followed a sort of misshapen circle around several craters. The geologists had explained before the mission that the moon's craters are like natural drill holes, that the ferocious energy of meteorite impacts had blasted chunks of rock from the crust and scattered them across the moon. Each crater of the Snowman was nothing less than a ready-made excavation into lunar history. By visiting different craters, the geologists hoped the moonwalkers would find out whether the lava flows that formed this region of the Ocean of Storms varied in age and composition. On earth, a geologist would have spent many days or even weeks on this kind of exploration, but the timeline gave Conrad and Bean a bit less than 2½ hours to get around the circle. Mindful of all they would try to accomplish, Conrad and Bean assembled a small assortment of gear—rock hammers, sample bags, core tubes, shovel, tongs, and maps—loaded them onto a portable tool carrier, and set off on a journey across the craters.

The first stop was the north rim of 360-foot Head crater, about 100 yards west of the LM. It didn't take long for the men to make a discovery. Bean looked at the places where Conrad's boots had dug into the gray soil and saw that they had uncovered a lighter gray, just underneath the surface. They could not hear the excited shouts of the geologists in the back room down the hall from mission control, but they knew they had found something significant. Before the mission, the geologists had told Conrad and Bean to look for evidence of a light-colored streak that was visible on the unmanned orbiter pictures—material that had probably been ejected from the impact that formed huge Copernicus crater, 230 miles to the north. Now it appeared they had found some. After the mission, this sample would let the geologists determine the age of the impact that had formed Copernicus, an important event in lunar history.

Conrad knew that he and Bean could easily have spent an hour at any one of these craters, but the geologists wanted as many different types of rocks as possible, and the clock was ticking. So he and Bean took off running, heading south to Bench crater.

Yesterday, during the first moonwalk, Bean had found that running on the moon was an experience all its own. He watched as Conrad skipped like a little boy and laughed with delight: "Wheee! Up one crater and over another. Does that look as good as it feels?"

Bean had discovered a better way. "Pete, bend and rock from side to side as you run. Like that. There you go." It wasn't really a run; it was more of a lope. Push off with one foot, shift your weight, and land on the other. Each step launched him into the air for long seconds, while he wondered whether he would land on a rock or in a pothole. To avoid this, he tried sticking his foot out to one side, but when he landed and pushed off again his uneven stance set him slowly rotating, giving him a new problem to deal with. He soon learned to anticipate each new step while he was still airborne, shifting his weight as he landed and immediately pushing off, setting up a rhythm, as if he were bounding across a rocky stream. It felt strange, and it was demanding—he couldn't take his eyes off the ground very often—but that only made it more fun. And the moon afforded him a luxury unknown to any runner on earth, a chance to relax in the midst of each step. It seemed, to Bean's amazement, that he would never get tired.

10:59 P.M.

"By the way," Conrad radioed Houston, "this is the smartest idea we've come up with. This map just works great out here." It worked great for Conrad, but most of the time Bean didn't have the foggiest idea where he was: from the surface, really big craters like Head and Bench looked nothing like the big, circular bowls on the map.

Bean tried to navigate a few times on his own and gave up. It was beyond him how Conrad, running ahead of him, managed to study the map even as he bounded along.

While Conrad led the way, Bean scanned the ground for something interesting. It wasn't easy to do field geology at a full gallop—and on the moon, it wasn't much easier standing still. Everything was so dust-covered that only the most subtle variations in texture were visible. Even the rocks looked almost exactly alike, until he held one right up to his faceplate. Then he could see hints of what lay beneath the dusty coating, the green glint of olivine crystals or a chunky white grain of feldspar. He could even see tiny pits, actually craters made by micrometeorites, peppering the sides of the rock, and he was telling Houston all about it when—"Hey, Al, quit baloneying and give me a hand"—Conrad cut him off. Bean understood; the geologists were going to have the rest of their lives to study these rocks; why waste time talking about them? And in the back of his mind Conrad was thinking, "Got to get around the circle." The spindly white probe—in Conrad's mind the mission's finish line—was waiting for them. Fortunately, Ed Gibson had good news: The moonwalk was being extended by half an hour.

11:31 P.M.

"You know what I feel like, Al?"

"What?"

"Did you ever see those pictures of giraffes running in slow motion?"

Bean laughed; he knew what Conrad was talking about. He watched Conrad's feet as he ran along beside him. Each time his boots hit the surface they sprayed a small shower of dust, sailing out on perfect trajectories.

As they ran by the southern rim of Bench crater, they were beginning to feel the strain of their adventure. The entire run to the Surveyor crater was slightly uphill, and they could feel it. Their mouths were as dry as the desert from breathing pure oxygen. "Tell you one thing I'd go for," Conrad said, "is a good drink of ice water."

To make matters more difficult for Bean, the tool carrier was getting heavy with rocks. He had to hold it up to his chest while he ran, to keep the rocks from bouncing out. This meant he was constantly fighting the arms of the suit. Before the mission he'd been so conscientious about exercise—every day he'd run a couple of miles on the hard beach at the Cape, pounding his legs into shape. And it turned out that most of the work was in the arms and especially the hands. He told Ed Gibson to pass on a message to Fred Haise, the lunar module pilot for Apollo 13, to do hand exercises.

Meanwhile, in Houston, the flight surgeon saw Conrad's and Bean's heart rates soar to 160 beats per minute. Ed Gibson passed on a message from the surgeon to the lunar surface, in code: "Pete and Al, we'd like an E-M-U check." EMU stood for Extravehicular Mobility Unit, in other words, the space suit and backpack. Ostensibly this was a request for each man to read out his oxygen supply and report any anomalies in suit pressure or the like, but in reality it let Gibson tell the men to cool it for a couple of minutes, without letting the whole world know about it. Conrad and Bean got the message. And they read out their oxygen quantity and the rest of it, but they couldn't bring themselves to stand around and do nothing. An interesting rock caught their eye, and they set about collecting it. Then they were on the run again. Everyone in mission control could hear them huffing and puffing.

The two men ran onward in their long, gliding, slow-motion-giraffe strides under the black dome of the lunar sky. Bean looked up as he ran, above the solar glare, and found a glittering crescent set in the velvet, exactly where it had been the day before. He looked down at his feet again, watching out for potholes and rocks, then leaned his head back for another glance. For just a moment, he silenced the voice in his head that always called him back to work, to the rule of the clock. And he talked to himself: *This is the moon. That is the earth. I'm really here.*

Suddenly, Bean felt his ears pop—his suit must be losing pressure! Heart pounding, he stopped and checked the gauge on his wrist. No change; it must have been some kind of weird transient. Only after the flight would the engineers decide that as Bean ran in the light gravity his body had bounced against the oxygen outflow port within his suit, closing it momentarily and causing a slight *increase* in suit pressure. But just now, that taste of fear was enough to put Bean's thoughts right back on the traverse. After that, he stopped looking up for a while.

Thursday, November 20
12:15 A.M., Houston time

Standing on the rim of the Surveyor crater, Conrad and Bean rested for a moment. Just beyond the far wall they could see *Intrepid,* like a tiny replica. To their right, 300 feet away, sat Surveyor 3, gleaming in the morning sunlight. Antennas and sensors still reached upward from its tubular frame, just as they had on April 20, 1967, when the craft thumped onto the moon amid blasts from its braking rockets. Like *Intrepid,* it had the strange organic look of a craft designed for the void of space. Fuel tanks and batteries protruded from its open frame. This unlikely robot had made history when its mechanical claw, under command from earth, had scraped the skin of the moon and given scientists their first real information on the physical nature of the lunar soil. It had lasted just fifteen days, until the onset of lunar darkness. It was a relic now, but it was about to make one more contribution to space exploration.

There was some concern, both on the moon and in mission control, that Surveyor 3 might welcome its first visitors by sliding down the crater wall on top of them. For that reason Conrad and Bean had decided to approach from the side, following the contour of the sloping wall. This was a relatively old crater, and the geologists had warned them to expect a soft, thick dust blanket here, but the men were relieved to find firm ground that gave good footing. With slow, careful steps they closed in. They could see, to their surprise, that the Surveyor wasn't white any more, but a light tan. Had it been baked by the sun, or had Conrad and Bean sprayed it with dust when they landed? They'd investigate when they got closer, and they'd take pictures of it from every angle, and more pictures of the soil and rocks nearby, so the scientists could compare them with the Surveyor's own images from thirty-one months earlier. They'd collect some of those rocks, and then Conrad would snip off pieces of the craft to take home. But before they poked, prodded, and otherwise cannibalized the probe, there was something Al Bean and Pete Conrad wanted to do.

Before the mission, they'd had one of the support crew go out and buy an automatic timer for the Hasselblad, a little spring-loaded gadget. Conrad and Bean's idea was that they would mount the camera on the tool carrier and then pose, side by side, next to the Surveyor. It would only take a minute for them to fire off a few shots—saluting, waving, shaking hands, whatever—and Conrad was sure that when they got home one of those pictures would end up on the cover of *Life* magazine. He couldn't wait until everybody asked, Who took the picture?

Conrad had managed to smuggle the timer in the pocket of his space suit. He'd remembered to bring it into the LM with him, and just before they'd headed out on the traverse that morning, he'd dropped it into the tool carrier—which was now full of rocks and tenacious lunar dust. Bean rummaged in the bag for a moment, looking for a glint of chrome, but saw only grime. The only solution was to take all of the rocks out of the tool carrier. They couldn't talk about it; the whole world would know what they were up to. So they made hand signals. While Conrad held the tool carrier, Bean rummaged among the samples, each in their little Teflon baggies. He wondered if he should just put the tool carrier down, get on his knees, and lay the rocks on the ground, but he worried he'd never get them all back in the bag if he did. After a couple of minutes, Bean realized the timer was buried inside the bag, lost in the dust. He said quietly, "Forget it."

12:38 A.M.

"Okay, Houston," said Conrad. "I'm jiggling it. The Surveyor is firmly planted here. That's no problem." Having assured mission control, Conrad wielded a pair of cutting shears. The engineers wanted some samples of metal tubing, to see how it had been affected by thirty-one months on the surface. But what Conrad wanted most of all was the Surveyor's TV camera. With all its circuitry and moving parts, it was the real prize. Conrad wrestled with the shears as he tried to cut through the camera's support struts. "Okay, two more tubes on that TV camera and that baby's ours." Another cut. "There's one," Conrad said. Now Bean gripped the camera with his gloves, and at last: "It's ours!" Conrad cackled. That TV camera was his trophy fish.

1:19 A.M.

Back at *Intrepid,* Conrad was emptying the tool carrier into the rock box when suddenly out fell the Hasselblad timer. He called to Bean, "I've got something for you."

"Just what we need," Bean said. He picked up the timer and threw it into the distance as hard as he could.

6:55 A.M.
5 days, 19 hours, 33 minutes Mission Elapsed Time

Pete Conrad was exhausted. He sat on the floor of *Intrepid*'s tiny cabin, leaning against the wall in his grimy space suit. As tired as he was, he couldn't believe he and Bean had climbed into the LM after only 4 hours outside, with all that extra oxygen—maybe a couple of hours' worth—still in their backpacks. He wasn't about to tell Houston how frustrated he was; before the flight he'd agreed that if mission control gave him one time extension during each moonwalk, he wouldn't ask for a second one. So he and Bean had some time to kill. They ate lunch, and then they started into the checklist for liftoff. They were about an hour ahead of schedule, and when they came to the T minus 30 minutes mark in the count, they stopped, because the next steps—pressurizing the fuel tanks, for example—would have to wait until the proper time. For now, Conrad rested.

Conrad looked at Bean. He seemed nervous. There was absolutely nothing to do but watch the clock, but Bean was fidgeting with something on one of the instrument panels. What was he thinking? Conrad remembered how he had felt on his own first spaceflight, Gemini 5. Crammed into that tiny spacecraft with Gordo Cooper after a week in orbit, circling the earth over and over,

Conrad found himself thinking about the retro rockets, which had been soaking in the frigid cold of space for days on end, longer than anybody had ever been in space before. Conrad was terrified that when it came time to fire the retros nothing would happen—in which case he was sure he would slit his wrists. When the retros fired right on schedule, he breathed a huge sigh of relief. Now, on the moon, he was sure Bean was going through the same drill about the ascent engine. For his own part, Conrad had no such worries this time around. He'd already been through three launches in his career; this was just another one in a different place. In any case, if Conrad had any lingering doubts, he put them aside when he saw his lunar module pilot. Finally Conrad said, "Beano, are you worried about the engine?" Bean answered a quiet "Yep" and Conrad tried a bit of humor: "Well, there's no sense worrying about it, Al, because if it don't work, we're just gonna become the first permanent monument to the space program." Conrad wasn't sure the joke made any difference in Bean's outlook. But an hour later, when the moment of truth arrived, there were no unpleasant surprises. And then, as the Ocean of Storms fell away in one last, spectacular view, Pete Conrad and Al Bean headed for a reunion with a best friend in lunar orbit.

Friday, November 21
Aboard *Yankee Clipper*, in lunar orbit

Bean was looking out the window. It was the first chance he'd had to relax and play tourist since arriving at the moon more than two days earlier. He thought about how strange it felt to orbit the moon. The strangest thing about it was the silence. There wasn't any engine noise, the way there always was in an airplane. And it felt odd to see the spacecraft fly in the same direction no matter how it was pointed. Orbiting the moon, Bean thought, was much more of a science fiction experience than walking on it had been.

Flying in space was better than Bean had ever imagined during those long years as a rookie. Yesterday, during the rendezvous, he'd been slaving away with the backup computer and the navigation charts while Conrad flew the lunar module. They had one more burn to do, and then they would have it made. And Conrad had said to him, "Why don't you just quit after this midcourse, and relax and enjoy it? You can take a minute and fly this vehicle." Startled by Pete's audacity, Bean wondered, wouldn't it put them off course? No, Conrad assured him, whatever digressions they made would be easy to correct. Bean was reluctant—surely mission control would know. Conrad laughed, "Not on the back side of the moon, they won't." Bean realized Conrad had planned this perfectly. And for a few minutes Bean had his hand at the crisp, responsive ascent stage. It was a moment that Bean would always remember as pure Pete Conrad, that in a small craft somewhere over the far side of the moon, he had taken the time to share with Bean a flying experience that even most astronauts would never know.

And then, Bean could see *Yankee Clipper* out ahead, growing slowly from a point of light into a gleaming spacecraft. He watched as the command module moved from sunlight into shadow and back again, thinking, *That looks so* neat. Dick Gordon did a perfect job on the docking—smooth as glass—and later, when he opened the tunnel to the LM, all he could see was two dim figures floating in a cloud of dust. He called down, "You guys ain't gonna mess up my nice clean spacecraft!" Before he would even let them into the command module, he made them take off their filthy suits. They passed them up through the tunnel and Gordon hurriedly zipped them into their stowage bags underneath the couches. Finally, when he and Conrad got up there, Gordon was happier than Bean had ever seen him. He was scurrying around, helping them stow their gear. He was offering them a drink of water. And in that moment, Bean was filled with a love for his crewmates that he had never experienced. Years later, Bean would say, his most special memories of the flight would not be about the moon or the earth; they would be about Pete Conrad and Dick Gordon.

Now Bean looked out at the bright, bleak cinder passing beyond his window. It was so utterly inhospitable. Everything in the universe has some function, Bean thought, but what is the function of the moon? Is it to make the tides? The earth would probably get along fine without them. Maybe it was as the geologists said: the moon is here to tell us the story that had been lost forever on our own planet. Maybe the moon would tell us where we came from. Bean didn't know the answer. As *Yankee Clipper* circled, Bean looked, and now and then, he wondered. He found himself thinking about the six-year journey that had gotten him here. He realized now, with his neck farther out than it had ever been, that life is too precious to spend it living by someone else's rules, even the unwritten ones of the Astronaut Office. He would be a good astronaut, but he would do it his way. As the moon bore silent witness, he told himself, "When I get back home—if I get back home—I'm going to live my life the way I want to."

On the way back to earth

Pete Conrad wasn't feeling well. For one thing, he had somehow managed to come down with a cold. "How did I get the world's greatest cold on the moon, for Chrissakes?" Bean answered, "Because you had the world's greatest LMP with you on the moon, who had a cold." Where Bean's cold had come from, he had no idea.

Then there was the rash on Conrad's chest—he'd had a reaction to something in the adhesive that stuck the biomedical sensors to his skin, and it raised itchy welts, like a bad case of poison ivy. It was almost the best moment of the flight when Houston finally told him he could take them off.

Bean was out of it, too—or at least it seemed that way to Conrad and Gordon. He was spending much of his time on the way back to earth sacked out in his sleeping bag. Not surprising—he'd peaked for the landing and the moonwalks, and the trip home was definitely an anticlimax. A couple of days ago Conrad would've been happy to fire up that SPS engine longer than the flight plan called for and come barrelling out of lunar orbit so fast that they would trim a day off the return. But Houston wasn't about to let them do that.

There was plenty of time to think on the way back, and Conrad did. He was proud of Apollo 12. They'd proven the pinpoint landing capability and accomplished their other objectives. Now astronauts could go to the places the scientists wanted them to. But along with the immense feelings of elation and satisfaction, there was something else, something Conrad hadn't expected. After seven years of eating, sleeping, breathing Apollo, he had finally had his mission, and it had been perfect, and now, just like that, it was all over. The flying had been brief but challenging; the sights had been the most spectacular of his life. And yet, going to the moon wasn't what Conrad had expected. Yes, it was spectacular, but it wasn't . . . *momentous*. Looking back on it now, he couldn't shake the feeling that it had been so much like the training that it was almost an anticlimax. Take away the weightlessness and the view, and he might as well have been in the simulator. Seven years ago, he had told himself that if he made it to the moon he wouldn't let it change him; now, he had no worries that it would. Conrad kept these thoughts to himself as *Yankee Clipper* headed home, and he had no idea whether his crewmates felt the same way. He was quite surprised when Al Bean turned to him and said, as if he could read his mind, "It's kind of like the song: Is that all there is?"

• • •

Sometime after he and Dick Gordon and Al Bean got out of quarantine, Pete Conrad got his crew together for a survival meeting of sorts. It was time to talk about their futures. For his own part, Conrad had not known how he might feel about the moon after the flight, but now that he was home, he already knew he wanted to go back. It was a spectacular place, and he had proved to himself that it was also a fine workplace. Before the flight, he'd always daydreamed that he and Gordon and Bean would each return to the moon, as commanders of their own missions: Gordon would land on Apollo 18, Bean on 19, and himself on 20. There had been talk about bringing along a one-man lunar flyer for those last two missions, to expand the range of an astronaut in search of discoveries. In his daydreams he and Bean would develop the flyer and then test it out on their missions. But the flyer had been canceled in favor of a four-wheeled lunar roving vehicle. And it was clear now that budget cuts would force NASA to cancel the final lunar landing, Apollo 20, to free up a Saturn V for its earth-orbit space station project. Apollo 18 and 19 didn't look much more secure. For all their jokes about staying together as a crew forever, Conrad, Gordon, and Bean knew they would have to break up.

Conrad planned to get out of Apollo and onto the space station project, because that's where the available seats were. He told his crew they should do the same, and Bean planned to take his advice. But Dick Gordon decided to stay on. Going by the pattern of Deke Slayton's crew rotation, he would probably be assigned as the backup commander for Apollo 15, and then command Apollo 18. Conrad reminded him that Apollo 18 could disappear, but Gordon said he would take his chances. Having come so close to his goal, he couldn't give up now.

CHAPTER 7

The Crown of an Astronaut's Career

APOLLO 13

I: A Change of Fortune

The new decade opened with a nation polarized by the war in Vietnam and a space agency unsure of its future. After a year of triumph, NASA faced a combination of public apathy and outright hostility for costly, high-tech government "boondoggles" like putting men on the moon. At the same time, NASA's leaders were on uncertain footing with the Nixon White House. Already, faced with the leanest NASA budget in nine years, Tom Paine had suspended production of the Saturn V, leaving enough boosters to fly missions through Apollo 20. But Apollo Applications, the project to launch a temporary space station made from Apollo "spare parts," was already going ahead, and now it became clear that a Saturn V would be needed to launch the station into earth orbit. In January 1970, Paine canceled Apollo 20, and before long there were signs that two more Apollo flights were in jeopardy.

Meanwhile, preparations went ahead for Apollo 13, set for a spring 1970 launch. But within NASA, the question was raised: Was it time to abandon the moon? In recent days none other than the MSC's director, Bob Gilruth, had privately called for an end to the moon landing program.

Some called Gilruth the father of manned spaceflight. No one had done more to make Apollo a reality, and no one had higher regard for the astronauts. To them, too, he had always been something of a father. Even more than most of the safety-conscious NASA managers, Gilruth had sweated every new mission—but he had always been willing to take the risks, provided they were worthwhile. But there were new challenges and new programs on the horizon. With two lunar landings accomplished, was it wise to risk men's lives—and NASA's future—again and again? Gilruth wasn't at all sure that it was. Stop now, he said, before we lose somebody.

NASA was a democratic organization in which such opinions were freely expressed, but Gilruth's entreaties did not change the course of Apollo. Everyone knew how risky a venture it was, and sometimes it seemed as though they were tempting fate with each new mission. And no one realized this more than the astronauts. Any of them, if asked, would have said that it was just a matter of time before some hidden flaw in the system, some unnoticed mistake, that the quality control checks hadn't caught would come around to bite them. Spaceflight had always been a game of probability. But Jim Lovell, commander of Apollo 13, had no reason to suspect, as he and his crew headed moonward on an April evening, that the odds would finally turn against them that night.

Monday, April 13, 1970
8:56 P.M., Houston time

For the first time in weeks, Marilyn Lovell could finally relax. She was sitting in the VIP viewing room that looked out over mission control, watching a telecast from her husband's moonbound spacecraft, and what she saw cheered her. There was Jim, floating in the command module—the image was shadowy, and Jim's face was hard to make out at times—but to Marilyn it could not have been more heartening. Here was a portrait of a man in his element. Everything was going beautifully. Jim and his lunar module pilot, a rookie named Fred Haise, had just given a televised tour of their lander, *Aquarius,* and had displayed some of their gear—their space helmets, the hammocks they would sleep on, and special bags for drinking water that they would wear inside their space suits while they walked on the moon. Now, back in the command module *Odyssey,* Jim was acting as emcee.

"We might give you a quick shot of our entertainment aboard the spacecraft," he said, holding up a small tape recorder. He brought it close to his communications microphone and turned it on; light piano-combo music filled the airwaves. Jim had no way of knowing that his words were going no further than the NASA centers; after two flawless lunar landing missions the networks weren't going to cut into the Doris Day show or the other prime-time programs to carry yet another telecast from space. Marilyn was here in the control center because it was the only place for her to watch.

"It's interesting," Jim said as he turned off the music, "to see that tape recorder floating there playing the theme from *2001: A Space Odyssey*. And of course, our tapes wouldn't be complete without 'Aquarius.' "

Marilyn was grateful things were going so well, grateful for the chance to feel good again. In the weeks before the flight she had felt an inexplicable sense of foreboding. Jim had made three flights before this one, including the first trip around the moon, but she had never been this anxious before any of those missions. It wasn't superstition, though she could confess to having had a twinge of that when she learned, the previous August, that Jim would command Apollo number 13. It was more a feeling that he had been lucky too long and that it was only a matter of time before his luck would run out. Still, she had not raised any objections, because she knew how much this flight meant to him. At last, he was going to do the thing he had wanted to do since he'd become an astronaut eight years before; he was going to land on the moon.

Last week, Marilyn had gone to the Cape to see Jim, and she had planned to return to Houston to watch the launch. But when it came time to leave, she found that she could not. So she stayed, to bid him farewell the way she had on Apollo 8: silently, while his rocket rose on a pillar of flame and disappeared into the sky. But as Jim had confided to her during a moonlit walk along the beach, Apollo 13 would be his last mission.

Also in the viewing room this Monday evening was a quiet young astronaut named Ken Mattingly. A member of the fifth astronaut group, he was thirty-four years old. Tall and thin, his brown, crew-cut hair almost gone, Mattingly was perhaps the most private man in the Astronaut Office. Even the other astronauts would not claim to know him well. But they would not have been surprised to learn that this evening Mattingly was in the depths of the worst depression of his life. Just a week ago, Mattingly had been the command module pilot on Apollo 13. Now he was a spectator.

The weekend before launch Charlie Duke, the backup lunar module pilot, came down with a case of German measles. He'd caught it from the child of a friend. The NASA doctors said Duke wasn't contagious and his illness had been incubating for two weeks, during which time he'd been to meetings and meals with Lovell, Mattingly, and Haise. Any one of them might also come down with German measles in the next two weeks. With less than a week until launch and the culmination of nine months of preparation, Lovell's crew were at the mercy of the doctors, who made daily blood tests and waited for the disease to show itself. By Wednesday the doctors had decided that Lovell and Haise were probably immune, but they could not be sure about Mattingly. His blood tests, they said, showed that he might already be fighting off the disease and could develop full-blown symptoms during the mission. By Thursday, Chuck Berry was recommending that Mattingly be pulled off the flight and that his backup, Jack Swigert, be sent in his place. And Mattingly was going through hell.

Like most pilots, Mattingly had an aversion to doctors. Pilots had a saying: There are only two ways you can walk out of a doctor's office—fine or grounded. More than one astronaut had sneaked off to a private physician, who was sworn to secrecy, rather than risk seeing a NASA doctor. And as the week wore on, Mattingly began to feel as though the doctors were not on his side. They were waking him up at 6 A.M. to draw blood. At 5 P.M. they drew more blood. Then they sent him to bed saying, "Now don't worry"—and then, the next morning, they'd wake him up again and say, "It doesn't look like you had a very good night's sleep." The whole process was enough to make anybody doubt his own sanity.

Aside from these hassles, Mattingly felt *absolutely fine.* But by Thursday, Swigert was being put through his paces in the command module simulator, and Deke Slayton was keeping a close eye on his work. Still, it was inconceivable to Mattingly that he would be left behind. He thought the doctors would either fill him up with so much gamma globulin he couldn't conceivably get sick, or else NASA would postpone the flight until the next launch window in May. His friends were consistently upbeat. If they believed his fate was sealed, they couldn't bring themselves to say so.

Friday, the day before launch, the decision was due. With Swigert getting all the simulator time, Mattingly had little to do, and he could only go running for so many miles on the space center's roadbeds. Slayton suggested he go flying. Mattingly drove to Patrick Air Force Base and did just that for a couple of hours, in a T-33. Afterward, he considered breaking the pre-mission medical quarantine to cheer himself up with some Dunkin' Donuts—those and the barbecue sandwiches at Fat Boy's were his main weaknesses at the Cape—but decided against it. He was wearing his blue NASA flight suit; surely he'd be recognized. Heading back to the space center, he turned on the car radio in time to hear a newscast: "NASA has announced that it will replace one of the three astronauts of Apollo 13, Thomas K. Mattingly . . ."

Suddenly it was real. Mattingly was furious to find out this way, but he wasn't about to blame anyone, especially not Deke. By the time he had reached the crew quarters he'd accepted the news. Jim and Fred were still over in the simulator, and everyone else reacted to Mattingly's presence with an awkward silence. Mattingly didn't have any more idea of what to say than they did. Slayton asked him what he wanted to do, and he answered that he needed to get away before the launch. Fine, Slayton said; he could grab an airplane and fly home that evening. After dinner he wished Lovell, Haise, and Jack Swigert good luck. When he arrived at Patrick, a T-33 was on the ramp, ready and waiting. Flying through the early-spring night, Mattingly was surprised to find the normally laconic air-traffic controllers calling him on the radio and chatting up a storm. He realized they were trying to make sure that he was awake. When he landed at Ellington, the ground crew was already waiting to take care of his airplane; he had only to get in his car and go home. And he realized that Deke Slayton must have quietly arranged for all of this; it was just the kind of thing Deke would do, and never acknowledge.

Now, watching the television transmission from Apollo 13, Mattingly was still in the depression that had descended on him four days ago. A distraction would have helped, but he had nothing whatsoever to do except watch the progress of the mission he should have been on.

8:59 P.M.

Jim Lovell ended the telecast with a farewell that had the ring of a 1940s radio announcer. "This is the crew of Apollo 13 wishing everybody there a nice evening"—all he needed was a closing theme song—"and we're just about ready to close out our inspection of *Aquarius* and get back for a pleasant evening in *Odyssey*. Good night." The TV monitors went blank, and Marilyn left, pleased that her husband was having a good mission. Mattingly, meanwhile, remained in the viewing room, listening to the voices of his former crewmates. In a few minutes, he and a friend planned to go out and have a beer.

• • •

At age forty-two, Jim Lovell was the most traveled man alive. With three spaceflights under his belt, he had racked up 572 hours and nearly 7 million miles, more than any other astronaut or cosmonaut. For many men that would have been enough, but not for Jim Lovell. From the day he joined the astronaut corps in 1962, his ultimate goal had been to command a lunar mission. Command was even more important to Jim Lovell than landing on the moon, and most veteran astronauts felt the same way. When Borman turned down Deke Slayton's tentative offer to fly the first lunar landing it had everything to do with the fact that he had been the commander on Apollo 8. If Lovell had any disappointment about his commander's decision, it vanished when Slayton assigned him to lead Neil Armstrong's backup crew. Within weeks after Armstrong's team came back from the moon, Lovell and his crew were training for their own landing.

By the spring of 1970, most of Lovell's colleagues from the

second astronaut group had moved on to other things. Neil Armstrong had disappeared into the world of postflight P.R. that greeted him on his return from Apollo 11; it seemed unlikely he would fly again. Jim McDivitt had traded the demands of flying Apollo missions for the equally demanding job of preparing for them, as manager of the Apollo Spacecraft Program Office in Houston. Tom Stafford, though still on flight status, had replaced Al Shepard as chief of the Astronaut Office, and wasn't scheduled for another mission. And Frank Borman had begun a new life as a vice president with Eastern Airlines. One day Borman came by for a visit while Lovell was in the simulator, and he seemed glad to be free of the training grind. "Jim," he said, "aren't you tired of this? I wouldn't want to go through this again."

Lovell couldn't have felt more differently. This was his seventh time around, counting the stints on backup crews, and even now his appetite for spaceflight was undiminished. He himself described it as an addiction. He could have gone on until NASA said he was too old to fly any more, but he knew that when he came back from Apollo 13 he would face a long wait, perhaps several years, before he flew again. Well aware of the astronauts still waiting for their first flights, he decided he would not get back on line for a fifth. Apollo 13 would be a great finale to a long spaceflight career.

Like every commander, Lovell wanted his mission to stand out, but he couldn't see why people would remember the third lunar landing. And that was fine with him. He wanted badly to land on the moon, and he was glad for the chance to make a contribution to science. The Apollo 13 mission patch read "Ex Luna, Scientia" —From the moon, knowledge—and Lovell thought of that when he christened his lunar module *Aquarius,* after the god of the ancient Egyptians who brought life to the Nile Valley (not to mention the popular song from the Broadway musical *Hair*). The command module *Odyssey,* meanwhile, took its name not only from Homer's epic work but from Arthur C. Clarke's science fiction vision of space travel. When Lovell was back on earth, he would find irony in odyssey's dictionary definition: a long voyage with many changes of fortune.

9:08 P.M., Houston time
2 days, 7 hours, 54 minutes Mission Elapsed Time
Aboard Apollo 13

The TV show was over, and Lovell waited in the lower equipment bay for Haise to finish closing up *Aquarius* for the night. They were about to shoot some pictures of Comet Bennett, an icy celestial wanderer that had graced the spring skies. But first mission control wanted Jack Swigert to stir up the service module's tanks of cryogenic liquid hydrogen and oxygen. In zero g, the super-cold fluids tended to become stratified, making it difficult for the astronauts and controllers in Houston to get accurate quantity readings. To remedy the problem, each tank contained a fan that acted like an egg beater to stir the contents. Strapped into the left-hand couch, Swigert flipped the switches marked H_2 FANS and O_2 FANS and waited several seconds, then turned the fans off. A moment later there was a loud, dull bang.

At first Lovell thought he knew what it was. The LM's cabin-repressurization valve made a bang whenever it was activated, and Haise had enjoyed scaring his crewmates half to death with it. Looking into the tunnel, Lovell called out, "Fred, do you know what that noise was?" Haise's dark eyes registered surprise; he hadn't touched the noisy pressure valve. Suddenly the master alarm rang in their headsets.

In *Odyssey*'s left seat, Jack Swigert looked up at the caution-and-warning panel and was alarmed to see that the light labeled MAIN BUS B UNDERVOLT glowed red. Main Bus B was one of the command module's electrical junctions, and the warning light could only mean that something had disrupted the flow of power to the command module's systems, perhaps an electrical short; if so, it was very bad news. Swigert called to Lovell, "The MAIN B light is on."

The electrical system was Fred Haise's specialty, and now Swigert called out to him to come back to the command module. In the meantime he floated over to Haise's side and scanned the electrical gauges. To his surprise, the voltage and current readings looked normal, and so did the fuel cells. Maybe there had been some kind of momentary glitch. But the bang—he'd *felt* it. The whole spacecraft had shuddered. It had scared the hell out of him. With only slight urgency in his voice, he called mission control. "Okay, Houston; we've had a problem."

"This is Houston," said Jack Lousma. "Say again, please."

Lovell was back in the center couch now. He keyed his mike. "Houston, we've had a problem. We've had a MAIN B BUS UNDERVOLT."

"Roger, MAIN B UNDERVOLT." There was a pause. "Okay, stand by, 13; we're looking at it."

Swigert could see no clear explanation on *Odyssey*'s instrument panel. Something must have happened to the lunar module. He wondered if a meteorite had struck it. The chances were supposed to be infinitesimal, but it would only take one sand-sized particle, moving at 8 or 9 miles per second, to cause crippling damage. If *Aquarius* had been hit, it could be losing pressure this very second. Swigert glanced toward the open hatchway and had the same thought that comes to men in a submarine taking on water: seal off the damage. As soon as Haise reappeared, Lovell and Swigert tried to install the forward hatch at the mouth of the tunnel, but it would not close; in their haste they had misaligned it. Suddenly they realized there was no air leak; apparently, nothing was wrong with *Aquarius* at all. But as Fred Haise was now discovering, something was very wrong with *Odyssey*.

When he heard the bang, and then Swigert's call, Haise had dropped what he was doing in the LM and headed for the command module. As he floated through the tunnel he heard pinging and popping noises: the sound of metal flexing. The walls of the tunnel were bending as the two joined spacecraft rocked back and forth against each other. Something had jolted Apollo 13. He heard the sound of *Odyssey*'s thrusters firing in response. Quickly, Haise made his way to his seat on the right side of the command module and began to investigate. *Odyssey*'s electric power came from three chemical-power plants called fuel cells, housed in the service module. Each fuel cell mixed hydrogen and oxygen to produce water; the by-product of this reaction was electricity. Current from the three cells flowed through two junctions, or buses, called A and B, which distributed power to run hundreds of different components.

Everything in the spacecraft—the gauges on the instrument panel, the computer, the ball valves in the service module's SPS engine—ran on electricity. Haise checked the voltage on Bus B and the needle sank past the bottom of the scale; there wasn't even enough voltage to give a reading. As more warning lights came on Haise checked fuel cell number 3, the one that was supposed to be supplying power to Bus B, and saw that it was dead. The mission rules were printed on a card attached to the instrument panel, but Haise didn't even have to look at them. They were forbidden to go into lunar orbit unless all three fuel cells were working. Without being told, he knew the shattering disappointment of what had happened. They were not going to land on the moon. They had lost the mission.

But this was no time to dwell on that. Haise focused on *Odyssey*'s stricken electrical system. His moves came instinctively now, sharpened by hundreds of hours in the command module simulator. He began reconnecting *Odyssey*'s systems from the dead Bus B onto Bus A. Suddenly another red warning light flashed on—that bus too was starving for current. He checked fuel cell number 1; it was now dead too. Only one cell—number 2—was still producing current.

Lovell and Swigert, back in their couches, joined in the troubleshooting. The command ship and the attached lander had been set in motion by the bang and were still slowly turning. Swigert's best efforts didn't steady them, and he wondered whether some of the service module's maneuvering thrusters had stopped working. Meanwhile, Lovell checked the gauges for the oxygen tanks that fed the fuel cells and found strange readings: Just minutes ago they had both been full; now tank number 2 was completely empty, and tank 1's pressure had dropped to a third its normal reading and was still falling.

The dimensions of the problem grew with each passing moment. The three men had never seen so many seemingly unrelated malfunctions at once—maneuvering thrusters out, fuel cells dead, oxygen tanks losing pressure—even the computer had hiccupped during the bang. At their diabolical worst, the simulator instructors had never served up anything like this. There was an unwritten rule against simulating *multiple, unrelated, simultaneous* failures. In a spacecraft with thousands of separate components, the possible combinations of malfunctions would have been so numerous that it was impossible to train for them all. And everyone had always considered the odds of such a scenario very small.

In Houston, Gene Kranz was at the flight director's console, presiding over a team of flight controllers who could not believe what they were seeing. In the minutes since Swigert's first report of trouble, the situation aboard Apollo 13 had worsened on so many fronts that they suspected some kind of instrumentation failure had caused a slew of wrong readings. Even now, as they searched for some way out of the maze of problems, they wondered if they could believe their data. But that was because they were earthbound. They had not heard the bang, or felt it the way Swigert had, or sensed its reverberations as the spacecraft slowly turned in response. And if they could have seen what Jim Lovell was seeing, right at that moment, they would never have blamed it on faulty instrumentation.

Lovell glanced out the side window next to Jack Swigert and what he saw gave him a queasy feeling in the pit of his stomach. A huge sheet of gas streamed from *Odyssey*'s side, swirling in the sunlight like cigarette smoke. For the first time he realized the depths of the crisis. He keyed his mike. "It looks to me that we are venting something. We're venting something out into space. It's a gas of some sort."

The gas was oxygen. When Swigert turned on the fans, he unknowingly triggered an electrical short inside a tank full of supercold liquid oxygen. The bang that rocked Apollo 13 at 2 days, 7 hours, 54 minutes Mission Elapsed Time was the number 2 oxygen tank exploding. Every one of the failures that followed was a result. The jolt from the explosion caused the fuel cells' reactant valves to snap shut, cutting off their supply of oxygen and hydrogen and starving the electrical system. It also closed valves in the propellant lines that fed the maneuvering thrusters, making it very difficult for Swigert to steady the spacecraft, which was turning in response to the propulsive action of the venting gas. It caused the computer to stop in the middle of its work and suddenly restart itself. And the blast tore out part of the plumbing for the service module's remaining oxygen tank, allowing its contents to spill overboard.

Lovell and his crew did not know what had happened inside their service module, and they could not lay eyes on it. The one thing they could see, the smoky haze spewing from its side, didn't single out either a micrometeorite hit or an internal explosion as the cause. But it wasn't important to know exactly why this was happening: all that mattered was that *Odyssey* was swiftly dying. It was only a matter of time before the last oxygen tank was empty and the last fuel cell was dead. At that point *Odyssey*'s only source of electricity would be the command module's batteries, but they had to be reserved for reentry. For Jim Lovell and his crew, the disappointment of losing the mission paled before the realization that they were in danger of losing their lives.

• • •

It took Marilyn Lovell about fifteen minutes to drive from the Mission Control Center to her home in Timber Cove, and when she arrived she phoned to invite her close friends and neighbors, Betty and Bob Benware, to come over for a drink. They declined; they had just come from one of the many Apollo 13 parties thrown by the aerospace contractors, and now they were tired. After Marilyn hung up, the doorbell rang; Pete and Jane Conrad had dropped by on Pete's brand-new red Honda motorcycle. They hadn't been there very long when a call came in for Pete on the "hot line," the direct link to the control center that NASA always installed in the crew's homes during a mission. As Pete went into another room to take the call, Marilyn's phone rang. It was another neighbor, Jerry Hammick, who ran the recovery operations for Apollo.

"Marilyn," Hammick said, "I just want you to know that countries from all around the world have offered to help with the recovery efforts, even the Soviets."

Marilyn couldn't make any sense out of what he was saying. She guessed he had been at the same party that had left the Benwares so tired. "Jerry," she asked, "have you been drinking?"

"Hasn't anyone told you?"

"Told me what?"

"The mission's been scrubbed."

She felt a wave of disappointment. "They're not going to land on the moon?"

"There's no way they can, Marilyn."

Before she could find out any more, Marilyn looked up to see Pete Conrad at the top of the small staircase leading to the family room, motioning for her to hang up. His eyes were wide with alarm.

"Jerry," Marilyn said, "I'll have to talk to you later."

Pete came to her and quietly explained what he had just heard: "They've had a serious problem. They can't use the command module at all. . . ."

Suddenly the house was filled with her closest friends and neighbors. They stood in her living room in anguish. Someone turned on the television set and there was ABC's Jules Bergman, intoning the grim worst. He was all but saying the three astronauts were doomed. Marilyn felt the grip of panic. She hurried out of the living room, away from the crowd, and headed for the bedroom, and then into the master bathroom, locking the door behind her. In tears, she sank to her knees and prayed to God to bring Jim back to her alive.

9:38 P.M.

"Thirteen, we've got lots and lots of people working on this. We'll get you some dope as soon as we have it, and you'll be the first to know." Jack Lousma was doing the best he could to be reassuring.

"Oh, thank you," Lovell said quickly. His noticed his mouth was dry.

Even as Lousma spoke, Lovell and his crew were trying to come to grips with their situation. They were 200,000 miles from earth, getting farther away every second. If only there had been enough electrical power left to operate the service module for even another hour, they might have been able to make what was called a direct abort, firing the SPS engine for all it was worth and executing a big U-turn in space. But even assuming the engine itself hadn't been damaged by the explosion—and there was no way to know—they could not operate it without electricity to open the valves to the combustion chamber, swivel the engine nozzle, and steer the spacecraft through the burn. The SPS was useless, and its 40,000 pounds of propellants were just dead weight. The bottom line was, they were going to the moon—or more precisely, they were going around it.

That would not have been a problem had Apollo 13 been on the free-return trajectory that Apollos 8, 10, and 11 had followed. But when Apollo crews started heading for landing sites in the western regions of the moon, it became necessary to change the timing of their arrival in order to ensure the proper lighting conditions for the landing. The solution was to make a slight change in their approach to the moon. The new path, known as the hybrid trajectory, also had the advantage of saving fuel. The extra risk from giving up the free return was considered minimal. By firing the SPS to place them on the hybrid path, went the logic, the astronauts would demonstrate that the engine was healthy and would work properly to get them into and out of lunar orbit. The very act of throwing away their free ticket home would demonstrate that they didn't need it. As proof, Pete Conrad's crew had used the hybrid trajectory on Apollo 12 without mishap. It had always been assumed that any failure bad enough to cripple the SPS engine, with all its redundant parts—let alone one that could knock out an entire spacecraft—would kill the crew. No one had anticipated the situation that was shaping up this Monday night—three very live astronauts with a dead command module, heading for the moon.

Lovell and his crew knew that unless they did something to get back on the free return—and soon, before they went around the moon—they would miss the earth by some 45,000 miles. The only hope was at the other end of the tunnel: *Aquarius*. Its descent engine had far too little power for a direct abort, but it would suffice to get Apollo 13 back on the free return. Before they could fire the engine, however, they would first have to align *Aquarius*'s navigation platform with the stars, and one glance out the window told the men that would be all but impossible. The sky was filled with particles of debris from the explosion—in a cloud that observers on earth estimated to be 20 miles across—that caught the sunlight and shone like real stars. Furthermore, the LM's navigation telescope couldn't home in on stars automatically the way the command module's could. There was only one way to align the LM's platform: Lovell and Haise would have to copy the alignment from the command module's platform and feed the information into the lander's computer. And they would have to act fast: *Odyssey* was almost out of power. Already, Houston had instructed Swigert to turn off some of the systems; soon he would have to turn off the rest, including the platform and the computer.

Lovell and his crew had arrived at this basic plan when, after an hour and a half of troubleshooting, Jack Lousma radioed, "We're starting to think about the LM lifeboat."

"That's what we're thinking about, too," Swigert answered with an unwavering calm that typified the radio transmissions that had gone back and forth across the translunar gulf for the past 90 minutes. As Swigert spoke, the pressure in oxygen tank number 1 was slowly falling toward zero. Realizing that it was beyond saving, Haise left his seat in *Odyssey* and headed back through the tunnel to *Aquarius*.

• • •

When the emergency began, Ken Mattingly had left the viewing room and taken a seat next to Jack Lousma, where he plugged in a headset. He listened to the voices on the flight director's loop as Gene Kranz and his controllers grappled with the bewildering mess that was unfolding 200,000 miles out in space. As an expert on the command module, Mattingly knew it was serious as soon as he heard Lovell and Swigert report the electrical failures. The crisis brought him right out of his funk. His thoughts now were of his crewmates, and how he might help bring them back.

It wasn't hard to understand why Mattingly, and everyone else in mission control, reacted to the emergency with disbelief. In this business there were certain well-based assumptions: Things broke, but they never did physical damage to the spacecraft in the process. It went without saying that oxygen tanks didn't explode. Listening to the voices of Kranz's men on the flight director's loop, Mattingly could tell that no one had a handle on the problem. The situation changed when Glynn Lunney took over the flight director's chair.

A seasoned veteran at age thirty-three, Lunney had the uncanny ability to go right to the heart of the most complex problem. Now, as Mattingly listened, Lunney proceeded to give an inspiring performance. At first he, like Kranz, focused his efforts on trying to save the command module. When it became clear that this was impossible, Lunney put his LM experts to work coming up with a new checklist to let the astronauts power up *Aquarius* as quickly as possible. There were those in mission control, including one or two off-duty astronauts, who took one look at *Odyssey*'s oxygen readings and privately concluded that NASA had lost three good men. But Glynn Lunney didn't have time for any such thoughts. Mattingly heard him feeding rapid-fire questions and instructions to his controllers, keeping tabs on the dying command module, and checking the progress on the lunar module checklist. He and his men were now marching together toward a clear goal, to get Apollo 13 back on the free-return trajectory, and then home to earth.

Meanwhile, Gene Kranz organized controllers and specialists into teams to work on the rest of the recovery effort. How quickly should they bring Apollo 13 home? To answer that they would have to know how long *Aquarius*'s supplies would last. What about the command module? Could it be powered up after days in the cold of space? Unlike the crises that had threatened the first lunar landing and the launch of Apollo 12, this drama would play itself out over a span of days. Already the simulators were being cranked up so that astronauts could test new procedures devised by Kranz's teams. And across the country, thousands of NASA and contractor workers were mobilizing to save the three men in deep space.

11:13 P.M.

"Didn't think I'd be back so soon," Haise said as he set to work turning on the *Aquarius*'s systems. Time was critical now. As written, the shortest possible activation checklist took two hours, but there were only minutes of electrical power left in *Odyssey*. Jack Lousma passed up instructions from the LM experts, who were working furiously to pare the procedure down to the bare minimum. Meanwhile, inside *Odyssey*, where only the computer and the cabin lights were still turned on, Swigert had switched over to one of the command module's batteries, knowing that most of its power had to be saved for reentry.

While Haise worked, Lovell handled the critical transfer of the platform alignment from *Odyssey* to *Aquarius*. "Do it right," Haise cautioned his commander. "Take your time." To convert the data from *Odyssey*'s platform to *Aquarius*'s frame of reference, Lovell had to go through a fill-in-the-blanks conversion using a worksheet that was printed in the flight plan. He remembered that he had made errors when he'd done this during simulations, but he and Haise both knew he could not afford a mistake now. "I want you to double-check my arithmetic on these numbers," Lovell radioed to Lousma. Once Lovell received confirmation from earth, Haise entered the data into *Aquarius*'s computer, and moments later the platform was aligned with the heavens. Minutes later, after Haise had activated the LM's maneuvering thrusters, there was no longer any need to keep *Odyssey* alive on precious battery power. Lovell yelled up the tunnel to Swigert: "Shut her down!" Nine minutes before midnight, Swigert turned off the last system, and *Odyssey* fell dark. Sometime after midnight, Lovell and Haise were in the midst of their work when Swigert floated into the LM cabin, a forlorn expression on his face. They understood; he had just put a friend to sleep. Swigert looked at his crewmates and said, "It's up to you now."

After midnight, Tuesday, April 14

Even now, oxygen spewed from *Odyssey*'s side like blood from a harpooned whale. The escaping gas acted like a small rocket, fighting Lovell's efforts to stabilize the joined craft—which the astronauts called "the stack"—with *Aquarius*'s thrusters. Lovell soon found that trying to control the stack from the lander was strange and awkward, like steering a loaded wheelbarrow down the street with a long broom handle. When he nudged the hand controller the joined craft wobbled unpredictably. It was, Lovell would say later, like learning to fly all over again. And he had to learn fast, because if he let the spacecraft drift uncontrolled, there was a danger that one of *Aquarius*'s gyros would be immobilized—a condition called *gimbal lock* that would ruin the alignment of the navigation platform. With no way of sighting on stars, there would be no hope of realigning it. Mindful of this, Haise said, "Why the hell are we maneuvering at all now?" The strain of the moment was in his voice. "Are we still venting?"

"I can't take that doggone roll out," Lovell said. Throughout the next 2 hours Lovell wrestled with his unwieldy craft, as the time for the free-return maneuver approached. He wondered if *Aquarius* would be able to point them toward home, and whether it would last long enough to get them there. Lovell and his crew had become the first astronauts to face the very real possibility of dying in space. If the free-return maneuver wasn't successful, Apollo 13 would miss the earth by thousands of miles, and continue to circle endlessly in the void. Years later, people would ask him whether he carried suicide pills, and he would answer that in all his years with NASA he had never heard of such a thing. There wasn't any need for them in a spacecraft; all that was necessary to take one's own life was to open the hatch. And even if he knew he were doomed, he would continue to transmit data back to earth as long as the radio held out, or until he and his crew had succumbed to lack of oxygen. But Lovell was determined to avoid that kind of death. At all costs, he told himself, he and his crew must get back to earth, even if they did not survive. Better to burn up in the atmosphere, Lovell thought, than to become the first human beings never to return to their home planet.

2:40 A.M.

After an hour of preparation, Lovell and Haise were ready to fire *Aquarius*'s engine for the free-return maneuver. Apollo 13 was oriented correctly and held in position by the LM's computerized autopilot. Just let that computer keep working, Haise thought; he knew this burn required enough precision that orienting the spacecraft by hand might not be good enough. At one point during the preparations, Capcom Jack Lousma joked, "How do you like this sim?" Lovell answered wryly, "It's a beauty."

Heirs to the Original 7: The "New 9" on desert survival training, 1963. (From left) Neil Armstrong, Frank Borman, Pete Conrad, Jim Lovell, Jim McDivitt, Elliot See, Tom Stafford, Ed White, and John Young. Eight of them would fly on Gemini; Elliot See would die before he got the chance. At far right is their boss, Deke Slayton. *(NASA photo courtesy National Geographic)*

Last in line: Al Bean (far left), Bill Anders (second from left), Walt Cunningham (kneeling), and Rusty Schweickart (with sunglasses) study a telescopic image of the sun with scientists at Kitt Peak National Observatory in Arizona in 1964. For them, a seat on a space mission would be years away.
(Unless otherwise stated, all photographs are from NASA.)

ABOVE: The Apollo 1 crew: Gus Grissom, veteran of Mercury and Gemini; Ed White, the nation's first space walker; Roger Chaffee, a bright and energetic rookie.

LEFT: Apollo's darkest hour: the scorched wreckage of the Apollo 1 command module, hours after the fire that killed Grissom, White, and Chaffee on January 27, 1967.

ABOVE: Apollo 8 commander Frank Borman. A key player in the recovery from the Fire, Borman was determined to vindicate the Apollo spacecraft. When Slayton offered him a chance to fly it around the moon, he took it.

BELOW: December 21, 1968: Borman's crew leaves earth, riding the most powerful rocket in existence, the Saturn V.

LEFT: Four days before launch, Bill Anders (left) and Jim Lovell share a joke before entering the command module simulator (background).

BELOW: "Happy Birthday, Mother." Lovell and Borman (background) on TV, 140,000 miles from home.

ABOVE: Like the deserted battlefield of the final war: the lunar far side, seen from a distance of 69 miles.

BELOW: The most electrifying sight of all: earthrise, photographed by Bill Anders on Christmas Eve, 1968.

ABOVE: One of the most unforgiving flying machines ever built, the Lunar Landing Training Vehicle was essential to learning how to land on the moon.

ABOVE: April 1969: Neil Armstrong (right) and Buzz Aldrin clamber through a suited run-through of their moonwalk.

RIGHT: July 20, 1969: An hour into history's first moonwalk, Buzz Aldrin photographed his own boot and its imprint on the ancient lunar dust.

ABOVE: Posing before the flag, Aldrin sensed the audience of millions on earth—and the enormous distance separating them from himself and Armstrong.

LEFT: Inside the lunar module *Eagle*, Armstrong grins through whiskers and tired eyes after the moonwalk.

RIGHT: Meanwhile, in lunar orbit, Mike Collins keeps a solo vigil in the command module *Columbia* until his crewmates return.

Best friends: Pete Conrad and Dick Gordon, after their Gemini 11 mission in November 1966. Dick Gordon's inscription reads, "To Pete: After three days I thought you'd stop yelling at me. P.S., I'm still listening." *(Courtesy Pete Conrad)*

BELOW: After six years of waiting: Al Bean in the lunar module simulator, training to land on the moon as Pete Conrad's lunar module pilot.

LEFT: November 19, 1969: Bean takes his first step onto the Ocean of Storms.

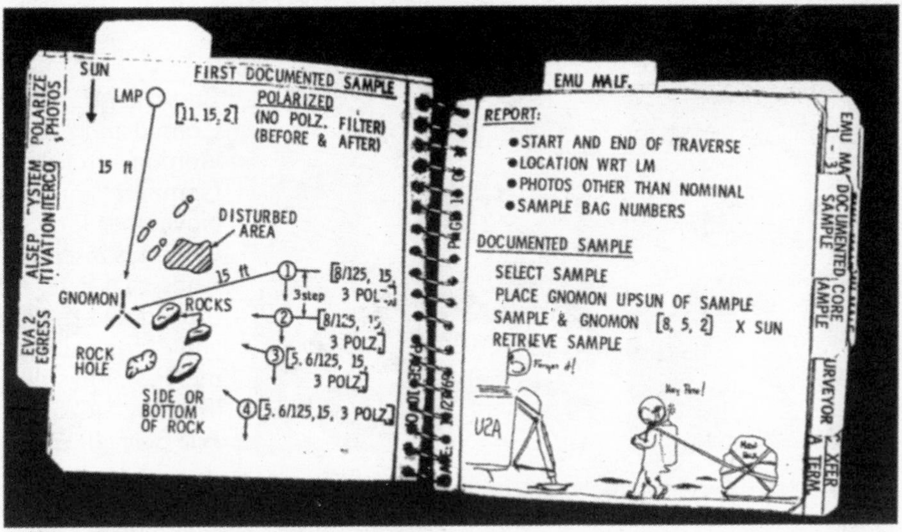

Two pages from Pete Conrad's cuff checklist for the moonwalks, complete with Snoopy cartoons. *(Courtesy Pete Conrad)*

BELOW: Visible proof of the pinpoint landing: Conrad examines Surveyor 3, perched on the slopes of a wide, old crater. Six hundred feet away, on the crater rim, the lunar module *Intrepid* looks like a tiny scale model. To its right is a collapsible communications antenna.

APOLLO 13

ABOVE LEFT: Ken Mattingly, grounded by NASA doctors three days before launch because of a suspected case of German measles, studies a flight plan in mission control. For now, Apollo 13 is still a normal mission.

ABOVE RIGHT: Soon after the explosion of an oxygen tank aboard Apollo 13, astronauts and flight controllers study data in mission control. From left: Tony England, flight controller Raymond Teague, Joe Engle, Gene Cernan, Ed Mitchell, Al Shepard, and Ron Evans.

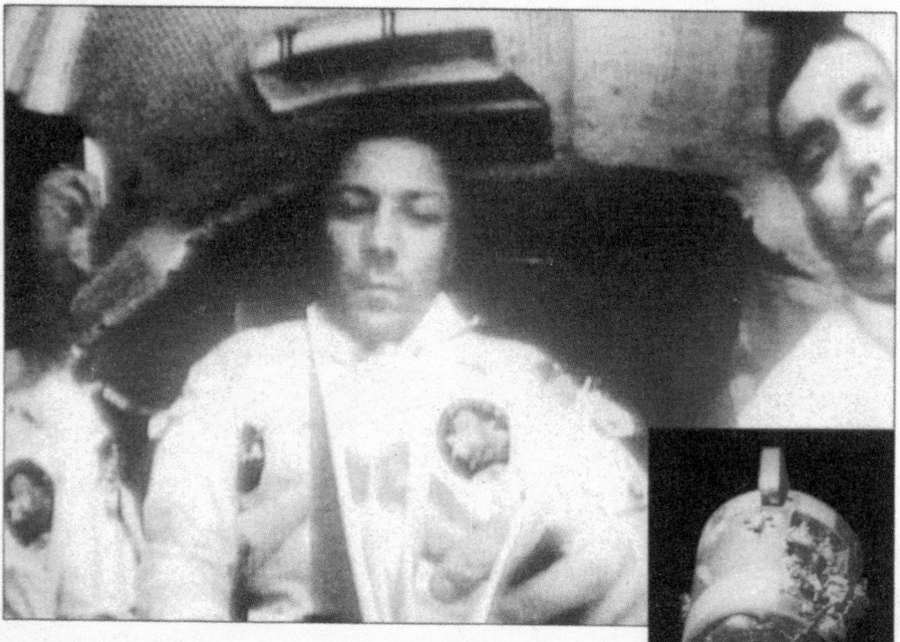

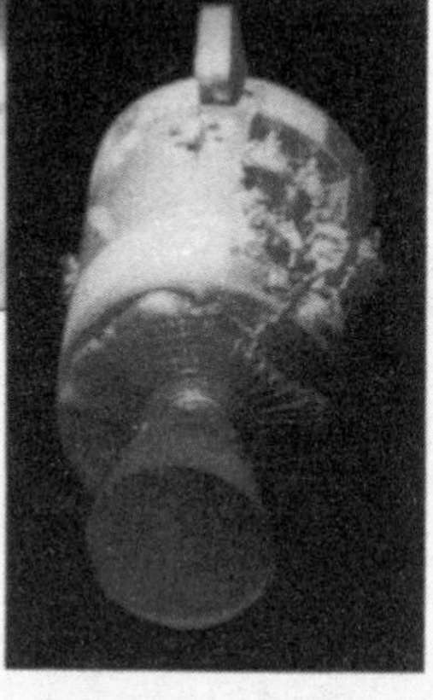

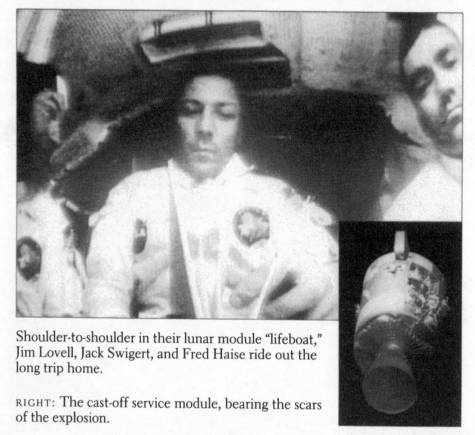

Shoulder-to-shoulder in their lunar module "lifeboat," Jim Lovell, Jack Swigert, and Fred Haise ride out the long trip home.

RIGHT: The cast-off service module, bearing the scars of the explosion.

Al Shepard (LEFT, with tool cart) and Ed Mitchell (BELOW, with map) during their walk up Cone crater on February 6, 1971. Minutes later they would be immersed in a tiring and frustrating search for the crater's rim.

BELOW: The end of a full-up mission: Stu Roosa, Al Shepard, and Ed Mitchell smile from inside their quarantine trailer.

The master teacher and some of his students: Caltech geologist Lee Silver (gesturing, in hat) points out features to the Apollo 15 prime and backup crews at 12,000 feet altitude in the San Juan mountains of Colorado. From left: Dick Gordon, Jim Irwin, Jack Schmitt (behind), Dave Scott, Silver, and USGS geologist Tim Hait. *(Photo by Joe Allen, courtesy Lee Silver)*

ABOVE: "Little Joe" Allen mans the Capcom console in mission control while Scott and Irwin explore the valley of Hadley-Apennine.

Three hundred feet up the flank of Hadley Delta mountain, Scott photographs a geologic find before collecting it.

With Hadley Rille as a backdrop, Dave Scott unloads supplies from the Rover near St. George crater on August 1, 1971.

ABOVE: The command ship *Endeavour*, with Al Worden inside, after three days of solo exploration in lunar orbit.

ABOVE RIGHT: One of Worden's cameras captured these two lunar mysteries: the unusually bright near-side crater Aristarchus, 27 miles across, and, to the right, Schröter's Valley.

Inside the Lunar Receiving Lab, Dave Scott gets a look at the Genesis Rock.

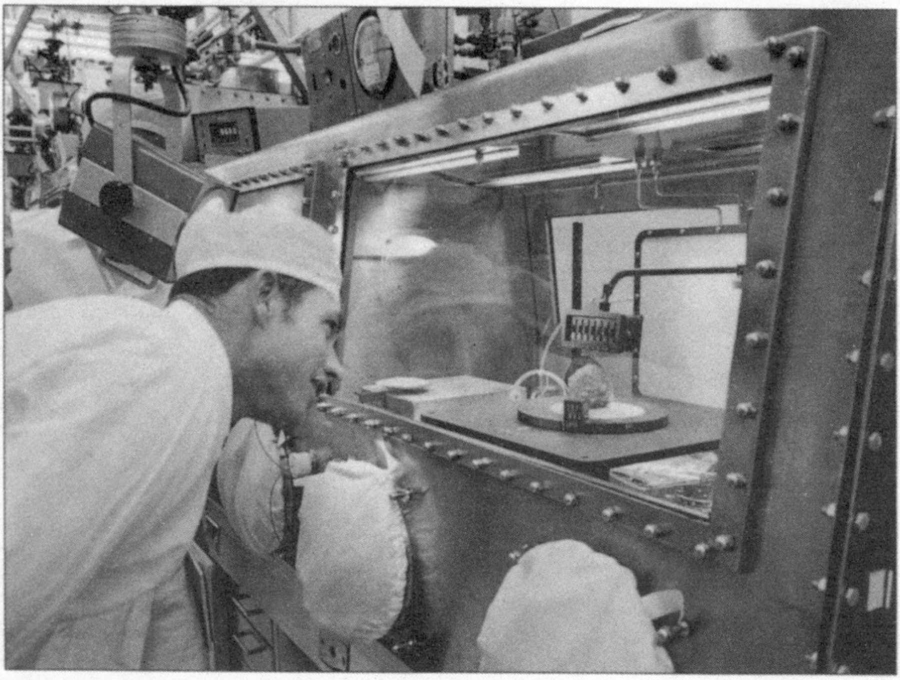

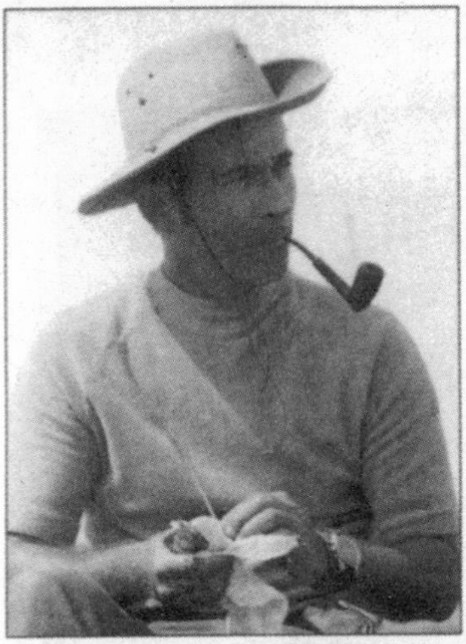

LEFT: John Young, whose intelligence was masked by his self-effacing wit and country-boy drawl, became the first man to journey twice into lunar orbit.

BELOW: Ready to explore the Descartes highlands, an ecstatic John Young salutes the flag while jumping three feet off the ground.

BELOW: At huge North Ray crater, Young gathers samples with a specially designed rake, looking for volcanic rocks but finding none.

RIGHT: Young approaches one of the more modest-sized boulders he and Charlie Duke sampled.

The full earth. Africa and the Arabian peninsula are clearly visible; Antarctica is at bottom.

BELOW: The first scientist on another world: Jack Schmitt, scoop in hand, prospects for samples at Camelot crater.

ABOVE: In the geology back room, Jim Lovell and Bill Muehlberger (seated) await the final moonwalk.

ABOVE: Televised by the Rover's remotely controlled camera, Schmitt goes about his work, high on the side of the North Massif. His face is visible through the clear inner visor of his space helmet.

LEFT: Ron Evans (left) studies the moon with his mentor, Farouk El-Baz.

"Talk about being a spaceman, this is it!" An exuberant Evans retrieves exposed film during his space walk 180,000 miles from earth.

The last moon voyagers: Gene Cernan, Ron Evans, and Jack Schmitt wave to recovery forces on the deck of the carrier U.S.S. *Ticonderoga*.

At 2:42 A.M. the descent engine rumbled to life at minimum power. Lovell waited 10 seconds, then slid the throttle to 40 percent. Lovell and Haise, standing, felt themselves pressed gently against the floor; Swigert settled onto the ascent engine cover. Twenty-one seconds later the computer shut down the engine automatically, precisely, perfectly. The burn was so good that Lovell and Haise didn't need to make any adjustments with the maneuvering thrusters. For the first time, the men let themselves believe they would make it back to earth. Another engine firing to hasten the homeward voyage still lay ahead, after Apollo 13 swung around the moon. But the worst was over—or so they thought.

II: The Moon Is a Harsh Mistress

Tuesday, April 14
3:30 A.M., Houston time
2 days, 14 hours, 17 minutes Mission Elapsed Time

Fred Haise looked at his watch and was amazed to see that more than 6 hours had passed since the accident. He had been so absorbed in getting the LM activated, then preparing for the burn, that he had been completely unaware of time. Hours had passed this way, in intense activity that was punctuated by breathing spells as he waited for mission control to verify his work. During those brief respites, Haise registered a swift succession of images beyond *Aquarius*'s triangular windows: a cloud of debris particles sparkling in the black sky, the moon looming before him, big and close. He could scarcely believe how thoroughly his reality had changed in such a short time. It seemed like only minutes since he and Lovell were putting on their television show in this same tiny cabin.

Fred Haise, who had been flying for NASA since 1959, applied for the astronaut corps not out of any particular desire to go into space, but because it made sense to him to go where his agency was concentrating most of its resources. He was a member of the fifth astronaut group, the ones who called themselves "The Original 19." There was more than a trace of irony in the nickname. They were the third string, the benchwarmers, the "Red Shirts" of the Astronaut Office. Their chances of getting a seat on an Apollo mission seemed slim to nonexistent. But the Nineteen took the lead in one very important arena: hardware. With most of the more senior astronauts already in training for Gemini or Apollo, many of the Nineteen were assigned to represent the Astronaut Office at contractors' plants. Some, like Jack Swigert and Ken Mattingly, were detailed to Downey, to help shepherd the first command modules through testing. Others, including Haise and a navy pilot named Edgar Mitchell, were assigned to Grumman, and their life became the lunar module. Haise all but lived at the factory. He developed the kind of intimacy with the craft that only a "company pilot" gets. He knew far more about the guts of the lander than anyone needed to know in order to fly it. He knew what the relays looked like; he knew which pins on the electrical connectors went to which systems. He knew just about every one of the odd, one-of-a-kind parts that Grumman fashioned at the height of the weight-saving crunch. More nights than he could count, Haise had curled up on the floor of a lunar module and catnapped while a test was delayed. After two years of this he felt he knew the machine almost like a part of himself. He had unwavering confidence in *Aquarius*, but he also knew that he and his crewmates were asking it to perform in ways it had not been designed for. From here on, they were on untried ground.

Aquarius was designed to support two men for 45 hours—a time span that included a systems checkout, the lunar landing, a 33-hour stay on the surface with two moonwalks, and the rendezvous with the command module. Now, Lovell's crew would ask it to support three men for perhaps twice as long—the journey back to earth would take anywhere from 77 to 100 hours, depending on what kind of engine firing the men made after rounding the moon. The situation was pure black and white: *Aquarius* would have to function at less than half its normal ration of supplies, or Lovell, Swigert, and Haise would not be alive when they reached earth. Immediately after he and Lovell returned to the LM, Haise began doing mental calculations of oxygen, power, water, and other consumables. Later, when there was a lull, he got out his books, where he had data on every one of the lander's systems, along with actual flight performance from previous lunar modules. Would *Aquarius*'s reserves be enough? Haise didn't even bother calculating oxygen. He knew without checking that there was more than enough for the trip home. First there were the oxygen tanks in the lander's descent and ascent stages. Then, the oxygen in the backpacks he and Lovell would have used on the moon. And finally, a set of emergency bottles in the command module. In all, there was enough oxygen to keep three men alive for 8 or 10 days—in other words, more than twice as much as they needed.

Haise's worry was electrical power. Unlike the command module, the LM had no fuel cells and relied solely on batteries. Fully charged, *Aquarius*'s batteries were good for roughly 2 days of normal operation. It was obvious that the only way to stretch them to 4 days was to turn off most of the LM's systems—cabin lights, gauges, and after the next rocket firing was over, even the computer—and he knew mission control would have them do just that. Haise went down through the list of the items he knew they *wouldn't* turn off: radio, 1.29 amps; environmental control system, 5.85 amps. . . . Haise figured it would take about 18 or 20 amps to keep *Aquarius* going at a bare minimum—about as much as a 1970-vintage refrigerator and color television would consume together, and less than half the 50 amps the LM used in normal operation. *Aquarius*'s batteries could supply 20 amps for 100 hours—enough to see them home, and at best, perhaps a full day's worth more than they'd need.

But battery power alone would not keep the LM running. Electronic gear gets hot, and it must be cooled to keep functioning. The most precious substance aboard *Aquarius* wasn't oxygen or food; it was cooling water, and Haise found that the 338 pounds stored in *Aquarius*'s tanks weren't enough to last the trip. By his calculations, *Aquarius* was probably going to run out of water about 5 hours before reentry. That would have been an ominous prediction if not for an important bit of data from Apollo 11. Before Neil Armstrong and his crew cast off *Eagle's* ascent stage in lunar orbit, they began an experiment: they left everything inside the lander operating, but deliberately turned off the water supply. *Eagle* was a good soldier; it transmitted valuable data right up until it overheated and died. If *Aquarius* was as good a ship as *Eagle*, and Haise was sure that it was, then even if the water ran out, *Aquarius*'s systems would keep running for about 8 hours more. That left 3 hours to spare. When Haise had finished his calculations, he knew that if nothing else went wrong, they would probably make it. And when mission control called with their own assessment of the situation, Haise realized that the LM experts, more than 200,000 miles away, had reached the same conclusions.

With the free-return maneuver out of the way, Haise found himself exhausted and emotionally drained. The normal flight plan had gone out the window after the emergency, and with it, normal rest periods. It had been more than a full day since the men had slept, and Houston recommended they set up watches. Lovell decided that he and Swigert would stand watch in the LM while Haise slept in the command module; that way, either himself or Haise would always be available to take care of *Aquarius*. "I'll wake you up," Lovell said, and Haise headed for *Odyssey*.

6:24 A.M.

Jack Swigert and Jim Lovell floated in *Aquarius* while Haise slept in *Odyssey*. Every hour, Lovell had to use *Aquarius*'s thrusters to rotate the stack 90 degrees, so that the sun's warmth would be evenly distributed on the hull. Normally, this was done by the command module computer, but the LM's computer had no software to perform this slow-motion-barbecue spin; Lovell had to perform the maneuvers by hand. Even this simple task was made difficult by the unwieldy craft, but at length he managed it. Now, as the stack turned, the moon slid into view. Already it was close enough to see craters with the unaided eye.

"That old moon's getting bigger and bigger, Jack."

Jim Lovell and Jack Swigert did not know each other very well. They had arrived at NASA four years apart, and as a backup crewman, Swigert had not worked closely with Lovell until the day he took Ken Mattingly's place. In some ways Swigert and Mattingly were opposites. Both were bachelors, but that was where the similarity ended. Mattingly's whole life seemed to revolve around the program. He appeared to have no social life whatsoever; he was an ascetic monk at the monastery of Apollo. Once, when a reporter asked him whether he had any plans to get married, Mattingly blushed and jingled the change in his pocket.

Swigert, meanwhile, was a world-class bachelor who pursued his calling with the same methodical, fastidious approach he brought to test flying. His date book included codes that signified single, divorced, or widowed. His success with women was at once legendary and utterly mysterious. More than one astronaut who teased Swigert about the way he dressed—he wore white socks with a suit and tie; his feet were flat and the sides of his shoes wore out before the soles—had to will his jaw shut when Swigert showed up with a drop-dead-gorgeous date. They knew Swigert was doing just as well wherever his T-38 took him. But those who worked with Swigert knew his dedication to Apollo. After the Fire, when it was time to get down to rewriting checklists and procedures, Swigert had been tireless. His battle cry was persistence: Even after NASA turned him down twice, for the second and third astronaut selections, Swigert was back to do battle with the ink blots and the treadmill in 1966; this time he won.

When talk about replacing Mattingly started, Lovell spoke to Tom Paine to try and stop it. He had nothing against Jack Swigert, but Mattingly knew the command module down to the last detail, and Lovell had come to depend on him. He told the NASA administrator that if Mattingly got sick it would probably be on the way home, with no impact on the mission. But Paine refused to overturn Chuck Berry's recommendation. It wasn't worth the risk, he said, especially since an illness would bring NASA a great deal of negative publicity. Lovell realized he had no choice but to accept the situation.

But taking Swigert onto the crew was no simple matter. It was true that Swigert was a command module expert, having lived with the craft at North American the way Haise had done with the LM at Grumman. When it came to the emergency procedures for command module malfunctions, Swigert literally wrote the book. He was one of the only astronauts who actually went to Deke Slayton and asked to be a command module pilot.

The problem was that Swigert had been training with John Young and Charlie Duke for eight months. He'd forged a close working relationship with them; he was used to their techniques. Now, with launch only days away, he was switching crews. Would he be able to mesh with Lovell and Haise in the critical phases of the mission? During an emergency rendezvous there would be no chance to explain verbal shorthand like, "You've got the next burn." Lovell told Tom Paine he'd agree to the switch on one condition: This shotgun marriage would have to be tested in the simulator.

And so in the last days of their training, just when they should have been taking it a little easier, Lovell and Haise were in the simulator for hours on end, running through practice launches and rendezvous with Swigert. By Friday, they knew the marriage was going to work. But by launch day, the sadness they felt for Ken Mattingly and the annoyance with the daily blood tests and other medical hassles of the past week all but robbed them of the excitement of going. When the three men arrived at the White Room, sealed in their space suits, the usual exuberance of the closeout technicians was gone; they were subdued and nervous. Still, Lovell knew, if he and Haise had been through ups and downs, that was nothing compared to what Swigert had been through.

Any astronaut who had ever served on a backup crew knew that the experience is an emotional roller coaster. All through the long months of training, the backup man keeps himself motivated with the hope that he might get to go instead of the prime crewman. About a month before the flight, he realizes he's not going after all. A week before the launch of Apollo 13, Jack Swigert was in the same position as every backup crew member before him: he had been mentally and emotionally backed down from the mission for weeks. His biggest worry was finding hotel rooms on the beach for the prime crew's launch guests. Then Charlie Duke got German measles.

And as it turned out, Swigert had fit in extremely well, Lovell thought. In the first two days of the flight, while everything was still going normally, Swigert had proved the benefits of standardized training. He handled his role with professionalism and skill. As much as Lovell hated to lose Mattingly, he had to admit that things were probably going to work out okay—until Monday night. Lovell could only imagine what Swigert was thinking now.

Since midnight Tuesday, Jack Swigert had been a man without a spacecraft. As long as they were in *Aquarius*, Swigert would not do any flying; he had barely set foot inside the LM simulator. He was a spectator, perched atop the can-shaped ascent engine cover in the back of the cabin, while Lovell and Haise worked. The lander seemed so unsettlingly delicate. At one point he was about to take a picture of the moon and Haise cautioned him not to hit the window with the camera; it might break! And he heard strange, distressing sounds. The LM whined and squealed; it gurgled. He asked Lovell and Haise, "Is there something wrong with the environmental control system?" They reassured him: those sounds are normal; the time to worry is when you *don't* hear them.

Now, Swigert watched as Lovell struggled to turn the spacecraft. Through the windows, he could see the moon as a pockmarked half-circle; then the earth, tiny and radiant. It had looked small before the bang; now it seemed even smaller. It was hard for Swigert to accept the fact that the best way to get back there was to keep going away from it. About 12 hours from now, at 6:30 this evening, Apollo 13 would swing around the moon and make its closest approach, or pericynthion, above the far side. Two hours later, according to the plan devised by mission control, the men would fire up *Aquarius*'s descent engine once more. Much would depend on that firing, which was known as the PC + 2 burn, short for "pericynthion plus 2 hours." Without it, the trip home would take a total of 4 days; with it, that time could be trimmed to as little as 2½ days. The increased margins on *Aquarius*'s supplies could mean the difference between making it and not making it. In addition, it would shift the splashdown point from the Indian Ocean, where the navy had few ships, to the South Pacific, where the recovery force was now stationed. But Swigert, with less confidence in this unfamiliar machine than in his own, looked ahead to the PC + 2 burn with some anxiety. He asked Lovell, "Do they think we'll have any trouble doing the next burn . . . ?"

"Probably not, if everything holds together," was Lovell's assessment. "We've already made one burn." But then, in a matter-of-fact tone, he added, "Well, Jack, this is going to be touch and go." He seemed to be talking about the entire journey home, not just the next burn.

Tuesday morning
Timber Cove, Texas

For Marilyn Lovell, the past twelve hours had been a vigil by the squawk box. The men had turned off their radio's amplifier to save power, and sometimes their words were nearly drowned out by static. It helped to have some of the other astronauts in the house, including Pete Conrad, explain to her what was going on. And she took some measure of comfort from the presence of friends, including Susan Borman, Jane Conrad, and Betty Benware. Marilyn's memory of these next hours would be hazy; she would be told later that she lit one cigarette after another, setting them down here and there while friends followed behind putting them out.

In the morning, she sent young Jeffrey, age four, off to nursery school as if nothing were wrong, but her two daughters stayed home. She had already talked by phone to her son James, fifteen, who was at a Wisconsin military academy. Earlier, Marilyn had worried about Jim's mother, who had recently suffered a stroke and was in a nearby nursing home; if she found out about the emergency, Marilyn feared it might kill her. So Marilyn called the home to have the television in her room turned off. But in the Lovell house, the TV was on, and ABC News's gloomy prognosis so upset Marilyn's younger daughter, eleven-year-old Susan, that the girl burst into tears. "He's coming back," Marilyn reassured her. "There's just no reason why he isn't coming back." Marilyn knew it was a white lie. She'd asked some of the NASA people about the odds of saving the men, and they'd been honest with her: 10 percent. But she also knew that she could not give in to her terror, imagining awful scenarios the way she had when Jim was a test pilot at Patuxent River.

Besides, other people had already done that for her. Last year, she and Jim had gone to the premier of a movie called *Marooned* in Houston. In the movie, three Apollo astronauts are stranded in earth orbit after their engine fails to ignite for retrofire. Before long, mission control has run out of ideas, and the head of NASA's manned space program, played by Gregory Peck, is spelling out the doom scenario to a closed-door meeting of his deputies: "If necessary, on Wednesday morning, the President will issue a message to the nation emphasizing the courage and determination of the crew, and their final wish—that the program be continued, without pause." But the chief astronaut, played by David Janssen, will have none of it. He slams the table and barks, "I don't give a damn about the next mission!" Before long, he's convinced NASA to launch a rescue mission with himself as the pilot. And the rescue succeeds —but not before the mission commander, whose name is Jim, is killed in a space-walking attempt to fix the engine. Marilyn did not find *Marooned* very entertaining.

But the movie was far from Marilyn's thoughts. She knew she must put fear out of her mind, or she would not last through this ordeal. After Father Raish came by to conduct a private Mass, her spirits were lifted, and she carried herself with a calm that amazed her friends. Susan Borman said, "I hope the navy's proud of you."

Meanwhile, the squawk box relayed the voices of Jim and his crew, amid static. The men were on hot-mike, and every once in a while it was possible to hear them talking to one another. Their voices did not show any signs of strain. She knew Jim's attitude about panic: Where would it get him? After you bounce off the walls a few times, Jim would say, you're right back where you started.

Right now, Jim was telling Jack Swigert to fill up some juice bags with drinking water from the command module. And then, it was possible to hear Jack greet a returning Fred Haise: "Come on in, Fred-o. Did you sleep good?" Not long after, Jim Lovell could be heard clearly saying, with some fatigue in his voice, "Lookit, I gotta get some sleep."

12:42 P.M.

Jim Lovell was back in *Aquarius* after only 3 hours. Unable to stop thinking about the situation, he hadn't slept. Had they missed anything? Could they find a way to stretch the LM's supplies even further? He'd seen Haise's numbers, but he had a hard time believing them. He wondered about the forecasts coming from mission control—they were a little too positive, a little too vague, and it occurred to Lovell that things might be worse than they were willing to admit. Lovell's thoughts turned to the PC + 2 burn, set for about 7½ hours from now. At that point, some 20 hours would have elapsed since he and Haise had aligned *Aquarius*'s platform. Lovell knew that there was a chance the gyros could have drifted out of alignment, perhaps only slightly, but still enough to affect the accuracy of the maneuver. There was no way to check it against the stars; the sky beyond *Aquarius*'s windows was still a blizzard of debris particles. The only chance to see stars would be during the half-hour that Apollo 13 flew through the moon's shadow. Lovell knew he might eat up those precious minutes simply maneuvering the stack to bring stars into sight of the LM's telescope. There had to be some other way.

Houston had figured out an answer, and at 1:42 P.M., a now-healthy Charlie Duke took the Capcom mike to read up the procedure, and Lovell didn't like what he heard. He and Haise were to test the alignment by sighting on the one star they could see, the sun. Mission control would send up a set of coordinates, which Lovell would use to steer the craft into position. If the alignment was good, the sun would show up in the eyepiece of the navigation telescope. It didn't have to be perfect, they said; even a 1 degree error in alignment would be okay—as long as the telescope cross-hairs were touching the sun's disk. Lovell agreed, but he had real doubts it would work. But the alternative—sighting on stars during the brief minutes in the lunar shadow—he liked even less.

It was around 3:00 P.M. when Lovell wrestled Apollo 13 into position. The task was even more difficult now, because the attitude indicator, or 8-ball, had been turned off to save power, and judging orientation from the computer's readouts was awkward. In Houston, flight director Gerry Griffin waited with great anxiety; he knew this was one of the most critical moments of the flight. Lovell wished Griffin and his controllers could look over his shoulder, but they told him now that they were having trouble locking up on *Aquar-*

ius's telemetry. Lovell would have to do the best he could without them.

Now the solar glare filled *Aquarius*'s windows. While Lovell sighted through the telescope, Haise checked the readouts. Lovell made a correction to the attitude, then moved aside and let Haise look. Lovell asked anxiously, "What have you got?"

Through the dark protective filter, Haise saw a bright arc. "Upper right corner of the sun."

"We've *got* it," Lovell exclaimed. In Houston, the flight controllers cheered, and Gerry Griffin's hands trembled as he wrote the successful result in his log.

Nearly three hours later, Apollo 13 flew into the moon's shadow. Outside, where the sky was suddenly filled with stars, the men could see dark silhouettes which they realized were debris from their explosion. Somewhere out ahead was the moon. Lovell knew he would not see it nearly as well this time as he had on Apollo 8, for Apollo 13 would come no closer than 136 miles. The encounter would be a fleeting one, for they would be moving so fast it would take just half an hour to round the far side, and only 20 minutes of that would be in sunlight. For Jim Lovell, the small, lifeless world that had been his goal for eight years was now simply a way point along a troubled journey. At 6:21 P.M. his headset filled with static as Apollo 13 flew out of contact with earth.

6:29 P.M.

With blinding suddenness the sun reappeared—reassuringly, it was exactly when Houston had predicted—and up in *Odyssey* Jack Swigert and Fred Haise were glued to the windows. Nothing could have prepared them for a sight as alien as the far side of the moon, and for the first time since the accident they forgot their life-or-death situation. As the craft sped past the cratered ball, Swigert and Haise fired off pictures.

"If we don't get this burn off," Lovell told them impatiently, "you won't get your pictures developed."

"Relax, Jim," Haise said. "You've been here before, and we

haven't." Like a billiard ball in bank shot, Apollo 13 rounded the moon, its path bent by the lunar gravity. Soon it would be pointing back toward earth. And now Lovell joined his crew at the windows, pointing out places that only he had seen before, until Houston was on the radio once more. As the full moon shrank into the distance, Lovell and Haise went back to work.

8:40 P.M.

Because of Lovell's anxiety to get ready for the PC + 2 burn, he and Haise finished powering up the LM almost an hour ahead of schedule. Lovell thought of all the power they were wasting just sitting there waiting, and he chided himself for jumping the gun. Finally, he and Haise stood at *Aquarius*'s controls, ready for the maneuver. Because the command module was attached, a situation the LM's software wasn't written to accommodate, Lovell had to fire the engine by hand. When the time came he pushed the ENGINE START button, waited 5 seconds, and slid the throttle to 40 percent. "We're burning," he told Houston. Twenty-one seconds later, the computer throttled up to full power. Lovell was amazed at how smooth and quiet it was; this engine was a beauty. He wished he were using it to land on the moon, right then.

• • •

For the first time since the accident, Jack Swigert's spirits lifted. Every second brought them closer to home instead of farther away from it. If *Aquarius* held out for another 62 hours—and based on mission control's latest estimates, there seemed no doubt that it would—they would make it back to earth. At the same time, Swigert knew, everything they had accomplished so far would be for naught if *Odyssey* didn't function properly when they got there. *Aquarius* was their lifeboat, but only *Odyssey* could ferry them through the earth's atmosphere to splashdown. Right now the command module was nothing more than an inert shell. Swigert called it the "upstairs bedroom." No one had ever tried to revive a dead command module after more than three and a half days of forced hibernation in space; mission control would have to invent the necessary procedures. Swigert had already spent some time thinking it over in his own mind—first align the computer, then get the entry monitor system going. . . . But the checklist was far too massive for him to devise on his own. He would need to have it in writing, carefully scripted, down to the last circuit breaker. And it would have to work right the first time; there would be no postponing reentry. Swigert knew only too well that under normal circumstances it took three months to write a reentry checklist, test it in the simulator, and work the bugs out. The people in Houston had less than two days left to finish this one.

But in the back rooms of Building 30, down the hall from mission control, the reentry checklist was just one of the many efforts which consumed engineers and mission planners. Ken Mattingly was in the thick of it, going around on a freelance basis, wherever the action was and wherever he could be of use. In each room, teams of specialists huddled over schematics and scribbled calculations on blackboards, facing problems like how to conserve power, or water, or how to align the guidance system without a computer. For more than three years, they had been thinking of "what-if" situations and what to do about them. To Mattingly, the most amazing thing was that nearly every solution the teams were coming up with had already been thought of, and sometimes even tested, in previous missions. Firing the LM's engines while it was docked to the command module, for example, was one of the engineering tests crammed into the ten-day Apollo 9 flight.

Then there was the carbon dioxide problem. The LM was equipped with round canisters of lithium hydroxide to scrub exhaled carbon dioxide out of the cabin air. But there were only enough canisters to handle two men for about two days—not three men for more than three days. Carbon dioxide buildup would kill them as as surely as anything else. First would come bad headaches; then their hearts would begin to race. Finally they would grow drowsy and drift off into a sleep from which they would never wake up.

The command module had plenty of canisters, but they were square, the wrong shape for the LM's environmental control system. Unless some way could be found to use command module canisters in the LM, Lovell's crew would suffocate from their own breath. Someone remembered a simulation from Apollo 8 in which the cabin fans had failed, leaving the air inside the command module dangerously stagnant. And the engineers had figured out that in principle it would be possible to build a makeshift air purifier using only items that were readily available in the spacecraft—a plastic cover from a checklist book, a plastic storage bag, and so on. Now, eighteen months later, they would have to build this contraption for real, and make it work in an entirely different spacecraft. Then they would have to figure out how to describe it so the crew could build it without a blueprint or a picture of the finished product. That alone had kept a team from Deke Slayton's Crew Systems division busy since the first hours after the accident. It was one thing to have the pieces of Apollo 13's recovery in hand; it was another to put them together and make them work.

On the rare occasions when Mattingly stepped outside, he saw the whole space center lit up with activity. The lights were on for twenty-four hours a day in the offices of anyone who had anything to do with Apollo. At the Grumman plant on Long Island, North American in Downey, and contractors around the country, people showed up at their offices after the first reports of trouble, and they did not leave. Around the world, Apollo 13 had drawn the attention of more people than any other space mission except the first lunar landing. Headlines from Toronto to Tokyo blared the message that Apollo was in danger. News bulletins cut into radio and TV programs to report new developments. Crowds gathered anywhere there was a television set. In Rome, Pope Paul VI prayed for the astronauts' return before an audience of ten thousand people. At a religious festival in India, pilgrims numbering ten times as many offered prayers for the astronauts. Jim Lovell had no way of knowing that when he'd imagined Apollo 13 would be a mere footnote to history, he'd been very wrong.

10:16 P.M.

"We've gone a hell of a long time without any sleep," Lovell said to his crew, his voice thick with fatigue. Still on hot-mike, he added,

"I didn't get hardly any sleep last night at all." After the PC+2 burn there had been much to do, including getting the joined craft back into a thermal-control spin, still a difficult and frustrating task. Now that was done, but there were other things to worry about. He and Haise were powering down the LM, and at mission control's request, they were about to turn off the guidance system and the computer. It was the last thing he ever imagined he would do intentionally aboard a spacecraft heading for the moon. When the time came to pull the circuit breaker he double-checked with Capcom Vance Brand. If there were any more midcourse burns to make, Houston would have to come up with a way of making them without the computer. When Brand said they were working on just that, Lovell kept his skepticism to himself.

Swigert, meanwhile, was doing some worrying of his own. Which control system, he asked Brand, would Houston want him to use for reentry? Mission control had asked him to make a check on one of *Odyssey*'s electrical buses—did they think it was okay? Suddenly Swigert heard a familiar voice.

"We think you guys are in great shape all around," Deke Slayton said. "Why don't you quit worrying and go to sleep?"

"Well, I think we might just do that," replied Jim Lovell, his tone suddenly more energetic. "Or part of us will."

Wednesday, April 15
12:44 A.M., Houston time
3 days, 11 hours, 31 minutes Mission Elapsed Time

The moon had dwindled steadily since the PC+2 burn; now, some 19,000 miles away, it was about the size of a golf ball at arm's length. But that was still close enough to see craters with the unaided eye, and Fred Haise, alone in *Aquarius* while his crewmates slept in the "upstairs bedroom," took some time to look at it. Near the terminator, just emerging from night, he could barely make out the highland region called Fra Mauro, where he and Lovell were to land. And if it was a strange and alien place, then it had become only too familiar to Haise. As backup to Buzz Aldrin on Apollo 11 he'd mastered the technical and physical demands of getting to the surface of the moon and working there. By the time he began preparing for Apollo 13, Haise was a trained geologic observer who was as excited about the science of his mission as the flying. But the moon beyond *Aquarius*'s windows was the face of disappointment.

At least he could feel good about *Aquarius*. Right now, with only the environmental control system and the radio going, the lander was drawing just a bit more than 12 amps—far less than Haise had figured on. When this flight was over, people would really know what a lunar module could do. *Aquarius* was the reason Haise had been able to sleep for five hours the morning after the accident. He'd trusted his calculations, and it turned out he'd been right. Lovell hadn't found it as easy to accept Haise's results, but Haise knew commanders tended to worry.

With almost nothing to do himself—at the moment, his only job was switching from one antenna to another as the spacecraft slowly turned—Haise's thoughts turned to Houston. From what the Capcoms had been saying, he could just imagine the massive, round-the-clock effort by the flight controllers and engineers. Apollo 13 had become their mission, their test, far more than it was his. According to Jack Lousma, things were going so well that they were already planning for splashdown. He asked Haise, "How would you like to spend a week on an aircraft carrier getting back?"

"If I can get on an aircraft carrier," Haise answered, "I don't care how long it takes, Jack."

III: The Chill of Space

Wednesday, April 15
3:04 A.M., Houston time
3 days, 13 hours, 51 minutes Mission Elapsed Time

"Fred is being relieved now. He went back to get some rest. This is Lovell here who's got the duty."

"Gee whiz," said a surprised Jack Lousma, "you got up kind of early, didn't you?"

In fact, Lovell had returned to *Aquarius* after little more than

3 hours. In part, it was because he knew Haise was tired; in part, it was because he could not stop thinking about what lay ahead. But mostly, it was because it had become very cold in the command module. Yesterday morning, when he and Swigert had tried to sleep there, sunlight had shafted through the windows, keeping them awake, and they had put up the window shades. Normally that wouldn't have been a mistake. But with *Odyssey*'s systems shut down the shades removed the only other source of heat—sunlight —from the cabin. Now it was as wet and cold as a damp cellar. Moisture from the men's breath and perspiration had condensed into a thin film of water droplets all over the walls, the instrument panel, the windows. Lovell tried to rest there after midnight Wednesday, but the chill penetrated the flimsy fabric of his sleeping bag. It was remarkable, though, how weightlessness affected things: He found that if he stayed perfectly still, his own body heat built up around him like a tenuous blanket, because there was no convection to carry it away. But his slightest motion dispelled this warmth and let in the cold. Finally he gave up trying to sleep.

Back in *Aquarius*, Lovell had something new to worry about: Jack Lousma had informed him that Apollo 13 was straying off course for reentry. As a veteran of one lunar reentry, Lovell knew only too well the narrow cone of safety the command module would have to fly through. Something was making Apollo 13's flight path too shallow. Lousma said a midcourse correction burn was planned for about 9 P.M. Wednesday evening. Lovell couldn't imagine how he would align the spacecraft properly for the burn with no computer or guidance system, but Lousma said they were working on it.

7:22 A.M.

Even the carbon dioxide problem hadn't escaped Fred Haise when he did his calculations on *Aquarius*'s consumables. But when he totaled up the number of lithium hydroxide canisters and decided there were enough, he only figured on two men; he forgot that Swigert had to breathe too. Fortunately the engineers on the ground had been on top of the problem, and they'd radioed word about their plan to have Haise and his crewmates construct an air purifier. It would be like building a model airplane, they said, but it would end up looking more like a mailbox. By Wednesday morning, Lovell and Swigert had gathered together the necessary materials: two lithium hydroxide canisters from *Odyssey,* a couple of pieces of cardboard from a flight plan book, a couple of plastic storage bags, and the stuff Haise called space-age bailing wire, gray tape. While Haise slept in *Odyssey,* Lovell and Swigert listened as Capcom Joe Kerwin described the procedure.

"Now then," Kerwin began, "we want you to take the tape and cut out two pieces about three feet long, or a good arm's length, and what we want you to do with them is make two belts around the sides of the canister. . . ." It was an elegant creation. Lovell and Swigert were told to encase each canister in one of the plastic bags, affixing the bag with tape. Before that, however, they were to use the cardboard to fashion a little archway over the top of the canister to keep the bag from getting sucked against the canister's inlet screens. "The next step is to stop up the bypass hole in the bottom of the canister. . . . We recommend that you either use a wet-wipe or cut off a piece of a sock and stuff it in there. . . ." With all the multimillion-dollar equipment aboard Apollo 13, survival had come down to cardboard, gray tape, and socks.

It took about an hour for Lovell and Swigert to finish two of the makeshift air purifiers, and they did resemble a mailbox—in fact, they looked exactly like the air purifier that astronauts and engineers had put together on earth. Later, after Haise installed one of them on the end of an oxygen hose, the carbon dioxide level fell toward normal. Looking back on it, the ingenuity and teamwork embodied in that "mailbox" would amaze him, as much as anything on the flight.

1:34 P.M.

"Jack, for your information," said Joe Kerwin brightly, "FIDO tells me that we are in the earth's sphere of influence and we're starting to accelerate."

It was a piece of news Swigert was glad to hear. "I thought it was about time we crossed," he told Kerwin. "Thank you. We're on our way back home."

In Houston, Gene Kranz and his White Team of controllers were racing the clock to come up with the reentry checklist. Nothing would draw more concentrated effort. Kranz split his people into three separate groups. One would write instructions for everything that should happen in the last four hours of the flight: powering up the command module, casting off the dead service module, letting go of the LM, and finally the reentry itself. Another group was to take care of translating the checklist into a form the astronauts could use. A third team would make sure that nothing called for in the checklist exceeded the slim reserves of battery power and water. At the top of their list was the daunting requirement to make the command module's batteries last three times longer than their normal 45 minutes. It was all but impossible—but they would have to figure out some way. Before long the White Team was aided by a "Tiger Team" of off-duty flight controllers. And Ken Mattingly hunkered down with them. They didn't have any time to waste; reentry was only 2 days away.

Wednesday evening

It was time to put Apollo 13 in position for the midcourse correction. Mission control had instructed Lovell to orient the spacecraft so that the sun shone directly through the LM's small overhead window. Then, looking through his forward-facing window, he would spin the stack around until he saw the earth, now a blue and white crescent, directly ahead. Then he would sight on the earth through a special gunsight, normally used for rendezvous maneuvers, and turn the spacecraft until the horns of the crescent were sitting on the crosshairs. Then, according to mission control, he would be in position to make the burn.

Earlier, when Lovell heard about this technique, he had a flash of recognition. Someone had dreamed it up during a "what-if" session before Apollo 8. Since then it had been taken out of the flight plan books as unlikely, and had been long forgotten—except, that is, by the experts in mission control. Lovell never would have dreamed, back then, that he would have to stake his life on such a scheme.

Around 9 P.M. Haise copied data for the burn; almost 90 minutes later Lovell had wrestled the stack into position. He informed Houston that he was ready for the burn and added that he hoped the guys in the back room knew what they were doing. When the moment came to fire *Aquarius*'s descent engine, all three men played a part. Lovell turned on the engine and controlled the LM's roll; Haise kept the stack oriented in pitch. And Jack Swigert kept his eye on the clock and called out the start and stop times. Fourteen seconds after it began, the midcourse burn was over.

"Nice work," radioed Jack Lousma.

"Let's hope it was," Lovell answered.

Thursday, April 16
3:21 A.M., Houston time
4 days, 14 hours, 8 minutes Mission Elapsed Time

Lovell was alone on watch while Swigert and Haise slept. There would be no sleep for Haise in the cold atmosphere of *Odyssey*—they had started calling it "the refrigerator"—so he had brought down one of the sleeping bags. Now he was a strange sight, wrapped in the bag, his weightless body suspended upside down with his feet up in the tunnel and his head above the engine cover. To keep from floating away, he'd hooked the loop on the bag's zipper to the hatch handle. Swigert was down on the floor just beneath the instrument panel; he'd wrapped a restraint harness around his arm to anchor himself.

Lovell had found that he could close his eyes and doze off for 15 minutes, floating before *Aquarius*'s controls, and wake up feeling a little refreshed. He didn't feel nearly as exhausted now as he had at the end of those 20 hours in lunar orbit on Apollo 8. The difference was adrenaline; the constant demands of keeping his crew and his spacecraft functioning kept him going.

At least now, the end was in sight. From mission control, Jack

Lousma had good news about *Aquarius*'s water: It was holding out so well at the present rate that there was enough to last 20 hours beyond reentry. "Your luck is really hanging in there," Lousma said.

Luck: Lovell had to smile at the irony of the word. He wasn't a superstitious man, and so when he found out he'd be flying Apollo 13 he didn't think anything of it. In fact, some of his Italian friends were happy for him; they said 13 was a very lucky number. So much for lucky numbers. And even though, after the accident, he'd wondered, *Why me?*—just as Swigert must have—he had to concede that, after 572 hours in space, he was a good target for the law of averages. What he hadn't realized at the time was how lucky he'd really been. After the accident he'd told Swigert and Haise, "It couldn't have happened at a worse time." After all, the explosion occurred just at the point where Apollo 13 was too far from earth to turn around. But he was wrong, Lovell knew now. If the explosion had happened in lunar orbit, getting out of orbit with only *Aquarius*'s engines would have been a much iffier proposition. If it had happened when the LM was on the surface, there would have been no recovery. He and Haise would be doomed men. And Jack Swigert would be dead, circling the moon in a command module without power, or an engine, or oxygen. Lovell realized it couldn't have happened at a better time.

Unlike Neil Armstrong and Pete Conrad, Jim Lovell had not received any pre-flight promise from Tom Paine about letting him and his crew fly the next mission if they had to abort. Another team of astronauts would have that chance. But when would that be? At the very least, Lovell suspected, there would be a hiatus of months while NASA tried to figure out what went wrong. At worst—if they didn't make it back—it could mean the end of the Apollo program. Yesterday, during an idle moment, Lovell had said to Jack Swigert, "I'm afraid this is going to be the last lunar mission for a long time." What he didn't know then was that the comment was heard live, on hot-mike. Tom Paine, waiting out the ordeal, went before the press to explain that, yes, there would be an investigation into Apollo 13, but after that, most definitely, NASA was going back to the moon.

• • •

By Thursday afternoon, the cold had begun to invade *Aquarius*. The command module, again, was even colder; at one point Lovell went up to *Odyssey* to get some hot dogs out of the pantry, and they were practically frozen. The men had donned a second set of cotton long johns, and they discussed the idea of putting on their space suits to stay warm, but had decided against it. They wouldn't be able to turn on the suit fans, because they ate up too much power. Without any ventilation they'd overheat and perspire, then run the risk of getting seriously chilled if they had to take them off. But their long johns provided little warmth, and the Teflon-coated fabric of their coveralls was cold to the touch.

Jack Swigert, meanwhile, had his own troubles. Almost a day earlier he'd been in his usual place, straddling the ascent engine cover, when he realized his feet were immersed in a puddle of water; the LM's drinking-water dispenser had been leaking. Swigert's feet were soaked, and soon, after several trips back and forth to the command module, half-frozen. He'd been rubbing them ever since to get them dry. This morning, Lovell and Haise had broken out their lunar boots to keep their own feet warm, and Haise had offered his pair to Swigert. But he declined; he wanted to keep rubbing his feet.

The afternoon dragged on with the three men shoulder to shoulder in *Aquarius*, trying to stay warm. Mission control had promised to read up the lengthy checklist for tomorrow's reentry, but it wasn't ready yet. The cold only made the waiting seem longer.

Around 6 P.M., with no checklist yet in sight, the conversation in *Aquarius* turned to Ken Mattingly. If he were sick, he ought to be breaking out with red spots by now. Before the flight Lovell had worked out a code with the Capcoms so he could find out Mattingly's condition covertly. Now he asked Vance Brand, "Are the flowers in bloom in Houston?"

"No, not yet," Brand replied. "Still must be winter."

"Suspicions confirmed," Lovell said.

"I doubt if they will be blooming even Saturday when you return," Brand ventured.

Not only was Mattingly healthy, but he'd been working like

mad with Gene Kranz's controllers to put together the reentry checklist. At 6:30, Vance Brand radioed that it was nearly ready and that Swigert should get ready to copy it down. "He'll need a lot of paper," Brand said. By 7:19, Swigert was still waiting, and Lovell could no longer hide his irritation. They couldn't just wait around up here, he told Brand; they had to have the procedures in time to study them, "and then we're got to get the people to sleep." Finally, around 7:30 P.M., Mattingly entered mission control, checklist in hand, and sat down at the Capcom's console to talk to the man whose fate had been so strangely entwined with his own. No words of greeting passed between them, no ironic remarks. It was all business.

"Hello, *Aquarius;* Houston. How do you read?"

"Okay. Very good, Ken."

"Okay. Let me take it from the top, here." There followed a conversation that lasted the better part of 2 hours, as Mattingly read every switch setting, every keystroke of the computer, the steps that would bring *Odyssey* back to life and ready for reentry. After every line, Swigert repeated what he had heard. Around 9:15, it was time to copy the LM checklist. Mattingly assured Swigert that both checklists had been tested together in the simulators, and he added, "We think we've got all the little surprises ironed out for you."

"I hope so," Swigert said, "because tomorrow is examination time."

Friday, April 17
1:41 A.M., Houston time
5 days, 12 hours, 28 minutes Mission Elapsed Time

"Fred, are you sleeping?"

"Go ahead."

Mission control had planned to let the crew of Apollo 13 sleep through the night. But early Friday morning Haise was awakened by Jack Lousma, who had a minor change in the checklist for him to copy. Soon it was clear that Jack Swigert was also awake. For the benefit of the doctors, who were keeping tabs on the astronauts' rest periods, Lousma asked, "How much sleep did you get, Jack?"

"I guess maybe two or three hours."

"You plan to get any more?"

"Well, if I get everything done, I'll try, but I tell you, it's almost impossible to sleep. All of us have that same problem. It's just too cold. . . ."

By now, *Aquarius* too was like a damp cave. The temperature hovered in the mid-forties. The LM's environmental control system, designed for warmer temperatures and only two men, was overloaded with moisture, not only from the exhalations of the three men, but in the wet, frigid air that drifted in from *Odyssey*. Big globs of water, shimmering in zero g, clung to every bend in the exposed plumbing and the wire harnesses. The windows were clouded with moisture. The men had to wipe off the instrument panel to read a gauge. The stack had strayed so far from its normal orientation—the thermal-control spin had deteriorated badly—that the sun no longer appeared in *Aquarius*'s windows; it was shining down on the service module's engine bell. The men waited out these last hours in near-darkness, studying the checklist by flashlight, rubbing their hands together.

Years later, Ken Mattingly would look back on the performance of Lovell and his crew with amazement. For more than three days they had been living only for survival, in miserable conditions, with nothing exciting to break up the long hours of waiting. And yet, they never lost their temper with each other, or with mission control. And if Jim Lovell got on the radio every once in a while and let his frustration hang out a little, he was still the epitome of restraint. If Apollo 13 had to happen to any spacecraft commander, Mattingly would say later, there wasn't anyone could have handled it better than Jim Lovell.

2:06 A.M.

"Hey, Jim, while you're up and things are nice and quiet, let me give you a couple of things to think about. . . ." Deke Slayton sounded almost as tired as the man he was talking to. "I know none of you are sleeping worth a damn because it's so cold, and you might want to dig out the medical kit . . . and pull out a couple of Dexedrines apiece. . . ."

"We might consider it," Lovell answered; privately he worried that the letdown from the drug would leave them more tired than they already were.

"Wish we could figure out a way to get a hot cup of coffee up to you," Slayton said. "It'd probably taste pretty good about now, wouldn't it?"

"Yeah, it sure would. You don't realize how cold this thing becomes. . . ."

Hearing this, Jack Lousma offered his own words of encouragement: "Hang in there. It won't be long now."

2:35 A.M.

"Okay, Skipper. We figured out a way for you to keep warm." It was Jack Lousma's voice. "We decided to start powering you up now."

"Sounds good," Lovell said, "and you're sure we have plenty of electrical power to do this?"

"That's affirmative." In fact, *Aquarius* had come through so well that from here on in, there would be twice as much power and water as the men would need. Lousma's go-ahead came none too soon; as he set to work turning on the systems, Haise looked at his window and thought he could almost see frost.

By 3:15, Lovell and Haise had brought *Aquarius* fully to life, and Lovell told Lousma it was warming up a little. Lousma answered, "Duck blinds are always warmer, Jim, when the birds are flying."

4:59 A.M.

Apollo 13 was only 55,000 miles away from earth, heading toward it at 6,100 miles per hour. When Lovell steered the stack into position for a new midcourse burn, the planet was a crescent of impressive size and growing larger by the minute. But even now,

Lovell's crew could not rest easy. Something was still causing their angle of attack to shallow slightly. Throughout the past couple of days, the men had seen more spurts of gas from the service module—suddenly the constellations would be lost in a flurry of sparkles—but Houston assured them that couldn't be the cause. Whatever the force bending Apollo 13's trajectory, it had to be corrected. All that was needed was a gentle, 21-second nudge from *Aquarius*'s maneuvering thrusters. But Lovell was so tired that he called up the computer program for the descent engine instead. In Houston, sharp-eyed controllers caught the mistake in plenty of time.

Tiredness was not the only cause of Lovell's mistake. Ever since Houston had told him that water was tight—confirming Haise's numbers—Lovell had decided to ration drinking water. The men had made good use of the water left in the command module's drinking tank, but Lovell had been particularly stringent about his own water consumption. The situation was compounded by the fact that in zero g, astronauts generally don't feel thirsty as often as they do on the ground. Though he did not realize it, he was becoming dehydrated. He would later find out that the loss of electrolytes made him more prone to errors.

Fred Haise had his own problem: he was sick, and all because of a misunderstanding with mission control. After the accident, with *Odyssey* powered down, the men couldn't use the normal system of dumping urine overboard; it required electricity to heat the line and keep it from freezing. Instead, they'd used an alternate line, which dumped urine through a fitting in the command module's side hatch. But mission control told them that the resulting swarm of ice droplets around the spacecraft was ruining their Doppler tracking data; please don't dump any more urine. They neglected to say, "for the time being." Lovell and his crew interpreted the request as a permanent ban. The misunderstanding created an immediate problem of where to store urine, and for nearly three days they'd been using every suitable bag and container, stashing them in the LM's storage cabinets. But it also led, indirectly, to Haise's illness. Mission control's request came at a time when the men were very busy. The easiest way to store urine, they decided, was to use the collection bags they normally wore inside their space suits. They kept them on while they worked, for some number of hours, until the pace of events slackened. That saved some time, but it also meant they were bathing themselves in urine—and that made them vulnerable to infection. By Thursday evening, Haise had developed a urinary tract infection, although he did not know it. He knew only that he was experiencing a burning pain when he urinated. And on Friday morning, when Haise went up to *Odyssey* to go to the bathroom, he bumped up against the walls with his bare skin. The cold metal drained the warmth out of him. When he returned to *Aquarius* he was shivering badly. He zipped himself into a sleeping bag and floated before the instrument panel. Swigert felt his head; he didn't seem to have a fever. The shivering stopped after a couple of hours, but for the first time in the flight, Haise felt truly exhausted.

7:15 A.M.

After the midcourse burn, it was time to cast off the stricken service module. Jack Swigert had been in *Odyssey* for more than two hours, where he had begun to turn on some of the systems, drawing power from *Aquarius*—another bit of ingenuity from Houston. To guard against a lethal mistake, Swigert had put tape over the switch marked "LM JETTISON." Down in *Aquarius*, Lovell pulsed the thrusters to impart momentum, and yelled, "Fire!" Swigert hit the switch marked "SM SEP" and there was a bang of pyrotechnic bolts, and the service module was away. Lovell immediately swung the joined LM and command module around—with the massive service module gone, the craft suddenly responded crisply to his commands—and through the overhead window he saw the service module, tumbling slowly as it departed. For the first time he could see what had happened to Apollo 13.

"There's one whole side of that spacecraft missing! Look out there, would you?" Instead of the small puncture wound Lovell had imagined, the explosion had blown off an entire panel from the skin of the service module. Inside he could see tanks, pipes, and other equipment amid a tangle of torn Mylar insulation. At the base of the cylindrical module, they could see that the big communications antenna had been bent out of position. The men fired off pictures until the service module was a speck in the distance.

In Houston, flight controllers were chilled by the verbal picture of destruction. As to the cause of the accident, that would be for the review board to determine. But Ken Mattingly had a good idea of what they would find; he had been directly involved. Back in March, the launch teams at the Cape had conducted the Countdown Demonstration Test, which included filling the spacecraft's tanks with cryogens and then emptying them again. When it came time to drain the service module's oxygen tanks, technicians were unable to remove the super-cold fluid from tank number 2. For eleven days, they studied the problem. Mattingly went to briefings by the engineers, who had worked around the clock to find out what was wrong. They explained that the tank had originally been scheduled for Apollo 10, but had been removed for modifications and replaced. Sometime after that, the tank had been accidentally dropped a distance of two inches—a minor jolt, but enough to damage the tube assembly used to fill and empty the tank. That problem, never completely fixed, was the reason for the present difficulties. But there was an alternate procedure, they said, that would work around the problem. They would turn on the tank's built-in heaters and warm the liquid oxygen until it boiled off through the tank's relief valve. They asked Mattingly whether that sounded okay; he agreed.

But neither Mattingly nor anyone else knew that the heater inside the tank had a design flaw. Back in 1965, the entire Apollo spacecraft, which was intended to operate at 28 volts of electricity, was upgraded to accept 65 volts from ground test equipment. Everything, that is, except the thermostat inside the service module's oxygen tanks. It was still rated for only 28 volts; no one caught the error. It was that thermostat that was supposed to shut off the heaters when the temperature inside the tank reached 80 degrees Fahrenheit. During the draining operation, the thermostat was activated, but the excess voltage caused an arc that welded its electrical contacts shut. The heaters stayed on for 8 solid hours, long enough for the temperature to reach 1,000 degrees. The technician monitoring the test was unaware of the dangerous condition because his temperature gauge went no higher than 85 degrees. Inside tank number 2, the intense heat baked and cracked Teflon insulation covering wires on a nearby motor; the motor controlled the fan inside the tank.

Nothing more happened until 2 days, 7 hours, and 54 minutes into the flight of Apollo 13, when Jack Swigert turned on the fan at mission control's request. At that point, the review board would later decide, there was an electrical arc inside tank number 2. In moments a fire had started inside the tank, fed by the generous supply of oxygen. Seconds later, the sudden surge of pressure blew off the tank's dome, flooding the surroundings with oxygen. When the pressure reached 30 pounds per square inch, the panel blew away in a jolt that probably also severed the plumbing for tank number 1, crippling Apollo 13.

In the test flight business, big mistakes are often the result of many little ones. Mattingly would always carry with him the knowledge that he had signed off on Apollo 13's damaged oxygen tank. But he was by no means the only one who had failed to stop the process that led to the accident; a number of managers and engineers had also said, "Go ahead." So had Jim Lovell.

9:00 A.M.

A strange and unprecedented combination, lunar module and command module, sped earthward. Inside *Odyssey*'s dark, frigid cabin Jack Swigert and Fred Haise faced the moment of truth. With just 2½ hours to go before reentry, it was finally time to bring the command module fully back to life. They knew the systems must be even colder than it felt inside the cabin. No one knew if the navigation system would function again after such a prolonged cold soak. The design specifications said it wouldn't work. They hoped the specs were wrong.

Before they could do anything they had to wipe off the gauges, because there was water all over the instrument panel. They knew there must be water *behind* the panel too, clinging to the insulation around electrical connections, perhaps ready to short-circuit as soon as they began pushing in circuit breakers. Swigert and Haise had no choice but to charge ahead, hoping they didn't see sparks or smell smoke. With relief, they realized their fears were unfounded. Undoubtedly the modifications made to both the command module and the LM after the Fire had prevented this potential disaster. And every one of *Odyssey*'s systems revived in perfect condition, even the computer. Swigert was so glad to have his command module back that he didn't even need Dexedrine.

But for Haise's friend, lunar module number 7, time had nearly run out. Swigert and Haise were in their couches in *Odyssey* while Lovell, now crowded in *Aquarius* among bags of unneeded gear, steered the joined craft into position. When Lovell was back in the command module, the men sealed the hatches between the two craft. When Swigert hit the switch there was a loud bang and a jolt far more violent than the oxygen tank explosion had been. *Odyssey* pitched perilously close to gimbal lock, but Swigert quickly steadied the ship. Then Haise watched the departing LM, legs extended, poised for landing on alien soil, but instead headed for a reentry it would not survive. He wished there had been some way to bring it home and put it in a museum. Joe Kerwin spoke for all of them: "Farewell, *Aquarius,* and we thank you."

11:43 A.M.

Odyssey sped through darkness toward its rendezvous with earth. Inside, Lovell, Swigert, and Haise were strapped in their couches, exchanging last-minute bits of information with Houston. After three days of constant background noise from *Aquarius*'s fans and pumps, the command module seemed unnaturally quiet. But now, with everything working perfectly, there was still a troubling unknown in Fred Haise's mind: the condition of the heat shield. Even before they saw the service module, they had talked among themselves about the possibility that it had been cracked by the explosion. Of course, that resin-filled honeycomb structure was extremely tough. And a small crack would probably seal itself as the heat shield began to ablate in the heat of reentry. A serious crack was another matter. But soon Haise was far too busy to think about it.

Now, with just 7 minutes to go until reentry, Joe Kerwin radioed the welcome news that *Odyssey*'s guidance system was perfect. Swigert suspected that when it was all over the trajectory experts in mission control would have a great party, and he told Houston he wished he could go to it. Kerwin relayed an answer from someone in the control room: "We'll cover for you guys, and if Jack's got any phone numbers he wants us to call, pass them down."

With three minutes to go, Lovell and his crew saw a tiny and distant moon, and as they watched, it blinked out, sinking behind the dark horizon exactly when mission control had predicted. And then, the blackness of space gave way to the first, soft light of ionized gas, a glow known only to the returning space traveler.

Timber Cove, Texas

A crowd had gathered with Marilyn Lovell to watch the recovery on television. In the past three and a half days she had hardly left her house. Yesterday, she had lunched with Fred Haise's wife, Mary—who was very pregnant—at Deke Slayton's house. And earlier this morning, she had gone to get Jim's mother from the nursing home so that she could watch too. For the past two days, her spirits rose steadily as the reports on Apollo 13 improved. Still, there had been difficult moments. She knew Jim wasn't getting much rest. At one point, someone sent word that Dr. Berry thought it might be a good idea if she went down to mission control and talked to Jim directly, to convince him to go to sleep. She passed word back that she just couldn't do that. She knew that if she tried to talk to him, she would fall apart.

Then, earlier this morning, she had been sitting by the squawk box when she heard Jim say, "There's one whole side of that spacecraft missing!" And she panicked. With no NASA people around to ask, she called Pete Conrad, who said, "I'll check on it, I'll check on it"—and soon she was reassured. But she would not rest easy until she saw the command module coming down on its parachutes.

Now Marilyn sat in front of the TV among her close friends, with little Jeffrey at her side. Just minutes ago she had heard the last conversations between Jack Swigert and mission control. "Everybody says you're looking great," Kerwin had said. "Welcome home."

"Thank you," Swigert had said. And then there was silence from Apollo 13. Long minutes passed as she waited anxiously for the communications blackout to end. She heard the NASA public affairs commentator say, "About thirty seconds to go for blackout. Less than ten seconds. We will attempt to contact Apollo 13 through one of the [Apollo Range Instrumentation Aircraft]."

And there was silence. Half a minute went by; nothing. Now it was a full minute. Suddenly: "We've had a report that the ARIA 4 aircraft has acquisition of signal." Now she heard Joe Kerwin's voice. "*Odyssey,* Houston. Standing by."

Long seconds passed. Finally, there was Jack Swigert: "Okay, Joe." And still, she could not relax; there were the parachutes yet to come. She was not alone in her tension; no one, either in mission control or inside the command module, knew whether the electronics to deploy the chutes were still good.

On television, the broadcast from the recovery ship showed morning in the South Pacific, a calm sea under a patchwork of fair-weather clouds. She heard Kerwin radio to Apollo 13 that the weather was good. The public affairs officer counted down for the deployment of the drogue parachutes. Less than 2 minutes; now 1 minute; now 30 seconds. For interminable seconds it was hard to tell what was happening. Then, the TV camera found the command module: a small dark cone floating out of the clouds on three beautiful orange parachutes. The room erupted in cheers and Marilyn hugged Jeffrey so tightly that he cried out.

Meanwhile, at the space center, cheers and applause thundered through mission control. Everyone shook hands and lit up cigars—including Ken Mattingly, who smiled and shook hands with Chuck Berry. And inside *Odyssey,* Jim Lovell remembered his ton-of-bricks splashdown from Apollo 8 and said to his crew, "Gentlemen, be prepared for a hard landing." But *Odyssey* must have caught the crest of a descending wave, for it fell gently onto a calm Pacific Ocean.

As they waited for the swimmers to arrive, Lovell, Swigert, and

Haise lay in their couches, amazed that even now, after the heat of reentry, *Odyssey* was still so cold that they could see their breath. When the swimmers opened the hatch, a great cloud of frosty air issued into the tropical morning. The chill of space would linger inside *Odyssey* long after Lovell, Swigert, and Haise were safely aboard the aircraft carrier *Iwo Jima,* recounting their ordeal for the doctors, and savoring the fresh, warm air of earth.

• • •

Marilyn Lovell had never seen a party like the one that enveloped her house at splashdown; nor had she seen so many champagne bottles emptied. People were leaning against the wall, unable to move. Marilyn lost her voice, but not before she went out to greet the reporters who had kept a patient vigil through the past four days. She jokingly gave them the old standby—"Happy, thrilled, and proud"—and one of them asked her what the past four days had been like. "It's been a nightmare," she told them. "I have never experienced anything like this in my life and I never hope to experience it again." That wasn't the kind of talk usually acceptable for an astronaut wife, but now for the first time since Jim became a test pilot, she had no reason to hide anything.

• • •

The day after Lovell's crew splashed down Richard Nixon came to the space center and presented the Presidential Medal of Freedom to Gene Kranz, Glynn Lunney, and the other flight directors of Apollo 13. Tom Paine was there too, sitting on the stage, watching as Nixon praised the ingenuity and courage of the NASA and industry personnel who had saved the lives of the three astronauts. For many in the audience, Apollo 13 would always be NASA's finest hour.

But the irony was that even now, Paine's vision for the space program was crumbling. In March the White House had issued a statement saying that space would no longer hold such a high position in the list of national priorities. There would be no expensive new programs, especially not in the area of manned spaceflight. There would be no money spent on any effort linked to the goal of sending humans to Mars. By summer, NASA's goal of a space station and reusable space shuttle was coming under fire in the Senate, where Walter Mondale protested, "I believe it would be unconscionable to embark on a project of such staggering cost when many of our citizens are malnourished, when our rivers and lakes are polluted, and when our cities and rural areas are dying." He asked, "What are our values? What do we think is more important?" Despite this and other outcries, Congress approved the NASA budget by a slim margin at the end of July. The next day Tom Paine surprised his colleagues at NASA by resigning as administrator to take a management job in industry. He would leave to his successor the coming battles for support—not from Congress, but from Richard Nixon.

But it was already clear that Nixon had rejected the vision proposed by the Space Task Group. By September, only the space shuttle had any chance of being approved—and that was still uncertain at best. As presidential adviser John Ehrlichman told historian John Logsden years later, inflation and other priorities were more important; the budget just wasn't big enough to do everything NASA wanted. Furthermore, Nixon was unwilling to pay for a Mars effort that would not bear political fruit until after he left office. But Apollo was another matter. After Apollo 13 some of Nixon's advisers, wary that a space disaster would do him political harm, had urged him to cancel the rest of the moon missions, but he resisted. According to Ehrlichman, the moon program held magic for Nixon for one reason: he liked heroes. To him the astronauts represented the best the country could produce. It was good for the nation, Nixon believed, to have heroes.

CHAPTER 8

The Story of a Full-up Mission

APOLLO 14

I: Big Al Flies Again

Some memories are so bright that the passage of time cannot dim them. For Alan Bartlett Shepard, the brightest was May 5, 1961: On that Friday morning, while a nation watched and listened, he lay inside a tiny spacecraft called *Freedom 7* atop a Redstone booster and waited to be hurled into space. The space program had suffered more than its share of failures, and images of exploding rockets were burned into the national psyche. The suspense was almost unbearable. This launch would give a badly needed boost to national morale and prestige—if Shepard wasn't killed. As for Shepard himself, he had a different worry: he didn't want to screw up.

Inside *Freedom 7*, Shepard fought nervousness by concentrating on his work. For hours the count was held up by one problem after another—first the weather, then technical difficulties—and Shepard's mood turned impatient. With less than three minutes to go, minor trouble with the Redstone halted the count yet again. Over the radio Shepard could hear the engineers in the blockhouse, anxiously debating whether to postpone the launch, and what he said was the stuff of folklore: "I'm cooler than you are. Why don't you fix your little problem and light this candle?"

At 9:30 A.M. fire erupted from the Redstone and Shepard rose from the east coast of Florida and soared into a sunlit infinity. Fifteen minutes later, having arced like a guided cannonball to an altitude of 115 miles, *Freedom* 7 splashed down in the Atlantic, and a nation rejoiced. Within days Shepard stood in the White House rose garden to receive NASA's distinguished service medal from John F. Kennedy. Afterward, he rode in an open car down Pennsylvania Avenue, to the adulation of thousands. Sitting next to him, Vice President Lyndon Johnson was amazed. "Where did all these people come from? You're a famous man, Shepard."

Yes, he was a national hero, but the most precious thing Shepard gained that day wasn't glory—the true pilot, he would say years later, never does it for the fame—but the satisfaction of having been first. Fierce competition for the first Mercury flight—and with it, the chance to become the first man in space—had been an undercurrent in everything the Original 7 did. But the choice was up to their boss, Bob Gilruth, the soft-spoken head of the Space Task Group who thought of the astronauts as "his boys." One day Gilruth asked each of the Seven to conduct a peer vote: "If you yourself cannot be the one to make the first flight," he said, "who do you think it should be?" Shepard wondered if this was some kind of stunt dreamed up by the shrinks who had been part of the Seven's lives since the astronaut selection. There was actually a duty psychologist, Bob Voas, assigned to them; Shepard never did trust him.

Shepard had no idea whether the peer vote was important, but he didn't worry about it. There were no bad pilots among the Seven, Shepard knew, although NASA had overlooked some discrepancies in their piloting backgrounds in its zeal to select perfect mental and physical specimens. But he had no doubts about where he stood—and never had. That supreme self-confidence was with him when he graduated from the Naval Academy in 1944, ready to end the conflict in the Pacific single-handed. After the war he began his aviation career, flying off carriers, and moved on to become a test pilot at Pax River. It was that quiet arrogance that Shepard, more than any of his colleagues, brought to the business of being an astronaut.

Still, Shepard doubted he would be chosen. He figured the first flight would go to his main rival: John Glenn. Among the Seven it was Glenn—charismatic and devout, with the Ohio small-town background and the sunny, freckle-faced look of an all-American—who most embodied the hero the country was looking for in its astronauts. At the Original 7's first press conference—which had scared some of them half to death—Glenn had been a natural. For an image-conscious NASA, Shepard figured, Glenn was the obvious choice. And Glenn, with his own résumé of test-fight accomplishments, was as determined as any of them to be first.

Around mid-January 1961, Gilruth came to see the Seven in their office and announced that after much consideration, the pilot for the first flight would be Alan Shepard. Grissom would get the second flight; Glenn would back up both flights. As Gilruth spoke, Shepard was looking at the floor, and long seconds passed before he looked up to see the other six coming over to congratulate him. It wasn't hard to guess that their smiles concealed great disappointment, but Shepard couldn't have known that John Glenn was so angry that he later considered going to Jim Webb to try to get the decision changed, then thought better of it. NASA kept the selection a secret, saying only that one of the three—Glenn, Grissom, or Shepard—would make the flight.

Then came the tough part. Looking back, the months leading up to the flight had been a true test, especially putting up with the forecasters of doom. They warned that a man would not survive a space mission, that he would succumb to the g forces of launch and reentry, that weightlessness would wreak havoc on his vision, or his sense of orientation, or his bodily functions. Some psychologists were predicting that the astronaut would experience a profound sense of separation from earth—a "breakoff phenomenon"—and be rendered helpless. Nor were these pronouncements coming from the fringe. In April, representatives of the President's Scientific Advisory Committee went down to the NASA offices in Langley, Virginia, and voiced their trepidation about the flight. Shepard, Glenn, and Grissom all volunteered to take extra runs on the monstrous centrifuge at Johnsville, just to prove to the skeptics that their fears were unfounded. Meanwhile, within NASA, Mercury planners looked ahead confidently. A chimpanzee named Ham had already made the trip and had come back in good shape. However, because of some minor problems with the booster, NASA postponed Shepard's March launch until early May, to allow time for one more unmanned test.

Then, early in the morning of April 12, Shepard's hopes of being the first man in space evaporated as the world learned that a twenty-seven-year-old Russian pilot named Yuri Gagarin had orbited the earth. Shepard fumed that NASA hadn't seized the chance to send him up in March, but there was nothing to do about it now. Shepard and everyone else connected with Project Mercury were caught up in preparations for the launch. After the May 2 attempt was scrubbed because of the weather, NASA could no longer conceal the pilot's identity. By the time he climbed atop the Redstone on May 5, the eyes of the nation were on him.

So were John Kennedy's. When Shepard flew, the president was already weighing the decision of whether to attempt a moon program. Shepard's flight put him over the edge. Shepard would always believe that the immense outpouring of pride that greeted his flight made a great impression on Kennedy. If a suborbital hop from Florida to Bermuda could so energize the nation, imagine what would happen if America put a man on the moon. By May 25, less than three weeks after Shepard's triumphant ride down Pennsylvania Avenue, Kennedy had made his decision. That day, as he addressed a joint session of Congress, his closest aides could tell he was nervous—by the way he kept playing with the pages of his speech, creasing them, smoothing them. They sensed that Kennedy wasn't sure he was doing the right thing, to ask for such an enormous sum of money for something so audacious. But none of that showed in his voice as he spoke the words that would mark the genesis of Apollo.

When Shepard next saw Kennedy, it was in February 1962, after John Glenn's orbital mission, which drew so much public celebration that it overshadowed the reaction to his own flight. Aboard Air Force One, flying from West Palm Beach to Washington, Glenn, Shepard, and Grissom sat with Kennedy undisturbed for most of the 90-minute flight. They talked about Glenn's experiences, and about the moon program. Aides came in and out of the forward cabin, but Kennedy wasn't interested in them. He didn't want to talk about anything but space. That day, Shepard saw the man's pioneer spirit. Some said the moon decision had been motivated by political concerns, but to Shepard it had been an act of statesmanship, not politics. And it was clear to him, inside Air Force One, that the lure of space exploration was very real to John Kennedy. The following September, when Kennedy delivered his stirring speech at Rice University stadium, Shepard saw a president who was behind Apollo all the way.

By that time, Shepard was already thinking about getting into space again. His own flight had been so brief—15 minutes and 28 seconds—and so busy—two minutes for this test, a minute for that one—that it had merely whetted his appetite. For raw thrills, nothing could top it, but it hadn't been long enough for him to really show what he could do. He wanted the next flight he could get, though he would have to wait his turn on the rotation. That came after Gordon Cooper circled the earth for 34 hours in May 1963; Shepard made a bid for a three-day mission. But Jim Webb, anxious to get on with Gemini, was very much against prolonging Mercury, with the time, energy, and expense that would entail. With Webb's knowledge, Shepard took his case to Kennedy himself, but in the end, Webb prevailed.

Shepard's disappointment was short-lived. He was still at the top of the rotation. Gemini was coming, and after that, he could look forward to Apollo. Soon Shepard was assigned as the commander of the first Gemini mission, with Tom Stafford as his copilot. But that summer Shepard's fortunes changed abruptly. At home one day, he was suddenly swept with dizziness, nausea, and vomiting. When he recovered, he thought he was coming down with the flu, but within days he had a second attack, and then a third. Reluctantly he took his problem to the NASA doctors. Their diagnosis was Ménière's syndrome, an inner ear disorder characterized by a buildup of excess fluid in the semicircular canal, causing impaired balance and attacks of vertigo. Medications would control the vertigo, the doctors said, but the prognosis was bad. There was no cure for Ménière's syndrome, and unless Shepard was among the 25 percent for whom the disease clears up on its own, he would never make another spaceflight. At first, Shepard was optimistic; he told Stafford, "We're going to fight this thing." But soon the pair were taken off the flight. Stafford was eventually reassigned to a Gemini mission, but Shepard had no recourse. He was forbidden even to fly an airplane by himself. For an astronaut, it was like a death sentence.

For six years, Shepard lived with the stigma of being grounded. He never gave any thought to leaving the astronaut corps. He still hoped the disease might go away spontaneously. In the meantime, Deke Slayton had his hands full trying to run the newly created Flight Crew Operations Directorate, and he asked Shepard to take over the Astronaut Office. Through Gemini and into Apollo, Shepard handled the humdrum administrative affairs, coordinating training and travel schedules, approving interview requests, and the like. Knowing the collection of egos under him, the prima donnas who balked at serving on a backup crew, Shepard felt compelled to use a firm hand. It meant he had few friends in the office, but that never seemed to bother him; in fact, it was better for morale not to give even the appearance of favoritism.

A Chief Astronaut who could not fly was like a warrior-king who could not fight, and Shepard's frustration was magnified by his role in preparing other astronauts for their missions. It began with the secret crew-selection process, of which Shepard was always a key part. Then, as each crew trained, he followed their progress at the Cape. On launch morning he ate breakfast with them, and then he did the hardest thing of all: he walked them out the door of the crew quarters and watched as they climbed into the transfer van and rode away, heading for the launch pad.

But the care and feeding of astronauts did little to satisfy Shepard's appetite for a challenge. To fill the void, he turned to business. Even before he was an astronaut, Shepard had dabbled in real estate and made a respectable sum of money. The Mercury era had left him well connected to the Houston social and business scene, and he was getting some expert advice from Leo De'Orsey, the lawyer who guided the Original 7 through the *Life* magazine contract and other unfamiliar waters. Shepard logged his share of failures, but also successes, especially with his lucrative part-ownership of Houston's Bay Town bank. Within a few years he had amassed a small empire that at one time or another included partnerships in shopping centers, hotels, and other assorted enterprises around the country. He bought a $200,000 house in Houston's opulent River Oaks district. If Shepard's escapades drew some private envy and resentment both inside and outside the Astronaut Office, no one could fault him for the way he looked after the astronauts' interests within the space center. And while he was sometimes known to put in half-days in the Astronaut Office and spend the rest on outside business, one of his Mercury colleagues would later say, "Al never let anybody down in that job. That was the beauty of Al Shepard; he could do both." On the fifth anniversary of Shepard's flight the Astronaut Office threw a party for him, and they showed a movie called *How to Succeed in Business Without Really Flying Very Much*, a slight alteration of the title of a very popular Broadway play. By that time, at age forty-two, Shepard was on the way to becoming a millionaire.

But none of this mattered to Shepard compared with making another spaceflight. By the spring of 1968, his astronaut career was slipping through his fingers. His condition had worsened; he still had attacks of vertigo, and he was nearly deaf in his left ear. But Tom Stafford had heard about a Los Angeles ear surgeon named William House who had devised a delicate, risky operation to help Ménière's sufferers. The procedure involved implanting a small silicone tube in the ear to drain the excess fluid from the semicircular canal, through the mastoid bone, to the top of the spinal column. Shepard went to see House, who warned that it would take some time to know whether the operation would be a success; he gave no guarantees. In the summer of 1968 Shepard secretly checked into a Los Angeles hospital as Victor Poulos (a name chosen by the admissions nurse, who was Greek). Shepard had ruled out the surgery until he had nothing left to lose; that time had come.

• • •

In the summer of 1968, Stuart Roosa didn't have the luxury of worrying whether he would ever make a second spaceflight; he was still waiting for his first, and things didn't look good. Like his colleagues in the Original 19, he'd been immersed in rookie's labors for the better part of two years. A thin, red-headed Oklahoma native, Roosa was thirty-five, but he seemed not to have aged a day since he was a smoke jumper for the U.S. Forest Service in Oregon, fifteen years earlier. On their first date, he had told his future wife, Joan, that he was going to be an engineering test pilot. But if there had been no Edwards and no NASA, Stu Roosa would probably have gone to Vietnam, for he had spent much of his adult life preparing to fly in combat. As a newly minted air force fighter pilot he'd spent a year in gunnery training. Then, in 1955, he was assigned to the 510 fighter-bomber squadron in Langley, Virginia. The 510's official designation was special weapons; in plain English that meant nuclear bombs. The 510 trained for the unthinkable, a full-scale nuclear war with the Soviet Union. The plan, which remained secret for years, called for the pilots to deploy out of bases in West Germany. Flying their F-100's over Russia, they would drop down to 50 or 100 feet to avoid the enemy's radar. At carefully chosen points, each man would arm his weapon and execute a series of precise maneuvers: Pull back on the stick, watch the airspeed like a hawk, and at exactly the right moment, with the jet zooming almost straight up, release the bomb, which would then follow a two-mile-long ballistic arc to the target. Meanwhile, the pilots would head in the other direction as fast as their jets could take them, to be safely out of range of the ensuing destruction. The planes did not carry enough fuel to reach their targets inside Russia and make it back. The pilots planned to eject, hundreds of miles from the base, and then, making use of special training in escape and evasion, walk the rest of the way.

Decades after the fact, it may be difficult to understand why Stu Roosa, like the other men in the 510, was not only prepared to carry out this mission, but half-wished he would get a chance to. But in 1955, at the height of the Cold War, the threat of nuclear war was a grim fact of life. Communism was the enemy, and Roosa's desire to defend his country from it was exceeded only by his love of flying. He would never forget the first time he was cleared to take off on his own as an air force cadet. In the open canopy, with the sun on his face and the roar of the engine filling his senses, he began to sing the air force song: *Off we go, into the wild blue yon-*

der. . . . If that sounded like *Life* magazine's image of an astronaut, then in fact, Roosa was it, more than many of his colleagues; he was a straitlaced, conservative family man with a soldier's devotion to his country. And when he became a member of the Original 19 in 1966, he was ready to dedicate his energies to the "peaceful war" that was Project Apollo.

No sooner had he arrived at NASA than he attended the party the astronauts threw for the fifth anniversary of Al Shepard's Mercury flight. He looked around and felt awed in spite of himself. He turned to his wife, Joan, and said, "Do you know that the first person on the moon is in this room tonight?"

Roosa had no illusions that that person might be himself or any of the Original 19. He was just one more in a sea of faces at the Monday-morning pilots' meetings. He and his friend Charlie Duke were assigned to cover the Saturn boosters; to them it felt like a backwater. But as the time for the first Apollo missions neared, the astronauts began to show more interest in Roosa's work. And then, one day in the fall of 1968, he was at North American when Al Shepard called him to say, "We want you to be on the support crew for Apollo 9." Then he added, "Just be patient. I've got something in the works." Shepard's tone was so matter-of-fact that the comment barely registered; Roosa was just glad that he was finally moving out of the pack.

Roosa's assignment was to help coordinate the activities in mission control, and he logged months of simulations with the flight controllers. When Apollo 9 flew, Roosa was more involved than anyone expected. After Rusty Schweickart suffered motion sickness, Roosa was part of the scramble to reorganize the schedule and save as many of the mission objectives as possible. He knew more about the flight plan than any astronaut who was still on earth, and he was at the Capcom's mike almost every minute that McDivitt's crew was awake. When the other controllers left at the end of their eight-hour shifts, Roosa just stayed where he was. After it was all over, Deke Slayton told him, "You were more on top of it than the flight directors were." Just as importantly, Chris Kraft took notice—and Roosa had always felt Kraft had a great deal of influence on an astronaut's career. But Roosa still had no sense of where he stood, and he didn't know what to make of it when one day Deke asked him, "Has Al talked to you about your new assignment?"

Al Shepard had always been a mysterious, intimidating figure to Roosa. Going to Al Shepard's office was as nerve-wracking as going to see the General—Roosa always made sure his thoughts were organized and his hair was combed. But around the fall of 1968, Roosa had begun to notice some subtle changes in Big Al. At the pilots' meetings, Shepard showed an unusual interest in the tech-talk—not just whether a man was on top of his work, but the details of what he was saying—some new change in the spacecraft or a mission plan. For the first time since anybody could remember, he started showing up at the gym. Every now and then Roosa would see him in mission control, talking to flight controllers. Roosa had no idea what was going on, until one day in April, when he was summoned to Shepard's office, along with Ed Mitchell. When they got there, Shepard was grinning broadly. "If you guys don't mind flying with an old retread, we're the prime crew of Apollo 13."

Roosa couldn't believe what he was hearing—no rookie had ever gone straight onto an Apollo prime crew without serving as a backup. He said to Shepard, "Did you say prime?"

Shepard flashed him a steely glance. "I said prime."

The surgery had worked. By the spring of 1969, the doctors had pronounced Shepard whole. He was back on flight status, and ready to take any mission he could get. He would have been happy to fly in earth orbit on Apollo Applications, if that's where Deke wanted him.

At a small press conference, a reporter asked Shepard who would take his place helping Slayton with the crew assignments, and Shepard said he guessed Deke would have to do it all by himself, and then, barely containing a laugh, he added, "Of course, my ol' buddy Deke, good ol' Deke . . ." and the place broke up. Someone asked, "Will you campaign?" Shepard replied, "You're not supposed to." But then, he didn't have to. His last act as Slayton's coauthor on crew selections was something no other astronaut could have pulled off: he recommended himself for the very next flight available. In the spring of 1969, that was Apollo 13. Slayton obliged him.

Word got around the Astronaut Office that Alan Shepard was now the commander of Apollo 13, and the reactions were mixed. Behind closed doors, more than one of the pilots expressed resentment that Shepard could walk right onto a prime crew without doing his turn as a backup. In reality, Shepard's ego ruled that out; nor was there any danger he might have to accept middle or right seat. Others realized that it was pointless to say that Al Shepard was cutting in line; he had been in line from the day he arrived. And as far as Deke Slayton was concerned, if Shepard wasn't qualified, no one was.

Only one other member of the Original 7 was still on flight status; for him, the news of Shepard's assignment was grim. For years now, Gordo Cooper's star had steadily fallen. No one would argue that he had flown a spectacular Mercury mission—by some accounts, the best of the series. But after that, he had begun to slack off. He had always raised casual to new heights; it was Cooper who, perched atop a fully fueled Atlas booster, waiting out a hold in the count, fell asleep. His strap-it-on-and-go attitude was legendary, but unfortunately he seemed to carry it into training. On his Gemini 5 flight the other astronauts had to goad him into the simulator. No one who witnessed that episode was surprised when his next assignment was a dead-end stint on the backup crew of Gemini 12. When Slayton gave him another backup post for Apollo 10, some wondered whether Cooper would ever fly again.

Cooper didn't help himself by entering a twenty-four-hour road race at Daytona, while he was in training. Slayton was no happier about that than a movie studio would be about a major star doing his own stunts—too much was invested in him to subject him to unnecessary risk. When Slayton pulled him from the race, Cooper bitched to the press, "They ought to hire tiddlywinks players as astronauts." In any case, Slayton had named him to back up Apollo 10 with the idea that he'd probably be able to fill in for Tom Stafford if necessary, but he was not at all sure that he'd follow the rotation and give Cooper command of Apollo 13. When Shepard became available, a few weeks before Apollo 10 flew, Slayton's choice was easy. Cooper would later hold an angry press conference, saying that Shepard had unfairly edged him out of his mission, but the truth was that Cooper had let the moon slip beyond his reach. And so Alan Shepard would become the only one of the Original 7 to reach the moon.

But not on Apollo 13. For the first time since he had begun selecting crews, Slayton's assignment was vetoed—by George Mueller, head of manned spaceflight at NASA Headquarters. Mueller insisted that after so long on the bench, Shepard needed more time to train. Slayton argued that Shepard had been in training for 13 since the spring and was making good progress. Shepard himself knew he had some catching up to do. But he'd had a head start; even while he'd been grounded, he had been learning the basics of the command module and the LM, and like every rookie astronaut he'd been grabbing simulator time whenever he could get it. He felt confident he'd be ready in time for Apollo 13. But Mueller held firm. Slayton had no choice but to swap Shepard's disappointed crew with Jim Lovell's, who had just finished backing up Apollo 11. One day in August, Slayton took Lovell aside and asked him whether he and his team could be ready for 13; and Jim Lovell—happily surprised at the offer—said yes.

• • •

For Alan Shepard the events of April 1970 had special irony. In the years after the Fire, it had occurred to him that being grounded might have saved his life; he might well have been the commander of Apollo 1 instead of Gus Grissom. Now he had been saved from the ordeal of Apollo 13 by George Mueller.

Shepard wasn't alone in such thoughts. Stu Roosa had been in mission control when Apollo 13 became a struggle to save the lives of three men. Initially he had been one of those who thought there was no hope for Lovell's crew, but over the next four days he'd witnessed the most impressive recovery in NASA's history. In Houston, as the accident investigators issued their report and engineers went to work on safeguards for later missions, the flight controllers savored a new confidence. Whatever problems might come up, they knew they would handle them. They were eager to take on Apollo 14, which had been postponed from July of 1970 until the end of the year.

But in the wake of their masterful rescue—which NASA called "a successful failure"—Apollo once again faded in the national consciousness. Even during the ordeal of Apollo 13, amid global concern, the country's attention had been divided. While NASA struggled to save Lovell's crew, a professor at Duke University encountered one of his students and asked, "Do you think they'll get them back?" The student responded by talking about the American troops in Vietnam. At the beginning of May, the U.S. invasion of Cambodia intensified the campus unrest over the war. Days later, during a demonstration at Kent State University, four students were killed by young, nervous National Guardsmen. By summer, the nation's conflict over the war in Vietnam had deepened.

It was also a troubled time for NASA. The relative austerity in the space budget continued to affect the program. At the Cape, there were bumper stickers that read, "Apollo 14: One Giant Leap for Unemployment." At the end of August, faced with additional cuts, acting NASA administrator George Low, who had taken over after Paine's departure, canceled two more Apollo missions. In the Astronaut Office, morale skidded as many saw their chances to go to the moon slip away.

And the astronaut image was showing signs of wear. It had been a year since Donn Eisele left his wife for another woman. For years astronauts had kept shaky marriages together because no one knew how the alternative would affect a spaceflight career. If the Eiseles' split stunned the Astronaut Office, it tore apart the community of wives, many of whom were forced to admit, at least privately, what they had carefully ignored for so many years: Their husbands were playing around. The Astronaut Wives Club, which had been the nucleus for social gatherings, was so shaken over whether to invite Eisele's new wife that it ultimately dissolved. Eisele left NASA without flying again, and some wondered whether his divorce had been the reason. But at the end of 1969, Al Worden, assigned to Apollo 15, showed that an astronaut could end his marriage and remain on a space crew.

For the most part, Stu Roosa had little time to think about these events, or anything else besides Apollo 14. It was true that he had pilot friends in Vietnam, and as always, he felt a little guilty about not flying alongside them. When one of them came home and expressed anger over the way the war was being handled and bitterness at the dissent at home, Roosa shared his distress. But he was more upset over what was happening to Apollo; he could not believe the nation was abandoning the most magnificent exploration in its history. The impact on Roosa was very real. He had known that as command module pilot of Apollo 14, assuming he stayed in the rotation, he would eventually command Apollo 20—but his hopes of landing on the moon had died in January 1970, when Apollo 20 was canceled.

But Roosa's personal concerns paled beside the demands of his mission. In addition to the normal training workload, he had to look after the modifications to the command and service modules. And Shepard had done something that impressed Roosa greatly: he'd let Roosa handle it. He didn't micro-manage; he wasn't afraid to delegate. He had faith in his crew, and Roosa had every faith in Shepard. The man was *very* sharp, a quick study, and in Roosa's opinion, the most competent astronaut in the office. He was also probably the most complicated man Roosa had ever known. But one thing was certain: this wasn't the same Big Al who had lorded over the Astronaut Office. When he came on flight status, he stepped down gracefully to the trench work of a mission commander. He was genuinely pleasant to work for. Roosa introduced him to his parents and they were charmed by him. Once Shepard accepted a person into his inner circle he could be surprisingly open. Of course, he could still turn to ice without warning—which is to say, he was still Al Shepard—but he never directed that anger at his crew. And when it came time to make up a mission patch, Shepard sketched a design that showed an astronaut pin flying from the earth to the moon, to convey that the entire Astronaut Office was going along, in spirit. In his own way, Shepard was the best commander

Roosa could have asked for; he nicknamed him "Fearless Leader."

Ed Mitchell, the man who would land on the moon with Shepard, had a doctorate from MIT and test pilot credentials from the space school at Edwards. In the Astronaut Office, it was his intellectual bent that set him apart from some of the other pilots, along with a certain hard edge. By his own admission, he had an impatient streak, and when angered he was capable of outbursts of temper. But when Mitchell spoke, it was with the soft voice of a midnight FM-radio announcer. Roosa shared an office with him and used to wonder how anybody at the other end of a phone conversation with him could understand what he was saying. But Roosa respected him as a professional. What really mattered was that Mitchell do his job in the command module, and that he did very well. And from what Roosa could tell, Mitchell—who matched Fred Haise in his knowledge of the LM—seemed to be carrying the load for Shepard with the lander's systems, letting his commander concentrate on the piloting.

Shepard, Roosa, and Mitchell trained for nineteen months, longer than any crew before them, but even now Roosa did not really think of them as friends. Al Shepard wasn't Pete Conrad, and you wouldn't see the crew of Apollo 14 going out for a beer together at the end of the day. And Roosa couldn't have cared less. What really mattered—and what they shared with every crew before them—was a burning desire for what Roosa called a full-up mission, that is, every objective on the flight plan accomplished. By January 1971, NASA was ready to send Shepard's crew to the moon to complete the mission Lovell's team had been denied. Apollo 14 would be the first flight devoted solely to the scientific exploration of the moon. Using the pinpoint landing technique proven on Apollo 12, Shepard and Mitchell would make the first landing in the lunar uplands, at a place called Fra Mauro. While they made two moonwalks, including a climb to the rim of the 1,100-foot-diameter Cone crater, Roosa would survey the moon from orbit with a special high-resolution camera called the Hycon. Roosa sensed that this time, a full-up mission had special importance: NASA couldn't afford another failure. In 1961 Shepard had been the man of the hour; as far as Roosa was concerned, there wasn't anyone better to entrust with the future of the space program than Alan Shepard.

There were some people who wondered why, at age forty-seven, having acquired fame, wealth, and status as an American hero, Alan Shepard would risk his life to go to the moon. The passage of ten years had deepened the lines around his mouth and his blue, slightly bugged eyes, but the toothy grin was the same as it had been on the cover of *Life* magazine in 1961. And though the military crew cut was gone—his hair now fell partway across his forehead—his reasons for wanting to fly in space had not changed. In Shepard's mind, to say that he was brave, to call him a hero, was hopelessly incomplete: he was a supreme test pilot, and nothing mattered more to him than getting a chance to prove it. A decade ago, lying atop that gleaming Redstone rocket, he had watched as technicians sealed him inside *Freedom 7*, had and steeled himself: *Okay, buster, you volunteered for this thing. . . .* On January 31, 1971, Shepard led his crew to Pad 39-A, where a booster a hundred times more powerful was waiting for them. There was more thrust in the command module's escape rocket than there had been in the Redstone. His space-flight experience was surpassed the moment Apollo 14 reached earth orbit. Two hours and twenty minutes later, when the Saturn's third stage ignited, Shepard was heading for the goal he'd kept in sight for nearly a decade—not the moon, but the chance to show what he was made of. Years later Shepard would say that he always believed in pushing out the frontier; it was nice that the moon happened to get in the way.

II: To the Promised Land

Sunday, January 31, 1971
7:41 P.M., Houston time
4 hours, 38 minutes Mission Elapsed Time

Apollo 14 almost ended before it could begin. As launch time neared Shepard's crew had looked ahead to their mission with an almost arrogant confidence: With the intensive scrutiny after Apollo 13, nothing should interfere with their mission. But now, only two hours after a perfect Translunar Injection burn, Shepard, Roosa, and Mitchell found themselves waiting for word from Houston, wondering whether they could continue. The trouble had begun shortly after Stu Roosa cut loose from the Saturn third stage and prepared to dock the command module *Kitty Hawk* with the lander *Antares*. In the past nineteen months he had simulated the maneuver so many times that he used less fuel than any command module pilot before him. Now he was doing it for real, and he wanted that fuel record. Floating above the center seat, Shepard peered through the hatch window as if he were hanging from a ledge by his fingers. He could see *Antares* dead ahead, and he told Roosa, "Gonna break the record, man."

Slowly, Roosa closed in. At last the two ships met, but a moment later they drifted apart. Roosa was mystified. He decided he hadn't come in fast enough to trigger the docking latches. He would have to back off and come in a little faster, and he knew that would cost him more fuel. "There goes the record," he said mournfully. Soon his disappointment turned to concern as the second attempt failed. Something was wrong with the docking mechanism. Over the next hour and a half, in consultation with Houston, Roosa made two more attempts without success. No one had to say what they all knew: if they couldn't link up with *Antares,* the mission was over.

Shepard's crew wasn't about to give up. The culprit might be something as simple as a tiny piece of debris on the mechanism; if so, it should eventually disappear. But they couldn't afford to wait indefinitely; in a few hours the Saturn's third stage was going to vent its extra fuel, and when that happened they had better be at a safe distance. If Houston didn't come up with a fix, Shepard thought, they would have to take matters into their own hands—literally. After putting on helmets and gloves they would depressurize the cabin, open the command module's top hatch, and bring the probe inside; there they might be able to fix it. And if they couldn't, Shepard had another idea. After Roosa steered *Kitty Hawk* back to *Antares,* Shepard would reach through the tunnel and pull the two ships together by hand; when the ships met, he was sure the docking latches would engage automatically. His only worry, one shared by his crew, was that the managers in the back row of mission control would not let them risk it. And five days from now, there would be another linkup in lunar orbit when Shepard and Mitchell returned from the surface. Trouble then would force the men to make an emergency space walk to the command module. They'd trained for every conceivable emergency, including that one. But would Houston accept that possibility now, when it would be far less hazardous to abort the flight?

"Hey, Stu, this is Geno." It was Gene Cernan's voice from mission control. "We've got one more idea down here. . . ." Cernan told Roosa he should fire *Kitty Hawk*'s thrusters to hold the command module against the LM. As he did so, Shepard would flip the switch to retract the docking probe out of the way. If the two craft were lined up properly, Cernan said, the contact might just trigger the docking latches. One hour and 42 minutes after his first attempt, Roosa once more steered a course for *Antares*.

"About six feet out," Mitchell said. "About two feet."

"Here we go," Roosa said.

They felt the ships touch. "Okay," Roosa said, "retract."

Shepard hit the switch and waited. "Nothing happened."

"Nothing?"

"I don't know." Suddenly Shepard saw the indicators on the panel in front of him go from gray to a pattern of stripes. "I got barber pole," he called. At that moment he heard the telltale "ripple-bang" of twelve docking latches snapping shut. Shepard announced, "We got a hard dock."

An hour later the joined *Kitty Hawk* and *Antares* were drifting away from the third stage on a course for the moon, and Shepard, Roosa, and Mitchell settled down to finish up what was becoming a long first day in space. It had been a close call; it would not be their last.

Monday, February 1, 1971
7:37 A.M., Houston time
16 hours, 35 minutes Mission Elapsed Time

In the dark, floating in *Kitty Hawk*'s left-hand seat, a tired Stu Roosa tried to settle in for the night. Like those who had come before him, Roosa found it difficult to adapt to sleeping in zero g; he wished he could lay his head on a pillow. Forty-five minutes later, still awake, Roosa noticed a light coming from underneath the right-hand couch, where Ed Mitchell was in his sleeping bag. Roosa assumed Mitchell had turned on his flashlight because he'd gotten tangled in a strap. He could not have guessed the real reason Mitchell was awake, that he was conducting his own private experiment in extrasensory perception, unknown to anyone except a handful of people on earth.

Edgar Dean Mitchell grew up as a rancher's son amid the dust storms of the Texas panhandle and eastern New Mexico. His father was a man of nature who had a great love for his animals; his mother was a fundamentalist Baptist. As a boy, sitting beside his parents in church on Sunday mornings, Mitchell was confronted by an inner conflict. The preachers in those small churches were the fire-breathing, Bible-waving kind, and the younger Mitchell found himself doubting the existence of the God they praised and the Satan they warned of. By the time he reached high school he had concluded that the Creation story was allegory, not literally true as the preachers insisted. Their stern admonitions against social and sexual transgressions, even dancing with girls, repelled him. As much as he respected his parents and the church he was raised in, as much as he sensed a great body of truth in its teachings, he could not follow that path. His strong interest in math and science only deepened his sense of irreconcilable conflict, for neither side seemed to acknowledge the other. From then on, Mitchell hungered for resolution, and he carried into adulthood a desire to understand the nature of the universe. It was still with him when he became an astronaut.

When the press asked Mitchell why he wanted to go to the moon, he told them it was the logical extension of everything he'd been doing as a fighter pilot and then as a test pilot. He was also quick to mention his scientific curiosity, and in truth, the greatest lure was the chance to explore the unknown. When he arrived at NASA the Nineteen had little hope of flying on Apollo, but Mitchell had his sights on a more distant goal: to captain the first manned expedition to Mars. By 1970 it was clear that NASA would have to abandon for the time being any plans of sending people to the Red Planet, but by then Mitchell was already training for Apollo 14. And even as he looked forward to probing the secrets of Fra Mauro, he saw a chance to explore a frontier even more mysterious than the moon: the nature of consciousness.

Mitchell was probably the only astronaut who missed the presence of psychologists in the space program. NASA had engineers to handle the technology of flying to the moon, and scientists to puzzle over its geologic riddles, but no one to unlock the inner experiences of the men who had been there. About three weeks before the flight, a chance conversation inspired Mitchell to take advantage of the fact that he would become one of the few human beings to leave the planet. As some of his colleagues knew, Mitchell had long been fascinated by the study of psychic phenomena, for which neither science nor religion offered a satisfying explanation. He'd become acquainted with a couple of surgeons in Florida who shared his interest. Together they wondered, was it possible to transmit thoughts across a hundred thousand miles of space? In the midst of the all-consuming preparation, Mitchell told them, "Line up some people and we'll do a little experiment on the flight."

And so they did. Each night of the trip to and from the moon, Mitchell planned to perform the experiment, waiting until forty-five minutes past the start of the sleep period, when he had privacy and quiet. He kept his plan a secret from NASA, knowing that the agency would be completely unreceptive to the idea. He said nothing about it to his crewmates. The test subjects had also agreed to keep quiet. And Mitchell wasn't worried about what would happen if someone found out; with all the canceled missions he was already certain that Apollo 14 would be his only spaceflight.

Now, floating in his sleeping bag, Mitchell pulled out a small clipboard bearing a table of random numbers. Each number designated one of the standard symbols used in ESP experiments: a circle, a square, a set of wavy lines, a cross, a star. Mitchell chose a number and then, with intense concentration, imagined the corresponding symbol for several seconds. He repeated the process several times, with different numbers, knowing that on earth, four men were sitting in silence, trying to see the pictures in their own minds. After several minutes of this, Mitchell put the paper away and closed his eyes.

Had Roosa known of Mitchell's activities, he wouldn't have been too surprised. In Roosa's opinion, *Kitty Hawk* was carrying three people who, within the narrow military-test-pilot filter, were as different from one another as it was possible to be. On the circle of astronauts, they spanned a full 360 degrees. Deke Slayton always said he could put the three most divergent personalities in the Astronaut Office on the same crew and they would do just fine. As far as Roosa was concerned, the Apollo 14 crew proved him right.

Friday, February 5
1:33 A.M., Houston time
4 days, 11 hours, 10 minutes Mission Elapsed Time

Ed Mitchell tapped on the instrument panel with a flashlight as *Antares* drifted through darkness over the near side of the moon. Only ninety minutes away from Powered Descent, the mission of Apollo 14 was once again in jeopardy. He and Shepard had been checking the lander's guidance software when engineers in mission control detected that the computer was receiving an errant signal from the abort button. They guessed that a tiny ball of solder was floating around within the switch and closing a contact, and sure enough, when Mitchell tapped on the panel, the signal disappeared. For the moment it had no effect, but if it came up when Shepard and Mitchell lit the descent engine, the computer, not knowing any better, would read the mistaken signal and automatically abort the landing. Since Apollo 13, mission control had trained for every malfunction in the book—and there literally was a book, with procedures for any conceivable emergency—but not this one.

Once more there was nothing to do but wait for mission control to come up with a solution. In Houston, and in Cambridge, at MIT, computer programmers began a feverish effort to work around the malfunction. They would have to tell the computer not to accept any signal from the switch at all, to "lock out" the erroneous command. But that would mean rewriting a portion of the LM computer's software—right away. At MIT a young programmer named Don Eyles went to work; he was done before Shepard and Mitchell went behind the moon for the last time before the scheduled Powered Descent. By the time *Antares* reemerged, the programmers had streamlined and improved the fix, and relayed it to Houston.

With only minutes to go before ignition, Mitchell keyed in the necessary changes, telling himself, "It's just like a simulation," a good trick to stay calm. He and Shepard had been through some horrible scenarios in the simulator; by comparison, this was almost a leisurely operation. But when the descent engine lit at 3:05 A.M., Shepard and Mitchell held their breath. The fix worked. As the engine rumbled silently on its long brake, Mitchell keyed in a few additional changes, and they kept going. Once again the experts on earth had saved the mission.

3:09 A.M.

Four minutes later, as *Antares* descended through 32,000 feet, everything still looked good. The next step was for the landing radar to lock onto the echoes of its own signals bouncing off the surface of the moon. But on the computer display, caution and warning lights glowed, signalling that the radar had still not locked on. "C'mon, radar," Mitchell said quietly, "let's have the lock-on." Still no change. "*C'mon*, radar."

"Go at five," Fred Haise radioed. Six minutes into the burn now. Velocity was good; altitude estimates by both the primary and backup computers were in agreement. But still no radar. Shepard and Mitchell knew that if the radar didn't come in by 10,000 feet the mission rules specified a mandatory abort.

Now Haise told them, "We'd like you to cycle the landing radar breaker." Shepard pulled the circuit breaker out and pushed it back in again. "Okay," he radioed, "it's cycled." *Antares* was down to 22,000 feet now, and Mitchell could no longer hide his urgency: "*Come on.*"

Suddenly the caution lights went out, and the radar signals began to come in. Within seconds the men could see that its data were good, and Houston confirmed that fact. Once more, with help from mission control, they had made a narrow escape. And when *Antares* cleared 8,000 feet and pitched over, Shepard and Mitchell were electrified. Cone crater was dead ahead, right where it should be. "Fat as a goose," Shepard said, and proceeded to steer *Antares* to a smooth touchdown, closer to his target than any other landing in the Apollo program.

When the dust had settled and they were on the moon to stay, Mitchell thought back to their close call, when *Antares* had been flying beautifully and only the balky radar and the mission rules had stood in their way. He asked his commander, "What would you have done?" Shepard wasn't about to say what he was thinking, that he would probably have rewritten the mission rules on the spot. Radar or no radar, he would have continued as long as everything still looked good. To Mitchell he said, simply, "You'll never know."

8:54 A.M.

It is impossible to look up at the full moon and not notice Mare Imbrium, the huge dark splotch that forms the Man in the Moon's right eye. Down where his right cheekbone would be, the lunar module *Antares* rested on a region of hills and craters called Fra Mauro. The roughness of the place surprised Shepard and Mitchell, who had expected to see something resembling the *mare* plains visited by their predecessors. Instead it bore the scars of eons of cosmic bombardment. Shepard was filled with the realization that time here was measured in billions of years rather than the thousands of years spanned by human history. Nevertheless, just before 9 A.M., when Shepard stepped off *Antares*'s foil-covered footpad, the words he spoke were not for the timeless moon, or even for the history books, but to mark the end of his own personal odyssey:

"It's been a long way, but we're here."

Finding his balance, Shepard made his way to one side of *Antares* and cast his gaze to the east. There, beneath the solar glare, was the broad rise of Cone crater. Tomorrow he and Mitchell would climb to its summit in search of geologic treasure. He told Houston the way would be clear; he could see that even from where he stood. Shepard realized that, finally, everything was going well. After all they had been through, he felt sure he would have his full-up mission. He took a moment to lean back so that he could look up into the black sky, and near the zenith his gaze found a small and lovely blue-and-white crescent. Suddenly he was overcome by the beauty of the earth, by the undeniable majesty of Fra Mauro, and by his own feeling of relief. Standing on the gray dust of this promised land, Shepard cried. For several long moments, while the checklist went unnoticed, his tears flowed in spite of himself.

III: Solo

Saturday, February 6
12:26 A.M., Houston time
5 days, 9 hours, 53 minutes Mission Elapsed Time

Stu Roosa hadn't slept well, but when *Kitty Hawk* came around from the far side of the moon on his fourteenth revolution alone, he lied about it and told Capcom Ron Evans he'd gotten six hours of good sleep. In fact he'd had a fraction of that, and none of it good, but he didn't want anyone on the ground to worry about him. And thanks to a steady flow of adrenaline, he felt well. But as he brought *Kitty Hawk*'s systems back up for another day, Roosa took a deep breath. Yesterday had been long; today would be even longer. His solo voyage was turning out to be a forty-one-hour fire drill.

Roosa was pleased at how well prepared he'd been for this whole business, so well that on the trip out—to his surprise—he did not experience awe or amazement at seeing the earth shrink to the size of a marble: after hearing five Apollo crews talk about the dwindling earth, there wasn't much impact when he saw it for himself. It wasn't until Apollo 14 was more than halfway to the moon that he sensed the immense distance; it came not by looking but listening to the long moment of silence between his call to Houston and the response. As for the conversations themselves—well, the air-to-ground transcripts from this mission were going to be just plain dull. The talk was all technical, and that didn't surprise Roosa one bit, not with this crew. There were no weather reports from space, no lighthearted exchanges with mission control. It was thirty-six hours into the mission before they said anything about the view outside, but—they weren't there to talk about the view.

Roosa had thought he knew what to expect from the moon, too, but nothing had prepared him for what he saw just before Lunar Orbit Insertion, when a thin but enormous crescent moon loomed beyond the windows. On the tape player, just by chance, was one of Roosa's favorite hymns, "How Great Thou Art." He couldn't have asked for better background music:

When I in awesome wonder
Consider all the works Thy hands have made
I see the stars, I hear the mighty thunder
Thy power throughout the universe displayed
Then sings my soul, my Saviour God to Thee;
How great Thou art, how great Thou art.

Gazing down on the moon from 69 miles, Roosa wished he could take an orbit to do nothing but look at it. But his mission—the first intensive program of scientific observation in lunar orbit—was far too demanding for that. It was largely the mission Ken Mattingly had meant to carry out; like him, Roosa had spent much of the past nineteen months working with the geologists. Each of his assigned craters was a geologic world unto itself, with its own peculiar features, its own riddles. His Solo Book, crammed with photographic tasks, was as demanding as the one Mattingly had prepared for Apollo 13. On top of all this, he had the work of a command module pilot: landmark tracking, maintaining the systems, firing the SPS engine to adjust his orbit. It was a grueling workload, and time seemed to be slipping through his fingers, mostly because the single most important piece of gear for his geologic reconnaissance, an automatic, large-format camera called the Hycon (another holdover from Apollo 13), had conked out.

The Hycon was a huge thing; the lens alone was as big as the 18-inch hatch window. It had a motorized film transport and exposure controls and its own timer. With it, Roosa had planned to take photographs that would show features less than 7 feet across. Just hours after he and Shepard and Mitchell arrived in lunar orbit, Roosa unpacked the Hycon and mounted it at the hatch window. It breezed through about 140 frames; then it began making a distressing *clack-clack-clack-clack-clack,* until he turned it off. Since then, he'd spent much of his time troubleshooting. He unplugged and reconnected every cable again and again. He unmounted it and peered into the lens with a flashlight while it clacked away. Houston offered troubleshooting advice, but nothing seemed to help, and he'd already lost so much time that he was way behind schedule.

Roosa was too busy to notice his isolation—most of the time. The exceptions came after sunset. As *Kitty Hawk* drifted silently through the realm of earthlight, the cabin cooled slightly—just a few degrees at first, but that was enough that the environmental control system could not remove all the moisture in the air. Then the spacecraft went out of radio contact and into total, unyielding darkness. He liked the solitude, but he couldn't deny a feeling of loneliness. His only company were the stars that filled the sky, except where the moon blotted out even these distant companions. While Roosa worked the air turned clammy, and suddenly it seemed he could *feel* the darkness. He knew, in that moment, what it was to be utterly alone.

After nearly a half hour of this—longer than most previous missions, because the landing site was well to the west, and most of the near side was in sunlight—something remarkable happened. In a finger-snap, the cabin was flooded with sunlight, and it was such a glorious feeling of renewal, even rebirth, that Roosa said to himself, "You know, we're really not creatures of darkness." Roosa realized where he was and felt great and small at the same time. It was the moon that made him feel big: He'd made it all the way out here. And it was the sight of the earth—now, for Roosa, an object of undeniable wonder and nostalgia—that made him feel small.

Only once, near the end of the first day, did Roosa have a brief respite from his flight plan. It happened after the last experiment of the day, which called for him to take pictures of the earthlit part of the moon using high-speed film. When the time came, on his eighth solo orbit, he pointed *Kitty Hawk* straight at the moon and turned out all the cabin lights. Mounted in his rendezvous window was the Hasselblad camera. All was silent except for the soft whine of the cabin fans, which he had long ago tuned out, and the steady *click-click-click* of a timer, telling him when to snap pictures. Below, shadows of morning lengthened as *Kitty Hawk* drifted westward, making the Ocean of Storms jagged and forbidding. Every ridge, hollow and pinhole was sharp. One large crater straddled the terminator, half of its rim in brilliant sunlight and half lost in the night. Even as *Kitty Hawk* crossed into blackness, he kept firing off pictures. After a few moments his eyes adapted to the darkness and the moon reappeared, bathed in soft, blue light. He photographed the earthlit ground for several minutes, and then the experiment was over, and his flight plan for the day was finished. No more experiments, just grab a bite to eat, take care of some housekeeping, and go to bed. But for now, Roosa let *Kitty Hawk* drift, its windows pointed at the moon, and allowed himself the luxury of just looking.

Not a single light glowed within the command module. Roosa floated in total darkness, except for the window in front of his face, which framed a strange and eerie tableau of craters, plains, and ridges, rendered in soft tones of bluish gray. His eyes registered remarkable details. Now he began to see *shadows*, cast by the earth, as deep as space itself. As *Kitty Hawk* coasted westward, while the earth sank progressively lower in the moon's sky, the shadows lengthened, and the moonscape became a maze of strange patterns of blackness and blue-gray land. As *Kitty Hawk* approached the realm of total darkness, Roosa saw a jumble of shapes that played tricks on his vision. Like a child looking under the bed with a flashlight, he saw the dragons of his own imagination. It was a spooky, marvelous encounter with something unknown—in himself as well as in the moon. Then *Kitty Hawk* was engulfed in total darkness. Roosa turned up the cabin lights, and began his last chores before closing up the command module for the night.

There were no moments like that on this second day alone. From the time he woke up and plunged into the troubleshooting on the Hycon once more, he was on the run. He realized he had put more into that Solo Book than he could possibly accomplish, and that he had nobody to blame but himself. Before the flight, he couldn't say no to the scientists who suggested another crater to photograph; he was here for them. His time in lunar orbit was so precious that he couldn't bear to see any of it wasted, but too much of it had gone to troubleshooting the Hycon. Around 5:00 A.M. mission control decided to abandon the camera altogether. That meant Roosa would have to step in and, using a Hasselblad with a 500 mm lens, take detailed photos of a place in the moon's central highlands called Descartes, which was under consideration as the landing site for Apollo 16. If Roosa did nothing else, he had to return to earth with pictures of Descartes that would be good enough to plan a lunar landing. To keep the landing site in the field of view of the telephoto lens, Roosa would have to turn *Kitty Hawk* precisely as he sped past, firing off pictures at the same time. So much confidence had been placed in the Hycon that Roosa hadn't practiced this contingency procedure very often, but it would have to work.

Every now and then, as he pressed on, Roosa got word of his crewmates on the surface and their own efforts on behalf of the geologists. He felt no envy; they had their mission, and he had his. He wondered if they were having an easier time than he was.

IV: The Climb

Pieces of the moon—chunks, chips, slices, grains—had been in the hands of lunar scientists for eighteen months, slowly giving up their secrets. Prior to Apollo 11, one researcher had stated, with possible exaggeration, that if so much as a single gram of lunar material were brought to earth the scientific understanding of the moon would increase one-millionfold. Two lunar landings later, 122 pounds of rock and dust were inside the sample vault at the Lunar Receiving Laboratory and in laboratories around the world, where the scientists were like children in an unending Christmas morning.

Before the first lunar samples arrived, scientists were divided into two main camps in their predictions about the moon. Nobel laureate chemist Harold Urey believed the moon was a chunk of debris left over from the formation of the solar system. He subscribed to the impact origin of the moon's craters. Looking up at the dark *maria* he envisioned collisions so violent that they created pools of molten rock which froze into smooth plains. Aside from the occasional new crater, Urey believed, the moon had remained cold and unchanged for four and a half billion years, a primordial fossil. Urey and his followers were known as "cold mooners." Other scientists—the "hot mooners"—believed that all lunar features, from craters to mountains to *maria,* had been formed by volcanic activity. Some even claimed to have spotted the glow of escaping volcanic gases through their telescopes. Both camps defended their positions with an almost religious fervor.

By the time of the first lunar landing, most geologists had taken positions somewhere between the two, envisioning a moon whose surface had been shaped both by impacts and volcanism. They had used telescopic observations and unmanned probe images to construct geologic maps of the moon and work out a rough sequence of lunar geologic history. But neither the geologists nor their astronomer colleagues had reached a consensus. On the most fundamental unknown, the moon's origin, three main theories vied for support: The "fission" theory stated that the moon had been torn from the infant earth; the "capture" theory held that it had formed in another part of the solar system and had later wandered into the earth's gravitational clutches; the "co-accretion" theory maintained that the earth and moon had coalesced separately out of the same cloud of gas and dust. But until the samples arrived, these ideas—and almost any reasoned speculation about the moon—were up for grabs. On the eve of Apollo 11, even a theorist who maintained that the *maria* had once been real seas, full of water, not lava, made his voice heard.

But the rocks from Tranquillity Base said otherwise. In the instant when a handful of geologists inside the Lunar Receiving Lab beheld a gray slab and recognized it as the volcanic rock called basalt, the cold-mooners—whether or not they could yet admit it—had lost. The Sea of Tranquillity was a plain of congealed lava, and that meant the moon had once been geologically alive. Remarkably, lunar basalts resembled the earth's in most respects; under the microscope their crystals of feldspar, pyroxene, and other familiar minerals looked just like their terrestrial counterparts. But there were some important differences. For one thing, they were pristine; unlike earth rocks, they had never been subjected to wind or rain. In fact, not a trace of water showed up in the lunar samples, not even within their molecular structure; nor was any organic material discovered. The geochemists' probings revealed other discrepancies: lunar basalt contained more titanium than earth's but less sodium and other volatile elements—a finding that raised immediate difficulties for advocates of the fission theory. The riddle of the moon's origin wasn't going to be solved after a single lunar landing.

But there was basalt on the moon, and the importance of that discovery could not be overstated. On earth, basalts are formed by partial melting within the layer called the mantle, whose rocks are rich in iron and magnesium. By inference, the moon was not a primitive body, but at some point had become hot enough for its interior to melt and separate into layers of differing composition, including a crust, a mantle like earth's, and perhaps an iron-rich core. More than just a relic from the early solar system, the moon was a world with its own evolutionary story to tell—if the scientists could be clever enough to decipher it. They took a giant step when geochemists assayed radioactive isotopes in the Apollo 11 rocks and found that the samples were 3.65 billion years old. That age, which dated the epoch when the Tranquillity lavas erupted, was the first hard nail of truth on which the geologists could hang their timeline of lunar history. Contrary to what Harold Urey believed, the *maria* were not primordial. Geochemists had already established from meteorites that the birth of the earth and moon had happened 4.6 billion years ago, nearly a full eon before the Tranquillity lavas poured forth. But some hot-mooners had to face the surprising truth that the moon's volcanic activity took place far earlier than they had thought.

The geologists had barely had time to probe the Apollo 11 rocks when Apollo 12's haul arrived. The *mare* basalts from the Ocean of Storms were a good 400 million years younger than those from Tranquillity Base, confirming the geologists' belief that the *maria* did not all form at the same time. That was consistent with geologists' theories about an era of *mare* volcanism that may have spanned a few billion years. The Apollo 12 samples also differed in composition from the Apollo 11 rocks and from one another. That implied that the source of the *mare* basalts, the lunar mantle, must also vary in composition from one part of the moon to another.

Meanwhile, the dust of the moon was telling its own story. It bore the tracks of cosmic rays and was laden with subatomic particles from the sun for the astronomers and solar physicists to study. For the geologists, it sparkled with tiny beads of glass. Some of these, the geologists deduced, had formed by volcanic eruptions, possibly from deep within the moon. Others had likely sprayed out as molten droplets from the tremendous heat of meteorite impacts. In all, the lunar samples from these two landings would have been enough to keep scientists busy for years.

But they were not enough. A glance at the full moon reveals that the *maria* are surrounded by bright-colored highlands, otherwise known as the *terrae*, whose composition was still almost totally unknown. The relatively smooth *maria* had been the safest choice for the first landings, but after the success of Apollo 12's pinpoint landing, NASA was ready to send astronauts to the more difficult terrain of the lunar highlands. There, at a place called Fra Mauro, Shepard and Mitchell were ending an uneasy night in the lunar module *Antares*.

Friday, February 5
8:02 P.M. Houston time
5 days, 5 hours, 39 minutes Mission Elapsed Time

"Are you awake?" whispered Shepard.

In the bottom hammock Mitchell answered in a hushed voice, "Hell, yes, I'm awake."

"Did you hear that?"

"Hell, yes, I heard that."

An unfamiliar sound had jolted the men out of a fitful slumber. It had been hard enough to fall asleep wearing space suits, but they were even more uncomfortable because Shepard had put *Antares* down with one footpad in a small crater, and the whole craft was tilted noticeably to one side. Now, as they lay in the darkened cabin, the strange noise made them wonder whether the lander was resting on solid ground. Shepard whispered, "You don't suppose this damn thing is tipping over?"

After a momentary realization—Why were they *whispering?*—the two men scrambled out of their hammocks, Shepard practically falling onto Mitchell in the process, and raised the window shades—and realized that *Antares* was still firmly perched on the rolling hills of Fra Mauro. For a few more hours they tried to sleep, without success. It didn't matter; they were so anxious to begin the second moonwalk that they felt no exhaustion.

Just past 2 A.M. on Saturday, February 6, Shepard and Mitchell emerged from *Antares*. Ahead lay the climb up Cone crater and a search for geologic treasure. They prepared for the traverse by loading supplies onto a two-wheeled cart called the MET (for modular equipment transporter), which carried geology tools, sample bags, magazines of film, and other gear. It also carried a message from Gene Cernan and the backup crew. Before the flight, Cernan's crew had devised a mocking version of the Apollo 14 mission patch featuring the Road Runner and Wile E. Coyote cartoon characters. On it, the Coyote, representing Shepard's crew, reaches the moon only to find the Road Runner—Cernan's crew—is already there, waving a "First Team" banner. Along the border was printed the Road

Runner's trademark "Beep Beep." The *real* message was unfit for publication: *Watch your ass—we're right behind you.* During the flight Shepard's crew discovered Road Runner patches in every notebook and storage locker in their two spacecraft. Even on the lunar surface, they couldn't escape: There, on the MET, was another "Beep Beep."

At 2:51 A.M. the men left *Antares* and headed to the east, where Cone crater rose into the glare of the morning sun. While Shepard pulled the MET, Mitchell studied a photo map. Their route would take them past a number of large craters, then onto Cone's flank. Reaching the upper slopes, they would head northeast along a broad ridge, following it right to the crater's edge. But almost immediately, he and Shepard had trouble spotting the craters they used as checkpoints. The place was a sea of hummocks, like sand dunes, with depressions in between. Some of them were 10 or 15 feet deep. Under the brilliant sun and black sky it was like an alien, rock-strewn Sahara—and just as difficult to navigate. He found himself looking at the map thinking, "That next crater ought to be 100 meters away," but it was nowhere in sight. Even a large crater could be so well hidden from his view that he wouldn't spot it until he was right next to it.

With some care he and Shepard managed to locate their first sampling stop, where they collected a handful of rocks, snapped photographs, and took readings with a portable magnetometer. After more walking and deliberating they found their second stop. They lingered there just five minutes, long enough for Shepard to pick up a single rock as a grab sample. Then, as Mitchell took his turn pulling the MET, he gave the rallying cry: "To the top of Cone crater."

In Houston, Mitchell's words were heard by Capcom Fred Haise, who had once planned to make this climb with Jim Lovell. Now he served as Shepard and Mitchell's link with mission control, and with a back room full of eager geologists. As he listened, Haise followed along on his own photo map; he could also check the men's position by glancing at one of the big screens at the front of the control room. On the board, Haise saw that the men should be near the sloping side of Cone crater. Then came Mitchell's voice: "We're starting uphill. . . ."

The ground on Cone's flank was firmer, but there were more rocks, and Mitchell was forced to slow down as he threaded a winding course among the craters. Every time the MET's wheels hit a rock it lurched upward in slow motion, and he worried it would tip over and scatter equipment and samples across the landscape. Shepard finally grabbed the back of the cart and the two men carried it, while Shepard jokingly muttered "Left, right, left, right," like a foot soldier.

From earth, Haise radioed, "There are two guys sitting next to me who kinda figured you'd end up carrying it up." He didn't have to explain what he meant. Gene Cernan and his lunar module pilot, Joe Engle, had bet Shepard and Mitchell a case of scotch that they wouldn't make it to the top of Cone as long as they had to drag the MET with them. But they were determined. The view into the 1,100-foot-wide pit would be spectacular. Furthermore, the scientists had told them, the deepest rocks blasted out of Cone would lie at the rim itself. With or without the MET, they would get there. Anything less, in Mitchell's mind, would be less than a full-up mission. Just ahead, the ground sloped upward in what was surely the last rise before the summit. But the climb was far more tiring than they expected. The stiffness of their pressurized suits fought every step.

As Shepard and Mitchell took a much needed rest they stole a moment to look behind them at the bright, undulating plains of Fra Mauro. Tracks from the MET's two small rubber tires stretched like shiny ribbons down the hillside, into the broad valley where *Antares* rested like a tiny scale model. Already the men were more than twice as far from their LM as any previous moonwalkers. Like their predecessors, Shepard and Mitchell found that the lack of familiar landmarks and the unreal clarity of the scene made it almost impossible to judge distances, and that only made navigating more difficult. But the climb was almost over now. From this high place, Shepard and Mitchell savored the anticipation of victory.

One thing puzzled Mitchell—if they were nearly at the rim of the crater, it didn't look anything like he expected. On one of the field trips with the geologists they'd visited a nuclear explosion crater in Nevada, and that thing had boulders the size of small cars around its rim. But there were no such rocks here. Seconds later, as he and Shepard reached the top of the rise, he knew why.

"We haven't reached the rim yet," Shepard said, his voice betraying little of his surprise. All they had done was climb over a ridge; out ahead, Cone's flank rose into the distance. Suddenly Mitchell wasn't at all sure where they were. He told Haise, "Our positions are all in doubt."

4:09 A.M.

Down the hall from mission control in the Science Operations Room, a thirty-nine-year-old geologist named Gordon Swann listened to the transmissions from Fra Mauro. Since the early 1960s, Swann had brought his considerable energy and geologic expertise to the work of planning lunar exploration for the U.S. Geological Survey. Raised in western Colorado, Swann had the humor, sensitivity and political savvy to be an ideal leader of the Apollo 14 field geology investigation. Swann had helped devise the traverse Shepard and Mitchell were now struggling to complete. Over the years he had gotten to know most of the astronauts on geology trips, and had made some good friends. But he never managed to get close to Shepard and Mitchell. Before the flight, Shepard had told him, "I guess you realize rocks and geology aren't too big with me, but I'll try and do a good job for you."

"I can understand that," Swann had answered. "I'm not too big on aeronautical engineering."

Shepard replied, "I guess we understand each other."

Now that he was on the moon, Shepard seemed to be confirming his promise. He was making his own geologic observations; a little earlier he'd even corrected Mitchell's description of a glass-splattered rock. Swann wasn't surprised, though, that he and Mitchell were having trouble navigating. He knew well from Conrad and Bean's experience that it was one thing to see a feature on an orbital photo and another to recognize it standing on the moon, with no obvious topographic clues. The solution was training. Swann had offered to brief Shepard and Mitchell on how to spot landmarks, and they had invited him to the Cape a few weeks before launch. But when Swann met with them in the crew quarters they seemed unconcerned. "We'll have the maps," they told him. "And you guys will be in the back room, telling us where to go." Swann did his best to get his message across, but the truth was that he and the other geologists couldn't do as well at navigating, from a quarter-million miles away, as the men who were on the moon. And now, no one in the back room knew exactly where Shepard and Mitchell were.

In his headset, Swann could hear the sound of heavy breathing; the climb was taking its toll. A few minutes ago Shepard's heart rate had hit 150, prompting a flight surgeon in mission control to request that they stop for a rest. One of the doctors called back to the Science Support Room and asked, "How important is it to get to the top of Cone crater?"

The question was fielded to Swann, but he didn't want to answer. Cone crater was a natural excavation into the Fra Mauro, and the closer Shepard and Mitchell got to the rim, the deeper the source of the rocks they would pick up. The deepest rocks, and perhaps the most important, would lie near the crater's edge. Getting to the rim was desirable, but not paramount—but Swann didn't want to say that. If he downplayed the rim, he feared, the doctors would call Shepard and Mitchell back. If he recommended to keep going, he knew the managers in the back row of mission control might veto the request. Swann hoped the men would push on as long as they could—and as long as mission control would let them. So he filibustered.

Even as he spoke, Swann listened to the voices from the moon. Shepard was arguing to spend more time on collecting samples. He had spotted some boulders up ahead which he felt sure were deep ejecta from the crater. He wanted to sample them and then turn back.

Swann could hear that Mitchell wasn't happy about that idea. "Oh, let's give it a whirl! We can't stop without looking into Cone crater," he told Shepard. "We've lost everything if we don't get there." Swann could understand the younger man's frustration. No one had ever visited a 370-yard lunar crater, and after coming all this way it was only natural that Mitchell wanted to look into it. But Swann doubted he would see anything of scientific importance; the unmanned orbiter photos had showed no signs of exposed layers or other features.

Just then, Fred Haise called back to ask for a verdict, and everyone in the room agreed that Shepard and Mitchell had come close enough. Swann heard Haise radio, "The word from the back room is they'd like you to consider where you are the edge of Cone crater."

Mitchell answered, "Think you're finks." But Haise responded with a bit of leeway. If they thought they could reach the rim soon, it was their decision. Mitchell wasn't giving up. And mission control was extending the moonwalk by half an hour. Swann smiled when he heard Shepard say, "We'll press on a little farther, Houston. And keep your eye on the time."

4:17 A.M.

Shepard and Mitchell came to another rise and stopped to rest again. All around them, the ground was littered with rocks that must have rained out of the sky like artillery fire after the impact that formed Cone crater. Two hundred and thirty feet below, the valley was awash in sunlight. Mitchell was sure that somewhere close by, among the rocks, was the crater.

"Deke says he'll cover the bet if you'll drop the MET," Haise said. But Mitchell rejected that idea; they would need their tools when they reached the rim. "The MET's not slowing us down," he said gamely. "It's just a question of time. We'll get there."

Following his commander, Mitchell watched him pull the tool cart. He realized Shepard wasn't heading in the right direction.

"Al? Head left. It's right up there."

"Yeah. I'm going there."

They pushed onward, dodging boulders. Both men were breathing hard. Again they stopped.

"We're right in the middle of the boulder field on the west rim," Shepard radioed. "We haven't quite reached the rim yet."

When Mitchell heard those words he realized Shepard thought they were farther to the north than they really were. No wonder he'd been heading toward the rise straight ahead; he thought they were just west of the crater. Mitchell was sure they were south of it. He pulled the map from its holder on the MET. Yes, he could see where they were now. If they headed north—off to the left—they'd reach the rim. He went over to Shepard and showed him the map.

"Look," he said, his words interrupted by heavy breathing. "Let me show you something. . . . We're down *here*. We've got to go *there*."

Mitchell pulled. The MET caught a boulder and almost tipped over, but he saved it. Now the grade flattened out; they'd reached the point of maximum elevation. But a minute and a half later the rim was still nowhere in sight. Mitchell realized that once more, the moon had fooled them. Intently, he studied the map. "This big boulder, Al, that stands out bigger than anything else—we oughta be able to *see* it." They were so close. If only they had a little more time. But that was something they had just run out of.

"Okay, Ed and Al." Haise's voice was matter-of-fact, but his words meant the quest was over. They'd eaten into the half-hour extension. They couldn't take any more time looking; they had to start sampling.

Shepard and Mitchell gathered some rocks from the place where they had stopped, and then they headed down the slope, to the northwest, toward the strange, white rocks Shepard had noticed earlier. The geologists had said the rocks of the lunar highlands would be different, and these surely were, unlike any of the samples they'd seen in the LRL. Wielding a geology hammer, Mitchell approached the boulders, which were 4 or 5 feet high. Streaks of gray and brilliant white ran through them like pulled taffy. At the first strike an outer surface crumbled away. Mitchell hit the rock again, harder, and knocked off a chip about the size of an egg. Clumsy in his gloves, he struggled to place it in a small Teflon sample bag, rolled up the top, and placed it in the rock bag on the MET. Already it was time to leave.

On the way down Shepard and Mitchell made up for some of the time lost in the climb. Now they could really fly, bounding downhill in giant, slow-motion leaps, sailing over rocks as they went. They had no trouble navigating; fortunately their last station stop had been near a large boulder. They spotted the big rock while they were still up on the slope and kept it in sight all the way down.

Three hours after they had left *Antares* Shepard and Mitchell returned with their precious cargo of rocks and film. By now they were both very tired, but Shepard had something to take care of before the moonwalk ended, a little "gotcha" for the TV audience. He'd dreamed it up one day during training, when Bob Hope came for a tour of the space center. Shepard and Deke Slayton were escorting Hope, and they took him over to the one-sixth g rig they used to simulate walking on the moon. Hope carried a golf club with him everywhere he went as if it were a pacifier, and he refused to let go of it even when they had him on the rig. There he was, bouncing around with that golf club in his hand, and that's when Shepard, also an avid golfer, said to himself, *There has to be a way. . . .*

Now Shepard stood before the TV camera. "In my left hand," Shepard announced, "I have a little white pellet that's familiar to millions of Americans . . ." In his right hand, he held the handle from the contingency sample collector, now slightly modified: it had a genuine six-iron at the end of it. His pressurized suit was so stiff that he had to swing his makeshift club one-handed. His first swing missed, and on the second he shanked the ball. Dropping another ball to the dust, he swung once more and made contact, and as the ball sailed away into the black sky, arcing over the craters in slow motion, Shepard announced, "Miles and miles and miles!"

Saturday, February 6
7:53 P.M., Houston time
6 days, 4 hours, 50 minutes Mission Elapsed Time

The cratered sphere filled the window, hanging in blackness. *Kitty Hawk* was climbing away from the moon on the Great Elevator Ride. Ed Mitchell felt a weariness that was not just physical exhaustion but a longing so profound as to be felt in the body. He did not want this moment to end. It wasn't merely the view that was so powerful, it was the *idea:* He and Shepard had just been down there. They had walked on another planet. Mitchell took a long, last look; he knew he would never return.

By the next day the three men were still recovering from the grueling day before. Now the quiet time of the mission had arrived, three days with little to do but keep the spacecraft operating and catch up on sleep. The pressure was off. Later Mitchell would look back and realize he was in the ideal condition for what was about to happen to him.

Beyond the command module's windows, a cloud-covered, crescent earth cast its light in the blackness. By now it was a familiar sight, but every once in a while Mitchell stopped to look at it. Gradually, as he worked and glanced at the bright crescent, he was filled with a quiet euphoria, great tranquillity, and an overpowering sense of *understanding.* It was as if he had suddenly begun to hear a new language, one being spoken by the universe itself. No longer did the earth or anything in the universe seem to be random. There was a sense of order, of worlds and stars and galaxies moving in harmony. In one moment he was a detached observer; in the next he could see that he was definitely a part of it all.

As he worked within the command module he had a sense of being outside himself, as if someone else's hands—way down there—were turning knobs, flipping switches. He found himself glancing over at Shepard and Roosa, looking for some glimmer in their eyes, some sign that they were sharing any part of this awakening, but they seemed the same as they had always been. He said nothing.

Several times during the trip home, the feeling returned, triggered each time by the sight of his home planet. Mitchell knew he had been enlightened, but in a way he did not understand, and with an impact that even now, he did not fully sense. In time, it would overshadow even walking on the moon. In it, Mitchell would try to find the seeds of resolution he had longed for all his life. He did not guess, as *Kitty Hawk* coasted toward earth, that Stu Roosa, and even Al Shepard had tasted their own moments of personal discovery. In ways they would never talk about, at least not to each other, this 360-degree crew had that in common.

• • •

When Shepard's crew arrived at the Lunar Receiving Laboratory to wait out the rest of their quarantine, they brought along a haul of rocks and photographs that made the geologists ecstatic. But their enthusiasm was tempered by the fact that Shepard and Mitchell hadn't documented their finds as well as the scientists had expected. On the moon, Mitchell radioed the geologists that one of the rocks he collected would be easy to recognize because it had "a definite shape." But as one geologist told a reporter, all rocks have a definite shape, and in the LRL, Mitchell could not identify this one among the other samples. And although he and Shepard had trained to photograph rocks in place before collecting them, they had only documented a few samples in this manner. In the weeks to come, the scientists would have to try to locate the rest on the astronauts' panoramic photos.

Meanwhile, using standard triangulation methods, some of the scientists had analyzed the pictures from Shepard and Mitchell's climb up Cone crater and had plotted the path of the two men. The results showed that Shepard and Mitchell had come within only 65 *feet* of the rim. Mitchell was furious when he heard. Ironically, he and Shepard had come the closest after they abandoned the search, when they walked down the slope to the strange white rocks. If they had continued another 20 yards to the northwest, they would have been staring into the enormous pit. Shepard and Mitchell had already paid up on the bet with Cernan and Engle, but the geologists did them one better: they sent a case of scotch into quarantine. As far as they were concerned, 65 feet was close enough. Gordon Swann got a laugh from Shepard when he said during a debriefing, "You weren't lost, and you didn't know it!"

One morning around the same time, Shepard made his own discovery. He was reading the paper at breakfast, when he noticed an article: "Astronaut Does ESP Experiment on Moon Flight." Shepard laughed. "Hey, Ed," he called out, "did you see this? This is the funniest goddamn thing I ever saw."

Mitchell came over to the breakfast table and looked at the article. "I did it, boss," he said.

Shepard stared back at him in silence. He could barely believe Mitchell had managed to get away with such a stunt without him finding out about it. If he'd known about it before the flight, he would have hit the ceiling—*Not on this mission, you're not*—but right now he was feeling pretty damn good. He'd had his full-up mission. He could afford to be magnanimous. Why let a little thing like this spoil a good breakfast?

• • •

Not long after he got out of quarantine, Stu Roosa went to Downey to thank the workers at North American who had built the Apollo 14 command module. In a comment that rightfully belonged not only to the astronauts but to mission control, one manager told Roosa, "You saved the space program."

But within the scientific community, that news was greeted with skepticism. Ever since the first lunar landing, NASA had been criticized for failing to include scientists on Apollo missions. On the day Apollo 14 arrived in lunar orbit, a leading figure in lunar science, Caltech geologist Eugene Shoemaker, was in London to provide televised commentary on the mission for the BBC. At a press conference with an embittered Gordon Cooper, Shoemaker blasted NASA for doing a "completely miserable job" in integrating scientific goals into the moon program. To Shoemaker, who had always believed that the purpose of sending humans into space was to make discoveries, the potential of this $24 billion undertaking was being wasted.

NASA was well aware of the criticisms voiced by Shoemaker and other scientists. Only three more lunar landings remained, but they had been planned to answer the scientists' frustrations. With those missions, Apollo would figuratively and literally reach its greatest heights.

BOOK THREE

CHAPTER 9

The Scientist

The words that John Kennedy spoke before Congress in May 1961 said nothing about going to the moon for science. There was nothing about extending the lifetime of a lunar module to three days so that astronauts could make three moonwalks instead of two, or upgrading the backpacks to nearly double the length of each trip outside. Certainly there was nothing that even hinted at developing a battery-powered car that could be folded up in the side of the lunar module like a toy in a matchbox and then, once unloaded on the lunar surface, take the astronauts miles into the distance until their lander was only a speck in the wilderness. And yet before Apollo was finished, all of this would happen, and it would happen in the name of scientific exploration. Even now, with forces already at work to bring an end to the moon program, Apollo was about to hit its stride. With these new innovations, astronauts would progress from visiting the moon to living there. And the final teams of moon voyagers, with trained eyes and hands, would visit some of the most spectacular places on the moon for the first extended scientific expeditions to another world.

For one man in particular, NASA's plans for the last Apollo missions were crucial. His name was Harrison Hagan Schmitt, known to friends as Jack. Dark-eyed and dark-haired, he was given to strong opinions and awful puns. He was also the first geologist-astronaut, and his ambition was to be the first scientist to practice his profession on another world.

Schmitt hadn't planned to be either a geologist or an astronaut. As a teenager in Silver City, New Mexico, he spent his weekends and summers assisting his father, a respected mining geologist, in the field. When he entered Caltech in the fall of 1953 it was not to study geology but, like most Caltech undergrads, in hopes of becoming the great physicist of his generation. After half a semester as a physics major, however, he knew it would be otherwise. The next year, Schmitt moved to the more familiar ground of his father's profession. When John Kennedy launched the Apollo program, Schmitt was a graduate student at Harvard, where his nickname was Bull Schmitt. The joke among the grad students was that of course the first man on the moon ought to be a geologist, but Schmitt showed only a casual interest in the new space program.

In spring of 1964, Schmitt was completing a postdoctoral fellowship, but the job prospects, both in his specialty of economic geology (which includes the study of ore deposits) and in academia, were dim. But some of Schmitt's friends from Caltech and Harvard had gone on to work for Eugene Shoemaker, who had established the U.S. Geological Survey's new astrogeology branch in Flagstaff, Arizona. Their reports encouraged Schmitt, and he wrote to Shoemaker, hoping there might be an opening for him.

Gene Shoemaker was one of the brilliant scientists of his generation. He'd made his mark at a young age with a study of Meteor crater, the 4,000-foot-wide hole in the Arizona desert that is one of the freshest and most recent impact craters on earth. Shoemaker made the first comprehensive geologic analysis of the crater and, in the process, reconstructed the awesome event that produced it: Fifty thousand years ago a 150-foot iron meteorite slammed into the plains and exploded with the force of a twenty-megaton nuclear bomb. No wonder the entrepreneurs and scientists who had searched for the remains of the intruder had failed: most of it had vaporized in an instant. And Meteor crater was *tiny*, insignificant compared to the hundreds of lunar craters that can be spotted with no more than a pair of binoculars. The moon's pockmarked face testifies to eons of bombardment by asteroids, leftover debris from the formation of the solar system. Shoemaker knew that the earth had endured the same terrible onslaught; anyone who wanted to see what our own planet looked like billions of years ago need only look at the moon. But the earth's ancient craters, along with most other signs of that era, have long been erased by the relentless forces of geologic activity. The surface is constantly reshaped by mountain building, volcanoes, glaciers, and erosion by wind and water. The moon, which died geologically while the earth was still young, remains a museum world. When Shoemaker looked at the moon, he saw a 4.6-billion-year history book waiting to be read.

By 1964, at age thirty-six, Shoemaker had put together from scratch, nearly single-handedly, a new scientific discipline called lunar geology. At that time the words "lunar geology" barely had any meaning to most people, but Shoemaker had made the first detailed geologic map of part of the moon, and had established a system for delineating lunar geologic time. Working from telescopic photos, Shoemaker and his USGS colleagues had staked out parcels of moonscape the way Survey geologists will tackle the corner of a state. To the Survey's map list, brimming with titles like "Geologic Map of the Sawtooth Ridge Quadrangle of Montana," they added titles such as, "Geologic Map of the Copernicus Quadrangle of the Moon."

But Shoemaker's ambitions were even more far-reaching than his telescope. In 1948, as a twenty-year-old Caltech geology student, he'd read of experiments with captured V-2 rockets and foreseen the coming of space exploration; he also knew that he wanted to be part of it. Since then, he had built his career on being the first scientist on the moon. By 1962, with the moon a national goal, Shoemaker's efforts went into high gear. And if NASA wasn't taking scientists into the astronaut corps, Shoemaker would do what he could to change that. But his dream was abruptly shattered in 1963 when he was diagnosed with Addison's disease. Even then, Shoemaker continued to pressure NASA to take scientists into the astronaut corps. Whoever he turned out to be, Shoemaker knew that the first geologist on the moon—the man who would go where Shoemaker so longed to go—would have an unprecedented chance for scientific discovery.

Schmitt's letter crossed in the mail with an invitation from Shoemaker to join the work under way at Flagstaff. Schmitt became part of an effort to figure out techniques for field geology on the moon. He had been at Flagstaff only a few months when, in October 1964, Shoemaker's efforts paid off: NASA announced it would select scientist-astronauts. The National Academy of Sciences would screen applicants and submit a list of recommendations to NASA. Shoemaker immediately encouraged Schmitt and two others on his staff to apply. In all, more than a thousand scientists sent their names to the Academy, far fewer than the selection committee had hoped. It appeared that most of the country's top young scientists weren't willing to gamble their careers on what they saw as a slim chance of going into space. The Academy had hoped to cull fifty or sixty names, from which NASA might select a dozen or so astronauts, but in the end they sent only sixteen names to the space agency. Three of Shoemaker's people, including Schmitt, made the list; NASA rejected all three. In fact, none of the geologists recommended by the Academy passed NASA's physical.

Shoemaker was astonished. To select scientist-astronauts for a moon program and not include a single geologist was beyond his comprehension. Of all the geologists, Schmitt had come the closest to passing, and Shoemaker persevered on his behalf. None other than Randy Lovelace, the country's top aerospace physician, reviewed Schmitt's case and pronounced him fit to fly. When NASA announced the names of six new scientist-astronauts in June 1965, Schmitt was on the list.

• • •

They were outsiders from the moment they arrived. The astronauts knew about the pressure from the National Academy—the lobbying in Washington, the letters and phone calls. And they couldn't believe Deke would have taken on a bunch of scientists if they hadn't been forced down NASA's throat. Behind their office doors, they wondered what it would accomplish. Nothing personal; but what were a bunch of scientists going to contribute? In the race to the moon, science was excess baggage.

However the pilots may have felt about them, the scientists faced a very real and crucial hurdle: they had to learn to fly jets. Two of them, physician Joe Kerwin and astronomer Curt Michel, were already qualified jet pilots. Physicist Owen Garriott had a private pilot's license, but the other three had never flown an airplane of any kind. These men—physician Duane Graveline, physicist Ed Gibson, Garriott, and Schmitt—headed to Williams Air Force Base in Arizona for a year of pilot training.

Within three weeks Graveline had left the astronaut program for personal reasons. The three who persevered at Williams, meanwhile, were soon immersed in learning to fly. After thirty hours of basic flying instruction in a Cessna 172 they moved on to jets, first the modest T-37; then, at last, the sleek, high-performance T-38. Competing against younger cadets, they graduated in the upper third of their class, with Gibson and Garriott very near the top.

But that hardly seemed to matter in the summer of 1966, as Schmitt, Gibson, and Garriott returned to the Astronaut Office with their wings, to the visible surprise of some of the pilots. In their absence, NASA had selected the Original 19, who now joined the five scientists for classroom studies. When classes ended, the scientists lost all illusions of moving ahead. The Nineteen were assigned to support crews; the scientists—who ostensibly had more seniority—weren't. No one had to come out and tell them what was suddenly obvious: the pecking order didn't apply to them. They were never in line. At this low point, each of the scientists faced his own private decision; each decided to tough it out.

• • •

It was almost beyond belief to the pilots when NASA selected eleven more scientist-astronauts in the summer of 1967. Still, they had a good idea what brought it on. They knew the tension between NASA and the scientific community had not abated. They probably did not know that at NASA Headquarters, George Mueller had been pressuring Deke Slayton to get "manned up" for the Apollo Applications earth-orbit and lunar missions that would follow the Apollo landings. What Mueller didn't seem to recognize—or perhaps didn't want to believe—was that most of those missions were disappearing from the NASA budget even as the new scientists were being screened. And on the day eleven bright-eyed young scientists reported to Houston for their first day as astronauts, they didn't get the welcome they expected. Shepard's politely worded greeting boiled down to this: "We can't use you, and if you had any brains, you'd leave." They didn't even get their own offices; they were corralled into one big area that became known as Boys' Town. Once the shock wore off, the new scientists did their best to take their predicament in stride. They hung in through flight training—once more, some graduating in the top of their classes—and classroom instruction, and then went after whatever assignments were available. If they had to be the jetsam of the astronaut corps, they could at least do it with a sense of humor; they dubbed themselves the XS-11.

It was clear from the beginning that none of the XS-11 had any hope of getting a seat on a moon flight. Even Apollo Applications, whose raison d'être was science, was a long shot, since Schmitt's group was ahead of them in line for those seats. As it would turn out, the XS-11 would have to wait until the 1980s to fly in space, by which time some of them would be in their fifties. By comparison, Schmitt's group had it good.

And you would not have known Schmitt was an underdog if you had seen him in action around 1967. Slayton assigned him as the astronaut representative on the ALSEP experiments, and under his own initiative he branched out to cover the lunar module's descent stage and its cargo of gear—geology tools, television cameras, other experiments. And these were just the beginning. His bachelor apartment, across from the Manned Spacecraft Center, became a center for round-table discussions on lunar exploration. One evening Schmitt would host his geologist colleagues, the next it would be an assemblage of top-level Apollo managers. Mission strategies and goals, rock-collecting techniques, new theories of the moon's geology, and other issues were under discussion.

None of the other scientists matched Schmitt's involvement in Apollo. He was almost fanatical in his focus. If he had any life outside of Apollo—which some doubted—he never talked about it. Geologist friends would call to invite him to dinner on a Saturday night to find out he was spending the weekend working. Even when they coaxed him out for a break, he talked shop. There was a bar at the Nassau Bay Hilton called the Seville Club, otherwise known as the Boom Boom Room for the loud bands that played there. For one brief moment in the sixties the Boom Boom Room had topless waitresses. One afternoon a few of the geologists dragged Schmitt there for a beer, and he spent the time in lively discourse on strategies for lunar exploration. If Dick Gordon, for example, had done that, they would have checked for a pulse.

Schmitt also seemed oblivious to pressure. Aside from the scrutiny from other astronauts, there were those outside of NASA who watched him even more intently. Now that he was an astronaut, some scientists expected him to be the scientific community's inside man. But Schmitt knew his strength was that he could play both sides of the issue. To the scientific community he was a friendly representative of the astronaut corps, and he worked hard to represent the astronauts' positions and to educate the scientists in the realities of spaceflight. And to the astronauts, he was a voice of encouragement to do more science, but without the unrealistic expectations that outside scientists often brought. And increasingly, Schmitt was doing what few of his colleagues on the outside could: he was spurring the pilots' interest in geology.

That wasn't easy; most of the pilots had little or no enthusiasm for geology classes, and with good reason. Schmitt saw the problem soon after he arrived. The astronauts were being lectured to, served up a hopelessly dull plate of chemical formulae and arcane schemes for classifying rocks, the kind of material that had almost put Schmitt to sleep when he was a Caltech freshman. And when it came right down to it, geology classes were just extra work for men who already had their hands full. Schmitt saw a chance to improve things with a training program that was tailored to the specific tasks of lunar exploration. But first he would have to get Al Shepard's approval; by the fall of 1967, he had it.

A year later Schmitt, like everyone at the space center, was caught up in the tremendous push to send Apollo 8 to the moon.

Frank Borman had his hands full getting ready, and asked Schmitt to put together a flight plan for the 20 hours in lunar orbit. From then on, Schmitt became Apollo 8's unofficial scientist-astronaut. When Bill Anders wanted extra preparation for his role as the first geologic observer in lunar orbit Schmitt was eager to give it. Borman never missed a chance to give Anders a jab about that, and at press conferences he would say, "Well, Bill Anders is the scientist; let him explain that." Borman didn't actually pronounce "scientist" as if it meant second-class citizen, but he didn't have to.

By the time of Apollo 10, Schmitt understood that his efforts would go much farther if he could win the support of the mission commander. Thankfully, Tom Stafford was interested, and he brought his crew to several briefings with the geologists. Frank Borman's team had called the moon a black-and-white world, raising new questions about its true colors; Stafford was eager to settle the issue, and to take whatever pictures and make whatever observations the geologists wanted, as long as they didn't interfere with the mission. Not that Stafford was that interested in the moon; more than anything else, Schmitt realized, Stafford wanted his mission to stand out, and if he and his crew could settle a few lunar mysteries, it wouldn't hurt.

Unfortunately, Stafford's crew didn't have much time for science; of course, neither did Neil Armstrong's. Schmitt understood that on the first landing science had to take a back seat—a fact that Shoemaker and some of his colleagues didn't seem to understand. They protested bitterly that Apollo 11's lunar module pilot wasn't a scientist. In his most naive moments, Schmitt could agree with them—"If they were really smart, they'd put a geologist on the first landing"—despite the fact that he had barely seen the inside of a simulator. Deke Slayton obviously didn't have any such delusions; he would say, "A dead scientist on the moon wouldn't do anybody any good." A scientist on a spaceflight had to be an astronaut first and a scientist second. And in his more realistic moments, Schmitt had to agree.

More importantly, some of the scientists had underestimated the astronauts; Neil Armstrong's work on Apollo 11 made that clear. For years, the geologists had the feeling that Armstrong was genuinely interested, and that he was picking up more than most of the other pilots. They were right; Armstrong turned in an excellent performance on the moon. In the postflight debriefings he was full of detailed comments on what he had seen, and he made clear the potential for a scientific observer on the moon.

With that in mind, Schmitt and other geologists—including a friend from his days at Flagstaff, Gordon Swann—eagerly began training Pete Conrad and Alan Bean for Apollo 12. By the time they left earth, Conrad and Bean were excellent geologic observers. But on the moon, to Schmitt's great surprise, they seemed to avoid talking about what they saw in detailed geologic terms. At one point Conrad mentioned a rock that had a glint of green in it, the color of a ginger ale bottle. Schmitt realized immediately that Conrad was talking about the mineral olivine—and he knew Conrad knew it too.

Schmitt was convinced that one reason for Conrad's reluctance had come from some of the geologists. They had been listening when Buzz Aldrin radioed a description of a sparkling rock at Tranquillity Base which he said looked like "some sort of biotite." The mineral biotite is a form of mica, and mica can only form where there is water. The moon has no water, and hence no mica. Behind their closed office doors these scientists complained, "Why try to teach them anything?" But these scientists missed the point. Aldrin didn't say it was biotite; he said it *looked* like biotite—he even hedged his bet by adding, "We'll leave that for later analysis." He was doing exactly as he'd been trained, describing what he saw as best he could. But these few scientists raised such an outcry that word got back to Conrad and Bean during training. After that episode, Schmitt could understand why Pete Conrad didn't want to risk making a "dumb shit" mistake on the moon, with the world listening.

But for Pete Conrad, geology was not at the top of the list. Until July 1969, he was still preparing to fly the first landing in case Apollo 11 failed; even afterward, he was absorbed in perfecting the pinpoint landing, visiting the Surveyor, and so on. But that changed with Apollo 13. For the first time, the objective was geologic exploration. For Apollo 13, Schmitt redoubled his efforts. He knew what was needed more than anything was a professional teacher, someone who could relate to the pilots and inspire them, who could do what those dull classroom lectures could not. In the summer of 1969, Schmitt called on his friends at Harvard and Caltech for help. Gene Shoemaker suggested that Schmitt contact a man who had been one of Schmitt's professors. His name was Lee Silver.

Leon T. Silver was a scientist of great standing, known not only as a skilled and thorough field geologist but a superb geochemical analyst. He had made his career roaming the desert and mountain country of the southwestern United States and trying to decipher the earliest portion of the geologic record. This was the expanse of time geologists call the Precambrian, everything that happened prior to the explosion of multicellular life some 570 million years ago. The Precambrian is truly the *terra incognita* of geology. Reading the story in these rocks is difficult; often they have been crushed, or melted and re-formed, so that their original character is all but unrecognizable. Sometimes it is accomplishment enough simply to determine how old such rocks are. Geologists do that by reading one of several isotopic "clocks" contained in them, and in the late 1950s Silver made some pivotal refinements to the method, specifically for measuring the decay of uranium into lead. That work won him high honors, including membership in the National Academy of Sciences. In the 1960s Silver, with his grad student Mike Duke, turned his attention to a certain class of meteorites that they felt might very well resemble what astronauts would find on the moon. By the time the first lunar samples were on earth, Silver, at age forty-four, was one of the scientists named to study them. Like Shoemaker, he was eager to see a trained geologic observer walk on the moon.

Schmitt knew Silver well. At Caltech Silver was known above all as a gifted teacher of field geology. It is hard physical work to go into the field, and after ten or fifteen years most geologists hand over their classes to someone younger, but Silver never did. He was a vigorous intruder in the wilderness, his thinning, golden red hair concealed under a floppy hat, his sunburned nose protruding from a pair of sunglasses. It was his boundless energy and enthusiasm that made the difference to his students; after one lecture they knew that they had encountered a man who would make an imprint on them. Schmitt was certain he could do the same for the astronauts.

But would the astronauts give Silver a chance? Schmitt sought out Apollo 13's Fred Haise, whom he knew to be one of the most enthusiastic geology students. With Haise's help, Schmitt arranged a meeting between Haise, Lovell, and Silver at a coffee shop in Cocoa Beach. Silver made a proposal: he would take Lovell and Haise and their backups, John Young and Charlie Duke, to a place he'd picked, for a trial run. "I'm personally convinced it's not going to be a waste of your time," he told Lovell and Haise. "But you've got to convince yourselves. I'm willing to put the time in; are you?"

It was up to Lovell; Silver could tell that he was still skeptical. "We'll give you this one trip," Lovell said, "and if it works out, we'll see about doing it again." In late September, Silver loaded a party of seven—including Lovell and Haise, Young and Duke, Schmitt, and a field assistant—into one Caltech carryall and headed for southern California's Orocopia Mountains. There was nothing official about it. They paid their expenses out of their own pockets and took the trip out of their vacation time. But in that place—a testing place, for all of them—the astronauts entered a new realm, as students with a master.

The Orocopia Mountains rise up from the hottest desert in the country. They stand naked, devoid of vegetation, in unrelenting 100-degree heat. They are a geologist's paradise, brimming with textbook examples of geologic structures. Variegated layers of rock, in shades of red, yellow, orange, white, brown, are draped and folded, angled against one another, cut by faults, twisted by ancient spasms, now frozen in a time exposure of planetary history. Arriving in this geologic wonderland, Silver led the astronauts to a little half-valley that looked out on the bare hills. Silver turned and pointed to a gnarled ironwood tree a few yards away. "That tree is the LM," he said. "You've just landed. I want you to look out the window of the lunar module and describe what you see."

One by one, Lovell, Haise, Young, and Duke tried their hand. With rough verbal sketches, they rendered the lines of hills that stood before them. Silver listened; then he coaxed them: What about the layers in that mountainside? What about the texture of that rock—how would you describe that? With impressive speed, the pilots caught on, and Silver pushed them to do even more. Soon they were proving to themselves the very thing Silver wanted most to teach them: years of test flying had already given them the skills they needed to be excellent scientific observers.

For eight days Silver led his new students through a kind of geology boot camp: up at dawn, out into the field after breakfast, working right up to dinner time. When nightfall came, Silver would lead them by flashlight back over the area they studied. They talked geology from the time they awoke until they went to bed. When the experiment was over, Lovell agreed that it had been a good experience, and that they would schedule regular field trips. Young and Duke, who were looking forward to their own explorations on Apollo 16, were sold on the whole program. Silver had four astronauts hooked. Meanwhile, Lovell and Haise emerged from the desert, grungy and unshaven, in time to attend the annual meeting of the Society of Experimental Test Pilots in Los Angeles. They walked into the lobby of the Beverly Hilton hotel looking like a couple of outlaws and approached the desk, and as the clerk stared in surprise Lovell said, "This is a stickup."

• • •

While Silver was working miracles in the California desert with commanders and lunar module pilots, other scientists fought their own battle for the command module pilots. The breakthrough came with Farouk El-Baz, a young, exuberant Egyptian-born geologist with powers of persuasion. He worked for Bellcomm, the Washington-based think tank hired by NASA Headquarters as a kind of scientific "Tiger Team," that helped tackle anything from picking out landing sites to acting as a liaison with the scientific community. One of the first things El-Baz did after joining Bellcomm in 1967 was to sift through all 4,322 pictures from the unmanned Lunar Orbiter probes and list every dome, crater, ridge, bump, and knob he could identify. From that exhaustive survey, which took three months, El-Baz formulated a list of sixteen candidate sites which together offered an example of every type of feature on the face of the moon. That effort turned El-Baz into a walking lunar data bank.

By 1970, El-Baz had already done some limited work with the lunar crews, including Tom Stafford's. But as for a real program of scientific observation, there was none. Mike Collins had little time to think about the moon when he was getting ready to go there; Dick Gordon didn't have much more. But for Apollo 13, with its emphasis on science, there seemed to be a new opportunity. El-Baz talked to space center geologist Mike McEwen, who told him he might be able to arrange a meeting with someone El-Baz had never met, Ken Mattingly.

When Mattingly heard that a geologist named Farouk El-Baz wanted to talk to him about the moon, he thought it was a joke. "*Farouk El-Baz?* Have we run out of geologists in this country?" Face to face with El-Baz, Mattingly grudgingly agreed to a briefing at the Cape. El-Baz found the conference room ahead of time and covered the walls with spectacular Lunar Orbiter panoramas, marked to show the orbital path of Apollo 13. When Mattingly arrived, with Lovell and Haise, El-Baz got to the point. "Anyone can look," he told them, "but few really see. We don't know very much about the moon. You have a chance to help us know more." Then he took the men on a tour of the moon, reciting from his mental data bank of landmarks and lunar mysteries, describing the lay of the land and sneaking in a good bit of geology along the way. He could tell Lovell and Mattingly were still apprehensive, but Haise was smiling with enthusiasm. Then they asked questions, and they listened to the answers. They began to make up nicknames for funny-looking craters they would use as landmarks. The meeting was supposed to go two hours; it went five. When it ended, Lovell agreed to schedule repeats. In time, El-Baz was meeting with Mattingly regularly, preparing him for a solo mission of scientific observation.

And so the command module pilots had their scientific mentor. Like Lee Silver, El-Baz had the infectious enthusiasm necessary to spark curiosity in the astronauts. And he had independently made the same discovery Silver had: once the astronauts called science their mission, their competitive, perfectionist energy did the rest.

Some time later, as El-Baz stood in the lobby of Building 4 at the Manned Spacecraft Center, a thin, redheaded man with a boyish face approached him. "Hey, you're Farouk El-Baz, aren't you? I'm Stu Roosa. I want you to make me smarter than Ken Mattingly!"

• • •

Had fate not intervened in the fortunes of Apollo 13's newly trained scientific observers, Lovell, Mattingly, and Haise would probably have set a new standard for the astronauts who followed. The geologists sweated out Apollo 13's return along with everyone else, but when it was all over, they mourned the mission they had lost. Unfortunately, the prospects for Apollo 14 weren't so bright. Alan Shepard and Ed Mitchell were among the smartest men in the Astronaut Office. There was no doubt that if they had really wanted to, they could have met or even exceeded Lovell and Haise's performance. But that wasn't what Shepard had in mind. One NASA geologist who was assigned to train Big Al was warned by his colleagues, "Don't be alarmed when he acts like this isn't very serious." During briefings, Shepard would tell jokes to whoever was sitting next to him, including his lunar module pilot. Whatever Mitchell's enthusiasm, it was apparently so dampened by Shepard's disinterest that another geologist later termed Apollo 14 "the nadir of my efforts." Some believed that if Shepard and Mitchell had gone in more prepared, they would have found Cone crater.

Looking back, Schmitt would allow as how Shepard had other things on his mind, like getting up to speed after a ten-year absence from spaceflight, and the fact that he was about to fly the first flight after a near-disaster. Whether his attitude would have been any different under other circumstances, no one could say. In any case, the geologists learned a lesson with Apollo 14: the commander sets the tone.

All this time, Jack Schmitt waged his own personal battle to win a seat on a lunar mission. Of course, as the only geologist, Schmitt knew he had an edge over the other scientists; if any of them were going to the moon, it would be himself. And his work with the lunar crews had won him greater standing in the astronaut office. Schmitt's biggest worry was that Shepard or Slayton might disqualify him on technical grounds. His edge as a geologist would mean nothing if he didn't measure up as an astronaut. And so Schmitt became a simulator hound. And when Fred Haise needed a warm body to help with the Crew Compartment Fit and Function tests at Grumman, Schmitt was happy to volunteer, again and again. By December 1969, he was feeling more confident that if he got the chance, he would not be found wanting. The newspapers were reporting that he would soon be named to a crew. Soon after the new year began, Shepard called Schmitt into his office and told him the same thing. He added that Schmitt should begin stealing simulator time, unaware that he had been doing that all along. By March 1970, it was official: Jack Schmitt was the backup lunar module pilot for Apollo 15. Now, working as an insider, Schmitt would help spark a competitive energy among the prime and backup teams of astronauts for the mission that would at last begin to realize Apollo's potential for scientific exploration.

CHAPTER 10

A Fire to Be Lighted

May 1970

On the highway that leads from Los Angeles to the Sonoran desert, a mud-splattered truck many field seasons beyond ever being clean again sped east with Lee Silver at the wheel, and a couple of rented four-wheel-drive vehicles following behind. Two hours from Pasadena, they pulled off the road into the Orocopia Mountains. Silver led his party on foot to a small valley strewn with cobbles and boulders. When winter rains were especially heavy, this dry wash became the mouth of a river several miles long and a hundred yards wide. These rocks were the legacy of a hundred thousand rainy winters, rolled and nudged from the surrounding hills, an inch or two at a time.

"Get me the suite," Silver said quickly. "You've got ten minutes." Silver watched as Dave Scott, Jim Irwin, Dick Gordon, and Jack Schmitt hunted among the boulders. The challenge for Silver's new students was to capture the variety of this place, from the typical to the exotic, in about a dozen hand-sized specimens. In geologic terms, such an assortment is called a *suite* of rocks. But in this case, Silver had asked the impossible. Dave Scott would always remember this as an exercise in frustration. There was too much variety here to be captured in just twelve samples, in just ten minutes. But that was the point; it wouldn't be any easier on the moon. In the spring of 1970, Silver was determined that picking up a suite of rocks would become as second-nature to these astronauts as flying the lunar module. And he had no doubts that with Dave Scott, his efforts would pay great dividends.

Even in a pack of overachievers like the astronaut corps, David Randolph Scott stood out. He seemed to have come straight from Central Casting, a six-footer with all-American good looks and built like a decathlon champion. In some circles there was a joke that if NASA ever came out with an astronaut recruiting poster, Dave Scott should be on it. Born on Randolph Air Force Base (hence his middle name), Scott was a general's son who graduated near the top of his class at West Point, then became an air force fighter pilot. He projected an admirable combination of charm and enthusiasm; he was also an able salesman. Other astronauts knew Scott as serious and businesslike, and something of a straight arrow, but even those who didn't get along with him put him at the top.

It was no accident that Scott was the first of the Fourteen to fly in space, but that experience, as copilot on Gemini 8's aborted tumble in earth orbit, left him with an albatross around his neck. Scott had lost the chance to make a two-hour space walk, and with it the chance to make his mark. Not until March 1969, when he turned in a flawless performance as Jim McDivitt's command module pilot on Apollo 9, did Scott feel he had shown the world what he could do.

Early on, Scott had shown more enthusiasm for geology than most of the pilots. He'd long harbored an interest in archaeology; as a fighter pilot stationed in Tripoli he'd visited the ruins of Roman cities in the Libyan desert. When the astronauts hiked into the Grand Canyon on one of their first field trips, Scott saw nearly two billion years of history written in twisted metamorphic rocks and perfectly exposed strata of limestone, sandstone, and shale. For the first time he understood what it meant to talk about *geologic* time, in which millennia are reduced to moments. And each outing brought new spectaculars. When Scott fell in behind Pete Conrad for geology training on Apollo 12, he wasn't there just because it was part of the mission. At his home in El Lago, Scott proudly displayed his rock collection in a specially made wooden cabinet. And when Jack Schmitt suggested that he consider a program of geology training with Lee Silver, Scott needed little convincing. He and Irwin met with Silver the day Apollo 13 was launched, and a month later, in May 1970, Silver returned to the Orocopias with his new students. Once again, it was a testing period for both sides, but when it was over Scott gave his okay. For the first time, geology was fully integrated into the official training plan.

By the time Lee Silver got to him, Scott was already poised to give his all for science. In the first days of training for Apollo 15, he told his crew, command module pilot Al Worden and lunar module pilot Jim Irwin, that their goal was to come back from the moon with the maximum amount of scientific data possible. More than any mission commander before him, Scott could afford to make that pledge. He'd come off Apollo 12 with the gut-level knowledge that he could land a lunar module. Even earlier, Apollo 9 had made him a rendezvous expert. He'd formed close working relationships with the flight controllers and knew what the conglomerate brain of mission control could do in a crisis. He didn't sweat free-return trajectories or reentry targeting. Like every Apollo commander, he wanted his mission to stand out, and he would do that with scientific achievement. But if Scott was a man who strove hard to make his mark, he was also ready to take Lee Silver as his mentor. Silver made him want to learn, to understand what the rocks could tell. He was hooked.

• • •

Scott's backup, Dick Gordon, was equally enthusiastic, but not for the same reasons. He couldn't profess any innate love of geology, though he liked Lee Silver immensely. For Gordon, the only thing that mattered was crossing those last 69 miles, commanding his own crew, flying his own lunar module to a touchdown, and leaving his footprints in lunar dust. And if geology was part of that bargain, then that was fine with him.

Gordon was happy to have Schmitt as his lunar module pilot —once he was sure Schmitt wouldn't get him killed. That worry faded quickly enough. Schmitt was by no means the best pilot Gordon had ever flown with, but he was certainly adequate. If Schmitt had trouble with anything, it was with the commander-subordinate relationship; that was something of an adjustment for the strong-willed geologist who tended to pursue his own ideas without submitting them to his commander. Gordon was ideally suited to the task of keeping Schmitt under control, and over time, the test pilot and the scientist became a team. Their rapport was real, in the air and at the field site. Gordon complemented Schmitt's exhaustive eye for detail with a talent for seeing the broad scope of things. If Jack Schmitt would reach the moon on Apollo 18, then he would have an excellent field partner. The geologists could hardly wait.

Schmitt proved himself too in the simulator, not only as a systems man, but by learning to make a successful lunar landing. One day the instructors decided to let him show it. During a descent they failed Gordon's controls, but instead of letting Schmitt take over, Gordon waved his lunar module pilot aside and landed with his controls while Schmitt laughed. After more than six years, Dick Gordon wasn't about to let someone else land his spacecraft on the moon—even in the simulator. After the run was over he emerged from the simulator and stormed past the instructors' consoles.

"Tried to get me to let Jack land, didn't you?" He kept walking. "It didn't work, did it? It never will." And he was out of the room.

• • •

The unpleasant possibility that had hung over Dick Gordon since he returned from Apollo 12 finally came to pass in the summer of 1970 when Apollo 18 was canceled. The news hit Gordon hard. He didn't say much, but when he did, once or twice, Lee Silver tried to commiserate. And being an optimist, he didn't stay down for long. As for Jack Schmitt, he seemed to be too absorbed in the training itself to be crestfallen. But then, if Schmitt felt disappointment, he wasn't the kind to show it. Gordon, Schmitt, and command module pilot Vance Brand agreed that they would not give up without a fight. It was a long shot, but perhaps Slayton would assign them to Apollo 17. In the meantime, they would work to be the best crew around, so that Slayton's decision would be an easy one.

At the same time, the cancellation reshaped the fortunes of Dave Scott and his crew. Until that time, Apollo 15 had been slated as the last of the so-called H-missions—the limited exploration missions in which astronauts stayed no more than 33 hours on the moon, traveled on foot, and took two moonwalks. But NASA, anxious to increase the scientific yield of the final three landings, redesignated Apollo 15 as the first of the so-called J-missions. In the works well before Apollo 11, the J-missions were designed to push Apollo's capabilities to the limit. They would feature an upgraded lunar module, allowing three-day stays on the surface, and a long-duration backpack that would extend each moonwalk to as much as seven hours. The J-mission crews would explore places of greater geologic complexity. And they would make the most extensive lunar traverses yet, thanks to a battery-powered car called the Lunar Roving Vehicle—or, as it was known, the Rover. Scott could not have been more delighted.

It remained to choose a destination for Apollo 15, and in mid-September the Site Selection Board convened in Houston. It had never been easy for the scientists to agree on landing sites, and this time was no exception. The various subcommittees and working groups had considered several alternatives, and had narrowed the choices to two. One group favored Marius Hills, a collection of low, dome-shaped rises that scientists believed might be small, young volcanoes. The other choice, a place called Hadley, lay along the shore of Mare Imbrium, the Sea of Rains.

Like most lunar *maria,* Imbrium's lava plains sit within a giant crater called an impact basin, ringed by mountains. Along Mare Imbrium's shore lies one of the moon's great ranges, called the Apennines, whose peaks tower as much as 3 miles above the *mare.* The geologists suspected that these mountains were blocks of a primordial crust that had been thrust upward by the tremendous force of the Imbrium impact. The chunks of basalt from Tranquillity Base and the Ocean of Storms had taken geologists back to the era of *mare* volcanism. The Apennines promised to open a window on an even earlier time, perhaps all the way back to the moon's birth.

Putting a lunar module down among the Apennines was out of the question, but the lava plains that formed the valley floor were a ready-made landing strip. Judging from the topographic maps, it should be possible for a team of the astronauts to drive partway up the side of an 11,000-foot mountain called Hadley Delta and hunt for samples; a single piece of primordial crust would justify the mission. The second lure, and the one that gave the valley its name, was a sinuous channel called Hadley Rille, a mile wide and thirteen hundred feet deep; it was one of a dozen such lunar features. Many geologists suspected it had once been a river of a hellish sort, brimming with broiling lava. The first visitors to a lunar rille might be able to solve the mystery. And there were enough other attractions at Hadley—unusual craters, for example, that looked to be volcanic—to make it a geologic bonanza.

The scientists were deadlocked. Although the final decision rested with Apollo program director Rocco Petrone, there was an unwritten rule that the choice had to have a thumbs-up from the mission commander. Dave Scott was in the room, and to him, the choice was easy—not only for science, but for something else he saw as essential: grandeur. It was good for the human spirit, Scott believed, to explore beautiful places. He told the scientists that although he felt confident he could land at either site, he strongly preferred Hadley; his words tipped the balance. The small valley at the foot of the Apennines became the target for the expedition.

• • •

For as long as he could remember, the mountains had been a part of Jim Irwin. Long before he knew the grand solitude that belongs to the jet pilot, he had found peace and exhilaration in high country. Some of the happiest times of his childhood had been while exploring the hills around Pittsburgh and the Wasatch mountains near Salt Lake City. As a young man in Colorado Springs he learned every hidden trail up Pikes Peak. On vacations the Irwin family scaled the rounded granite towers of Yosemite and reached the lofty summit of Mount Whitney. Before he soared to even greater heights, Irwin had stood atop many of the major peaks in the country. In the astronaut years, the mountains had been his refuge. When Houston's unbroken flatness and brown water oppressed the eye, when the steamy Gulf air was all but unbreathable, he and his wife, Mary, would escape with the children to the Rockies, to hike in the clean, rarefied air. How happy Irwin was, then, to discover that he would fly the first mission to the lunar mountains. Now age forty, Irwin was one of the oldest members of the Original 19 (and two years Scott's senior), and there was a quiet self-confidence about him, a maturity, that set him apart; perhaps almost losing his life and livelihood had reordered Irwin's sense of what mattered. In 1961 he shattered both of his legs in an airplane crash while teaching a student to fly. For a while the doctors feared he might lose a leg to gangrene. But Irwin fought his way back to flying and built himself into a figure of physical strength. He was probably among the more religious men in the Astronaut Office, but most of his colleagues would not have known that about him. He was as amiable as any of them, but he did not call any of them close friends. His best friends were his wife and children. For a test pilot, he was decidedly unflamboyant. No lunar module pilot before him had so completely avoided competing with his commander. In his own way, Irwin seemed ideally matched to Scott. Where Scott could be rigid, Irwin was deferential. When he did disagree, it was always with tact. Years later, Scott would say that Jim Irwin was probably the only person who could have gone to the moon with him; only Irwin could have put up with his authoritarian style of command. In truth, Irwin admired his commander's intelligence and energy. He was sure that if they were to send Dave Scott off to the moon all by himself that most of the mission would be accomplished. There was no question about it; Apollo 15 was Dave Scott's mission.

November 1970
San Gabriel Mountains, California

The rocks were white; they lay in a jumble of slabs and chunks on the floor of a tiny canyon. They had fallen here from much higher up; Scott could see that now. Silver explained that the San Gabriels had pushed up in a spasm of mountain building 20 million years ago. The upheaval had been rapid, in geologic terms, lasting perhaps only a few million years, and even as the rocks were pushed upward they were fractured by the tremendous stresses of the event. Much more recently, these white rocks had slid off the face of the mountains, whole layers like flecks of peeling paint, into this tiny canyon.

According to the geologists' best guess, this chunky white rock, called anorthosite, was the stuff of the moon's primordial crust. One of the unmanned Surveyor landers had radioed back data suggesting the existence of this rock in the highlands near the crater Tycho. And tiny pieces of anorthosite had turned up among the coal-dark dust from Tranquillity Base and the Ocean of Storms; because of their different composition, it was obvious to the geologists that these misfits had been hurled there from some distant place—namely the highlands (including the Apennine mountains) that lay beyond the shores of the *maria*. One look at the full moon, and the contrast between the dark gray *maria* and the white highlands, spoke to the sense of the idea.

Explaining why most of the moon's crust should be composed of anorthosite led some geologists to an extraordinary scenario. Within the infant satellite, they proposed, there was so much heat that the entire outer shell became an ocean of molten rock. As this "magma ocean" cooled, minerals crystallized. The heavier species, including the iron- and magnesium-rich crystals, sank to the bottom. The lighter crystals, specifically, the mineral plagioclase, which is the main component of anorthosite, floated to the top. In bringing Scott and Irwin to the San Gabriels, Silver's hope was that if they found anorthosite in the Apennines, they would recognize it from having seen these rocks.

It didn't matter how much Scott and Irwin understood about the San Gabriels, or even about the moon's magma ocean—though they seemed to take in just about everything Silver told them. What mattered, on this trip and the others that had become a regular part of the training schedule, was their ability to describe accurately what was in front of them. At first glance, the white blocks seemed to be a random jumble. But the more Scott and Irwin looked, the more they could see order—or *organization*, as Silver liked to say. Now they could see a band of darker rock running through the fragments, broken but still coherent, like the pattern on a shattered dinner plate. This was the kind of detail that would speak volumes, transmitted across the earth-moon airwaves. To Pete Conrad—

"Why talk about the rocks when they'll have the rest of their lives to study them?"—Silver would have answered with a single word: context. To a field geologist, as much as a sociologist, context is crucial. Rocks, like people, are a lot easier to understand when you know where they came from and who they grew up with.

To be sure, when it came to context, the moon didn't make things easy. On the moon, small rocks are hot potatoes. They begin life as high-velocity projectiles fleeing the blast of a meteorite impact. Once they come to rest, perhaps many miles from their source, they're then likely to be bounced around by later impacts. Over the eons, a cobble-size specimen might be kicked and nudged hundreds of miles away from the place where it formed. But the bigger a rock is, the harder it is to move. The larger the boulder, the more likely to have formed near the spot where it is found, simply because any jolt strong enough to move it would probably have broken it into pieces. As for an outcrop of true bedrock, that was something no astronaut had yet identified. The promise of outcrops, up on the slopes of the mountains, and perhaps at the rille, was one of Hadley's lures.

All the geologists were impressed with how fast the astronauts caught on to the intricacies of lunar field geology. What amazed Silver was that Scott could find time for geology at all. It would one day occur to Lee Silver that the massive effort mobilized for Apollo 15 was like a military invasion. Bob Gilruth, Chris Kraft, Jim McDivitt, and Rocco Petrone were the generals, each with his own theater of operations, and troops of engineers, flight controllers, crew systems specialists, and of course, contractors around the country, numbering in the tens of thousands. Silver was a captain somewhere in the infantry. There were enough changes and additions to the hardware and the mission design to make Apollo 15 seem almost like a new program. On Long Island, the Grumman people had outfitted Scott's lunar module with improvements such as bigger fuel tanks, extra batteries, and even a bigger exhaust nozzle for the descent engine, to give the LM extra thrust needed to carry the Rover and other gear to the surface. At the Boeing Company in Seattle, engineers wrestled with the fledging Lunar Roving Vehicle. Meanwhile, in Houston, trajectory specialists hammered out the steeper descent path required for Scott's lunar module to come in over the Apennine mountains.

For all of this and more, Dave Scott was the point man. All of these developments competed for his attention. And yet, Scott never turned away the scientists' concerns. He was their advocate with Petrone and the other generals in this invasion. The geologists could hardly believe the resistance that came up whenever they mentioned a new piece of gear. They wanted Scott and Irwin to carry a telephoto lens to take pictures of the mountains and the rille; Petrone and McDivitt fought it. Silver devised a rake to sift small rock samples from the soil; they fought that too. It was understandable; every pound in that lunar module was worth precious seconds of hover time; it was a matter of safety. But here again, Scott was the geologists' best ally. He would reach into his shirt pocket and pull out an index card already brimming with "action items" and add one more, and he'd knit his brow with a kind of mock concern and say to Silver, "We'll work it, Professor."

Most of all, Scott made time for the field trips. The training had intensified since the spring. By summer, the trips were no longer merely teaching exercises; in his drive to get ready, Scott pushed Silver to make them more like true simulations, and the men wore backpacks, cameras, and radios. By November, a training version of the Rover was added. Everything they did was like an actual mission. They followed traverses designed by the geologists. They used aerial photo maps that had been blurred until they matched the level of detail of the best photos of Hadley. And they solved real geologic problems. The best measure of progress was the increasing sophistication of their geologic descriptions. Up on a hill, Joe Allen, the young scientist-astronaut who would serve as Capcom for the moonwalks, sat in a tent with a two-way radio. With him were geologists who were unfamiliar with the area, so that they depended on the astronauts' words.

Through the end of 1970 and into 1971, Silver and his students were taking a trip every month, sometimes two, and the dynamic that was shaping up was really impressive. The healthy competition that had first showed itself in the Orocopias with the Apollo 13 prime and backup crews was in full flower. Jack Schmitt was very much a part of that chemistry. By his very presence, he raised the level of discussion. On field trips, during the review sessions, Schmitt would notice a key detail that Scott and Irwin had missed. It was easy to see that Dick Gordon was benefiting from having Schmitt as a field partner, and they in turn gave Scott and Irwin something to shoot for. There wasn't any question that each was trying to outdo the other.

Field geology is as much an art as a science, altogether different from the rigid, structured world of the simulator. Scott liked the change. He liked the hard work of the field trips, and he liked the camaraderie. Up to now Scott had not known Jack Schmitt well, but he could honestly say that he had never resented his presence, or any of the scientist-astronauts', in the Astronaut Office. The program needed the diversity they brought—but it was also clear to Scott that letting them fly before the lunar landing had been mastered was out of the question. Now that Schmitt was a key player in Apollo 15, Scott was glad for his help, though he had little patience for Schmitt's jokes—or anything from anyone that felt like a waste of time. And years later, Scott would deny that he was competing with his backup crew. He would say that his motivation was simple: He wanted to please his Professor. But that was costing him. More than once on these trips, Silver heard Scott mention how little time he had with his children. His wife, Lurton, had signed up for an introductory geology class at the University of Houston, not only to help fill the time when Dave was away, but so that when he was home, she would have something to talk about.

By March, the team had jelled, and people from all levels of NASA were witnessing it. That month Scott convinced Gerry Griffin, who would be the flight director during the moonwalks, to join a trip to New Mexico's Rio Grande Gorge, so he would understand what Scott and Irwin would try to accomplish on the moon. In April, when they ventured into California's Coso Hills, there must have been thirty or forty people, including Rocco Petrone and other members of the NASA brass, who had come along to see just what Lee Silver and his students were up to. What they saw was an expeditionary force for a geologic assault on Hadley.

• • •

On the last day of June 1971, the space center was rocked by a reminder that even now, there was nothing routine about sending humans into space. Three Soviet cosmonauts were returning to earth aboard their Soyuz 11 spacecraft after spending twenty-four days in orbit aboard the Salyut 1 space station, shattering all space endurance records. The flight was nothing less than a milestone toward missions to the planets; Americans would not better the feat for another two years. Everything was fine as the Soyuz hurtled toward reentry under automatic control. Minutes later, recovery helicopters in Kazakhstan spotted the craft descending under an orange-and-white parachute, but were unable to contact the men by radio. After touchdown, ground crews excitedly hurried to the ship and opened the hatch. To their horror, they found the men dead, still strapped into their seats. Later, they determined that the cosmonauts had perished from a sudden air leak just before reentry.

In Houston, the deaths sent a ripple of concern through the space center. With less than a month to go before launch, managers debated changing the flight plan to have Scott's crew don their space suits for reentry. Scott vetoed the idea; just because the Soyuz had a problem didn't mean that Apollo was suddenly suspect; his faith in his spacecraft was unshaken. Meanwhile, outside NASA's world, the question that had dogged the moon program for two years still lingered: Why send people back to the moon at all? The previous fall, the Soviets had scored two landings with unmanned probes. In September, Luna 16 had brought back a sample of lunar soil. Then, in November, a wheeled robot called Lunakhod 1 was dispatched to roam the plains near the moon's Bay of Rainbows, several hundred miles from where Scott and Irwin were slated to land. Gene Shoemaker was now saying it would be better if NASA spent its money on unmanned missions, if they insisted on not sending scientists to the moon. In Scott's mind, the argument was misguided. He and Worden and Irwin would go to the moon with trained eyes and trained minds. No robot could do that. Lunakhod couldn't see what they would see, speak the words they would say, feel what they would feel, or come back to tell about it. Hadley was waiting for him and Irwin, and they would meet it with the human's capacities to probe, to take advantage of the unexpected, to make discoveries. This wasn't about technology, not anymore. It was the curiosity that now burned inside him, which Lee Silver had helped to ignite, that Apollo was about. In Scott's mind, the first all-up scientific expedition to the moon would be a testament to one of Scott's favorite quotes, from Plutarch: "The mind is not a vessel to be filled but a fire to be lighted."

CHAPTER 11

To the Mountains of the Moon

APOLLO 15

I: "Exploration at Its Greatest"

Of all the dreams that might have visited an eighteenth-century sea captain, could he have envisioned a spaceship sailing a sunlit void in perfect silence, circling the moon? Could he have imagined its commander, a man who, like him, was skilled in pilotage and navigation, but whose energies were fired by the promise of scientific discovery? No doubt if by magic the two men could have stood face to face in the wood-paneled den of some timeless Explorer's Club, they would have had much to say to one another. For his part, Dave Scott had read up on Captain James Cook, the Englishman whose voyages of discovery preceded Apollo 15 by two centuries. In 1768 Cook ventured to the unknown reaches of the South Pacific on the first true scientific expedition, aboard the ship he christened Endeavour. *After the mercantile journeys of Columbus, Drake, and other pathfinders, Cook followed a new lure: knowledge. Though his attempt to witness a rare transit of Venus early in the journey was foiled by haze, Cook went on to map the unexplored lands of Australia and New Zealand. He returned to England in 1771, exactly two hundred years before Apollo 15 left earth.*

If the parallels between these explorations were clear to Dave Scott, he was equally aware of the differences. In two centuries the scope of exploration had changed dramatically, and with it the role of the explorer. Cook had played a part in everything from commission-

ing the boat to finding a crew to charting a course. Getting to the moon required the combined efforts of hundreds of thousands of people for the better part of a decade. Cook and his crew were completely on their own from the time they left port until they returned; the crew of Apollo 15 could talk to mission control almost every step of the way. Scott had no illusions about being a modern-day Cook—there simply were none—but as the command module Endeavour *slipped into lunar orbit on July 29, 1971, he could still take inspiration from Cook's words: "I had the ambition to go not only farther than man had gone before, but to go as far as it was possible to go."*

Friday, July 30, 1971
5:13 P.M., Houston time
4 days, 8 hours, 39 minutes Mission Elapsed Time

Riding the invisible flame of its descent engine, the lunar module *Falcon,* laden with scientific cargo, cleared the crests of the Apennine mountains and headed for the surface. With less than a minute to go until pitchover, Dave Scott readied himself for the ultimate flying challenge of his career. Just the night before he had told Jim Irwin, "I'm ready. I'm ready to put that baby in there right now." But now, as *Falcon* cleared 9,000 feet, Scott looked to his left and saw something that took him by surprise: the bright flank of Hadley Delta mountain, rising above him into the black sky. He had never seen this view in the simulator. For a brief moment Scott lost the feeling of powered flight and had the unreal sensation of floating slowly past the mountain. Craning to look through the triangular window for a glimpse of the land ahead, Scott saw no sign of Hadley Rille. He was sure they were off target. And moments later, Ed Mitchell in Houston radioed that there was an error in the lander's flight path; they were coming in 3,000 feet to the south.

Suddenly, right on schedule, *Falcon* pitched over, and Scott was confronted by a landscape he did not recognize. Craters he had been memorizing for months—Index, Salyut and a dozen others that were so clear on the Lunar Orbiter photographs—were nowhere to be seen. Instead he saw a nearly featureless, sun-drenched plain. But now, in the distance, he could see Hadley Rille, as distinct as a trowel's cut through wet clay. There wasn't time to go hunting for any other landmarks. Scott used the canyon as his marker and aimed a good distance short of it. He steered to the right, taking out the error in the automatic guidance, and brought *Falcon* in on instinct. Lower now, he could begin to see features he recognized, and at last, a good landing area. In his earphones, he heard the calm voice of his lunar module pilot with a steady stream of altitude readings and LPD angles: "One thousand feet. Four-five. Nine hundred. Four-five."

Scott gave himself a silent reminder: *Bring it on down.* He'd studied the landings of those who went before him, and he knew that for some reason, perhaps out of an unwillingness to really trust their own landing radar data—and undoubtedly, because it was so hard to judge one's altitude above a moonscape by eye—Conrad and Shepard had unwittingly descended along a stair-step pattern that used up fuel. Scott wanted to avoid that.

At 150 feet he began his final, vertical descent. Immediately he saw streaks of dust blow across the pockmarked ground. At 60 feet a dust storm raged beneath him, completely obscuring the surface. He turned his gaze away from the window and continued on instruments, just as he had done many times in the simulator. He slowed *Falcon* to a near hover, and as Irwin read off the diminishing altitude Scott waited for the Contact Light, mindful of *Falcon*'s bigger exhaust nozzle, which increased the necessity to shut down the engine at the moment of touchdown. When Scott saw the blue glow of the Contact Light, his hand went immediately to the ENGINE STOP button. Several thousand pounds heavier than previous landers, *Falcon* fell to the moon with a firm thud that rattled every piece of gear in the cabin and brought an exclamation of surprise from Irwin: "*Man!*"

The ship pitched backward slightly and tipped to one side. Then there was stillness. "Okay, Houston," Scott announced, "the *Falcon* is on the Plain at Hadley." As the dust settled the men saw that they had come down in the right place: straight ahead stood Bennet Peak, christened for trajectory expert Floyd Bennet, who helped devise the landing profile that brought them safely over the mountains. Somewhere, much nearer, was Hadley Rille, but it was hidden by the hummocky terrain. From now on, Scott told Irwin, they would call this place Hadley Base.

Scott and Irwin would not go outside now; they would take their first moonwalk tomorrow, after a good night's rest. That was unlike any mission before theirs, but Scott had realized from the first that with 67 hours on the surface, he and Irwin could not do business the way previous crews had. There was an important difference between staying on the moon and *living* there. To perform at their best they would need to maintain their normal day-night cycle, and so, the previous autumn, Scott pushed to change the flight plan. Not only was the new schedule in tune with his and Irwin's circadian rhythms, it opened up some time for something Scott really wanted: a reconnaissance. Lee Silver had taught him: when you arrive at a new field site, go to a high place and look around. Scott was ready to do just that, and the high place was the lunar module's top hatch.

7:22 P.M.

As the last wisps of oxygen left *Falcon*'s cabin, Irwin stood by while Scott opened the top hatch, climbed onto the ascent engine cover, and stood up, so that his helmeted head and shoulders stuck up above the LM's gleaming metal structure. From this high vantage, Scott beheld a scene that was at once alien and uncannily familiar. All around him, beyond the undulating *mare*, stood the rounded peaks of the Apennines. Nothing, not months of study, not even the view from orbit, had prepared him for the majesty of the lunar mountains. Their smooth, bright forms were arrayed in fluid sculpture under the black sky. Their slopes were virginal, marred by only an occasional small crater. For eons they had stood unchanged; now, gazing on their ancient beauty, Scott was all but overwhelmed.

Immediately, he spotted familiar features. To the northeast, a slender crescent of sunlight delineated the ridge crest of Mount Hadley, almost 3 miles high, the bulk of the great mountain still cloaked in shadow. Due east, beneath the sun's obliterating glare, stood the hills he and Irwin had christened the Swann Range. Looking southeast, Scott spotted the lone rocky pinnacle they had named for their professor: Silver Spur. None of these places were accessible to him and Irwin, but to the south lay their destination for the first two moonwalks: Hadley Delta mountain. Over 11,000 feet high, Hadley Delta would have held its own among the Rockies; here it towered imposingly above the *mare*, its steep sides topped by a wide, flat crest. Before the mission, some of the geologists had warned that Hadley Delta's flank might be littered with boulders; no one could predict whether he and Irwin would be able to drive there. Now, thankfully, Scott couldn't see any boulders at all. Tomorrow, he and Irwin would drive onto the western slope to St. George crater, a gouge the size of twenty-seven football fields in the mountainside. Farther along the flank to the east was a much smaller pockmark named Spur, one of the prime targets for the second traverse on Sunday. Monday's plans included Hadley Rille, and while the canyon was still out of view, Scott could see another destination for the final moonwalk, an intriguing collection of craters called the North Complex. Scott would have been amazed to find so many spectacular features in one place on earth, let alone another world.

Scott aimed a telephoto lens at Silver Spur, the North Complex, and a handful of other lures. He rattled off descriptions, speaking not only to Joe Allen in mission control, but to Lee Silver, Gordon Swann, and the rest of the scientists in the geology back room. Then, half an hour after emerging into the sunlight, Scott withdrew into the safety of his lunar module, sure that he and Irwin would not find this place wanting.

10:32 P.M.

Jim Irwin let his remarkably light body settle into his Beta-cloth hammock. He was grateful to be in one-sixth g, glad to be free from the mild but persistent disorientation he'd felt in weightlessness during the trip out to the moon. He could not have wished for a more comfortable bed; his 25-pound form barely sank into the cloth strip. For the first time, no bulky space suit would compromise an astronaut's sleep on the moon. The newly designed suits not only offered more mobility, to aid the work on the surface, but were easier to put on and take off, affording him and Scott the luxury of stripping down to their long johns and placing the suits, like stowaways, in the back of the tiny cabin. Before the flight, they'd trained to sleep by spending a night in the simulator, with tapes of the LM's coolant pumps and machinery playing in the background. Now, just in case, they wore earplugs. Neither man talked as he lay in the darkened cabin, knowing it was important to fall asleep as quickly as possible.

For just a few minutes, Irwin was alone with his thoughts. He felt lucky to be here; not too long ago he had doubted this day would come. Less than a year ago, he had been so troubled that he had considered resigning from the crew. The grind of training had stressed his marriage almost to the breaking point. His family life had been turned into a series of long separations interrupted by weekends at home that were spoiled by arguments and cold silences. He wanted to focus his entire being on the mission, but his troubles at home distracted him.

It was different for Scott; he and Lurton had an air force marriage, and she understood the priorities. She'd been through the training grind with Dave four times before this, counting his stints on backup crews, and by now the family was used to his long absences. But Mary Irwin had never been exposed to military life before she met her husband. As an astronaut's wife she distrusted the fame that went along with the job. She kept her distance from the other wives, forgoing the afternoon get-togethers and the social events of the Astronaut Wives Club. But unlike her husband, Mary Irwin was no loner, and she sought support at a neighborhood church. She and Jim belonged to different Christian sects, and their religious differences were driving yet another wedge between them. By Christmas, with the mission less than eight months away, Mary was on the brink of divorcing him.

More than once, Irwin had considered going to Deke Slayton and asking to be taken off the crew, but each time he'd thought better of the idea. Instead, he took the problem to his commander. He could not have known that his crewmate, Al Worden, had done the same thing when he faced the breakup of his own marriage in 1969. Do what you need to do, Scott had told Worden, promising to talk to Slayton on his behalf. And in fact, Worden stayed on the crew after his divorce. But when Irwin went to Dave Scott, his advice was to wait it out. "Everybody goes through it," Scott said. "It'll change." And it did change; the problems at home began to ease in the last few months before the flight. Relieved, Irwin was free to devote his entire energies to getting ready. Meanwhile, as launch day approached, it was Scott who seemed to grow more irritable, more tight, under the pressure of command.

Last Monday morning, as Irwin lay inside *Endeavour* and felt the Saturn V rise from the earth, he experienced a feeling of release so strong that he almost shed tears: no one could call him back now. Minutes later, when *Endeavour* reached orbit, Irwin was amazed to look out and see the crescent moon, perfectly framed in his window. He was sure it was a good omen.

Now, before he closed his eyes, Irwin did what he always did on earth; he prayed. He gave thanks that he and Scott were here, that everything was working so well. He did not think of his family now, or of events at home; he would say later that he had divorced the earth. He prayed for the success of the mission they were about to begin. Then, Irwin fell into a sound sleep, his best sleep of the whole flight.

Saturday, July 31
8:29 A.M., Houston time
4 days, 23 hours, 55 minutes Mission Elapsed Time

If Dave Scott's heart pounded as he hopped down *Falcon*'s ladder, it was not really at the thought of becoming the seventh man to walk on the moon; he would say years later that he had already accepted being here. His elation, as a newly minted lunar field geologist, was at arriving at the best field site he could have imagined. As he took his first steps at Hadley Base his words, composed with history in mind, were a preamble for what he and Irwin were about to do: "Man must explore. And this is exploration at its greatest."

The key to that exploration, Lunar Rover 1, was folded up against *Falcon*'s side like a toy in a matchbox. Within minutes, Irwin had joined Scott, and the two men set about bringing the Rover to the surface. They pulled on a pair of lanyards, and the chassis lowered slowly like a drawbridge. Suddenly wheels of wire mesh, with orange fenders, popped out from the corners. With a bit more pulling—interrupted when Irwin slipped and tumbled, laughing, onto the soft powder—the first manned, motorized vehicle was on the moon.

9:22 A.M.

It would have been a strange sight, had anyone been there to witness it: a four-wheeled craft heading out among the craters, bearing two space-suited men. The Rover was less a car than a spacecraft on wheels, with its own navigation computer, communications system, and cargo space for such essentials as maps, geology tools—and moon rocks. For all its sophisticated technology, there was still a hobby-shop look to this moon car, with its antennas and folding lawn-chair seats and tool racks. It had been designed, built and tested by Boeing in less than two years; the success of the mission rode on it.

Aboard Rover 1, the ride was more exciting than anyone, including Scott and Irwin, had banked on. Though the Rover averaged only 5 to 7 miles an hour, it seemed more than fast enough to the two men perched atop it. Every time the Rover hit a bump—and the "plain" at Hadley was all bumps and hollows—it took flight. Each new obstacle set one or two wheels off the ground for a long moment. As an added encumbrance, the steering on the Rover's front wheels was inexplicably out of commission, forcing Scott to rely on the rear-wheel steering alone. The ride was especially harrowing for Jim Irwin, who felt as though he were strapped onto a bucking bronco, especially when Scott had to swerve to avoid a crater. The transmissions from the Rover were punctuated every so often with a warning from Scott—"Hang on!"—followed by Irwin's grin-and-bear-it laughter.

Scott and Irwin were having as much trouble gauging distance on the airless, treeless moon as everyone had before them. But they were in no danger of losing their way, thanks to the Rover's navigation system; it clicked off readings of heading and distance as Scott drove. Irwin, meanwhile, followed along on the photo map.

Aside from the usual difficulty in recognizing landmarks, the maps were misleading: Many of the craters and bumps that looked so prominent on the over-enhanced photos were barely noticeable on the real moon. But one landmark was impossible to miss, and as the Rover headed southwest, over the undulating *mare,* both men suddenly caught sight of it: "There's the rille!" For a time, they followed the canyon's curving edge, then stopped at Elbow crater to gather samples. Soon they were driving again, heading uphill toward St. George; Scott could feel the difference.

11:10 A.M.

"Oh, look back there, Jim—look at that! We're up on a slope, Joe, and we're looking down into the valley, and—That is spectacular!" Standing beside the Rover, Scott and Irwin looked down the mountainside and saw Hadley Rille winding into the distance, its walls curving through shadows and sunlight. They could see boulders the size of houses strewn on the valley floor. Even as he and Irwin set to work, Scott was still reacting to the sight. It was ironic: he'd studied this place so thoroughly he almost felt he knew it in advance; he'd even driven over it in simulations. He'd looked forward to the work he was about to do. But there was one thing about Hadley he had never anticipated, its spectacular beauty.

11:18 A.M.
Science Operations Room, Manned Spacecraft Center

"We have a view of the rille that is absolutely *unearthly.*" Even as Joe Allen's voice sped moonward, it was heard by a roomful of exhilarated geologists in a small command center across the hall from mission control. Here in the Science Operations Room, otherwise known as the geology back room, all eyes were riveted to the television monitors suspended from the ceiling. The color pictures were the sharpest ever transmitted from the moon. Lee Silver felt an excitement that was hard to describe. This was not vicarious pleasure. In essence, he was a participant. He had given some thought, years ago, to the astronaut program, but had decided against it. For one thing, he was almost too old to apply; for another, he knew that, ironically, the total commitment necessary to go to the moon would likely rob him of the opportunity to study the rocks he brought back. Now he and the other geologists were here, not as spectators to thrill vicariously to their students' adventure, but to participate in the exploration itself.

All morning, Gordon Swann and his Surface Geology Team had followed the unfolding moonwalk with the precision of battlefield commanders. One man, wearing headphones, took notes on index cards, jotting down every rock collected, every picture taken, every comment pertaining to geology. Another listened to the astronauts' descriptions and made annotated sketches, which were then thrown onto the wall by an overhead projector. Another marked down the astronauts' positions on a map that was projected onto another wall. And along the walls were panoramic views of Hadley Base, captured in Polaroid pictures taken from the television monitors. Presiding over this group was another familiar face: Jim Lovell, who had taken charge of the geology back room at Dave Scott's request.

But there was another presence in the back room, and it was a reminder of discord. How to collect rocks on the moon had been a matter of heated debate within the scientific community. To the geochemists, a rock's context wasn't important; all that mattered was getting it back to their laboratories. The geochemists would have been happy to have the astronauts simply fill up a gunny sack with as many rocks as possible. Some even opposed giving them any geology training at all, afraid that the astronauts would go to the moon with the preconceptions of their instructors. On the eve of the mission they had warned the geologists in no uncertain terms that they better not have biased Scott and Irwin. Now, one of them sat in the back room, watching Silver and the rest of the geology team.

Silver had never been able to convince these scientists what he and the others were trying to do with Scott and Irwin. This was not an exercise in programming, but in learning. When Silver showed them anorthosite in the San Gabriels, he made a point of saying that no one knew what the moon's primordial crust was made of.

No one knew what they'd find. And the astronauts understood that.

Just now, the Rover's television camera was panning under remote control to show Scott and Irwin at work. There was Dave Scott, standing near a dark, sharp-edged boulder, about knee-high. Scott stood back, describing it before he disturbed it. Silver knew that Scott knew the importance of finding large boulders. Never before had the geologists had the luxury of knowing exactly where a sample had come from. This rock was big enough; it had probably been just where it was for a long time.

Scott and Irwin were in the process of scooping up some of the dust around the rock. There was glass splattered on top of the rock; Scott managed to pull some of it off. Then Scott drew his hammer and prepared to knock a piece off the boulder. He told Allen, "You know what we're going to do when we get through with this thing, Joe? We're going to roll it over and sample the soil beneath." That sample, which had been shielded from the effects of solar and cosmic radiation, would let the scientists determine how long the rock had been sitting there. And once he and Irwin had hammered off pieces of the boulder, which turned out to be composed of two different types of rock, Scott did just that. In this one stop, he and Irwin gathered samples that would meet the requirements of several different groups of lunar scientists. Scott had recognized the optimum collecting site on his own; none of the geochemists, Silver knew, could have prepared him to do that. And when he pushed over that boulder, he felled many of the unspoken doubts that had hung over the back room.

As Scott and Irwin headed back to *Falcon,* the geology team was on a high that would continue long after this first moonwalk was over. History would record that in the evening Dale Jackson of the USGS dined with Joe Allen, Dick Gordon, and several friends at a restaurant called Eric's Crown and Anchor, near the space center. Jackson made no attempt to contain his enthusiasm for Scott and Irwin's performance on the moon.

"Did you see those guys today? They got up there on the side of that mountain and found that boulder"—his booming voice filled the room—"and they sampled the soil around the rock, and then they knocked a piece off it, and then they rolled it over and got some of the soil underneath it!" By this time people at the next tables had turned around to look. Everyone heard Jackson proclaim, "Why, they did everything but fuck that rock!"

12:33 P.M., Houston time
Hadley Base

The sweet taste of fruit, a cool swig of water: sustenance for a long day's work. The fruit stick and the water bag inside Scott's suit was a reflection of the fact that no one had ever been at work in a space suit for so long. He and Irwin sorely needed the boost, because now, 4 hours after setting foot on the moon, they returned to *Falcon* to unload their ALSEP package and deploy it on the rolling plains. But before the work could even begin, Allen had some bad news. For reasons that would never be known, Scott's oxygen usage had been higher than expected—in part because of the demands of steering Rover 1 safely among the craters. The moonwalk would probably have to end early.

But for Scott, the hardest work of the day was still ahead. Apollo 15's ALSEP had a new experiment, a pair of thermometers designed to measure the amount of heat flowing from the interior. Scott's task was to bore a pair of 10-foot holes using a battery-powered drill, then insert the sensors. While Irwin set up other instruments, Scott activated the small, boxlike machine that had a handle on either side and a long tube extending from the bottom. Drilling proved to be an unexpected struggle. Shy of the 10 feet requested for the heat flow probes, the drill met resistance and would go no further.

The drill took its toll on Scott. Despite their greater freedom of movement, the new suits had one real weakness, the gloves. No one had been able to develop a pressure glove that was easy to operate. At rest, the glove assumed a position that was half opened, half closed. In order to grip an object like a hammer or a pair of tongs you had to exert constant pressure to keep the glove closed around it. After a few hours of this the forearms ached, but even worse was the damage it did to the fingers. Before the mission, in the interest of having as much dexterity as possible, Scott and Irwin had arranged to have the arms of their space suits shortened so that their fingers were right against the tips of the gloves. When they extended their arms or brought them in close, the rubber tips pressed against their fingernails. At first it was just a minor discomfort; after a while it felt like someone had hit their fingers with a hammer. But for the most part, Scott's excitement eclipsed his pain. And as this first excursion came to an end after some 6 hours, too short in Scott's mind, he stole a last chance to pick up samples. "Oh, look at what I got," he called to Irwin, who was already back in the lander. And then, treasure in hand, he ascended *Falcon*'s ladder, a geologist in a remote and spectacular place returning to the comfort of his field tent.

II: High Point

Sunday, August 1, 1971
Hadley Base

The mountains of the moon captivated Jim Irwin, just as he knew they would. That anticipation had been with him yesterday as he followed Scott down *Falcon*'s ladder. When he put his weight on the footpad, it tipped so suddenly that he went reeling backward, hanging on with one hand, until he was gazing almost straight up at the sparkling blue world he had left behind. Regaining his balance, he took his first buoyant steps into the valley, and could not suppress his elation. The Apennines were not gray or brown as he had expected, but golden in the light of lunar morning. Their rounded, dust-covered forms reminded him of ski slopes—Dollar and Half Dollar mountains at Sun Valley. He was surprised to find that a place others had called stark and desolate could feel so friendly. The mountains surrounded Hadley Base like a cradling hand. He felt at home on the moon.

But yesterday's exploration had taken a toll on Irwin. Not until he and Scott were back in the LM did Irwin feel the full measure of his exhaustion. Aside from the mental fatigue brought on by hours of intense concentration, and the pressure of finally doing the real thing instead of just simulating it, he was physically spent. The incessant glare from the sun and the bright landscape had given him a fierce headache. And he was parched; the water bag in his suit had refused to work and he'd gone more than seven hours without a drink. Worst of all was the pain in his fingers. By the end of the moonwalk they were so sore that when it came time to take off his gloves he had to ask Scott for help. When they had settled in for the evening, Irwin used a pair of scissors to trim his fingernails all the way back; that helped. He suggested his commander do the same, but Scott declined, reluctant to do anything that might compromise his dexterity.

After the moonwalk Irwin had been overtired, and he had not slept well. Then, this morning he and Scott were awakened early by a call from mission control, to check on a water leak in *Falcon*'s cabin. When they looked behind the ascent engine cover, they found a couple of gallons of water that had leaked through a broken filter. Irwin was glad that Grumman had waterproofed the LM's electronics. Thankfully, despite all of this, Irwin felt renewed as he and Scott prepared for the second traverse. Yesterday, they'd been denied the chance for real prospecting up on Hadley Delta; now they would have one more chance to find out what the mountains of the moon had to offer.

8:07 A.M., Houston time
5 days, 23 hours, 33 minutes Mission Elapsed Time
Science Operations Room

Rover 1 did its drunken gallop across the *mare*, heading south, and in the science back room, Lee Silver listened to Scott and Irwin's progress. As usual, Scott was going full throttle and they were making good time. They zipped past Dune crater, a yawning, block-rimmed hole in the *mare* that was their last landmark before Hadley Delta. Then they reported they were going up a slight incline; they were on the flank of the mountain. With Scott concentrating on his driving, Irwin was doing most of the talking, and Silver was glad for what he heard.

Only a few days before launch, after the last of a regular series of evening geology briefings at the Cape, Silver had said good-bye to his students. In the parking lot of the crew quarters, he wished them well. He felt admiration for them, and also concern, but they radiated confidence. He turned to Irwin. Though Irwin was just as adept as Scott, Silver knew, he usually deferred to his commander and said comparatively little. One day late in the training he'd taken Irwin aside and said, "Talk to me up there, Jim. Don't just follow Dave around." Now, saying his good-bye, Silver offered one more smiling admonition: "Now, Jim, we're going to hear from you up there, right?"

Irwin smiled back. "You bet."

Silver needn't have worried. Irwin was full of detailed, precise descriptions as the Rover bounced along. He brought a patient thoroughness to the explorations that complemented Scott's drive and exuberance. He frequently directed his commander's attention to something too important to ignore, but always as a suggestion: "Dave, shouldn't we get that rock over there?" And now, as Silver listened, Irwin made a new discovery: The side of Mount Hadley was covered with a slanting pattern of straight lines, as if a giant comb had been dragged across it. Everyone in the back room wondered excitedly whether the lines might help solve the riddle of how these mountains had formed; perhaps Mount Hadley was a block of layered bedrock that had been raised up by the violence of the Imbrium impact. For more clues, they would have to wait for the photographs. And when Irwin brought Scott's attention to the new find—"Oh, *yeah*," Scott said—Silver found himself smiling.

8:28 A.M.

The Rover climbed effortlessly; even now, heading directly up the slope, it was making 6 miles per hour. The plan was to work their way along the mountainside, sampling whatever geologic variety Hadley Delta had to offer. Scott knew what the geologists wanted—a crater that had acted as drill hole into the mountain, strewn with boulders torn from the flank. But there were no such craters in sight. There was a sameness to the terrain up here; everything was much more worn and subtle than the photographs had led them to believe. Angling along the contour of the mountain, Scott searched for a target, finally spotting a medium-sized crater. There weren't any boulders, but it would have to do.

When he hopped off the Rover, he almost fell over backward. The drive had been so easy that he had no idea the steepness of the slope they were on. Glancing at his feet, he saw to his surprise that his boots were half-buried in dust, and yet the wire-mesh wheels of the Rover, fully loaded with both of them in spacesuits, had penetrated only a fraction of an inch. He warned Irwin to be careful. Then he turned around, and what he saw almost knocked him off balance again. They were more than 300 feet up. Hadley Delta's broad flank swept downward and merged with a bright, undulating desert adorned with the brilliant white rims of craters. Beyond, the Apennines invaded the dome of space. The entire valley was one enormous, incredibly clear panorama: It was the face of pure Nature. And out in the middle of this pristine wilderness, a single artifact: his lunar module. *Falcon,* that big metal-and-foil bird, was a mere speck.

No scene could have conveyed more vividly the reach of this exploration—or the risk. If something had happened to the Rover now, Scott and Irwin would have faced a long and difficult walk to safety, for the LM was more than 3 miles away—much more than the distance covered by Shepard and Mitchell during their round trip to Cone crater. The whole question of how far to let two Rover-riding astronauts go had consumed hours of pre-mission deliberation. At no time could the men be allowed to drive farther than they could walk back with the amount of oxygen remaining in their backpacks. Because their oxygen supply dwindled as the moonwalk progressed, this *walkback limit* would be an ever-tightening circle.

Even if the Rover worked flawlessly—and so far, it had done nearly that—there was always the chance that a backpack would fail. In that case, the men would break out a set of hoses that would allow them to share cooling water. The man with the failed backpack would survive on his own emergency pack, which contained about an hour's worth of oxygen, and if necessary, his partner's emergency pack, allowing more than enough time for the Rover to race back to the lander. A more dire scenario was that both the

Rover *and* a backpack might break down. For a time, this remote possibility had so worried the managers that they considered writing the mission rules around it—a change that would have severely limited Scott and Irwin's explorations. In the end, NASA bought the risk of the double failure, knowing that if it came to pass, one of the astronauts would not make it back to the lander alive. But all of this was far from Scott's mind as his gaze lingered on the view. He felt a wave of excitement: everything was working. He joined Irwin and the two men headed down the slope to the crater.

The mountain was a difficult workplace. The stiffness of their suits hindered climbing; they were reduced to taking small, ineffective hops. With every step, the soft, thick dust fell away from their feet, as if they were walking on the side of a sand dune. A few steps left them nearly out of breath. They spent the better part of an hour here, winning only a few samples for their efforts. There just wasn't the variety they'd hoped for. They would make another try at Spur crater, but first Scott wanted to get to a boulder they'd spotted along the way.

"Man, I'd sure hate to have to climb up here!" said an amazed Scott as he struggled back up the slope, grateful that the Rover had done so well. "You'd *never* get here without this thing." And yet, they had barely begun to scale Hadley Delta. With no haze to block the view, its upper reaches were clearly visible, many thousands of feet beyond. Gazing on that bright frontier, Irwin silently wished they could go higher.

9:38 A.M.
Mission Operations Control Room

Joe Allen sat at the Capcom's console, studying a photo map of Hadley Delta. Every so often he made a mark showing the best estimate of where the Rover was. His brown hair, short on the sides but longer in front, fell across an unlined forehead. Allen listened intently to the voices of his friends on the moon; at the same time he kept an ear tuned to the conversations in the geology back room. He also had a TV monitor on which the scientists could write him notes—"We'd like a core-tube at this stop"—so that if a question came down from the moon, Allen would be ready with the answer. And it was Allen's job to take the myriad inputs to the astronauts, from the geologists, the flight director, and the flight controllers, and synthesize them into one coherent voice.

Joseph Percival Allen IV, an Iowa-born physicist who came to NASA with the XS-11, was sometimes known as Little Joe, because at five feet six inches he had usurped Pete Conrad by half an inch as the shortest man in the Astronaut Office. At thirty-four, he looked as young as the day he entered Yale to begin working toward his doctorate. In the minds of the geologists, Allen's assignment as mission scientist was one of the best things that happened to Apollo 15. Allen combined a youthful enthusiasm with a keen scientific mind and an innate sense of people. On the field trips, it was Allen who took the edge off Scott's driving intensity with a humorous remark. Now, as Capcom, Joe Allen was Scott and Irwin's link with the back room. No one but he could speak directly to Scott and Irwin; any requests, advice, or questions from the back room had to be spoken by Lovell over the loop to flight director Gerry Griffin and then, via Allen, to the moon. It was no accident that every link in this human chain had been out on the field trips at one time or another, and that each understood the scientific objectives of the mission. Allen in particular understood so well that often he did not even need to ask the back room what to say. He was more than just an ally, Lee Silver would say years later; he was a colleague.

Throughout the moonwalks, Allen had tried to convey a sense of support and optimism. He knew Scott and Irwin were working very hard, and he never missed an opportunity to ease the pressure with a comic remark, a private joke. At each new achievement, however minor, Allen voiced his approval: "*Extraordinary. Superb.*" Deke Slayton had come down on him once or twice for deviating from the spare style of communication that characterized standard pilots' radio discipline. Allen didn't let that bother him. There wasn't any operational reason not to let Scott and Irwin know that he was pulling for them.

He knew roughly where Scott and Irwin were: a few hundred meters east of Spur crater, and slightly upslope from it. Scott had already radioed that he didn't think it was worth going any farther to the east; they'd already seen as much variety as they were likely to see. They had stopped to sample a lone boulder that was apparently too good to pass up. Allen listened intently as Scott maneuvered the Rover, then stopped. Suddenly the radio link was scratchy, but through the static Allen could hear Scott's labored breathing. He heard Irwin say, "Gonna be a bear to get back up here, you know."

"Hey, troops," Allen said with contained urgency, "I'm not sure you should go downslope very far, if at all, from the Rover."

"No, it's not that far," came Scott's answer. Irwin added gamely, "I think we can sidestep back up."

There was no television from the Rover just now, but Allen could picture the steep, powdery slope. He also knew Scott and Irwin wouldn't give up a prize without a fight. And he could sense that the managers in the back row of mission control were getting worried.

Suddenly Scott called out that the Rover was beginning to slide down the hill. "The back wheel's off the ground," said Irwin. Their voices didn't convey the seriousness of the situation: If the Rover got away from them, it meant scrapping much of the mission, and a long walk back to the LM. Allen listened as Scott held on to the Rover to keep it from sliding down the hill. Now he heard Irwin talking about something he'd noticed on top of the boulder, some kind of green material, he was saying. Allen knew that was unusual, and he wasn't surprised when Irwin urged Scott to get a sample. The men traded places; now Irwin held the Rover while Scott made his way to the boulder.

"Use your best judgment here," Allen cautioned once more, sensing now that Scott and Irwin had the situation under control. Within minutes, having bagged a piece of their strange find, the men were driving again, heading west toward Spur crater. Time was running out; this would be the last stop on the mountain.

10:00 A.M.

Heading downslope once more, Scott and Irwin angled toward Spur crater and stopped next to the rim of the football-field-size pit. Spur looked promising. There weren't many big rocks, but there was great variety in the small fragments that littered the soil. Immediately Irwin noticed a white rock, perched atop a small, light-gray pinnacle, that was different from anything he'd seen. Irwin wanted to rush over to it, but Scott was already collecting another rock. Even as Irwin went to help him, he spotted something even more curious. It was more of that green material, like the coating on the boulder at the last stop. What could it be? Moon rocks were gray, or they were tan, or black, or white—but not sparkling light green. To Irwin, a man of Irish descent who was born on St. Patrick's Day and had shamrocks tucked away in the lunar module, a green rock was a special find. There was some debate about whether the color was real—Scott thought their gold-plated visors were playing tricks on them—but they picked it up anyway. It looked like a chunk of basalt, shot through with tiny holes where gas bubbles once frothed in molten lava. But it couldn't be basalt; it actually yielded to the pressure of Scott's gloved fingers, which left streaks in its surface. Not until the rock was unpacked in the Lunar Receiving Laboratory would the men see that it truly was green, and that it was actually made of tiny spheres of glass. Even among moon rocks, it was a rarity. And in time, the story it would tell, of eruptions from the hidden depths of the lunar interior, would hold geologists spellbound.

But just now, Scott and Irwin's attention was riveted to the strange, white rock Irwin had seen earlier. As they approached, they strategized about the best way to collect it. Irwin suggested lifting it off its dusty pedestal with the tongs. Scott did so, and he raised it to his faceplate to inspect the find, which was about the size of his fist. It looked fairly beaten up, and it was covered with dust, but as he grasped it with his gloved fingers some of the ancient coating wiped off, and he could see crystals: large, white crystals. Suddenly—

"Ahhh!"

"Oh, man!"

—both men realized what they had discovered. The rock was almost entirely plagioclase. Exposed to light for the first time in untold eons, its white crystals glinted in the sun, like the ones they had seen in the San Gabriels. This was surely a chunk of anorthosite, a piece of the primordial crust. Scott radioed the news.

"Guess what we just found. *Guess* what we just found. I think we found what we came for."

"Crystalline rock, huh?" prompted Irwin.

"Yes, sir," said Scott.

"*Yes, sir,*" repeated Joe Allen, like a brother at a revival meeting. After describing the rock in detail, Scott placed it in a special sample bag by itself. Of all the rocks brought back from the moon this one would be the most famous. To the geologists, it would be sample number 15415, but to the world it would be known by the name bestowed upon it by a reporter covering the mission at the space center: the Genesis Rock. In time, probing the treasure with electron beams in their laboratories, the geologists would peg the rock's age at 4.5 billion years. If the moon was any older than that, it wasn't much older; the solar system itself was thought to have formed only 100 million years earlier.

Scott couldn't hear the geologists' ecstatic reactions in the back room, but he didn't need to. And Spur had even more to offer. As they made their way along the crater rim they happened on one remarkable find after another, poking out of the dust. Scott sounded like a rare-coin collector who had stumbled across a treasure chest in the attic. "Oh, look at *this* one!" Scott told Allen what was already clear: "Joe, this crater is a gold mine!"

"And there might be diamonds in the next one," said Allen, a gentle reminder that Scott and Irwin could not linger here. They were getting close enough to their immutable walkback limit that Allen was telling them to press on. Now Allen radioed that Irwin's sample bag was about to come loose—everyone in mission control could see that on the big screen, and they worried that all those priceless samples might tumble into Spur crater. Precious minutes were lost as Scott cinched it up. With less than 15 minutes remaining, Scott and Irwin looked hopefully at a boulder that lay perhaps a dozen yards farther along Spur's rim, the only large rock in sight, and therefore the only rock for which the geologists could be sure of its place of origin. Undoubtedly it had been torn from the floor of Spur.

Just now, Allen passed up word from the back room: forget the boulder; they wanted them back at the Rover, using the rake to collect a bunch of walnut-sized fragments. Scott eyed the boulder. He told Irwin to go ahead and start raking; in the meantime he went into high gear. A short run along Spur's rim brought him to his quarry. Working on stolen time, he clicked off Hasselblad pictures of it from every angle. There wasn't time to hammer off a piece, but he spotted a small fragment in the dust—obviously knocked loose from the big rock—and got his sample after all. He hoped the geologists would be able to figure out where it came from by studying the pictures. Scott turned and ran back to join Irwin in his work. He hated to leave. It was a frustration he would feel again and again on the moon, just as he had that first day in the Orocopias with Silver: there was never enough time.

12:41 P.M.

Heading back to the LM, Irwin was glad to be traveling on more level ground. Up on the mountain, there had been times, driving along the contour of the steep slope, when Irwin was afraid the Rover would tip over. Now, pulled away from their explorations by the pressures of time and timeline, he and Scott made a brief stop at the LM, then revisited their ALSEP package, that tiny city of instruments sprawled in the dust. Scott had to summon more strength from his sore, tired hands to drill into the soil for the so-called deep core sample. While Scott drilled, Irwin dug a trench in the gray dust, took pictures of it, and used a device called a penetrometer to test the bearing strength of the trench walls and bottom. Irwin's tests would reveal much about the mechanical properties of lunar soil.

Scott's fingers were already so badly injured that they were beginning to turn black, and they were sore from cinching up sample bags all morning. Using the drill just made them worse. He had to bring his hands in close to his chest in order to squeeze the drill's trigger, and he could stand only a minute or so of the pressure against his nails before he had to back off, shaking his arms against the pain. But he kept at it. Now one core tube was in, then he added a second. He could feel the drill meet resistance, then penetrate more easily again. Soon he had drilled the full 10 feet down. It was the deepest lunar core sample ever; inside were millions of years worth of lunar history. Now he tried to pull the core out of the ground, but it wouldn't come. When Joe Allen told him to leave it until tomorrow, he was disappointed. Still, it was getting to be a long day, and he could feel it. By the time he and Irwin climbed back inside *Falcon,* they had been outside for more than seven hours. Today alone, they had driven almost 8 miles. And they had been rewarded for their hard work; it was a day in the field any geologist would have been proud of.

The secret of living on the moon, in Dave Scott's mind, came down to one thing: getting out of his suit. He and Irwin would have been miserable if they'd been sentenced to three full days inside them. By now, the whole process was routine: Off came the gloves, then the helmets, and then, one at a time, each man stepped into a white stowage bag and pulled it up over his legs to catch the rain of soot, and stripped to his long johns. Piled in the back of the cabin, the suits would dry out from almost nine hours worth of wear. Then came the chores: recharging the backpacks with oxygen and water, stowing the rocks they had collected that day, and attending to *Falcon*'s systems. And there was another routine that wasn't in the flight plan. Each day, Scott and Irwin paused in their work for a brief chat with their orbiting companion. The signal was broken up until Worden cleared the mountains, but then there was time to exchange news and pleasantries as *Endeavour* sped, starlike, overhead. Scott was glad to hear that Worden and his arsenal of scientific instruments were going like gangbusters. And then, before *Endeavour* drifted out of range, more mundane matters.

"Hey, Al," called Irwin, "throw my soap down, will you? And my spoon."

"You forget something, Jim?"

Irwin grinned through lunar grime. "I really need my soap."

• • •

Aboard *Endeavour*, everything was still quite clean and neat—except, perhaps, Al Worden, who had at this point gone a week without a shower or a shave and was happy to have Irwin's soap aboard. Like his friends at Hadley Base, Al Worden was too busy to care how grungy he was. He was, in effect, the commander of the first fully equipped orbiting lunar science platform. *Endeavour*'s service module was crammed to the hilt with high-powered cameras and sensors, and much of Worden's time was spent turning the instruments on and off as dictated by the flight plan. Two different sensors were designed to map the chemical composition of the surface. Another would try to detect evidence for volcanic gases seeping out of the moon during recent geologic time. A mass spectrometer was included to sniff out extremely tenuous gases that might surround the moon. There was even a tiny satellite, which Worden would release into lunar orbit in a couple of days, equipped with a magnetometer to stalk the elusive lunar magnetic field. And the two cameras were superb creations. One of them, based on declassified spy-satellite technology, could photograph details on the surface as small as a yard across. It was just possible, Worden thought, that this enormous haul of data would eclipse the knowledge gained from the rocks his crewmates were collecting down there amid the dust.

And to this list of sensing apparatus, add Worden himself, for like Scott and Irwin, he had given himself to the study of the moon. Worden, like Stu Roosa, had made Farouk El-Baz his mentor. For the first time, a program of visual observations from lunar orbit were an official mission objective. Worden not only admired the exuberant Egyptian—or the King, as the astronauts called him—he liked him as much as he did anyone in the space program, and found him an inspiring teacher. After the study sessions, they would go drinking together. "Now," Worden would say, "we're going to see what a White Russian can do to an Egyptian." It would always turn into a competition, and El-Baz would hang in there with a tenacity any fighter pilot would have applauded.

It was for the King that Worden spent dozens of hours flying over the mountains and deserts of the western United States, honing his observational skills, and many more hours studying the Lunar Orbiter photos. And now, in lunar orbit, each time *Endeavour* flew into sunlight, Worden was amazed to look down and see among the pockmarked chaos a crater he recognized. How strange, to be *here* and see something familiar.

No feature was more memorable than giant Tsiolkovsky crater, and for this and other targets Worden carried photographs with instructions for specific things El-Baz wanted him to watch out for. Looking down on Tsiolkovsky, he could see that it was packed with spectacular geology. The crater was so big that when he was directly overhead it nearly filled his view. The central peak was enormous, a great white mountain rising out of that floor of dark lava. Worden could see layers in the peak, as if it were a great slab that had somehow been turned on its end. Around it, he spotted boulders that must have been the size of a city block. Along the giant's rim, he could see what appeared to be rock avalanches that had cascaded away from the crater onto the surrounding hummocks.

Now he was over a place called Littrow, near the Sea of Serenity. Farouk had told him to look very carefully at Littrow, because in the unmanned photographs, there were craters, surrounded by halos of dark material, that resembled small volcanoes, or cinder cones, as Farouk called them. If that was true, they might represent an entirely different, and possibly much more recent, type of volcanic activity from the *maria*.

Sure enough, Worden could see the small, dark-haloed craters, and he told Capcom Karl Henize excitedly, "It looks like a whole field of small cinder cones down there." On the next orbit he would get another look, and he would report the craters were cone-shaped, and that he was even more convinced they were volcanic, and Henize would say, "Keep talking like that and we might end up going to Littrow sometime."

Like anyone who had ever been here before, Worden's mind was split between the operational tasks that filled his consciousness and the sights, sounds, and feelings that were even now being stored in his subconscious. Only during the orbital night did he have time to savor the experience. He had never believed that the solar system was the center of the universe in any sense; in the back of his mind, he'd always thought that the earth was just one planet among millions. But it wasn't until he found himself over the far side of the moon and saw all those stars—God, there were so many stars—that he was sure there was more to the universe than he had ever imagined. He knew that astronomers had debated the probabilities of intelligent life in the universe, but this light show was enough to convince Worden that it had to be all or nothing: Either there was no life out there, or the cosmos must be teeming with it.

Just before emerging from darkness, Worden looked out ahead and saw waves: thin, sinuous bands of light, suspended between the starry heavens and the unyielding darkness that was the moon. At first, Worden did not know what they were. Then he realized he was seeing the rays of the sun, which had not yet risen, glancing off the tops of distant mountains along the horizon. They were the opposite of shadows. The waves thickened until, without warning, night turned almost instantly to day. As his eyes adapted to the light, Worden searched once more among the craters for a familiar place, and for the places he would try to make a little less unknown.

• • •

Jim Irwin lay in his hammock, awaiting sleep. Before the mission, inspired by the prospect of visiting the lunar mountains, Irwin had felt a desire to hold a short religious observance on the moon, but when he mentioned the idea to his commander, Scott quickly turned him off. But ever since they had arrived on the moon, Irwin felt a spiritual quality about the place. Seemingly insoluble problems had come up, and each time they had been resolved. When he was deploying the ALSEP, he had a problem that had never happened in training; the cord for deploying the central station broke. Irwin prayed for guidance, and immediately he knew the solution was to get down on his hands and knees and pull the cord manually. Then there was the Rover's rear steering, which had been out of commission yesterday, and nothing they did could fix it. Inexplicably, when he and Scott came out this morning, it was working. He felt a glow inside him that whatever problem came up, they would solve it.

Once again Irwin took a moment to pray. He gave thanks that everything was going so well. He gave thanks too for the discovery of the white rock, so remarkably displayed on that pedestal of dust, as if it were being presented to them. Tomorrow would surely hold more discoveries, and before he closed his eyes, Irwin prayed that it might also bring him an opportunity to make his own small spiritual gesture. On the moon, a favorite biblical passage, a verse from Psalms had drifted through his thoughts like a refrain:

I will lift up mine eyes unto the hills:
from whence cometh my help?
My help cometh from the Lord.

He hoped that sometime during the next day's activities he would find the right moment to recite it. As he prayed he sensed that God was near him, even here.

III: The Spirit of Galileo

Joe Allen liked to think he was cut from the same cloth as the pilot-astronauts, but working for Dave Scott had been a lesson in humility. The man's stamina was simply amazing. It was beyond him how Scott could go through the long days of meetings and simulations without the slightest sign of wearing down, seemingly as productive and alert at ten o'clock at night as he had been at breakfast that morning. No question, it did wonders for his motivation to see his boss going at it so relentlessly, but sometimes it was all Allen could do to keep up. He was reminded of the climactic scene from one of his favorite movies, *The Hustler*. Jackie Gleason as Minnesota Fats is twenty-four hours into his pool-room confrontation with Paul Newman as Fast Eddie. Just as it seems that Eddie is about to clean the old master out of his last dollar, Fats takes time out. Going into the seedy little bathroom, he washes his face, cleans his fingernails, puts on a fresh shirt and a tie, and emerges into the smoke feeling like a new man. "Fast Eddie," he says, "let's shoot some pool." Fats goes on to win back all his money and win the match. Allen told

Scott about it one day, and after that, when Scott would see him flagging midway through a late-evening planning session, he would lean over and say, "C'mon Joe, let's shoot a little pool." Now, as Scott and Irwin finished suiting up for a final day of exploration, Allen asked Scott if he and Irwin might be ready to shoot a little pool themselves. The answer from the moon: "Joe, today's the day for a little pool." Now came the endurance lap.

Monday, August 2
4:07 A.M., Houston time
6 days, 19 hours, 32 minutes Mission Elapsed Time

Outside, the first men to spend three days on the moon saw the sun almost directly overhead, sending its fierce heat down on Hadley Base. Even within their suits, Scott and Irwin felt its warmth. They had been nearly two hours behind schedule getting to sleep last night and waking up this morning, and that had cost them. The liftoff time, slated for later this day, could not be changed. Allen had already told them that this third moonwalk would be shortened to only four or five hours. Scott and Irwin planned to visit Hadley Rille, and then, if there was time, the mysterious group of craters called the North Complex. A young Survey geologist named Jerry Schaber had advanced the idea that the North Complex might actually be a cluster of small volcanoes. Even if they were impact craters, Schaber said, they would still be worth the trip; the largest was almost 2,500 feet across, big enough to have punched through the frosting of *mare* basalt and bring up chunks of the more ancient rocks underneath. The North Complex hadn't been on the original traverse plan, but when Scott heard Schaber's idea, he personally made time for it. Solving that mystery was one of Scott and Irwin's fondest hopes for the mission. But they would have to wait; Allen was sending them to the ALSEP site, for a showdown with Scott's nemesis, the drill.

4:41 A.M.

It was waiting for them when they reached the ALSEP, sticking out of the ground just as Scott had left it the day before, when he had tried unsuccessfully to extract the deep core sample. Scott had already begun to wonder whether the core was worth the time and effort it was costing. Fighting their pressurized suits, the men bent over to grab the drill's handles and pulled as hard as they could. The drill did not budge. Scott gave a Herculean pull—"One, two, *threee*"—and it just barely moved.

In mission control Joe Allen listened with amazement. Scott and Irwin represented as much physical strength as any two men in the astronaut corps. In training, the drill had gone through Texas hardpan like so much butter. It had never caused any problems, going in or coming out; it was, as Scott would say, "a nuthin'." What Allen did not know—what no one knew—was the lunar soil had been so thoroughly compacted and tamped down by eons of micrometeorite rain that there was barely any room for the drill to penetrate. Once driven into the moon, it might as well have been in a vise.

"I don't think it's worth doing, Jim," Scott said. "We're not going to get it out." But Irwin wasn't ready to give up. He suggested that each of them hook an arm under one of the handles. That helped; they managed to pull the core about one-third of the way out. Now the men crouched down and each put a shoulder under a handle and tried to stand up: "One, two, *threeeee*—" More progress. Another push, then another. They were winning. Suddenly:

"There we go!" The drill flew upward, and as Irwin said, "We almost flew with it!"

But Scott's troubles were not over. The entire core sample was 10 feet long; before they could bring the core home they would have to dismantle it into sections. And for some unknown reason the vise from the Rover's tool kit refused to work properly. Scott realized with exasperation that it had been assembled backward; no wonder it wouldn't grip. Irwin, meanwhile, broke out a hand wrench. Eleven more minutes went by while they struggled to dismantle the sections. Scott's patience was dwindling. Every minute spent on this one sample was time lost for the explorations to come. The trip to the North Complex hung in the balance. "How many hours do you want to spend on this drill, Joe?"

Scott wasn't the only one losing patience. For a while now, Allen could hear voices on the flight director's loop, from McDivitt and the others in the back row of mission control, pushing flight director Gerry Griffin to abandon the deep core and get on with the next item on the timeline. Irwin was to take a movie of Scott driving the Rover for the engineers. But Griffin, who was fully committed to the science of Apollo 15, wasn't going to be diverted. He understood the value of the deep core, and as the managers pressured him he walked over to Allen's console and said quietly, "You worry about that core. I'll take care of the back row."

On the moon, the struggle continued. Scott and Irwin managed to disassemble the core, but only partially; that would have to do for now. Twenty-eight minutes after they began, Allen told them to move on. They'd pick it up on the way back to the LM at the end of the traverse. Scott wasn't sure where they'd put it in the command module, but he'd think of something; that deep core cost too much to leave on the moon. When the deep core was finally scrutinized in the Lunar Receiving Lab, it would teach what geologist Don Wilhelms would call "the lesson of the moon's antiquity and changelessness." Scientists would identify no less than forty-two separate layers of soil; the bottom layer had apparently remained undisturbed for half a billion years.

5:17 A.M.

With the deep core struggle behind them, Scott and Irwin were glad to be heading west toward Hadley Rille. They had expected a short, easy drive, but instead found themselves pitching over dunelike ridges and troughs. Here, Irwin thought, was a place where they could get lost. But even as *Falcon* disappeared over the horizon behind him, the first time astronauts had ventured out of sight of their lander, Irwin felt no anxiety, so familiar was this valley to him now. Impatient to reach the rille, he urged Scott to take the shortest route, but Scott methodically stuck to the heading Allen had given them, telling Irwin, "We're making good time." As they neared the canyon's edge, Scott's attention was diverted to a small exploration he could not pass up, a small fresh crater that would prove the youngest ever visited on the moon, a scant million years old.

5:57 A.M.

When Scott and Irwin finally stood at the edge of Hadley Rille, they were rewarded with a sight no one had expected. On the far wall, which was in full sunlight, distinct layers of rock poked through a mantle of dust, like the levels of some ancient civilization. They were surely lava flows. Over many millions of years, perhaps, a succession of outpourings had piled atop one another to build up the valley floor. This was the first—and only—time that Apollo astronauts would find records of the moon's volcanic life, not as fragments scattered around the rim of an impact crater, but in place, preserved from the day they were formed. This was true lunar bedrock. And there was more of it on this side of the rille. Scott and Irwin gathered their tools and went to work.

There was no sharp dropoff at the rim of Hadley Rille; it was more like the gentle shoulder of a hill, and thankfully, the ground was firm. Effortlessly, Scott continued past the rim and loped several yards down the slope. Even now he could not see the bottom; it was hidden from view beyond the curved flank. He turned and ascended once more. All around him and Irwin were big slabs of tan-colored basalt, shot through with holes from long-vacant gas bubbles; some of the boulders were scored by layers in miniature. But no sooner had Scott hammered off a chip than Joe Allen passed up word that he and Irwin would have to return to the Rover to collect a rake sample, and then it would be time to leave. On another day, Scott might have put up a fight; today he was too tired.

But in the back room, the geologists decided that the rille was worth more time. Urgently, Scott called Irwin to join him a little farther down the slope, where masses of darker rock waited. As they set to work, the men heard Joe Allen's voice once more.

"Out of sheer curiosity, how far back from . . . the edge of the rille are the two of you standing now?" Scott wasn't sure what he meant until Allen clarified it: "It looks like the two of you are standing on the edge of a precipice. . . ."

Scott managed to suppress a laugh. He could just imagine the back row of mission control—McDivitt, Kraft, Petrone, and the others, tied up in knots, their worst nightmare splashed on the big screen and multiplied on every TV monitor, because it looked like their boys on the moon were about to wander over the edge of a cliff. "Oh, gosh no," Scott reassured, "it slopes right on down here. . . ."

The extension amounted to all of eight minutes, enough time to gather a few more samples. Scott and Irwin headed back to the Rover wishing, once again, that they had more time. One thing was still true, and it would be true throughout the Apollo era: those on earth who carried the burden of responsibility for human lives would greet each new exploration with apprehension, while the men on the moon, confident and committed in their work, wanted only to push a little bit farther.

7:56 A.M.

Back at Hadley Base, Rover tracks converged on the foil-clad lunar module, and the ground was littered with gear and stowage bags attesting to the activity of the past three days. Beyond, the Apennines were bathed in sunlight, their slopes seemingly as soft and virginal as a meadow after a fresh snowfall. On the drive back, gazing at them, Jim Irwin had finally quoted the words of the Psalmist. But there had been no time for the North Complex; they would not solve that mystery after all. For all that they had accomplished, that loss was hard to take; Irwin felt their explorations were only half-finished. But right now, the most important thing was to get back into the LM with plenty of time to prepare for liftoff.

When everything was all packed up, Scott attended to a task that wasn't on the checklist. He had thought he might not get to it, but it was Joe Allen's idea, and he didn't want to disappoint him. He reached into a pocket on his suit and pulled out a falcon feather. Then, picking up his geology hammer, he loped back to where the Rover was parked and stood before the TV camera.

"In my left hand, I have a feather. In my right hand, a hammer." Centuries before, the story was told, Galileo Galilei had stood atop the Leaning Tower of Pisa and dropped two weights of different sizes, proving that gravity acts equally on all objects regardless of their mass. And where would be a better place to confirm his findings, Scott said, than on the moon? He held the hammer and the feather out in front of him, and let go. They fell slowly through the vacuum, side by side, and for one brief moment the spirit of the great Italian scientist was conjured on the airless ground of Hadley Base. When the hammer and the feather hit the dust at the same time, there was applause in mission control. "Nothing like a little science on the moon," Scott said.

8:08 A.M.

Alone, Scott drove Rover 1 to a small rise about 300 feet east of *Falcon*. From here, mission control would be able to aim the TV camera back at the LM and watch the liftoff, four hours from now. Scott made some final switch settings, and then, before he headed back, he pulled out a small red Bible and set it atop the Rover's control panel. If anyone should come this way again, he wanted them to understand who had left this machine here. Then he walked about twenty feet until he came to a small, subtle crater, reached down, and made a hollow in the dust. Into it he placed a small plaque bearing the names of fourteen men. The previous winter, at a New Year's party at Deke Slayton's house, Scott and Slayton had shared the hope that no more astronauts would ever be lost to the cause of space exploration. Scott resolved that he would find a way to memorialize those who had already been killed; maybe then, no more would die. He had never met some of the men remembered on this plaque, including the three cosmonauts of Soyuz 11. Others he had called friends. Next to the plaque, he placed a small aluminum figure, a stylized representation of a fallen astronaut.

Scott paused to take one last series of panoramic photos. Then, casting his eyes on the high slopes of a peak just to the south of Mount Hadley, he saw what looked like layers, and automatically began to describe them for the geologists. But Joe Allen cut him off; there wasn't time.

Scott took off in long, easy strides, heading across the rolling plains toward his lunar module. He could not deny a sense of loss, knowing that he would never return to this ultimate field site. But just now, Joe Allen was quoting the science fiction author Robert Heinlein: "We're ready for you to come again to the homes of men on the cool green hills of earth." Scott felt the chin strap of his communications carrier rasp against a week-old beard. He'd promised his children he wouldn't shave until he got home. After all: explorers always come home with a beard.

IV: The Final Selection

The first thing Jim Irwin did when he climbed into the life raft was dip his hand into the cool water of the Pacific and bring it to his face. He wore no respirator. There was no quarantine trailer waiting for him and his crewmates on the carrier *Okinawa*. Three landing crews had returned to earth healthy; not a trace of anything alive had turned up in the samples, and at last NASA could do away with quarantine without risking any political heat. And Irwin was very glad of that. To arrive from the dust of Hadley to the blue Pacific was the most wonderful sensation he could have imagined. He wanted only to let the sea air wash over him and the morning sun fill him with warmth. He would have stayed there in that raft, but all too soon he heard the roar of the helicopter about fifty feet above him and felt its wind. The recovery people lowered the net for him; Irwin climbed into it and ascended from the water back into the sky.

Aboard the helicopter, Irwin tried to come up with something to say when he arrived on the carrier deck. Dave Scott reminded him and Worden that when they stepped out of the helicopter they should all salute in unison. They had talked about this before, Irwin remembered; Scott didn't like the fact that the other crews hadn't saluted together. On this flight, they were going to salute like military officers. Suddenly they were landing, and when the door opened Scott was out first, and when he stood on the stairway he saluted immediately, without waiting for them.

On the carrier deck, Irwin felt unsteady and realized he hadn't cleared his ears in the command module on the way down, but those last five minutes had been the most hectic of the entire flight. One of the three parachutes had collapsed as soon as it was deployed, and the craft had descended faster than normal. On the way down Irwin had seen them have a near-miss with a helicopter. They'd hit the water with a solid impact, but everything was all right.

When it was his turn at the microphone, Irwin expressed gratitude for the trip, and to everyone on earth who supported them. He had been trained at the Naval Academy, he said, and he'd been on ships before, but he'd never been so glad to be on one as he was right now.

During the night Irwin slept fitfully; as he lay flat in his bunk he felt as if his body were tilted heads-down at a steep angle. At five in the morning he awoke to the din of the anchor chains up on the deck. Not long afterward he and his crewmates were getting the heroes' welcome at Hickam Air Force Base in Hawaii. By mid-morning they were airborne again, inside a big transport headed for Texas. They were strangely isolated in their small compartment, as if everyone were avoiding them, as if they weren't sure ending the quarantine was such a good idea after all. Probably it was just as well, because no sooner had the C-140 lumbered off the runway at Hickam than Scott pulled out a stack of first-day covers for them to sign; he had carried them on the flight in the pocket of his space suit. The covers were part of a deal with a stamp dealer who had approached Scott during training; eventually, some of them would be sold in order to set up trust funds for their children. Before the flight, Irwin had harbored reservations about the deal, but had decided not to say anything to Scott for fear of causing friction with his commander.

Irwin sat in silence, signing the envelopes and assembling in his mind the experiences of his twelve days in space. He thought of the wonderful mental clarity he'd had on the moon, the powerful sense of God's nearness. He thought of the discovery of the white rock. He could not imagine how his crewmates might have been affected, but here in this small cabin he knew his soul had been stirred. He was a nuts-and-bolts man who had come back with something he had never anticipated: the seed of spiritual awakening.

And when he was back home in Nassau Bay he found joy in the most ordinary things. Just to sit in a chair, to eat at a table without having to chase food into his mouth, the reassuring sense of up and down—he had a new appreciation for all these things. Life itself was cause for celebration. Irwin felt so renewed that he wondered if he'd discovered a cure for despondency, for those who have lost their appetite for life: lock them up for a while, deprive them of all earthly experience, then let them out into the world again.

But physically, Irwin wasn't quite right. He was still having difficulty adjusting to the pull of gravity, including some bouts of dizziness. He wasn't doing very well on the stress tests in the medical physicals. And the doctors had told him that during the flight, after he and Scott had returned to the command module, he'd had a problem with his heart. At the time, Irwin had no idea anything was wrong; he knew only that he was suddenly very tired. Unbeknownst to him, the doctors watching his EKG in mission control were alarmed to see a so-called bigeminy rhythm, in which both sides of the heart contract at once. Houston never said anything about it, nor about the fact that both he and Scott had also experienced minor heart irregularities during their time on the lunar surface. Both men were angry that no one from Chuck Berry's staff had informed them while they were in space. Now, on earth, the doctors were ascribing the problems to the rigors of training. During practice sessions for the moonwalks, Scott and Irwin had spent hours working in space suits in intense summer heat. To replenish lost fluids, they drank large amounts of an electrolyte solution, which tended to leach potassium from their systems. As a result, the doctors would conclude, they left earth with a potassium deficiency that was exacerbated by the stress of the mission, causing their subsequent heart irregularities. Irwin hoped that soon he would be back to normal.

Dave Scott was also tired, though not as much as his lunar module pilot. He wished he and his crew could have been locked up along with the moon rocks in the Lunar Receiving Laboratory. He wanted nothing more than to be put away for three weeks, with time to rest and write the mission reports and talk to the scientists. He would have loved to have time to look at the videotapes of the moonwalks and study the rocks with geologists. There wouldn't have been any pressure from the home front to get him back quickly; it would have been fine with Lurton. Before the mission, he'd pushed hard for it, but not hard enough. Instead, he was let out into the world for long days of debriefings. In his free moments he found himself signing photographs; it seemed that every general in the air force wanted one autographed specially for him. And at night, it was the time of the neighborhood party. Scott would have enjoyed it much more if he hadn't been so worn out. And still to come, there was the trip to Washington, the speeches before Congress. Then to New York and Los Angeles for talk shows and parades. A month after splashdown, instead of savoring the accomplishment, he would be completely exhausted.

To Al Worden, the debriefings were interminable. Day after day they rehashed the mission for the engineers, and then the operations people, and then the managers. By the second week of answering the same questions over and over again Worden was coming home physically drained—sometimes arriving to a party in full swing in his apartment. He couldn't wind down until well past midnight. He was so overtired he couldn't sleep. Alone in his darkened living room, memories of the flight, and the training, came into his head, and he began to write about them. The words seemed to be coming from somewhere else. They flowed out of him. He began to jot down ideas, most not even complete sentences, page after page of them. It all came back—seeing the universe revealed above the far side of the moon, the battered moon itself, a tiny crescent earth adrift in blackness. And then there was the spacewalk. In the middle of the voyage back to earth, Worden had ventured outside *Endeavour* to retrieve cassettes of exposed film from the scientific cameras on the side of the service module, and he remembered every moment:

A spacewalk
Is like
Being let out
At night
For a swim
By Moby Dick.

It was as if there were another debriefing he had to perform on himself, to give his mental computer a chance to play back the images and impressions he had stored up in space. It was the data the engineers didn't ask about. Over the weeks he rewrote and rearranged the lines until they were poems, the poetry of a man caught in the lingering pull of another world:

Quietly, like a night bird, floating, soaring, wingless
We glide from shore to shore, curving and falling but not quite touching;
Earth: a distant memory seen in an instant of repose. . . .
I glide upward, above the waves of the ocean moon. She is
forever moving just out of reach and I sail on,
never touching, only watching and wanting to know.

• • •

In August 1971, NASA was on a high. With Apollo 15, the agency had scored its biggest success since the first lunar landing. Within the scientific community, even skeptics applauded the mission. Gerry Wasserburg, a Caltech geochemist who had voiced his share of opposition to Lee Silver's astronaut training efforts, publicly named it "One of the most brilliant missions in space science ever flown." For Dave Scott, there was an unaccustomed gesture of praise from a fellow astronaut: a note from Alan Bean congratulating him on a superb performance. And throughout the space center there was the new realization that Apollo had reached maturity. Almost every one of Apollo 15's innovations, from the Rover to the upgraded lunar module to the training and planning efforts, had come through with barely a hitch. The way was clear for the final two missions to be even more ambitious.

And for six men, August was the moment of truth. The crew of Apollo 17, yet unnamed, was about to be announced. Six men eyed three seats on the final moon-bound command module. So did the scientific community. All through 1971 they had been turning up the heat on NASA, decrying the fact that NASA had a fully qualified scientist and wasn't sending him to the moon. Word of this filtered down to the Astronaut Office, where Dick Gordon had been making a hard run to snare Apollo 17 for his own. He knew very well that by Slayton's rotation it was Gene Cernan, who had backed up Shepard on 14, who was in line for the mission. When Cernan crashed his helicopter in the Banana River during training, Gordon had hoped Cernan had taken himself out of the running, but that little episode had just slid off Cernan's back. Now, as the decision neared, Gordon was betting that Schmitt would be his edge. And he had no qualms about politicking for the flight; he went to Shepard and Slayton and asked them to keep his crew together.

From the sidelines, the other astronauts watched and waited with great interest. There were two ways this could go, they realized. Even if Gordon didn't get the mission, Slayton could still take Schmitt and put him on Cernan's crew—in which case things were looking very bad for Cernan's lunar module pilot, Joe Engle. He was one of the most experienced test pilots among the Original 19, and, in Cernan's mind, one of the best pilots he'd ever flown with.

But Engle had a little of Gordon Cooper's strap-it-on-and-go attitude, and it had already hurt him. During Apollo 14, he had taken his role as backup lunar module pilot too casually. The computer and all its software routines was daunting enough to any astronaut, but Engle just wasn't applying himself. Tom Stafford, now the chief astronaut, was coming down on Cernan—"You're the commander of this goddamn crew; get him in gear." And Cernan was telling the simulator instructors, "Whatever you can do to help Joe, do it." But ultimately it was up to Engle to change, and he didn't seem to realize how carefully his performance was being scrutinized.

But in truth, Deke Slayton wasn't especially concerned. In his mind, only one thing mattered; it would have been unthinkable for NASA to have a geologist-astronaut and not send him to the moon.

August 12

It was Thursday evening, and at the King's Inn Hotel on NASA Road 1, Jerry Schaber was listening to the evening news. Back in 1965, Jack Schmitt had hired Schaber to work at the USGS in Flagstaff; over the years they had become friends. Schaber had just come through one of the most exhilarating experiences he could remember, watching Scott and Irwin explore Hadley; nothing, he would say later, could compare with seeing your friends go to the moon. But on this Thursday evening Schaber's thoughts were on Jack Schmitt. The crew announcement was due any day now. Schaber was pessimistic; he couldn't imagine how Schmitt could get the flight when a popular pilot like Joe Engle was in line for it. He'd just bet his roommate Jim Head five dollars that Schmitt wouldn't make it.

Suddenly, on television the reporter was saying that the crew of Apollo 17 would include geologist-astronaut Jack Schmitt. Schaber yelled to Head, "Guess what they just said! Jack's on Apollo 17!" Schaber picked up the phone to call Schmitt. "Hey, congratulations," Schaber said, "we just heard!" The voice at the other end was characteristically gruff.

"It's not true! I haven't heard anything from NASA. Until NASA tells me, I don't know anything about it."

"Well, can we come over and help you wait for the call?"

"I don't care."

In the Nassau Bay Apartments across from the Manned Spacecraft Center, Jack Schmitt was sitting around, by himself. One bedroom was still full of unopened boxes he'd brought with him from the Survey in 1965.

"Gee, Jack," said Head, "it's awful quiet around here for a guy who's going to the moon."

"It's not true," Schmitt said.

Head went over to the bar, opened his briefcase, and pulled out five small airline whiskey bottles.

"What's that for?" asked Schmitt.

"That's in case you get the right call," said Head.

Suddenly the phone rang. It was Schmitt's sister, calling from Silver City to congratulate him. Schmitt said, "It's not true!"

More waiting, and then the phone rang again. Suddenly Schmitt got very quiet and businesslike. Schaber and Head looked at each other. "Yes, sir," they heard Schmitt say. "Yes, sir, I'll do the best job I can." Schmitt hung up the phone, went over to the bar, and very calmly unscrewed the caps from three of the little bottles of whiskey, picked them up together, and slugged them down.

Schaber would always remember, with amazement, that no one from the space center called to congratulate Schmitt that night. He and Head ended up taking him out to a nearby Pizza Hut.

Meanwhile, in Acapulco, a vacationing Ron Evans was scuba diving with his daughter, Jaime, and when they came out of the water there was the rest of his family, and Gene Cernan and his family, all excited. Deke Slayton had just called with wonderful news. The children made a banner and strung it across the cabana; it said, "Apollo 17." But on both sides of the border there were pangs of sadness, as the ones who made it thought about the ones who didn't.

The next day, Jack Schmitt said to Dick Gordon, "Why don't you let me talk to Deke?"

"No, I've had my shot," Gordon said. "The decision's been made." Schmitt went to Slayton anyway and asked him not to break up the crew. He and Gordon had worked very well together, he said, they made a great team. But his effort was in vain; Slayton wasn't going to change his decision. In the days that followed, Schmitt also spoke with Joe Engle. It was an awkward time; Engle was visibly upset and Schmitt could tell he was bitter. But Engle handled himself as well as could be expected. He told a reporter that the toughest thing he could remember doing in a long time was explaining to his kids that he wasn't going to the moon.

CHAPTER 12

The Unexpected Moon

APOLLO 16

I: Luna Incognita

The moon that shone down on Houston in the spring of 1972 looked exactly the same as it had throughout human history. No telescope in existence could have picked up any sign that four teams of explorers had landed there, disturbed the stillness of the ages, and picked up rocks. But in the minds of the geologists who were analyzing those rocks (now totaling almost 385 pounds), the miles of photographic film taken on the surface and from orbit, and the rest of Apollo's burgeoning harvest of data, it was a world transformed. The moon of speculation was giving way to the moon of fact, and in many respects it was just as the geologists had theorized. The impact origin of most lunar craters, the formation of the *maria* by volcanic eruptions, and the existence of an anorthosite primordial crust had all been confirmed, though the details of these features would continue to be debated.

Knowing how lunar features formed was only one goal of the geologists; it was just as important to know when. With each new landing, the scientists had added more dates to their timeline of lunar history. Apollo 11 and 12 had given them the times of two different episodes of *mare* volcanism. And Apollo 14 gave them the best chance yet to look back before the *mare* eruptions to the moon's youth, and they saw an era of almost unimaginable violence. This was the period of heavy bombardment, when giant asteroids collided with the moon to form the impact basins. The bombardment came to an end when a chunk of debris the size of Rhode Island slammed into the moon, creating the Imbrium basin, 720 miles across, the largest and most spectacular crater on the lunar near side. Debris ejected from the impact appeared to have gouged, blanketed, or otherwise altered terrain for perhaps hundreds of miles around; the Fra Mauro hills were just one result. Even before Apollo, the geologists had realized that the date of that cataclysm —which reshaped so much of the near side—would be among the most important points on the timeline. With Apollo 14, they found it. The rocks from Cone crater gave a shaky but still usable date of 3.85 billion years. At Hadley Delta—which, along with the rest of the Apennines, formed the basin's rim—Scott and Irwin found rocks at Spur crater that confirmed this age. And thanks to the Genesis Rock, they could look back even further. Before Apollo no one had anticipated that the moon's primordial crust would be made of anorthosite, and many scientists still questioned the theory of a global magma ocean. But the idea was becoming harder and harder to refute; data from the sensors in Apollo 15's orbiting service module confirmed that anorthosite was a widespread component of the lunar highlands.

As the geologists were coming to understand the lunar surface, other scientists were lifting the veil of mystery from the interior. Including the ALSEP package at Hadley Base, there were now three fully equipped scientific stations scattered across the near side. Their seismometers were sending back data on tiny moonquakes caused by the pounding from small meteorites, and bigger ones triggered by the intentional crashes of spent lunar modules and Saturn boosters. For geophysicists, the shock waves from those quakes provided soundings of the moon's rocky depths. They revealed a crust some 40 miles thick, undoubtedly with variations from one region to another. Below the crust lies the lunar mantle, the source of the *mare* basalts. By analyzing those samples, the geochemists had determined that their source regions in the lunar mantle contained iron- and magnesium-rich rocks. By all signs, the mantle had cooled down enough to bring an end to the moon's volcanic activity eons ago. But was there a molten core farther down? Instruments had failed to detect a magnetic field, apart from traces of magnetic signatures preserved in some of the rocks. The earth's magnetic field is thought to arise from motions within a core made of hot, metallic fluid; by implication, the moon was probably fairly cool throughout.

This was where lunar research stood by the spring of 1972. There were many questions remaining, most notably about the composition and evolution of the highlands, the last great lunar unknown accessible to Apollo. Their lure was undeniable: The *mare* plains were the slate on which the last 3.6 billion years of lunar history was most clearly written, but only the battered highlands might contain a record of everything that came before. Sending astronauts to the ancient highlands had been one of the geologists' top priorities. The hills of Fra Mauro did not qualify, since their record of lunar history went no farther back than the Imbrium impact. To sample the oldest highlands, the astronauts would have to travel beyond the reach of Imbrium's extensive blanket of ejecta. One such landing site—one of the few that could be called a favorite of nearly all the geologists—was magnificent Tycho, the freshest large crater in the southern highlands and one of the most distinctive features on the lunar near side. A sharp unaided eye can spot Tycho from earth; its rays of ejecta stretch almost all the way across the full moon. On Lunar Orbiter photos giant boulders, torn from the depths of the highland crust, were visible along the rim of the 51-mile-diameter crater. Nearby, nestled among the hills, were ponds of once-molten rock that would make fine landing places for a lunar module. In January 1968, the final Surveyor probe had successfully touched down in one of these valleys. When it came time to choose a landing site for Apollo 16, the geologists pushed hard for Tycho. But their campaign hit a wall when NASA got a look at the hellishly rugged terrain that lay underneath the approach path to the landing site. Jim McDivitt, now the manager of the Apollo Spacecraft Program Office, saw the Lunar Orbiter photos and said flatly, "You will go to Tycho over my dead body." McDivitt's reluctance was compounded by the fact that Tycho was hundreds of miles away from the lunar equator. Getting there would take much more energy than previous missions; that would mean cutting down on Apollo 16's scientific payload. And more importantly, the trajectory necessary to reach Tycho was so far outside the free return that if the command module were disabled enroute, an Apollo 13–type rescue would stretch the LM's fuel reserves uncomfortably far. Tycho was a risk NASA would not take.

Instead, after some debate, the geologists chose a patch of hills west of the Sea of Nectar and not far from the 30-mile crater Descartes. Unlike Tycho, the Descartes highlands would probably not provide pieces of the ancient crust, which the geologists still hoped to sample on Apollo 17. But the geologists believed Descartes would yield an equally important prize, something no other place in the highlands seemed to offer: volcanic rocks.

The reason the geologists were willing to bank an entire mission on the promise of volcanic rocks was simple: they are the only direct means of knowing what has happened in a planet's interior. The *mare* basalts proved that the lunar mantle had once been partly molten, and that was tantamount to saying that the moon had been geologically alive. But those samples were only part of the story. The *maria* covered only 17 percent of the moon. Furthermore, the basalts brought back to earth had formed during a fairly narrow, 350-million-year window of the moon's 4.6-billion-year history. If volcanic rocks could be found in the highlands, the geologists hoped, they might unlock greater expanses of lunar history before and after the *mare* eruptions. In short, they would tell the geologists how full the moon's geologic life had been.

Many of the mappers who studied pictures of the Descartes highlands saw ample evidence for volcanism. To these geologists, the shapes of the mountains suggested that they had been formed from lava, but one of different composition than found in the *maria.* On earth, basalt lavas are relatively thin and runny; they spread across the land like oil on a table top. Lavas whose composition is more like granite, on the other hand, are thick, like toothpaste. The difference is silica content; because granite has more silica, it is more viscous. Not only would such lavas produce landforms like those at Descartes, they would be lighter in color than basalts, just as the lunar highlands are lighter than the *maria.* When the geologists looked at the Descartes mountains—elongate mountains, with rounded tops and irregular shapes—they saw silica-rich volcanic rocks, which had never been found on the moon. Perhaps the mountains were made of rhyolite, the volcanic equivalent of granite. Or maybe they were piles of silica-rich volcanic ash and cinders.

Between the mountains, meanwhile, were patches of smooth plains that resembled other lunar regions that the geologists called the Cayley formation; they too were almost certainly volcanic deposits. They might even be the highlands' ancient version of *mare* basalt, brightened due to the effects of meteorite bombardment over the eons. In any case, Descartes seemed to offer what the geologists had long hoped for, a window into the evolution of the highlands. The scientists who had mapped the area believed some of its features might date back to almost 4 billion years, while others had formed only 1 billion years ago—fully two aeons after the *mare* eruptions. Descartes was not only easy to reach, it appeared to lie beyond Imbrium's influence. For the geologists, Descartes was too good to pass up. And on April 16, 1972, when veteran moon voyager John Young and his crew—rookies Ken Mattingly and Charlie Duke—headed out of earth orbit, their primary goal was to make known Apollo's last *luna incognita.*

Thursday, April 20, 1972
1:27 P.M., Houston time
4 days, 1 hour, 33 minutes Mission Elapsed Time
Aboard the command module *Casper*

It's my fault. I've come all this way, and people have put all this time and money and effort into this, and I've managed to screw it up. Even as these thoughts flashed through Ken Mattingly's mind, he forced himself to focus on the job. Young and Duke were out there, drifting away from him, heading for their Powered Descent. And he was flying solo, getting ready to fire up his SPS engine and change *Casper*'s orbit from an ellipse with a high point of 69 miles and a low point of 9 miles—the so-called descent orbit—to a circle 69 miles above the moon. He'd routed electrical power to the engine, and turned on the gyros, and it was all normal, until he tested the secondary control system. There was a set of little thumbwheels that controlled the gimbal motors for the engine nozzle, and as soon as he touched the yaw thumbwheel, he felt *Casper* begin to shake. On the instrument panel in front of him, the 8-ball nodded back and forth. He pulled back his hand and the shaking stopped. He changed some switch settings and tried it again. More shaking. It felt like a train on a very bad track. He said aloud, "It's not gonna work." Once more, he switched settings and moved the thumbwheel. No change. He spoke to the empty cabin: "I be a sorry bird."

Was it his fault? He *knew* he was doing everything just the way he'd done it a hundred times in the simulator. And yet, how could something have gone wrong with this beautiful flying machine that had worked perfectly for four days? Back at the Cape, a few weeks before the launch, he'd gone to the pad just to look at the Saturn V. And he'd realized, as he gazed at the towering machine, that he barely knew what he was looking at. Sure, he understood the basic design, and he knew the parts and pieces he had to know. But there were, what—several *million* parts in the whole thing? And each one had been designed, fabricated, tested, and installed by *someone.* Standing there, he knew the scope of Apollo was beyond the grasp of any one mind.

He rode the elevator up to the place where the third stage met the spacecraft adapter section, and there, at the juncture, was an open hatchway. He climbed through until he was standing inside a great metallic ring lined with pipes and electrical lines and all kinds of components. The lone technician who was working in there was startled—"Who are you? Get out of here."—but once he understood that he was talking to one of the men who would ride this rocket, he was just as gracious as could be. He said to Mattingly, "You know, I can't imagine what it's going to be like for you. But I can tell you this: It won't fail because of what *I* do." Mattingly realized that the reason Apollo worked at all was because thousands of people had said to themselves, "It won't fail because of me." From that moment on, Mattingly had taken that statement as his credo. He'd told himself as launch day grew near, and then, lying on his back in the command module, just before the engines lit, *It won't fail because of me.* How could he not feel the weight of the mission on him now.

"Hey, *Orion*?"

"Go ahead, Ken." John Young's voice.

"I have an unstable yaw gimbal number two . . ."

"Oh, boy." John Young, the last active member of the New 9, was the most experienced man in the Astronaut Office. He'd even been here before, the first man to make two trips into lunar orbit. He knew more about spaceflight than anyone Mattingly could think of. If only he were here right now, Mattingly thought, he might know how to work around this. But he was several thousand feet away. He couldn't see the engine, he couldn't feel *Casper* shake when he touched the gimbal control. To Mattingly went the honor of being the first astronaut to face a crisis alone, over the far side of the moon. He asked his commander, "You got any quick ideas?"

"No, I sure don't."

More than most astronauts, Mattingly thought, John Young seemed mindful of the risks of his profession. Around the Astronaut Office, his memos were well known, sounding the alarm about some engineering problem he'd uncovered. He wouldn't rest until he knew every detail about the particular system or technique that worried him. And when he had learned all he could, then it was time to go fly—with his eyes wide open. That was the only way to handle this business; that was what made him so good. Maybe Young worried so much because he saw so clearly. But when it came down to the real question—Will you fly it?—John's answer would always be yes. And Mattingly could understand; he felt the same way. Maybe that's why he liked John Young so much.

Now Mattingly heard Charlie Duke's familiar drawl. "Hey, Ken, why don't you just stop it and start it again?"

"I've done that twice."

Mattingly knew his role as command module pilot. He was a truck driver: Get the big boys out to the moon, drop them off, and get them home. Secretly he'd always wished something would go wrong so he could show how good he was. Let the LM's ascent rocket give out before Young and Duke could reach him; he'd go get them. Let the computer fail on the way back to earth—he'd get them home with a pencil and a few directions from mission control. But not this. Failure, Mattingly had always thought, was supposed to be clear-cut: it works or it doesn't. This was a gray area. Was the engine control really broken? The nozzle moved, after all; it just shook a lot. Could he still make a burn with it? He had no idea. And as much as he wanted to do something to save the mission, he knew that if he went ahead and tried to make the burn and the engine *didn't* work right, he would have made a very bad decision.

Now he heard Duke again. "What do your rules say, Ken?" Mattingly knew what the mission rules said; there'd been a change about a week before launch, and Mattingly was so mad about a change that late that he almost didn't read it. Now he wished he hadn't read it. It said that all four thrust-vector-control circuits, primary and secondary, had to be working or else the "circ" burn was forbidden. The only thing to do was rendezvous with *Orion*, come around to the near side, and get word from Houston.

Mission control agreed; Young and Duke would have to take a wave-off on the landing attempt, rendezvous with Mattingly, and fly in formation until there was some answer. That would be difficult; the onboard computers weren't programmed for the subtle nuances of orbital mechanics required to bring together a command module and lunar module that were flying in the same orbit, less than a mile apart. Mattingly would have to brute-force it—just point *Casper* at the lunar module and fire the thrusters. That would cost lots of maneuvering fuel, but there wasn't any other choice.

The hours dragged on as *Casper* and *Orion* circled the moon in limbo, waiting for Houston's decision. Mattingly doubted they'd find a way out. If only they could have said, "You guys get back together; we'll figure this out and try again tomorrow." But that was out of the question; the sun was rising over the Descartes highlands, and by tomorrow the lighting conditions would be no good for a landing. They would have to solve this today or call off the mission.

"*Casper*, we'd like you to . . ." Jim Irwin was telling Mattingly to try the gimbal motors again; mission control would record the telemetry and show it to the SPS experts. He could picture what was going on in Houston right now—Apollo spacecraft program manager Jim McDivitt overseeing the troubleshooting effort; phone lines to Downey and to MIT; engineers poring over data. And he knew, after witnessing Apollo 13, that given enough time, the experts on earth could solve any problem. But they couldn't possibly have enough time on this one. Mattingly was sure the mission was over. And he could tell by the dejected tone in Young and Duke's voices that they thought so too.

The funny thing was, Mattingly had a feeling he knew what might be wrong. A long time ago, he'd heard about a test of the engine in which the cable that carried signals to and from the gimbal motors was a little too short; when the nozzle was pointed to one side the cable went taut and pulled some pins out of its electrical connector. Mattingly wondered if that's what had happened to the secondary control system. And if so, then how could he be sure the primary system was okay? All the control signals went through the same cable. No point in worrying Young and Duke about this; he kept the thought to himself. As *Casper* and *Orion* circled the moon, Mattingly wondered just how much of his spacecraft really worked. He knew that if they had to, he and his crewmates could use the lander's engine to get out of lunar orbit. Was that what they were getting ready to do? Was this mission really over?

Mission Operations Control Room

In the back row of mission control, Jim McDivitt waited for *Casper* and *Orion* to come back around from the far side of the moon for the fifteenth time. McDivitt had been the manager of the Apollo Spacecraft Program Office for three years now, having taken the job after coming home from Apollo 9. Often, during the missions, he had decided whether or not to continue after emergencies struck. On Apollo 12, it had been his recommendation to leave earth orbit after the lightning strike. Apollo 13 had been his ordeal as much as anyone's at NASA. On the flight Al Shepard had thought might be trouble-free, McDivitt had faced Apollo 14's problems with a balky docking mechanism, and then, hours away from Powered Descent, an errant abort signal. Now it was McDivitt's turn again, as Apollo 16 hung in the balance.

There were some astronauts who would keep flying as long as you let them—no one more than John Young—but not McDivitt. There were only four ways to get out of this business: Get grounded, get fired, get killed, or retire. In McDivitt's mind, the only good choice was the last one, but if he waited too long, it would be one of the other three. After commanding Apollo 9—which had been enough of a test for any astronaut—he knew he would probably command one of the later lunar landings, but he had never viewed going to the moon as an end in itself. He would have stayed to fly the first landing, but not the fourth. He had always wanted to manage a big program, preferably in government, where there would be more control. When the program office wanted him, he felt it would be the best use of his abilities.

Now, three years later, he had no regrets. But the job had taxed him in ways that flying never had. Tracking lunar modules and command modules as they went through manufacturing and testing, shipment to the Cape, and final checkout for launch, and making sure everything happened on schedule to meet each new launch window, was as complex as anything he'd ever done. He could fully appreciate now the awesome responsibilities faced by Kraft, Petrone, and the other Apollo managers.

He'd never forgotten that going to the moon was dangerous. But in his mind, NASA accepted that fact the moment the Saturn V left the pad. With each successive step—leaving earth orbit, getting out to the moon, going into lunar orbit, and finally landing—the danger increased, but so did the *investment,* in terms of risk, to get there. Now that Young's crew were circling the moon, McDivitt was prepared—as long as they were safe—to fight even harder to keep them going than he would have if they'd been in earth orbit. A short time ago McDivitt had looked at the strip-charts of the telemetry from *Casper,* and it appeared that despite the oscillations, steering signals were getting through to the engine. At Downey, engineers had been analyzing the data and feeding it into a mockup of the SPS. They concluded that if Mattingly had to use the secondary system, the engine might shake, but it would be controllable. Was their data enough? McDivitt needed an extra measure of confidence, and as he had many times in this job, he got it from his own experiences on Apollo 9. In one of the many what-if exercises of that engineering test flight, they'd caused the SPS engine to shake while it was being fired, and it had still performed well.

Just now, while *Casper* and *Orion* were behind the moon, McDivitt had met with Kraft and the other managers and told them it was safe to proceed. And once more, having taken the risk of sending men to the moon, NASA was about to make good on its investment.

5:55 P.M.

Mattingly couldn't believe it when Irwin called up and said they'd looked at the test data and figured out that everything would be okay. Whoever was responsible for this spectacular save, Mattingly was going to buy them a case of beer. But didn't they realize that the primary system might be out too? Well, if their judgment said go ahead, he wasn't about to argue, not after six years of working his tail off to get here. He marveled at the boldness that seemed to have filled the managers after more than a decade in this business. When he was back on earth, Mattingly would tell McDivitt that he couldn't believe he let the mission continue, not when both of those control lines went through the same cable. And McDivitt would tease, "We *didn't* know they went through the same cable—you're the only one who did! You're right; we *wouldn't* have let you land!"

8:23 P.M.
In the lunar module *Orion*

"Okay, eighty feet, down at three. Looking super. There's dust." *Orion* sank straight down toward the Descartes highlands, kicking up wisps of bright soil. John Young kept his gaze out the window while Charlie Duke gave him data. "Let her on down. Six percent; plenty fat." If the past several hours had been rough on Ken Mattingly, they'd been even more traumatic for Young and Duke. Before undocking, *Orion* had an antenna problem, and then, because of a failed regulator, a dangerous buildup of pressure in the tanks of maneuvering fuel that threatened to abort the flight until Houston suggested shunting propellant into the ascent engine tanks. And then came the wave-off. They circled the moon, feeling their mission slipping away from them. But now, six hours behind schedule, Young and Duke were going to have their landing. *Orion* was flying just like the LLTV at Ellington, and after edging past one last 50-foot crater, the lander thumped to a stop. Charlie Duke, never one to restrain his excitement, couldn't contain himself. "Wowwww! Whoa, man! . . . Old *Orion* is finally here, Houston! Fan*tas*tic!"

"Well," drawled Young, "we don't have to walk far to pick up rocks, Houston. We're among 'em."

Friday, April 21
10:57 A.M., Houston time
4 days, 23 hours, 3 minutes Mission Elapsed Time
Descartes highlands

"Hey, John, hurry up!"

"I'm hurrying," said John Young as he made his way down *Orion*'s ladder. And then, for a moment, as he stood at last in the lunar sun, he raised both fists in triumph. "There you are, our mysterious and unknown Descartes highland plains. Apollo 16 is gonna *change your image*." No one listening on earth knew how prophetic those words would turn out to be.

Young added, "I'm glad they got ol' Brer Rabbit, here, back in the briar patch where he belongs." Years later, Young would be reluctant to explain the quote, but it seems clear that Brer Rabbit is himself, and the briar patch is spaceflight. Even now, on his fourth space mission, he showed no signs of wanting to do anything else. But Young hardly looked the part of a seasoned spacecraft commander. One day a simulator instructor was showing a friend around the space center, and when he pointed out Young across the room—this rumpled little guy in a Ban-Lon shirt and jeans—the reaction was, "*He's* an astronaut?" Even people who worked for the NASA contractors didn't know what to make of him. He drawled his way through conversation and gave the impression he was still the quiet country boy who grew up in Orlando, Florida, back when it was mostly farmland. You could see them sizing him up: *This poor guy just isn't on top of it—I'll have to go slow here*—but before the day was over they learned differently. Young would sit through a presentation on the computer software without saying a word, and when the specialist had finished, he'd drawl, "Well, I don't know about the W-matrix or anything like that, but what gets me is, how come if that's true, when I do . . . this, I get . . . *this*?" And you could almost hear the floorboards falling away in the engineer's mind, as he thought it through: Hey, he's *right!* Some people saw the country-boy bit as an act; it wasn't. It was just John's way of getting the people around him to think a little harder about the problem. In any case, you only made the mistake once: John Young was no hayseed. Ken Mattingly would call Young one of the best-read people he had ever met. His sharp, intuitive approach to engineering problems was well known to his colleagues. Inside Young was an unwavering determination, an overriding sense of responsibility—to the space program, to the country, to his crew—and an almost childlike sense of wonder at the universe.

When it came to lunar field geology, Young had caught the spark from Lee Silver, beginning on that first trip to the Orocopias with Lovell and Haise, and he hadn't lost it. At that time, he and Charlie Duke had been looking forward to Apollo 16 as the first J-mission. In 1970 the pair were doing practice geology work at the USGS training site in Flagstaff when they found out they would lose that milestone to Dave Scott's crew; it was a tremendous disappointment. Now they were faced with the task of equaling or even surpassing Apollo 15's scientific haul, and Young was ready. "Oh, look at all those beautiful *rocks!*"

Science Operations Room
Manned Spacecraft Center

In the geology back room, William Muehlberger inspected a photomap of the Descartes highlands, ready to oversee his surface geology team. He was new at this game, and he'd had a very stressful twenty-four hours. The geologists had done their share of sweating during the wave-off. Then, during the night, they'd learned that because of the late landing, the managers wanted to cancel the third moonwalk, and Muehlberger had set his planning team to work on a position paper to convince them to keep it. They were up all night, and in the morning they made their case. The managers were persuaded, but the go-ahead would depend on whether Young and Duke could stretch their supplies of power, oxygen, and cooling water to allow it. Now, beneath his excitement that Young and Duke were finally about to face the mysteries of the Descartes highlands, Muehlberger worried that they might not have time to unravel them.

Muehlberger's tall and powerful form recalled his days on the football team at Caltech, where he had been a classmate of Lee Silver's. Unlike the disheveled, lived-in look of some geologists, Muehlberger was a picture of neatness; he was almost dapper. A professor at the University of Texas, Muehlberger had joined the Apollo geology effort after Apollo 14. For Apollo 16, he had taken over not only as chief of the Surface Geology Team, but for Young and Duke's geology training; Lee Silver had been too busy with his work on Apollo 15, not to mention his teaching load at Caltech. Now, seated at the back of the Science Operations Room, Muehlberger listened to the voices of his former students coming from the moon.

Since the preparations for Apollo 16 began Muehlberger had heard more about the volcanic interpretation of the Descartes highlands than he could remember. One of the unique aspects of Apollo's lunar explorations was that the geologists had to predict what the astronauts would find at each landing site long before the flight began. That wasn't the way Muehlberger or most other geologists were used to doing business, but NASA wanted the geologic objectives to be as carefully planned as any other aspect of the missions. And it would have been out of the question to convey any uncertainty to the managers. Had he or his colleagues stood up at a site selection meeting and said, "We're really not sure what we'll find," they would have weakened their position, and possibly even lost the chance to go to Descartes.

Just as on previous missions, the geologists had made their predictions about Descartes using the discipline of photogeology. In this effort, their basic tools were their eyes, their minds, and photographs of the moon, along with knowledge gained from unmanned probes and the previous Apollo landings. Muehlberger knew that there were scientists in other fields who had nothing but disdain for photogeology. To them, any research that based its conclusions on photographs, or even worse, visual observations, was second-rate at best. Even some geologists agreed with them. But neither they nor anyone else could offer an alternative; there wasn't one, save going there. And each new mission seemed to confirm photogeology's predictive powers. The last four landings had given their share of surprises, but they had also validated a lot of good, solid scientific effort.

But with Descartes, the geologists were at a disadvantage. The best pictures available were the ones Stu Roosa took when Apollo 14's Hycon camera failed. For all of Roosa's skill, the smallest details they showed were calculated at 66 feet across. The challenge of analyzing Roosa's pictures went to USGS geologists Don Elston and Gene Boudette, who subjected the pictures to the most intensive scrutiny imaginable. Magnifying them under stereographic plotters, Elston and Boudette charted details that weren't much bigger than the grains in the film, including boulders they said were about 16 feet across. After a heroic effort, their finished map was dotted with all kinds of volcanic features, from lava flows to cinder cones to explosion craters. Muehlberger was amazed. Sometimes, when he looked into Elston and Boudette's stereo plotters, he wondered why he couldn't see the tiny details they talked about. But he knew Elston and Boudette had been at this for years, and they seemed sure of their conclusions—in fact, out of all the geologists, they were the most enthusiastic about the volcanic hypothesis. Muehlberger had had second thoughts about that scenario; after all, some experts had expected volcanic rocks to turn up at Fra Mauro, and none did. But he never really questioned the interpretation. And if others on his team harbored doubts about volcanics in the Descartes highlands, they hadn't voiced them loud enough for him to hear.

The late landing had thrown everything into disarray. Already, they'd lost the chance to have Young and Duke take telephoto pictures of Stone Mountain in the early morning light, to find out whether it had a pattern of linear grooves like Mount Hadley, and whether it might be an illusion caused by lighting conditions. But today's excursion was packed with objectives, and as Muehlberger listened to the voices of his two students, he heard the most exuberant pair yet to reach the surface of the moon.

"Hot *dawg*, is this great! That first step on the lunar surface is *super*!" No surprise that Charlie Duke should be so excited; he was often that way, like an overgrown kid. On the field trips Muehlberger had heard him chattering away in his South Carolina drawl while John Young mumbled in the background in a steady stream of one-liners. The grind of training seemed to bring out the vaudeville in them. Now, on the moon, it was the John and Charlie Show all over again. Presently, Duke was doing his impression of W. C. Fields: "Here's the ol' photomap. . . . Just like training. . . . A picture of Hadley Rille. . . ."

Young, meanwhile, was setting up a telescopic camera to photograph ultraviolet radiation from stars, and was discovering how much fun it was to handle the instrument, which was heavy and cumbersome on earth, in lunar gravity. Young's normal reticence was gone.

"Look at that, Charlie! Look at me carry it! I'm carrying it over my *shoulder*! Ha ha ha!"

Soon the Rover's TV camera was on, and everyone on earth could see the pair in action. They had planted the American flag, and now Duke was composing a picture of his commander. "Hey, John, this is perfect, with the LM and the Rover and you and Stone Mountain. And the old Flag. Come on out here and give me a big navy salute." Young did just that—and he did it while jumping three feet off the ground.

1:07 P.M.
Science Operations Room

Mark Langseth, a tall man in his thirties with brown hair, wire-rimmed glasses, and long sideburns, had taken a seat next to Jim Lovell to watch Charlie Duke drill into the moon. It was customary, during an ALSEP deployment, for a Principal Investigator to take this seat while watching his experiment being deployed. Langseth, a geophysicist at Columbia University, had been working for six years to find out how much heat was flowing from the lunar interior.

His experiment had been aboard Apollo 13; it burned up, along with *Aquarius,* in the atmosphere. At Hadley Base, Dave Scott's best efforts hadn't succeeded in getting the thermometers down to the desired 10-foot depth; Langseth got his readings, but the temperatures were twice as high as he expected. Only new data from another location would let him interpret the Apollo 15 results. Langseth was especially eager to see whether the heat flow would be different in the highlands. And it looked as if he was going to get what he wanted. Duke had the benefit of a redesigned drill, and even the moon seemed to be cooperating.

"Look at that beauty go!" Duke raved. Within minutes the hole was drilled and the thermometer inserted to its full depth. Duke beamed, "Mark has his first one all the way in to the red mark on the Cayley Plains." Langseth beamed too.

"Outstanding," said Capcom Tony England. "The first one in the highlands."

To which John Young quipped, "Ask him what we're going to do if the temperature shows like it does at Hadley." Young was working on the ALSEP, standing among the sensitive instruments and their ribbons of electronic cable. The camera was pointed at him now, as he started to walk away from the Central Station. Langseth saw something wrong. He nudged Lovell and said, "Look, he's got the cable around his foot." Before Lovell could call mission control, everyone saw it pull free of the Central Station. Judging from the cable's position, Langseth thought it must belong to the seismometer. But that thought didn't last long.

"Charlie?" Young's voice was almost plaintive.

"What?"

"Something happened here."

"What happened?"

"I don't know. Here's a line that pulled loose."

Duke, who had been drilling the hole for the second heat-flow probe, stopped and looked toward Young. "Uh-oh."

Young sounded worried. "What is that? What line is it?"

"That's the heat flow. You've pulled it off." Langseth couldn't believe what he was hearing. Neither could John Young.

"I don't know how it happened. God almighty."

"Well," Duke said flatly, "I'm wasting my time."

"God, I'm sorry. I didn't—I didn't even know it." You could hear it in Young's voice; he really had no idea it was happening. Like any moonwalker, Young could not easily see his own feet as he walked; for one thing, the small chest-mounted control unit was in his way, and also, he had to lean unnaturally far forward to compensate for the mass of his backpack. And it was well known that the ALSEP cables developed a "memory" from being rolled up, and once unrolled they refused to lay flat in one-sixth g. That was something Young and Duke had noticed in training, and they'd warned the engineers about it. But that didn't matter now. Cocooned within his suit, Young never felt the ribbon snag on his boot, pull taut, and sever, crippling Langseth's experiment.

"Tell Mark we're sorry," Duke said. "Is there no way we can recover from that, Tony?"

"I'm sure we're working it," England said. It was true; even now, as a stunned Langseth absorbed what had happened, experts were coming together to see if it was realistic to try a repair.

3:04 P.M.
Science Operations Room

Bill Muehlberger studied the photomap, tracing Young and Duke's progress as they drove west on their first geology traverse. With only a couple of hours available, the men would just have time to drive to some relatively nearby craters and get their first samples of the Cayley Plains.

For months, Muehlberger and his team had speculated on what the Cayley Plains would be like. Today, in TV pictures that were even clearer than Apollo 15's transmissions, they were seeing it. "Old *Orion*" had set down in a small "inlet" between two mountains that were aligned roughly north-south, like the peninsulas of San Francisco. On the monitors in the back room, the place did not appear to be very different from Hadley Base. There was the same expanse of undulating, crater-pocked, gray desert. And in the background, there were mountains, not as distinctive a skyline as the Apennines, but similar in their rounded form. But if the geologists' ideas were correct, the similarities were deceptive; those mountains, like the plains, were made of volcanic rock.

But so far, Young and Duke hadn't found what everyone expected. The first rock Duke picked up, back at the ALSEP site, had not been a piece of rhyolite, or basalt, or any other volcanic rock. It had been a breccia—an impact rock. A breccia is a mixture of rock fragments and soil particles welded together by the enormous energy of a meteorite impact. Ever since Gene Shoemaker had found breccias at Meteor crater, geologists had expected to find them on the moon. Sure enough, minor amounts of these rocks turned up in the *maria*, and breccias had dominated the samples brought back from Fra Mauro. Some were universes in miniature, containing pieces of a host of separate rocks, occasionally even fragments of preexisting breccias. The information locked in one such sample can keep a geochemist, armed with an electron microprobe, busy for years.

Now, as the men drove west, Muehlberger could hear Duke describing some nearby boulders; he said he thought they were breccias too. Via Tony England, Muehlberger sent up a question: "Have you seen any rocks that you're certain aren't breccias?"

"Negative," Duke answered.

Well, some breccias were to be expected anywhere there were impact craters. And judging from Young and Duke's reports, there were more of them than anyone had expected. The term "plains" conjures up images of Kansan flatness. But at one point John Young said sarcastically, "Cayley *Plains*? There's nothin' 'plains' about this place." Instead, "there's just craters on top of craters." The craters made the driving rough, but even worse, Young couldn't see them as he headed west. With the sun at his back the landscape was as bright and featureless as a snowfield; all shadows were hidden by the objects that cast them. Muehlberger could hear him telling Duke, "This driving down-sun is murder." What worried Muehlberger—and Young—was that Elston and Boudette had mapped 10- to 16-foot-high scarps. So far, though, Young and Duke hadn't reported any.

Now Young and Duke were nearing their objective, a worn, 1,000-foot crater called Flag, that was big enough to have penetrated deep into bedrock. But how to sample the rocks it brought up? By chance, there was a much smaller (120-foot), more distinct crater, called Plum, right on Flag's rim; *its* rim was hoped to be littered with pieces of Flag's ejecta. Those rocks would show what the Cayley Plains were made of.

3:44 P.M.
Plum crater

At Plum, Young and Duke picked up one rock after another and described them all as breccias. Muehlberger's team, wondering if they'd misheard, sent a message to Tony England to find out if perhaps they'd found a volcanic rock after all. England didn't want to nag Young and Duke; he was sure that if they'd found one, they would have said something. Meanwhile, Young approached a boulder, hammer in hand, and pried off a piece. The rock was somewhat weak and friable, not hard, like basalt; everyone listening knew that such weakness is typical of rocks that have been through an explosion from a meteorite impact. England asked, "Do you think it's a breccia?"

"Yeah, it's a breccia, Houston," said Young. "Or a welded—" Muehlberger could hear the words forming on Young's lips: a welded ashflow tuff, which is a rock made of ash fragments welded together by a matrix of volcanic glass. Welded ashflow tuffs can look just like breccias, and they are rich in silica, just as the Cayley rocks were thought to be. Some of Muehlberger's team had predicted there would be welded ashflow tuffs at Descartes.

"No, that's not right," Young said. "It's a breccia."

In the back room, the question was getting harder to ignore: Where were the volcanic rocks?

4:01 P.M.

Ever since Young and Duke arrived at Plum, Muehlberger's team had eyed a rock sitting on the crater's east rim, near the Rover and its TV camera. They thought they could see a crystal shining through a layer of dust, and that meant it might be igneous. They forwarded the request to mission control.

"Are you sure you want a rock that big, Houston?" asked Young, mindful of the 200-pound limit on the samples *Orion* could carry. "That's twenty pounds of rock right there." It turned out closer to 26 pounds, and it was bigger than a football; about a quarter of it had been buried in the dust. To get it, Duke had to get down on his knees, roll the rock up the side of his leg, and then, clutching it to his suit, try to stand up without losing his balance. "If I fall into Plum crater getting this rock," Duke said with mock annoyance, "Muehlberger has *had* it." It would prove to be the biggest moon rock ever brought back to earth, and it would be christened "Big Muley," in honor of Bill Muehlberger. The rock was so covered with dust that the men couldn't tell what it was, but Duke ventured that it might be a breccia.

4:47 P.M.
Mission Operations Control Room

In mission control, Tony England listened to the communications loop as Young and Duke headed back toward *Orion* to end their first moonwalk. Charlie Duke had been using more cooling water than expected, and flight director Pete Frank had decided to trim the second and last geology stop of the day to only 19 minutes. Frustrated by the cutback, Young and Duke were thankful that the driving was easier as they headed east, because they could follow their tracks. At Buster crater, while Young took readings with a portable magnetometer, to record any remnant magnetism in the rocks of the Cayley Plains, Duke had enough time to gather several more samples. Every one was a breccia. And as he listened, England could not help but feel apprehension for the men at Descartes.

One month shy of his thirtieth birthday, with a clear, serious face and wire-rimmed glasses, England looked as though he might be a junior professor in a college science department. In fact he had a Ph.D. in geophysics, the only geoscientist in the astronaut corps besides Jack Schmitt. Unlike many others in the XS-11, who had immersed themselves in Skylab activities in hopes of improving their chances of flying, England had chosen to stay with Apollo. He knew that in doing so he was probably giving up his chance to fly anytime in the near future, but his greatest fascination lay with the moon and the planets. If NASA had not canceled the last three lunar landings, England might now be looking forward to going to the moon himself. Instead, as Apollo 16's mission scientist, this seat in mission control was as close as he would get.

There were times when the earth-moon gap seemed especially vast to England, and one of them had been when the heat-flow cable broke. England had not been looking at the screen until it was too late to warn Young. Young hadn't said a word about it since the mishap, but he didn't have to; England knew he must feel absolutely terrible. So must Duke; he had put in a great deal of time training with the drill, and he'd really been looking forward to doing a good job for Mark Langseth. Now, despite the exuberance they'd shown at Plum, Young and Duke were a subdued pair as they drove past the ALSEP. England wanted to say something to cheer them both up. "A day ago, it didn't look like we were even gonna land," he radioed, "and now we've sampled our first highlands. I feel pretty good about the science without the heat flow."

"Well, I know Mark's disappointed," Duke said, "and I sure am." His voice was flat with disappointment.

"Me too," Young said.

England worried about Young and Duke's morale even aside from the heat flow accident. The rocks were turning out to be so different from what they had all expected. Were Young and Duke doubting themselves? Were they telling themselves that if they didn't find volcanic rocks, they weren't doing a good job? England could only imagine how big the silence at the other end of the earth-moon communications link must feel; he tried to fill it with reassurance. When one of the astronauts made a discovery, England radioed his own enthusiasm—"*Outstanding!*" And when one of them made a joke, he keyed his mike to broadcast his own laughter. He tried to communicate what he could not say over the air: "Don't worry. You're seeing something we didn't expect. Don't try to make it fit. See what you see and document that; we're going to have to put together a whole new picture of this place because of what you do see."

But England need not have worried about Young and Duke. They had seen enough breccias on field trips, and in the Lunar Receiving Laboratory, to recognize them without hesitation on the moon. They had also seen more volcanic rocks than most field geologists, and there definitely weren't any in sight on the Cayley Plains. And they never doubted themselves. In fact, it occurred to Charlie Duke that the geologists might be doubting what they were hearing. He could imagine them saying to themselves, "My God, we wasted three years of training on them." He hoped that wasn't the case.

It wasn't. There were some perplexed scientists in the back room, and even a wounded ego or two, but if Muehlberger and his team were surprised by the descriptions from the moon, they did not doubt their veracity. Like Scott and Irwin before them, Young and Duke had won the geologists' confidence before they left earth. The simple fact that they were at the Descartes highlands, "field checking" the volcanic hypothesis, was exciting enough to overshadow the confusion of confronting the unexpected. Soon enough, they would have pieces of the Cayley Plains to study for themselves. On their monitors, Muehlberger could see Young, standing in *Orion*'s shadow, using the conveyor line to haul the day's rock box up to Duke. Would those samples disprove the volcanic hypothesis? Today's prospecting had turned up a few rocks that did not seem to be breccias, but they weren't clearly volcanic either. Maybe some welded ashflow tuffs would turn up after all. Or maybe the highlands were indeed covered by a surface layer of impact rocks—with so many craters around, that would not be surprising—and there were volcanic flows underneath. If so, the samples in that rock box, now in Duke's hands, would not give the answer. That would have to come from the mission's prime objective, huge North Ray crater, which the geologists hoped would be big enough to have blasted deep into bedrock. North Ray was shaping up as the single destination for a shortened third moonwalk. And tomorrow's excursion, to the flanks of Stone Mountain, would offer the first chance to probe the riddle of the Descartes mountains. That moonwalk was now threatened by the possibility that the astronauts might be called on to repair the broken heat flow experiment; Muehlberger hated to think of the time that might eat up. In any case, Young and Duke would follow the traverse plan just as they had trained. True, those traverses had been meticulously designed to investigate a volcanic terrain, but there wasn't any way to reframe them now, and even if there had been, no one could think of any better plan.

8:44 P.M.

Meanwhile, back in *Orion*, John Young was worried about orange juice. The NASA doctors had been quite alarmed by Scott and Irwin's heart irregularities on Apollo 15, which they attributed largely to a loss of potassium. To prevent a recurrence on Apollo 16, they added electrolyte to the food, mostly in the form of potassium-spiked orange juice, and they instructed Young, Mattingly, and Duke to drink as much of it as possible. Already that had caused a problem for Charlie Duke. Back in lunar orbit, preparing for the landing, Duke had discovered that every time he moved his head within his bubble helmet, the microphone on his communications hat triggered the valve on the drink bag just below his neck. Before long, Duke found himself staring at a big blob of orange juice floating in front of his face, just out of reach. When it touched the microphone boom it attached itself and migrated under the fabric of his hat, into his hair. By the time he could take off his helmet, after landing, he looked as if he'd been shampooing with orange juice.

But the steady diet of potassium-spiked juice had even more unpleasant effects. A few minutes ago, Tony England had passed on word from the medics that the EKG's looked great, and added, "Just push on the orange juice."

Young drawled, "Push on the orange juice and everything will be fine?"

"Yes, push on the orange juice. Roger."

"I'm gonna turn into a citrus product is what I'm gonna do."

England suppressed a laugh. "Oh, well, it's good for you, John."

But Young wasn't kidding. "Ever hear of acid stomach, Tony? I think I've got a pH factor of about three right now. Because of the orange juice." Young was fed up with the havoc being wreaked on his digestive tract. Up in the command module the three of them had all but overpowered the environmental control system with methane. So far they'd kept their complaints to themselves, but now, as Young and Duke finished their housekeeping chores for the night, they didn't realize that Young's mike button was stuck in the "on" position. What came down from the moon was a backstage look at the John and Charlie show.

"I got the farts again, Charlie. I don't know what gives 'em to me, I really don't. I think it's acid in the stomach, I really do."

"Prob'ly is," Duke said.

"I mean, I haven't eaten this much citrus fruit in twenty years. But I'll tell you one thing, in another twelve fuckin' days, I ain't never eatin' any more. And if they offer to serve me potassium with my breakfast, I'm gonna *throw up.* I like an occasional orange—really do. But I'll be damned if I'm gonna be buried in oranges."

England beeped his mike button several times to signal the performers, but they kept talking—no longer about oranges, but about the moonwalk they'd just finished. Finally, England spoke up: "*Orion,* Houston."

A crisp response from Young, still unaware: "Yes, sir!"

"Okay, John, [you] have a hot-mike."

Young, more subdued now: "How long—How long have we had that?" No one within the sound of Young's voice would forget the episode, including Deke Slayton. He had been sitting next to Tony England throughout the day, keeping watch on the mission, as he had with every important moment of every flight since Young and Gus Grissom flew the first Gemini mission. Now, as Young and Duke prepared for sleep, Slayton manned the Capcom mike for a short time. With his lined face and western shirt he looked like a cowboy who had seen many hard winters. When he spoke to *Orion,* the men on the moon were glad to hear his voice, and a brief chat ensued. Duke talked about the previous night, when he had lain awake, overcome by excitement and anticipation, while Young sawed wood. "Couldn't believe we'd go to sleep, Deke, but man, this guy John sleeps like a baby up here."

"It sounds like the best place in the world to sleep," Slayton said simply. "I wish I was with you."

Young answered, "We do too, boss."

II: "You Just Bit Off More Than You Can Chew"

Saturday, April 22
Manned Spacecraft Center

It had been a long night for Mark Langseth and his heat flow experiment. Shortly after the first moonwalk was finished, a task force had convened in a back room, including engineers from the Bendix Corporation, who built the experiment; Fred Haise, Apollo 16's backup commander; and a couple of people from the "Tiger Team," the squad of ever-ready engineers who were called into service to solve the knottiest hardware problems. Langseth was amazed by them; they soaked up information; they were inexhaustible; they were so confident that Langseth felt they could have surmounted any failure. The Tiger Team brought a five-gallon jug of ice cream and went to work. The cause of the accident became clear once the Bendix people described the connector that had attached the heat flow cable to the Central Station. Originally it had been hard-wired; but at a late date Bendix changed the design to a connector. Langseth never saw the connector they used, but now they told him that it had a sharp edge; the cable had probably broken because the connector had sheared it in two. Langseth wasn't angry at John Young; he was furious at Bendix.

But how to recover? The cable could be reinserted into the connector, but first the astronauts must have a way to strip off its insulation to expose the wires. What was needed was an abrasive of some kind, and it just so happened that a very good one was available: lunar rock. By morning (and many scoops of ice cream later) the team had come up with a plan. Young and Duke would bring the cable and the still-attached heat flow electronics box into the lunar module at the end of the second moonwalk. Inside the lander, they would wrap the end of the cable around the handle of the geology hammer and scrape off the insulation with a rock. Finally, they would cut a clean edge with the scissors. On the third moonwalk, they would reconnect the experiment to the Central Station. Fred Haise tested the procedure; it worked.

Around nine o'clock on Saturday morning, Langseth and the team went to the VIP room at the back of mission control to make their case to Chris Kraft and Rocco Petrone. A few other ALSEP scientists went along for moral support. No one needed to say that the astronauts would go along with the plan; fixing things was part of their being. And Langseth pointed out that a repair would demonstrate the value of having humans in space instead of machines (a point he hoped would play better than "I want my experiment fixed"). Kraft and Petrone listened, and they allowed as how it might work. But it was hardly a sure thing; there were forty-eight separate wires in that cable that had to make contact, and no one could guarantee the procedure would work on the moon. And even if it did work, it would cost a lot of time; the team estimated a total of an hour during the two moonwalks, plus another hour for the work inside *Orion*. Petrone said, "We're not going to do it." Langseth would say later that he did not question the decision. The astronauts' time was too precious to spend on fixing one experiment; they had the moon to contend with.

12:03 P.M., Houston time
6 days, 0 hours, 9 minutes Mission Elapsed Time
Descartes highlands

In one way, Stone Mountain was just like Hadley Delta: it was steeper than either Young or Duke had realized on the way up. Young parked the Rover in a small, subdued crater so that it wouldn't start sliding down the hill the way Scott and Irwin's had. When he turned to look back to the north, he saw the mountain drop away from him in a series of steep-sided ridges. Off to the west it looked really treacherous, even more steep than the slope they had driven up. Young said to himself, "You've just nearly bit off more than you can chew." But the view from this high vantage—like the one that had confronted Scott and Irwin—was dazzling. They were 500 feet above the valley floor, higher than any Apollo moonwalkers ever had been or would be again. And the most spectacular sight was South Ray crater, five times the size of a football field, its rim an absolutely *brilliant* white, seemingly as fresh as the day it was blasted out of the Cayley Plains. Rays of boulders—black boulders, white boulders—were sprayed across the landscape for miles in all directions. The boulders next to the rim looked to be 90 feet across. South Ray was a beautiful excavation into the Cayley, and Young had pushed hard to go there on one of the moonwalks—even to add a fourth excursion for the purpose, if necessary. But astronomers had made soundings with radar beams from earth, and when they analyzed the echoes from South Ray crater they determined there would be too many rocks for a Rover to gain safe passage. Now Young could see that they had overestimated the hazard. Still, on the way up here, he and Duke had driven across a rise with so many boulders that he feared one of them would break a wheel assembly off the Rover. Perhaps getting to South Ray would have been tough, but he felt sure they could have made it to its nearby, smaller counterpart Baby Ray. Under his excitement, Young felt the frustration of discoveries beyond his reach.

But if they could not visit South Ray, its rocky artillery fire was scattered across the southwestern face of Stone Mountain. Those rocks—clean and sharp-edged—set Stone apart from the bland slopes of Hadley Delta. And if the rocks really were South Ray ejecta, as Young suspected, then even though they were 500 feet up on Stone Mountain, they were picking up pieces of the Cayley Plains. "You know, John," said Duke, "with all these rocks here, I'm not sure we're getting Descartes."

"That's right; I'm not either." When these words reached the back room, they only confirmed what Muehlberger and his team had anticipated before the mission. Realizing that Young and Duke might find South Ray ejecta here, they'd chosen five craters, deep enough to have penetrated bedrock, for Young and Duke to visit; they were named the Cincos. If Young and Duke could get to one of the Cincos, they would probably find large boulders whose source could be confidently tied to the mountain. Young and Duke had been looking for the Cincos on the drive up, but they were never sure where they were. Ironically, the largest of the five, Cinco *a*, was only 40 yards from the place where they stopped. Young and Duke would never realize that it was just out of view behind a ridge.

There weren't any large boulders here, but Young wasn't giving up. "Okay, Houston. I'm digging an exploratory trench." According to the geologists' theory, the mountain could have formed from sticky lavas that oozed to the surface and congealed, or cinders that flowed across the landscape and then solidified. The rocks Young plucked from the trench were covered with dust, but as far as he could tell they looked just like what he had seen down on the Cayley Plains. "I wish I could say these rocks look different, Houston, but they don't."

Now Tony England sent word from the back room to look for another crater within walking distance that might yield samples of Descartes. Young made his way about 50 yards up the slope to a 60-foot-wide, block-strewn bowl. When he reached the rim, he realized these rocks too were invaders from South Ray. Young had another idea that showed his insight: He decided to rake the side of the crater facing *away* from South Ray, the side that would have been shadowed from the barrage of ejecta. Like a prospector panning for gold, Young stood on the crater's soft, powdery wall and hacked away with the rake, struggling to hold the tool with tired, aching hands. His difficult harvest was not what he expected; what looked like rocks turned out to be clods of dirt that fell apart in the sample bag.

Years later, Young would say that when you are on the moon, inside a pressurized space suit, with only six or seven hours outside, and no more than an hour at any given place, you just can't take time to try and see the big picture. You're supposed to be getting the samples and documenting them, according to the timeline. If you try to be anything more than a technician, you are doing a disservice to the scientists who sent you. There was no anxiety on Stone Mountain now, as Young and Duke came bounding down the hillside, laughing as they returned to the Rover. The tension was in the back room, where Muehlberger and his team knew they had one chance left to snare samples of the Descartes on Stone Mountain.

1:04 P.M.

After a nervous drive downslope, Young parked next to a 50-foot crater that seemed to be free of South Ray ejecta. Young again suggested that he and Duke work the slope that faced away from South Ray. After some effort—the men felt as though they were on the verge of tumbling into the crater—they culled from the crater wall some small whitish rocks that were not jagged like South Ray ejecta but rounded; to Young that meant they were older, having been eroded by micrometeorite rain. But even these were not clearly pieces of Descartes. He told Tony England, "I don't think this is going to be a simple problem, even after you get the rocks back. . . ."

But when the men climbed out of the crater and headed back to the Rover, they got the biggest surprise of the day; Young almost tripped over it. A shoe-sized white rock sparkled in the sunlight. "We're gonna get that one," said Young excitedly. "That's the first one I've seen here that I really believe is a crystalline rock." It seemed to be made entirely of plagioclase, much like the Genesis Rock, but with crystals that were tiny, like sugar. Like the Genesis Rock, it could be a surviving fragment of anorthosite from the primordial crust, but it had clearly been kicked around by eons of meteorite impacts. But it surely was not volcanic. And as Young and Duke headed down from Stone Mountain it seemed that on this mission there would be no moments of clear discovery like the one that had heralded the Genesis Rock, no excited announcements from the moon—*We found what we came for!* Instead, Young and Duke had found what was there. And in the back room, Muehlberger and his team felt it had been a good day. Young's ingenuity had greatly impressed them. Whatever the Descartes was made of, they knew Young and Duke had probably succeeded in collecting representative samples. The true nature of the Cayley and the Descartes was unfolding word by word, rock by rock.

III: ". . . Or Wherever Geologists Go"

No one who knew Ken Mattingly would accuse him of being a patient man, but in the last two years he had sorely needed patience, and even more, raw persistence. After getting bumped from Apollo 13 a week before launch, Mattingly dove into two more years of training, with no letup, for Apollo 16. Then, with just two weeks to go, the NASA doctors told him they'd found an irregular reading in one of his blood tests; he had an elevated level of bilirubin, an indicator of liver function. Physically, he felt fine; in fact, he was in superb condition. But as far as he could tell, the doctors were suggesting he might be coming down with hepatitis. Surely they understood that this was an impending *disaster*—and yet they offered him no advice on how to avert it. They just kept taking blood tests, waiting to see what would happen. For three or four miserable days, Mattingly feared another medical false alarm was going to steal his last chance to go to the moon. Then the doctors decided he was fine.

Needless to say, it had all been worth it. The past seven days had been the climax of his career, and even aside from the flying, he had been living through one unforgettable sight after another. It got so he didn't want to look at each new spectacular for fear of erasing the memories of what had come before. Then, after the trauma of the wave-off, he was finally ready to carry out his solo mission. He knew the moon so well that he didn't even need to look at a map. It had been Mattingly who had been first to sign on with Farouk El-Baz. But even now, after working with the man for almost three years—and he liked El-Baz a great deal—he could not honestly say that he was interested in geology, or in the moon. And he had told El-Baz that, up-front. But he was ready to do whatever El-Baz asked, not just because he admired him, but because he wanted to be something more than a truck driver. He wanted to make a contribution—hell, he wanted something to do.

But what was he going to contribute? Would he really be able to see anything with his own eyes that wasn't already in the thousands of Lunar Orbiter pictures, or the Apollo photos? Oh,

Farouk told him that there was no substitute for the human observer, but he wasn't unbiased, was he? After all, if there were no need for the human observer, then NASA wouldn't need Farouk!

But now, with three days in lunar orbit under his belt, he had to admit he could see more than he'd ever expected. What had looked like a pile of whitish rubble was now brimming with details. He found himself staying awake well past the start of his sleep periods because he didn't want to miss anything. Whatever the Cayley formation was made of, there was more of it all over the far side of the moon. To Mattingly, the bright, smooth highland plains didn't look volcanic; they seemed to consist of debris that had been shaken, like a bowl full of gravel, until the surface was relatively flat. He would tell the geologists all of this when he was back home. Still, Mattingly knew, you can't analyze the dimensions, or the brightness, or the color, of an impression. The real data would be the pictures: cameras don't make mistakes; he couldn't say the same for himself.

Sunday, April 23
10:39 A.M., Houston time
6 days, 22 hours, 45 minutes Mission Elapsed Time

Inside *Casper*, in the pitch blackness of an orbital night, Ken Mattingly heard his own voice, on tape, break the stillness. "In twenty seconds the DAC will go on and remain on until sunrise. Adjust the settings." Outside, the depths of space were crossed by fingers of light, streamers of the sun's outer atmosphere. It is this envelope of dimly glowing gas, the corona, that frames the moon's silhouette during a total solar eclipse on earth. For a glimpse of its cold, eerie light astronomers will travel halfway around the world, but Mattingly now saw the corona as only the space traveler could, in the last minutes of orbital night, while the sun still hid below the unseen horizon. It was Mattingly's task to capture the corona on film using the Data Acquisition Camera (DAC). And the tape—that was a matter of efficiency. Mattingly knew the inside of the command module so well that even in pitch darkness he could find his way around. And he knew that if he flicked on a flashlight to glance at a checklist, even for a moment, he'd ruin his night vision. So he'd spent an hour during the trip out to the moon reading his checklist into the portable tape recorder.

"Stand by for the start. Four, three, two, one, *start.*" Mattingly activated the camera and heard the steady click of advancing film. Data: that's what it was all about. Slowly the luminous coronal streamers brightened, and then pitch darkness yielded to blinding sunlight. And now came the best time of all. Even though he would have traded places with Young and Duke in an instant, he could not imagine how bounding across the surface of the moon was more fun than orbiting it alone. Charlie Duke liked country music, and he'd brought along a tape of "Grand Ole Opry" stars from the same guy who'd supplied Pete Conrad and Stu Roosa. He played it all the way out to the moon. Mattingly almost got to like country—almost. But now, with *Casper* to himself, Mattingly savored the music he loved. In Houston they could hear snatches of it in the background whenever he keyed his mike: a Mahler symphony, Holst's *The Planets,* and Berlioz's *Symphonie Fantastique.* ("It didn't sound as good as 'Ridin' Old Paint,' " teased Stu Roosa, "but I guess it'll do.") And here, over the far side, it was the perfect background music to the silent, unreal panorama that filled the windows. He went about his work in a carefully orchestrated, weightless ballet, tending the scientific instruments, taking pictures, and turning his eyes moonward in search of a new discovery for Farouk El-Baz.

"And there's old Mother Earth," Mattingly said aloud. "Man, that's a beauty, too. Never get tired of watching earthrise." Earth: the source of all his troubles. To no one's surprise Mattingly had put together a staggering solo flight plan and was bent on doing it all. He wouldn't waste time checking the mission clock for the timing of his tasks. Instead, whenever he was in radio contact, he would let the Capcom watch the systems and the clock for him, leaving him free to concentrate on the moon. When Houston needed him to turn an experiment on or change a switch setting, he was a happy robot. And if they'd only left him alone, he would have been glad to continue that way for another two days. But the engine problem, and then Young and Duke's late landing, seemed to have thrown all of mission control into a tailspin. For some reason—he still wasn't sure why—they were going to cut an entire day off the mission, one of the two days the three men had planned to spend studying the moon after the surface work was finished. That alone wreaked havoc on Mattingly's flight plan. He'd specifically talked to the flight directors before the mission, and they'd agreed that if something unexpected should interfere with the schedule, they wouldn't try to reschedule everything; they'd just lop off the parts that were affected. But the flight directors seemed to have forgotten all about that. Every time he turned around there was another revision to copy down. Even more frustrating, the flight planners were sending him procedures that hadn't been tested; they hadn't had time to check their own work. It was as if they'd never flown a mission before. They were ruining his ballet. He wanted to get on the radio and say, "Alright, you guys, knock it off." Instead he called up every ounce of self-restraint and kept his irritation to himself.

Even when he was out of their reach, over the far side, he fell prey to the same problem that had plagued his predecessors: too much to do. Before the flight, without even realizing it, he'd let extra tasks creep into the time reserved for eating, and he didn't realize what that would be like until now. As it was, just about the only time he managed to eat was when he had to stop to go to the bathroom. There he was, slurping down a plastic bag of juice while hooked up to the urine collection hose, with a fecal collection bag flypapered to his rear end. Not the image of the Intrepid Lunar Explorer. Mattingly always wanted to be the first man to go to Mars. If that trip was going to be anything like Apollo, he'd tell them to forget it.

Sunday, April 23
On the Cayley Plains

Like most astronauts, Charlie Duke would tell you that he rarely dreams at night, and remembers his dreams even less often. But Duke had a dream, six months before the flight, that he had no trouble remembering. In Hawaii on a geology trip, Duke came down with the flu and ran a high fever; at times he was almost delirious. Some of the wives had come along, and Duke's wife, Dotty, took care of him while his colleagues were out in the field. In a fever sleep, Duke saw himself and John Young on the moon, driving their Rover toward North Ray crater. They came up over a ridge and suddenly Duke spotted something that made his heart race. A set of tracks crossed the ground ahead. Young stopped the Rover and they got off to investigate. The imprints in the dust looked like those from the Rover, but they were definitely different. Duke asked mission control, "Can we follow the tracks?"

"Go ahead," was the reply from earth. The twin trails stretched eastward, and Young and Duke turned to follow them. They drove onward for miles, over hills and across craters, until finally, topping another rise, they saw it: a vehicle, looking amazingly like the Rover, stopped on the surface. Aboard were two figures in space suits. After calling Houston to announce their incredible discovery, Young and Duke climbed off the Rover and approached the two figures, motionless in their seats. When Duke reached the one in the right seat, he could not see into its helmet because of its opaque sun visor. He put out his hand and raised the visor and saw his own face. The one in the left seat was John Young's double. After taking pieces of the space suits and Rover at mission control's request, Young and Duke drove back to the LM and blasted off for home. The next thing Duke knew, he was on earth, presenting the samples to the scientists. The test results: The craft was 100,000 years old. Then he woke up. The dream was so vivid—not scary, just *real*—that Duke remembered it from then on, and as he descended to the real moon inside *Orion*, he glanced out to his right at North Ray crater, and scanned the ground not only for boulders—"Looks like we're gonna make it, John; there's not too many blocks up there"—but for a set of tracks.

10:32 A.M.
Science Operations Room

Consider the athletic standing of the Caltech football team (low), and understand that Bill Muehlberger had seen his share of tough games. By any objective measure, Apollo 16 had turned out to be one of them; in the contest between the geologists and the moon, the moon clearly had the upper hand. Yesterday, Muehlberger had ruefully realized that the astronomers' radar soundings of South Ray crater—so promising for samples from deep within the Cayley—had misled them all into thinking it was inaccessible. And today the clock was running out on their chances to find volcanic rocks in the Descartes highlands. The managers had trimmed Apollo 16's final moonwalk to only 5 hours—no extensions, lest they violate the lift-off time—leaving just enough time for a dash to North Ray crater, some 3 miles north of the lunar module. North Ray crater was a prime destination for many reasons. For one thing, no astronauts had yet reached the rim of a large, well-preserved crater to hunt for samples of deep bedrock. North Ray certainly fit the bill; not only was it fresh, but it was more than six-tenths of a mile across—nearly as big as Arizona's Meteor crater—and more than 650 feet deep. North Ray was big enough to have punched through any surface layer of soil and debris into the underlying Descartes rocks at the base of Smoky Mountain. The explosive impact that formed it had sprinkled boulders across the landscape like grains of sand. The ones on North Ray's rim might have come from hundreds of feet down, and some of them were so big that they showed up on the photo-maps. Young and Duke had already spotted a few of those boulders during the drive up. As he waited for them to reach the crater, Muehlberger was not grim. If his team had known a little confusion in the past two days, then their two surrogates on the moon had more than made up for that with their skill and insight. Muehlberger was proud of them.

10:37 A.M.

"Oh, spectacular! Just spectacular!" Young's excited voice filled the back room; he and Duke had reached the crater. Soon a TV picture appeared on the monitors, but Young and Duke weren't in it. Ed Fendell, the engineer who operated the Rover's camera from his seat in mission control, had aimed it at a blue-and-white crescent in a black sky. "C'mon, camera," Muehlberger said impatiently, "quit looking at the earth. Goddammit." At last the camera panned down to the sight of two tiny space-suited figures standing before a gigantic pit, which was so big that it extended well beyond the field of view. Everywhere, rocks poked through a mantle of dust. To some of those watching, including Tony England, this was the most nerve-racking moment of the mission. The crater walls plummeted steeply to a rocky floor 200 yards below, and the ground at the rim might be ready to give way at any moment. If one of the astronauts fell in, he would never get out. But the picture was deceptive, just as it had been with Scott and Irwin at Hadley Rille; the men were far from the edge. The geologists had hoped Young and Duke would be able to get close enough to the crater to see the bottom and look for exposed bedrock. Now they realized that was out of the question.

"That rascally rim—it slopes [toward the edge] about ten or fifteen degrees, which is the kind of slope I'm standing on right now," Young explained. "And then all of a sudden, in order to see to the bottom, I've got to walk another hundred yards down a twenty-five to thirty degree slope, and I don't think I'd better." Before the flight, Young had talked about bringing along a 100-foot tether so that one man could venture to the rim, or even part way into the crater, while the other stayed behind to anchor him. The tether never made it onto the stowage list, not only to save weight but because the idea made the managers too nervous. Now, without that line, Young wasn't about to let either of them get anywhere near the rim.

The men backed off and spent forty minutes taking pictures of the crater walls and collecting samples. While they worked, the TV camera panned the bright ground near the crater's eastern rim, and there, far in the distance, was a black shape. "Good Lord," said one of the geologists, "is that a boulder?" Indeed it was; Muehlberger could see it on his photomap, and it was dark, like the rocks on the crater floor. If Young and Duke couldn't see the bottom of North Ray, they could still bring back a piece of it—provided they got to the boulder in time. Little more than an hour was allotted for this stop; much of it was already gone.

Young and Duke had been eyeing the big boulder from the moment they arrived. Now, finally, Young said, "Okay, Charlie. Let's go back to the Rover. Put your bag on there and head out for the big rock." He cast his gaze toward the boulder and said with amazement, "Look at the *size* of that biggie. It may be further away than we think."

"No, it's not very far," promised Duke.

But by this time Young had had enough experience with trying to judge distance on the moon that his lunar module pilot's words didn't convince him. "Theoretically, huh? Like everything else around here: 'A couple of weeks later . . .' " The pair began to lope away from the Rover, Duke leading the way.

On the monitors, Young and Duke ran onward, getting smaller and smaller. It began to dawn on the geologists that this boulder was even bigger than they thought. Suddenly, from the back of the room, Jack Schmitt's voice brought an eruption of laughter: "And as our crew sinks slowly in the west . . ."

Still Young and Duke ran. On the monitors they were tiny. The geologists heard amazement in Duke's voice: "Look at the size of that *rock*!"

"They're not even *there* yet," Muehlberger said quietly. Then, at last, Young and Duke stood next to a wall of dark, rough stone. The boulder was as tall as a four-story building and twice as long.

"Well, Tony," said Duke, "that's your House Rock, right there."

"House Rock?" asked a voice in the back room.

"House Rock," Muehlberger repeated quietly, like Ahab sighting the White Whale. If anything was going to show what the Descartes highlands were made of, it was House Rock.

"Okay," Young advised, "we had to come down a pretty good slope to get to this rock, so we may have to leave early to get back."

Schmitt looked at Muehlberger. "Get ready to cut other time somewhere," he said. "They're going to be here awhile." But there weren't going to be any extensions here. The walkback limit was immutable.

"You've got about seventeen minutes before you'll have to drive off," Tony England radioed, "so we'll have to hustle with this."

House Rock was dark, like basalt, but it didn't take Young and Duke long to see that it wasn't basalt; it was an enormous breccia, with fragments that were more than 6 feet across. Young and Duke weren't about to say so, but here was the last nail in the coffin for the volcanic hypothesis, hammered in by a rock the size of a house.

Monday, April 24
11:13 P.M.
Aboard *Casper*, heading back to earth

Taken out of lunar orbit a day early—they were sure the managers were just too nervous to wait any longer to bring them home—Young, Mattingly, and Duke sped away from the moon. Like those who came before them, they recorded the unreal view with their TV camera, but the pictures went no further than mission control. Even during the moonwalks the networks didn't broadcast for more than a couple of hours at a time. Duke's parents had to go to mission control's VIP room to see their son at work on the moon.

At the Capcom's suggestion they turned the camera on themselves so that their wives could get a look at them on tape when they came in the next day. When it was Young's turn to be on camera, he held a smudged hand before the lens. "See that? Can you see the dirt under those fingernails? That's moon dust. You talk about two dirty people. It took us ten minutes before we could get Ken to open the door. And we're still that way." He wasn't exaggerating; even now, little gray pebbles drifted around the cabin, and packed away in the rock boxes were a whopping 207 pounds of samples. "Yeah," Young told his audience, "wait 'til you see some of those rocks."

But Muehlberger and his team couldn't wait. They were under some pressure from the media to explain the surprising findings. The nature of the Descartes was still an open question, but for the Cayley Plains, Muehlberger's team was throwing out the volcanic theory and resurrecting an older idea that went back to the cataclysmic blast that formed the Imbrium basin. The geologists envisioned fast-moving debris surging away from the newly formed basin and across the face of the moon. Reaching the Descartes highlands, more than 600 miles away, it filled the valleys between the mountains to form the Cayley Plains. When Tony England told Young and Duke about the theory of a "slosh" from Imbrium and asked whether they had any response, Young tried to be diplomatic. It was too early to be saying something like that, he said, too early to be jumping to conclusions about the geology when they hadn't even seen the rocks. He said simply, "It ain't good science." The geologists would simply have to be patient. The world's only experts on the Cayley and the Descartes were on their way home, and their credentials were under their fingernails.

Tuesday, April 25
2:43 P.M., Houston time

For Ken Mattingly, the understanding of what it means to be in space did not begin the moment he reached earth orbit, or saw his world shrink to the size of his thumb; it did not come to him in solitude above the far side of the moon. It came, instead, during the trip back to earth, when he opened *Casper*'s side hatch, heart pounding with excitement, and floated outside. His mission was to retrieve two canisters of exposed film from the side of the service module. Training for the spacewalk he'd spent hours immersed in a water tank, clambering around in a vacuum chamber, and frantically practicing his tasks in the KC-135, a converted cargo plane that created about half a minute of weightlessness by flying in a parabolic arc. By the time Mattingly climbed into his suit aboard *Casper* and readied for the trip outside, he had anticipated every aspect, except the experience of being in deep space.

The sun was so staggeringly bright that Mattingly immediately pulled down his gold-plated outer visor. He heard the reassuring whoosh of oxygen flowing into his suit through the 50-foot umbilical. He was completely outside now. On the silver and white skin of the cylindrical service module, he saw that here and there the paint was bubbled, from the heat of the maneuvering thrusters. The scientific instrument bay was near the other end of the cylinder, and he made his way there by "walking" with his hands along a handrail. It was effortless, just the way it had been in the zero g airplane. Except that it wasn't like the airplane, because everywhere he looked, beyond, around, and past the service module, there was *nothing* at all, and he realized that this machine he was holding onto—for nine days, his universe—was a speck in the void. He squeezed the handrail so tightly that if hadn't been for the gloves, he would say later, he would surely have left fingerprints.

Arriving at the scientific instrument bay, Mattingly slid his boots into a pair of slipper-like footholds and rested. Looking back along the bright hull he saw Charlie Duke standing in *Casper*'s hatch, tending the umbilical. Beyond Duke, just off the nose, a full moon glowed, 50,000 miles away. When he looked to his left, he saw a tiny crescent earth, 180,000 miles away. Looking at them through *Casper*'s windows, he had never sensed the emptiness that lay on the other side of the glass. And in there, he had seen stars: *Where were all the stars?* It was a three-dimensional abyss. Charlie Duke kept saying, "My God, it's *dark* out here!"—and each time, Mattingly laughed, but his heart raced.

Mattingly was sure the "disappearance" of the stars was due to his gold visor. The doctors had advised him to leave the reflector down, lest he be exposed to harmful solar radiation, but he couldn't stand it anymore. He blinked the visor open just long enough for the universe to show a familiar face: *There they are!* His work finished, Mattingly pulled *Casper*'s hatch shut, and his universe became once more a small spacecraft drifting toward earth.

• • •

When Charlie Duke stepped off the ramp at Ellington, there was NASA geologist Fred Hörz coming to greet him out of the crowd of well-wishers. Duke grinned, "Those were sure funny-looking volcanic rocks, Fred."

"What do you think, Charlie, are they impact rocks?"

"We'll leave that to you," Duke said.

Any uncertainty vanished when the samples were unpacked in the Lunar Receiving Lab. Nearly all the rocks were breccias; the rest were ancient chunks of anorthosite. No traces of volcanic rock turned up, save for a few fragments of *mare* basalt thrown in by a distant impact. And while the Apollo 16 rocks could not rule out that lava or cinders had erupted elsewhere in the highlands, they all but proved that none had flowed at Descartes. It was not the first time the geologists had misinterpreted photographs of the moon; it would not be the last. But that did not mean there was anything inherently wrong with photogeology. The error, one geologist would write years later, was that they had neglected to define more than one working hypothesis. It was clear now that Elston and Boudette had been so enthusiastic about highland volcanism that they had overlooked alternative explanations.

But why had so many geologists hopped onto the volcanic bandwagon? In part, they had been misled by a false assumption that had plagued them since the beginning, that the moon was somehow earthlike. The resemblance of lunar highland formations to terrestrial volcanic features had reinforced the long-held idea that the moon had enjoyed a long and rich history of volcanic activity. In the wake of Apollo 16 geologists would come to understand that the overwhelming force that had shaped the surface of the moon was not volcanism, but the violence of cosmic bombardment. In the years to come, they would often recall a comment Ken Mattingly made from lunar orbit: "Well, back to the drawing boards, or wherever geologists go."

But they would also realize that Apollo 16 was perhaps the greatest leap in understanding of the moon since the first lunar landing. As one of them would later write, ". . . the mission to the Descartes Highlands illustrated once again that science advances most when its predictions prove wrong." Deciphering the evolution of the highlands would be far more difficult than anyone had imagined, because instead of a well-defined progression of lava flows, the reality was near-chaos; the origin of any particular rock or hill might be traced to a crater miles away or perhaps a giant impact basin hundreds of miles away. Two decades later, geologists would still be debating whether one of those impacts—the Imbrium cataclysm—had left its imprint at Descartes. They would search the breccias for clues to the violence that had preceded Imbrium, when huge asteroids struck the moon and earth. And in Apollo 16's collection of anorthosites, the largest of all the Apollo missions, there would be glimpses into the nature of the moon's primordial crust. The geophysicists too were rewarded when, three weeks after Young and

Duke left the moon, Apollo 16's seismometer picked up the largest lunar impact ever recorded. Its reverberations showed that the crust at the Descartes highlands was more than 9 miles thicker than average. And finally, Ken Mattingly's photographs and observations, combined with the data from *Casper*'s scientific instruments, would help the scientists extend the lessons of the rocks to highland regions across the moon. Apollo 16 did more than change the image of the Descartes highlands. It reaffirmed the most basic reason for exploration. If they'd known so surely what was there, they wouldn't have needed to go to the moon to begin with. The point of going was to find out.

CHAPTER 13

The Last Men on the Moon

APOLLO 17

I: Sunrise at Midnight

Charlie Smith had seen more of the sweep of history than anyone in the United States. Born in Liberia, he'd been taken aboard a slave ship at the age of twelve and brought to Galveston, Texas, where he grew up on a white man's ranch. He'd toted a .45 since he was thirteen, and could tell tales of riding with Jesse James and Billy the Kid. In his adult life he had witnessed the invention of the telephone, the automobile, the airplane, television, the atomic bomb, and the microchip. In December 1972, Charlie Smith's age was given at one hundred and thirty years. As the oldest living American he was invited to the launch of Apollo 17, and so he and his seventy-year-old son Chester traveled from a central-Florida town called Bartow to the Kennedy Space Center. As dusk fell on December 6, they sat with dozens of other celebrities in the VIP bleachers near the Vehicle Assembly Building. Across the waters of the tidal basin the last of the moonships gleamed in the floodlights, transformed into something mythic. The light from Pad 39-A could be seen all around Brevard County, in Cocoa, and Eau Gallie, and Titusville, towns for whom the imminent departure of a team of moon voyagers had become almost commonplace. Only in the first decades of this century—when Charlie Smith was in his seventies, his eighties—had the exploration of this planet marked its last major objectives: the polar crossings, the Himalayan treks. As the scope of exploration had broadened, the pace of time had compressed. Less than twelve years had elapsed since Kennedy's decision to go to the moon, just four years since Apollo 8 lifted off from Pad 39-A. Now, at the same launch pad, the first Age of Lunar Exploration would enter its last act.

By now, of course, that was old news. In Houston, the space center was already gearing up for the Skylab flights that would begin in the spring, and beyond that was the joint mission with the Soviets, the Apollo-Soyuz Test Project. And then, in the late seventies if all went according to plan, would come the Space Shuttle. It tugged at the flight controllers even as they prepared for the last moon mission. It would launch like a rocket and land like an airplane. It would withstand the fires of reentry unscathed and be ready to fly again in a matter of weeks. It would be a technological marvel; it would be a true flying machine. It would have *wings*. Now was the time to move away from Apollo's throwaway lineage and make space economical, and that was the shuttle's main billing: the first reusable spacecraft. By the 1980s, NASA hoped, the shuttle would make getting to space so routine that the space program would turn a profit. It was nothing less than the future of spaceflight. Most of the Astronaut Office was already working on it. And other than the three men in that rocket, and their backup crew, there were no veterans of lunar missions who were still working on Apollo. Eleven of them had left the astronaut corps to begin Life After the Moon, with varying results. And then there was Deke Slayton. After ten years, he was finally back on flight status. He'd undergone a risky procedure at the Mayo Clinic in which a catheter was inserted through his veins and into his heart, so that the doctors could see once and for all whether it had been damaged—and he passed with flying colors, literally. There was talk he might be assigned to fly on the joint Soviet-American flight in 1975, by which time he would be fifty-one years old. He was already studying Russian.

But on the whole, it had not been a good year for astronauts. In the spring of 1972, it had come out that Dave Scott and his crew had carried four hundred unauthorized first-day covers to the moon and back. The men kept three hundred of the covers and gave the remaining hundred to a German stamp dealer named Horst Eiermann, with the understanding that they would be sold to collectors after the Apollo program was over, privately and with no publicity, with the three astronauts sharing equally in the profits. Each man stood to gain $8,000 by the deal, with which they planned to set up trust funds for their children. But Eiermann began selling the covers within weeks after Apollo 15 returned. Scott's crew immediately notified Eiermann that he had broken the terms of the agreement, that the deal was off. All three men refused to take any money. But word of the sales got into the European press, and it was only a matter of time before NASA found out. By the fall there were reports that envelopes were selling for $1,500 apiece. When rumors began circulating, Deke Slayton went out on a limb to defend his people, saying he didn't believe the stories. Then Slayton found out, from Jim Irwin, that they were true. He was furious. Congress was demanding an investigation into improper conduct by astronauts. NASA formally reprimanded Scott and his crew. By June, Irwin had announced that he was leaving NASA and retiring from the air force to launch a Baptist ministry called High Flight. Al Worden was transferred to NASA's Ames Research Center near San Francisco. And Dave Scott was moved to a management post within the Manned Spacecraft Center.

The other astronauts were divided in their reactions. Some saw it as simply a dumb mistake. Others thought Scott, as crew commander, should be court-martialed. To some, it was a gray area. Astronauts had sold their autographs, for example, and profited in other less dramatic ways from their fame. But this time, there was so much money involved, and it had all become so public. It had tarnished the astronaut corps. That it had all been done by earnest, straight-arrow Dave Scott, whose mission had been such a high point for Apollo, only made the shock greater. Years later, Scott would blame his bad judgment on the pressures of getting ready for a lunar landing mission.

Whatever the astronauts thought of the stamp affair, the damage was done. Within NASA, the people who had always felt the Astronaut Office had too much clout were determined to see their wings clipped. For better or for worse, the myth of the Perfect

Astronaut had crumbled. Now the public knew that astronauts didn't always follow the rules. Astronauts were fallible; they were human.

By December, the stamp affair was old news. In the VIP stands, the atmosphere was something like the last performance of a long-running stage play. Some of the stars of that production—several of the men who had crossed the translunar gulf, including Jim Lovell, Neil Armstrong, and, with special irony, Dick Gordon—were on hand to see the last of their comrades leave the earth. Nearby, more than a thousand newspeople filled the press site. And a crowd estimated at more than a million people jammed the roads and beaches. All of them were lured by the promise of a space spectacular: The particulars of Apollo 17's launch window required that this be the first—and in fact, the only—night launch of the Saturn V. And then there was Charlie Smith, to whom this whole thing was literally beyond belief. Wearing a Stetson and a string tie, and an Apollo 17 mission patch on his lapel, Smith gazed at the Saturn and said in a sly but paper-thin voice, "I see that's a rocket, but th' ain't nobody goin' t' no moon. Me, you, or anybody else."

And for a while, that night, it seemed that the old man was right. As the launch time of 9:53 P.M. approached, lightning danced among distant purple clouds, seemingly ready to threaten the launch but never doing so. Behind the party-din of conversation the loudspeakers carried the voice of the public information officer, Jack King, his flat monotone counting down the last hours and minutes. Vapor issued from the Saturn, its stages filling with super-cold propellants. Now the crowd quieted and the voice on the loudspeakers fully emerged, echoing in the calm, damp night, counting down, now five minutes to go, now one. Then, at thirty seconds, the steady cadence of numbers halted. "We have a cutoff," said the voice. "We have a cutoff."

Three and a half miles away, at the pinnacle of the rocket, Gene Cernan, Ron Evans, and Jack Schmitt waited tensely inside the command module *America*, all thoughts on the abort handle in Cernan's left hand. No other moment could have totally crystallized the responsibility that went along with command of a spacecraft and a $450 million lunar mission. Cernan had trained for Apollo 17 as the climax of his career; he was not going to be denied. If the Saturn's guidance system went out during the launch, he was prepared to fly it into orbit himself, using the 8-ball, the hand controller, and the stars—but let this machine get off the ground. Don't make Apollo 17 end with a twist of the abort handle.

There was no danger. The launch pad's automatic sequencer, which controlled each split-second event in the complex launch sequence, had failed to pressurize the Saturn's third stage. In the blockhouse 3½ miles away, the launch controllers noticed the problem and sent the necessary command directly to the booster. But they were too late. The sequencer, aware of its own error, didn't accept their action as a substitute; it stopped the count on its own.

For a time it seemed depressingly likely to Cernan and his crew that they would not launch that night, but soon there was welcome news. The launch teams would work around the malfunction; it would take a couple of hours. During the delay Evans fell asleep, and Cernan and Schmitt got on the air to complain about the snoring.

Far from the crowds and the VIP's, at the special viewing site reserved for astronauts, their families and friends, Jan Evans waited tensely with her children, Jon and Jaime. Barbara Cernan and her daughter, Tracy, were here too. All day Jan had been going on pure excitement; it had to be that way after all those parties, like the bash thrown by *Life* magazine, and the other celebrations that had been in full swing for days. The tension she felt now was not out of fear; Jan had too much faith in the hardware for that. It was out of the desire to see her husband do what he had longed to for the past six years. From this same grassy shore, she and Ron had watched most of the moon flights begin. They had heard thunder and shed tears, out of elation for their departing friends. And Ron had always said, "One day, that's going to be me."

Jan was as devoted to her husband as any astronaut wife. In 1965, when she and Ron had a house near Miramar Naval Air Station in San Diego, a letter arrived informing him that he had the qualifications to apply for the fifth group of astronauts, and asking whether he wished to volunteer. At that moment Ron was 8,000 miles away, flying sorties from the U.S.S. *Ticonderoga* off the coast of North Vietnam. The letter said all applications must be in within ten days. Jan sprang into action. She worked up her courage and called Deke Slayton herself, and told him, "Ron *definitely* volunteers." Slayton assured her they'd accept his application late.

Many months later, after he had gone home for the physicals and the interviews, and then rejoined his ship, Evans returned from a mission and got the message to report to the ready room. Wondering what he had done wrong, Evans arrived to find the captain and several of the other pilots gathered, and the captain read the letter informing Ron that he had been selected as an astronaut. His shipmates would say later, he just about floated out of his chair.

Hours ago, just before nightfall, Jan and the children had waited at the entrance to the crew quarters to say good-bye to Ron. First came Gene Cernan, emerging into the glare of the television lights and blowing a kiss to nine-year-old Tracy. Ron was right behind him, grinning, giving a little skip as he came out, like a kid. For a moment he stopped, and she put an arm around the bulk of his white space suit and planted a kiss on the clear bubble helmet, and watched him climb into the transfer van, followed by Jack Schmitt. Then the van headed off into the night.

Now, for Jan Evans, there was only waiting. It was past midnight now. Just as the delay seemed interminable, the count entered its final minutes, heading for a launch time of 12:33 A.M. Once more the voice on the loudspeaker counted down, and this time, it did not stop. At the 8-second mark there was an explosion of red flame underneath the Saturn that gave way to a river of incandescent white. The sight of it drew from the crowd one great *Ahhhhh!* as twin plumes of fire and yellow smoke issued into the dark sky on either side of the launch tower. At the moment of release, as the great rocket rose from the earth, the sky filled with a brilliant golden light, like the light of sunrise. As the crowd cheered, the announcer's voice, suddenly electric, was heard, "It's lighting up the sky, it's just like daylight here at the Kennedy Space Center . . ."

Then the noise hit. Up to now no one had been aware of the eerie silence, as the shock waves rolled across the tidal basin, arriving at the viewing sites just as the Saturn cleared the launch tower.

It was a rippling, ear-splitting staccato roar that pummelled the chests of the surprised spectators and shook the ground. It rattled houses all up and down the Cape. Far beyond, the glow from this second sun was visible across Florida and as far away as North Carolina. Jan Evans stood transfixed by the rocket carrying her husband, never taking her eyes off it, not even when she was suddenly aware of a great commotion in the water in front of her. Out of the corner of her eye she saw fish jumping into the golden light, so many that the water seemed to boil with their thrashing until, at last, the sonic wave crested. The rocket was now a glowing torch suspended high overhead, its flame fanning out in the rarefied air at the outer reaches of the atmosphere.

Now there was stillness again. She could hear Gene Cernan's voice, charged with adrenaline, echoing from the loudspeaker in curt exchanges with Houston. When the first stage fell away there was a sudden, brilliant flash. A moment later the escape tower departed and she heard a cry of delight from Ron: "*Aaa-ha!* There she goes!" But that was all; from now on she heard only the clipped, technical jargon of the mission. She did not cry, not this time; she was possessed of a wonderful, joyous calm. As she watched her husband's rocket dwindle to a bright star she felt that a part of her was going with him to share the adventure for which he had worked so long. He was way out over the Atlantic now, hundreds of miles away from her already. She strained to follow Apollo 17 into the night with her eyes, but before it reached orbit, it was gone.

Meanwhile, at the VIP bleachers, Charlie Smith, who had watched this extraordinary leaving with a steady gaze, said, "I see they goin' somewhere, but that don't mean nothin'."

• • •

Nine years of a man's life can go by in a flash; at least it sometimes seemed that way to Gene Cernan. In 1963, the year he and Barbara went to Houston—he couldn't believe he had been chosen over so many pilots with test pilot credentials—their daughter, Tracy, was born. When he flew Gemini 9 she was only three; she didn't understand what was happening. By the time he went to the moon on Apollo 10, she was six—old enough for him to take her outside one night a month before launch and try to explain things: The moon was very far away, where God is, and hardly anybody had ever been there, but he and Mr. Stafford and Mr. Young were going there. And when he came back, he took Tracy out under the moon once more, and she said, "Daddy, you went to the moon! And the moon is way, far away, where God is—and you went with Mr. Stafford and Mr. Young, and hardly anybody else has ever been there." Then she got quiet for a moment, and Cernan figured she was about to plumb the depths of six-year-old curiosity, and that his answer would be the makings of tomorrow's show-and-tell. What he forgot was that for a little girl whose next-door neighbors were astronauts, going to the moon wasn't a very big deal.

"Daddy?"

"What, Punk?"

"Now that you've been to the moon, when are you going to take me camping like you promised?"

Tracy Cernan was nine now, and she understood very well that her daddy was going back to the moon, and that this time he was going to walk on it. Probably, though, she didn't realize how much it meant that he was returning as a mission commander. That was even more important to Cernan than crossing the last 50,000 feet he had missed on Apollo 10; it was so important, in fact, that Cernan had actually turned down a chance to walk on the moon as John Young's lunar module pilot. It had been a calculated gamble; Slayton had not promised him another flight after Apollo 10, and even after he was assigned to back up 14, there was absolutely no guarantee that he and his crew would get their own flight. But in August 1971, his gamble had paid off. And if command was the greatest challenge he had faced in his career, then Cernan had grown into it. During his first two flights, Gemini 9 and Apollo 10, Cernan had been Tom Stafford's shadow; though he treated his crew like equals, Stafford was a natural leader and a strong commander who knew how to use his power. Cernan had learned much simply by being around him. And looking back on Apollo 10, he'd come to realize how little he knew then about his spacecraft, when he thought he had known so much. Now, at age thirty-eight, Cernan had matured into a seasoned astronaut.

In a sense, he was an amalgam of the broad spectrum of personalities in the Astronaut Office. Like most of his colleagues, his ego was among the healthiest in the country, but those who met him were always pleased to find him "just a regular guy." Cernan was one of the astronauts whom people who worked at the space center liked to introduce to their friends. Here was this charismatic six-foot figure with cool blue eyes, salt-and-pepper hair, and quiet poise—like a good PR man, Cernan could make a stranger feel immediately comfortable. He seemed at once to savor the hero's mantle and to disown it. To be sure, he enjoyed the social perks that went along with the job; he made the Houston party scene with as much enthusiasm as Wally Schirra and Jim Lovell ever had. Richard Nixon's vice president, Spiro Agnew, played golf with him and dined at his home. He and Barbara had the aura of a storybook couple.

If Cernan had stayed in the navy, his goal would have been to be a squadron commander, and with the Apollo 14 backup crew he had his squadron. It even had the squadron's spirit—what other team had come up with its own backup patch? They were what Ron Evans called a "crew-crew." They flew together; they played together. And they would've gone to the moon together, had it not been for the cancellation of Apollo 18. Cernan had seen that coming, he understood that NASA would have been foolish to have a geologist available for the last landing and not send him—especially considering the storm of criticism that would have unleashed. Still, Joe Engle was his friend and his crewman, and Cernan had fought to keep him on, all the way to NASA Headquarters. But it had come down to an ultimatum: accept the change or lose the mission. And in that episode, Cernan had learned the limits of the spacecraft commander's power.

Now that was ancient history. Cernan would be the first to acknowledge that Schmitt had earned his seat. He knew that many of the Old Heads thought putting a scientist on a lunar mission was too risky, that NASA had caved in to pressure from the scientific community, and that it was a mistake. Cernan didn't agree. Schmitt was not the greatest aviator the world had ever seen, but certainly adequate, and during the flight he wouldn't do any flying anyway.

More important, he was a damn good lunar module pilot. His geologic expertise, of course, was unquestioned. As to whether Schmitt would really be able to *use* that expertise on the moon, working to the timeline and encased in a pressurized space suit—that was something Cernan, and many astronauts, wondered about. But he was certain of one thing: he and Schmitt would stay on the moon the longest, collect the most rocks, take the most photographs, bring home the most data. They would make the last the best.

NASA's managers did not seem to share his spirit. To them, Apollo 17 was like a last-minute, all-or-nothing bet at the blackjack table—because if Cernan and his crew didn't come back from the moon, the shuttle and the rest of the space program would be in jeopardy. Some confided to Cernan, "If it were up to me I'd cancel this flight." Even Chris Kraft, who had succeeded Bob Gilruth as director of the Manned Spacecraft Center, was telling him, "Take off your white scarf. We don't want to lose anyone now. Don't take any chances out there; just get home alive." Understandable though it was, this kind of talk rankled Cernan. He sensed, as Neil Armstrong had in the months before Apollo 11, the damage that failure would cause the program and the national image. And as a mission commander, he was already conservative; he didn't need to hear Kraft telling him not to take unnecessary risks.

Whatever the mood in Houston, Cernan was concerned that elsewhere, Apollo's imminent end was hurting morale. It was true that at the Cape, where technicians readied his Saturn V for launch, the biggest job cuts were long past, but nine hundred more workers were due to be laid off after the mission. In Downey, California, the Rockwell teams were at work on command modules for the Skylab missions and the joint Soviet-American flight. But on Long Island, the Grumman people were literally working themselves out of a job; the entire lunar module program would be over when Apollo 17 left the moon. Cernan and Schmitt spent a lot of time at Grumman, and if morale was bad, they never saw it. It amazed Cernan how they worked as hard, if not harder, on the final lander; maybe he and Schmitt had something to do with that.

At the Cape and in Houston, Cernan's crew organized softball games with flight controllers and support teams. They threw parties. Years later, those who were involved would remember this Apollo crew for those times. They would also remember that Cernan, who had never been shy about public speaking, was in his element. Sometimes after a few drinks—and sometimes not even after a few drinks—he would stand up on a chair and launch into an oration that was part pep talk and part campaign speech. "Apollo 17 may be the last flight to the moon," Cernan would say, "but it's not the end. It's the end of the beginning." And for the most part, it worked, though some of his audience smiled to themselves at this mixture of ham and sincerity. Later they teased, "Gene, I'm not sure I've got it straight—is it the end or the beginning?"

• • •

Whatever you called it, Apollo 17 was the last roll of the dice for the geologists; only one chance remained to unravel the knotted ball of string called lunar evolution. Before the flight, when the press queried him on the value of yet another moon mission, Jack Schmitt liked to say that the moon had as much land area as the entire continent of Africa—and yet, astronauts had explored only a handful of acres. How could anyone think that an entire world might give up its secrets after only a half-dozen modest expeditions? Obviously, the choice of landing site for Apollo 17 was especially consequential, and when he was named to that mission Jack Schmitt had resurrected his goal of landing on the far side, on the dark lava floor of the crater Tsiolkovsky. It was a dream site, with not only a far side *mare* to sample—would it differ chemically from those on the near side?—but a gigantic central peak that offered samples from deep within the far-side crust. Needless to say, Schmitt didn't need to sell the idea to the geologists. But when he talked it up to his flight controller friends they half-jokingly responded, "Well, Jack, out of sight, out of mind." There would be no earth in the sky above Tsiolkovsky, and the only way to communicate would be via special relay satellites in lunar orbit. A pair of Tiros weather satellites were already available as off-the-shelf spacecraft; a single Titan booster could place both of them at a point about 30,000 miles above the far side—one of the so-called liberation points—where gravitational conditions were such that they would remain suspended over the surface, within sight of both Tsiolkovsky and earth.

Schmitt had first raised the idea late in the spring of 1970, as part of an effort to save Apollo 18 and 19: if he could show that those missions could reach places where the geologists had always wanted to go, maybe NASA would think twice about canceling them. But at the space center, it seemed everyone was too caught up in the recovery from Apollo 13 to consider such schemes. Now, in the fall of 1971, Schmitt pushed for Tsiolkovsky again—this time, to generate renewed public excitement for Apollo. But it was no easier to sell the far side now than it had been the year before. The satellites added a new level of complexity to an already risky enterprise. Most of all, there was no money for them. Cernan discussed the idea with George Low and Chris Kraft and was told it wasn't going to happen, but Schmitt kept pushing until Kraft told Schmitt to stop talking about it. Tsiolkovsky's secrets would have to wait for the next generation of lunar explorers.

No matter; the geologists had found a place that could only be described as the jewel in the crown of Apollo landing sites. It gained this status during Apollo 15, when Al Worden flew over the southeastern shore of Mare Serenitatis (the Sea of Serenity, which forms the Man in the Moon's left eye) and saw "a whole field of small cinder cones down there." His verbal pictures of tiny volcanic craters—whose conical shape seemed a dead giveaway—were a summons to exploration. But the real clincher came when the film from Apollo 15's panoramic camera was unreeled on light tables at the space center, and the geologists could see with their own eyes what had so excited the lone astronaut. Some of those craters lay within a small box-canyon ringed by the steep-sided peaks called the Taurus Mountains that formed the worn, ill-defined rim of the ancient basin. Seventeen miles to the northeast lay the crater Littrow. What was so striking about the valley of Taurus-Littrow was its floor, which was covered by some of the darkest material seen on the moon. It seemed entirely out of place, appearing as it did among the light-colored highland terrain. Whatever it was, it was widespread; the geologists found it all along the southeastern portion of

Mare Serenitatis. They even spotted patches of it on the high elevations of some of the mountains. How had it gotten there? On earth, that can happen when pockets of volcanic gas suddenly find release at the surface, spraying molten rock high into the sky. If the same thing had happened on a very intense scale on the moon, it could have blanketed a wide area with volcanic ash. The coating probably wasn't very thick; on Worden's pictures the rims of small impact craters were clearly visible poking up through the dark mantle. Because it was so dark, the geologists surmised the deposit hadn't been exposed for very long to micrometeorite rain; otherwise it would have become faded by being mixed with the soil. Perhaps it was only half a billion years old. After the disappointment of the Descartes highlands, here was new reason to hope that the moon hadn't died geologically so long ago after all.

At the same time, the valley walls—and in particular, two rocky prominences christened the North and South Massifs—might finally give a look back to the time before the Imbrium impact, whose influence had yet to be fully escaped. Everyone agreed that the Serenitatis basin had formed before Imbrium, and hoped the Massifs would contain the oldest lunar rocks yet found. And there were still other lures: a probable landslide at the base of the South Massif that would undoubtedly contain a rich variety of samples from the mountain; a scarp that resembled features called wrinkle ridges commonly seen on the *maria* but never visited; and the Sculptured Hills, whose knobby shape set them apart from the Massifs and sparked renewed hopes of highland volcanism. Geologically speaking, Taurus-Littrow was the most complex site yet, worthy of the full capabilities of the J-mission. Here, in one place, lay the possibility of exploring the beginning and the end of lunar history.

But getting there wouldn't be easy, primarily because the valley was only 4½ miles across, from one rocky wall to the other. The normal target ellipses—the ones that showed possible trajectory errors—spilled up onto a mountain on one side of the valley and a landslide at the other. But that was because the ellipses were very conservative. Perhaps the engineers had never gotten over the fact that Armstrong and Aldrin had landed 5 miles past their aim point, because the planners had never changed the size of the error ellipse, even though the next four lunar modules had set down within a few hundred feet of their targets. One of the men involved in the site selection process, Jack Sevier, made that point to the trajectory experts, and they soon came back with a smaller ellipse that fit neatly within the valley. Now it was up to the managers. At the site selection meeting, Kraft and McDivitt flashed a thumbs-up, as if to say, "It's the last one; let's go for broke."

In part, the geologists chose Taurus-Littrow because they knew Jack Schmitt would be there to explore it. But before he and Cernan could turn their combined geologic prowess on the valley, they had to get there. Schmitt had no unease about becoming the first of his chosen profession to practice on another world, but he knew he must not turn in anything less than an outstanding performance as a lunar module pilot—not just for the sake of the mission, or the lives of his crewmates, but because he was nothing less than a test case for all the scientist-astronauts who were waiting for their chances to fly. Before he could be a geologist, Schmitt would have to be an astronaut. And on the afternoon of December 11, when he and Cernan stood side by side at the controls of the lunar module *Challenger*, descending into the valley of Taurus-Littrow, the transmissions that came down on the earth-moon airwaves were the same jargon-rich shorthand that had been the soundtrack to each of the five previous lunar landings. Schmitt's concentration was total. Except for a glance at the moment of pitchover, he did not look outside. He did not see the dark valley opening before them, its white mountain sentinels casting long, pointed shadows. He did not see the smooth place, just near the crater Poppy (Tracy Cernan's nickname for her late grandfather) where Cernan planned to set down. When he heard Cernan say, "I don't need the numbers anymore," Schmitt knew his commander was getting ready to take over from the computer, but he did not know that there was a boulder field out ahead that Cernan had to maneuver around. At last, 60 feet above the moon, Schmitt took another brief look and reported, for the benefit of his listeners on earth, "Getting a little dust. . . . Very little dust." Then he returned to the instruments, unaware that *Challenger*'s spindly shadow had appeared on the dark plain, moving toward them, now passing under them. "Stand by for touchdown,"

Cernan called, and *Challenger* crept down the last 25 feet; as soon as Schmitt saw the blue contact light come on he called it out, and after a moment of free fall, he felt them come to a stop with a firm thud. Even as Cernan took a moment to exult, "Okay, Houston, the *Challenger* has landed!" Schmitt was reading checklists and flipping switches. Later, he would jokingly complain he had missed the whole landing.

Monday, December 11, 1972
6:08 P.M., Houston time
4 days, 21 hours, 15 minutes Mission Elapsed Time

"Hey, who's been tracking up my lunar surface?" The answer to Jack Schmitt's question was Gene Cernan; now, with characteristic irreverence, Jack Schmitt was climbing down the ladder to make his own tracks in ancient dust. And when he was at last standing on the valley floor—the end of a journey that began in Flagstaff eight years earlier—his thoughts were on the significance of the moment, not to himself but to humankind. What had brought Schmitt to NASA, and had fueled his fanatical drive toward lunar exploration, was his devotion to history. He didn't need to be told that he and Cernan would only scratch the surface, that the real exploration of the moon would be left to the ones who would follow them here. How many years or even decades away that was, Schmitt had no idea; he knew only that the space program was in serious trouble. But Schmitt had a kind of stubborn optimism; he would not give up hope.

"Boy," said Cernan, "your feet look like you just . . ."

Schmitt finished the thought. "Walked on the moon? Well, I tell you, Gene, I think the next generation ought to accept this as a challenge. Let's see them leave footsteps like these some day." And for a moment, as he took his first steps, Schmitt's gaze went to a landscape unlike any he had ever seen on earth. The Massifs rose like flattened pyramids into the velvet sky, their sides impossibly steep. They had neither the hewn-granite sharpness of the Rockies, nor the smooth, glacial roundness of Norway's fjords, where Schmitt had done the research for his doctoral thesis. He was looking at mountains that dated back almost to the formation of the solar system. Out on the dark plain, he saw rolling hills, their fluid forms littered with boulders ejected from the larger craters. In the near field he noticed small, rimless hollows in the soil, the center of each miniature crater marked by a dark spot of fused glass. At his feet, soil and broken rocks sparkled in the unfiltered sunlight, their textures and subtle hues amazingly vivid, their crystals iridescent. On the surface of the larger cobbles he saw small white spots where micrometeorites had blasted tiny craters. Everywhere he looked, on every scale, his eyes found new detail. He said, "A geologist's paradise if I ever saw one." Around him, light reflected from the distant mountains and nearby slopes penetrated the shadows. In the southwest, his gaze was arrested by a sparkling blue gibbous earth, hanging over the South Massif. He said nothing.

6:41 P.M.

It was only natural that Gene Cernan had come to know Jack Schmitt well; the two men had been living in each other's pockets for sixteen months. Cernan felt a kind of fatherly responsibility for him. He'd gotten used to Schmitt's outspoken manner, and his penchant for going to management with a pet issue; like Dick Gordon, Cernan had learned to keep Schmitt from getting out of hand. They'd developed a great working relationship, in the simulator and in the field. But Cernan could not say that he really knew what made Schmitt tick. Schmitt was hard to get at. As outspoken as he was, he kept his emotions well guarded. Cernan was probably one of the most gregarious astronauts, but Schmitt was undoubtedly one of the most private. Before simulator runs, Cernan and Evans would be gabbing with the instructors; Schmitt would be off in a corner with the training manuals.

Cernan could certainly tell that Schmitt was excited to be here. In between his expert descriptions—"The basic bright-colored rock type in the area looks very much like the cristobalite gabbros in the *mare* basalt suite . . ."—Schmitt was broadcasting his excitement in snatches of song (selections included "Bury me not on the lone prairie," "What is this crazy thing called love," and "We're off to see the wizard") and horrible puns. Informed by Capcom Bob Parker that his suit temperature was running a little high, Schmitt responded, "I'm just a hot geologist; that's all."

But underneath all that brashness, was Schmitt absorbing this experience on a personal level? During the past several days, Cernan had hoped Schmitt and Evans were taking it all in for themselves. The night launch was a sight to behold—even before they lifted off, as he could feel the Saturn V's engines coming to full power, Cernan glanced out through the little window in the boost protective cover and saw the glow—and though only he could see it, he said, "*Look at the light.*" In earth orbit, knowing they would be leaving only too soon, he'd made sure his crew saw the sunrise. And yesterday, just before the Lunar Orbit Insertion burn, Cernan had glanced out the command module's hatch window and had seen what he had never witnessed on Apollo 10: an enormous crescent moon bathed in sunlight, only 8,000 miles away and growing larger by the minute. In this moment the moon was neither the forbidding hole in the stars seen by Bill Anders, nor the ghostly earthlit sphere that had greeted Armstrong, Aldrin, and Collins. He remembered when, on Apollo 10, he and his crewmates were leaving the moon, climbing away so fast that one of them had said, "If we saw this coming in, we'd have to close our eyes." Now he was seeing it. They were making a dive-bombing run on the moon. Shading his eyes from the sun's glare, Cernan could see huge craters, some of which he recognized from three years before. The monocular brought him closer: He could see boulders on the crater walls, ridges curving over the bright horizon. He tried to convey to Evans and Schmitt how unusual it was to see this view. As they drew closer it grew to fantastic dimension, then disappeared into blackness as Apollo 17 headed for lunar orbit.

Cernan would not forget these sights, and he would talk about them for the rest of his life. In his core, he believed it was important to come home with more than just rocks. Now, loading the Rover with supplies, Cernan looked up from his work and saw the earth suspended above the South Massif. No one had ever seen it like this, so close to the horizon; that was because Taurus-Littrow was so far from the geographic center of the near side. On the way out to the moon, the sight of earth, like a gemstone suspended in dark water, had held him in awe; it was that sight that brought home the fact that there was nothing at all routine about going to the moon for a second time, the single experience that had made him say to himself, *My God, I'm out here again.* Just as he had on Apollo 10, he'd watched it turn on an invisible axis. And once again, he had consciously asked himself whether he understood what he was seeing. His overwhelming feeling was of a sense of purpose, a gut-level awareness that it was simply too beautiful to have happened by accident. He felt as if he were seeing earth as it had appeared in the moment before the Creation, in the mind's eye of God. Now that he was standing on the moon, immersed in the abyss of space—how could the sun shine on him so brightly and not dispel that darkness?—he felt as though he were looking out from within a dream, and the earth was his link with reality. To Cernan it was the most precious possession a man could hold in his memory. He had to say something.

"Oh, man—Hey, Jack, just stop. You owe yourself thirty seconds to look up over the South Massif and look at the earth."

"What? The *earth?!*"

"Just look up there."

"Aaah! You seen one earth, you've seen them all."

It was just like Jack Schmitt to answer his friend's display of emotion with feigned disgust. But in truth, Schmitt had been looking at the earth for three days. Planning to indulge his long-time fascination with meteorology, he'd arranged before the flight to be briefed by air force weather forecasters; he'd launched with satellite photos in the pocket of his space suit. During the trip out, from the moment he floated out of his sleeping bag, he was broadcasting his observations like a human weather satellite. His descriptions filled pages in the air-to-ground transcripts: "That tropical depression I saw earlier north of Borneo is now even more strongly developed, at the tail end of the front that stretches up toward Japan. It really looks like a humdinger from here. . . ." The last thing he needed to look at right now was the earth. He was here to focus on the moon, not because that was his mission, but because it was his profession. And he was impatient to get on with it.

On earth, Cernan and Schmitt had almost spent less time practicing their geologic activities than the purely mechanical tasks that would punctuate them. Paradoxically, that was for the benefit of the geology work: it didn't make sense, for example, to get on and off the Rover at half the pace that they could achieve with practice. By the same token, the men spent hour upon hour laying out their ALSEP experiments, so that they would take up as little of the first moonwalk as possible. But on the moon, predictably, those plans went awry. Fortunately, the heat flow experiment was set up without difficulty—giving Mark Langseth his last chance to confirm the puzzling readings at Hadley. But Schmitt was frustrated by another experiment, the Lunar Gravimeter, designed to show the existence of gravity waves but which now refused to operate. And then there was the lunar drill. Like Dave Scott before him, Gene Cernan was forced into a long, hard battle to obtain a deep core sample. Even the special long-handled jack designed to help Cernan extract the cores seemed to do no good. In mission control, flight director Gerry Griffin watched as Cernan, on his knees, wrestled with the jack. With Cernan's heart rate pushing 145, one controller warned Griffin, "He's eating into that oxygen," and urged that Schmitt help him, "or we're really going to be in trouble." Cernan was in danger of depleting the oxygen reserves designated for the first geology traverse. Schmitt went over to help pump the jack handle, launching his suited body into space and coming down horizontally on the handle, again and again, interrupted by a tumble that sent himself and pieces of stray equipment spinning into the dust—while Schmitt laughed. The flight controllers' concern aside, this was the first sign that something was different about Apollo 17. Both men, and especially Schmitt, took to their work with a kind of physical aggressiveness, even arrogance, rarely seen in any moonwalkers before them, and certainly not on the first day outside; they seemed to have no wariness about the risk of a damaged space suit. It wasn't recklessness—both men had a healthy respect for their situation—but the kind of confidence that comes with familiarity, as if they had done all of this many times before. And in a sense, they had, because five teams of astronauts had gone before them. Once again, the phenomenon of instantaneous evolution that had marked each Apollo mission, not only among the astronauts but the flight controllers and engineers, was at work: Cernan and Schmitt had absorbed the experiences of their predecessors, and now, on the moon, they were building on them.

Finally the moon's grip loosened and the men had their core. Much to Schmitt's frustration, they were an hour behind the timeline, and Cernan's oxygen consumption had forced Gerry Griffin to cut the upcoming geology traverse short; they would go only half as far from *Challenger* as planned. Both men had tired hands and sore fingers, but they showed no signs of slowing down. The Rover's TV camera caught Schmitt as he bounded back to the lander in crazy cartoon leaps, and once again the airwaves filled with song:

"*I was strolling on the moon one day—*"

Cernan joined in, "*In the merry, merry month of—*"

"*December—*"

"No, May," said Cernan. "May is the year of the month."

Schmitt laughed. "That's right, May. *When much to my surprise, a pair of bonny eyes . . .*"

"Sorry guys," said Capcom Bob Parker, unable to resist, "but today *may* be December." Corn on the moon. Jack Schmitt and Gene Cernan weren't letting anything get to them now; they were on an absolute high.

"*Da dada dada dada* deee *da dee . . .*"

Tuesday, December 12
1:12 A.M., Houston time
5 days, 4 hours, 19 minutes Mission Elapsed Time
Inside *Challenger*

The pungent odor of spent gunpowder—that is, the smell of moon dust—filled the cabin as Cernan and Schmitt took off their helmets and gloves. Schmitt knew that on earth, particles of dust are covered with a thin layer of air, but the dust in *Challenger*'s cabin had never been exposed to oxygen. Each grain was still chemically active, just as gunpowder is immediately after it has been set off, and it reacted with the nasal passages in the same way. Now, to his surprise, he was hit with an attack of hay fever.

It had been a long, hard day. Before the geology traverse, Cernan had accidentally caught his hammer on the Rover's right rear fender and before he realized it, most of the fender was gone. As a result, the entire geology traverse was accompanied by a spray of dust that shot skyward and rained down on the two men. They were so filthy that they had spent fifteen minutes at the bottom of the ladder trying to dust each other off.

Out of their suits now, the men ate dinner; even though the meal was cold (because the LM had no hot water) it was welcome. It had turned out to be a tougher day than either man had expected. Their hands and forearms ached, their fingers were bruised. Schmitt saw blood under Cernan's nails; no doubt partly due to working with the drill. None of this had dampened their spirits out on the surface; excitement had pushed all that discomfort into the background. Schmitt's only worry was whether he would be able to accomplish what he had come for. Dick Gordon had once offered him a bit of wisdom regarding spaceflight: "Time is relentless." And time had reined in him and Cernan today. Their single geology stop, at a medium-sized, block-strewn crater, had lasted only half an hour, just long enough to chip off pieces from a couple of boulders, and to drag the rake sampler through the hard ground for a bagful of cobbles and dust. He hadn't had time to do much observing; mostly he had used stolen moments during other work, and during the traverses, relaying data from his seat on the Rover as they bounced across the dark plains. He had hoped to shed some light on the origin of the dark mantle, but that had eluded him. He still did not know whether it was made of volcanic particles, as the mappers had suspected. Out on the surface, he'd begun to suspect that the mappers had been wrong, that the dark soil might simply be ground-up material from the boulders in the area, which were fairly rich in dark minerals. But the more he looked, the less sure he was—the boulders just weren't dark enough, and if you ground them up, the dust would be even lighter. He hoped there were some experts on earth who were puzzling through these questions in their laboratories. He would have been happy to talk through dinner about it, but mission control had already relayed the questions from the back room and seemed to be letting him and Cernan alone.

The moon was throwing him some curves—of course, that was what his profession was all about. The surprises made it fun. But Schmitt realized that he was in the scientifically vulnerable position of coming all the way to the moon and not learning very much. If something happened and the mission had to be cut short, he and Cernan would leave with a few pieces of basalt—whatever the dark mantle was, those rocks proved there were lava flows underneath it—along with a few bags of soil, a deep core, and a bit of data from a portable geophysics experiment designed to probe the structure of the rock layers beneath the dusty plain. But that wasn't very much, not compared to the secrets the valley of Taurus-Littrow still held.

II: Apollo at the Limit

Nothing was ordinary at Caltech, not even a practical joke. Tradition dictates that one morning each spring the graduating seniors disappear, leaving their doors locked with high-tech security measures as a challenge to the ingenuity of the underclassmen. On Senior Ditch Day, as it's called, the object is to break into each fortified room, and once inside—if a student is clever enough to get in—he can do whatever he wants. Over the years this has resulted in some diabolical pranks. A senior returned to his room to find that his furniture had been nailed to the ceiling. Or there was no room at all; the door had been plastered over, painted, and a light fixture installed. Or he opened his door to find a 7-foot weather balloon, full of water. This was not a simple problem. He couldn't move it. He definitely did not want to break it. What did the victim do? He broke it; there was a big flood. Jack Schmitt had witnessed many a Senior Ditch Day prank, and so had several people in mission control. They decided to awaken their slumbering colleague on the moon with a musical reminder of another hallowed Caltech ritual. Every morning during final exams, at 7 A.M., the students in each dorm would tie their stereo systems together, aim the speakers into the courtyard, and play Wagner's "Ride of the Valkyries" full blast. Anyone within a mile radius levitated out of his bed. Inside *Challenger*, the effect wasn't quite as stirring when Schmitt heard the familiar cadence of horns in his earphones, but it did wake him up. And he was glad to be awake. On earth, Schmitt always needed at least seven hours of sleep (unlike Cernan, who seemed to be able to get by on far less, putting in a long day of training even if he had partied the night before; Schmitt assumed the ability to survive on little sleep was an inborn trait of navy pilots who survived tours of duty on aircraft carriers). Now, even though he had only gotten six hours of intermittent sleep, Schmitt felt rested; thankfully, the soreness in his forearms had disappeared overnight. He would later speculate that the cardiovascular system was so much more efficient in one-sixth g that it cleansed the muscles of lactic acid and other waste products before they could cause any damage. But what lifted Schmitt's spirits most was simply that he and Cernan were still on the moon, that nothing had gone wrong to end the mission, that now, they would finally get a chance to explore the valley of Taurus-Littrow in earnest. Schmitt was actually more excited today, as he climbed down *Challenger*'s ladder, than he had been yesterday, walking on the moon for the first time. After repairing the Rover's broken fender with some maps, gray tape and clamps—a fix devised overnight in one of mission control's back rooms—he and Cernan headed for the slopes of the 7,500-foot South Massif.

• • •

There are quite a few places on earth where you can stand in one spot and see more than a mile of vertical relief; the Himalayas, and some parts of the Andes are among them. These mountains testify to the grandest forces on earth, the slow and steady action of plate tectonics. Since 1967 geologists had come to realize that the earth's rigid outer layer is divided into a set of interlocking plates that move along on the fluid upper mantle at a speed of a few centimeters per year, about the rate that your fingernails grow. As these plates collide, move apart, and slide past one another, the continents migrate and the ocean floors are recycled. In the process, most of the earth's major landforms are created. The crust beneath the Pacific Ocean is diving under the west coast of South America, causing massive eruptions at the surface and creating a chain of lofty volcanoes. India is crashing into Asia, pushing up the world's tallest mountain range in the process. But the moon has never had continents or plate tectonics, and the geologists who pondered the origin of the Taurus Mountains looked to the titanic forces unleashed by the collision of the Serenitatis asteroid with the moon. But the details of the cataclysm were still entirely beyond their grasp. Had the Massifs simply been thrust upward as unbroken pieces of crust, something like the great limestone blocks of Himalayas? That had been one working hypothesis for the Apennines, and when Scott and Irwin saw what locked like hundreds of thin layers in the side of Mount Hadley, it had seemed it might be the right one. But since then there had been some doubt about whether those layers were real or simply an optical illusion caused at certain angles of illumination. Another hypothesis was that the Apennines and the Massifs were simply piles of ejecta from their respective impact basins. However they had formed, their majesty assured that Taurus-Littrow would be remembered as one of the most spectacular places ever visited by human beings. Because the Serenitatis basin had formed before the Imbrium impact, samples from the Taurus Mountains—beginning with the South Massif today, and continuing with the North Massif tomorrow—were the highest priority of the mission. For Cernan and Schmitt, getting to the South Massif meant driving more than 5 miles, the longest traverse ever made on the moon. An hour was budgeted for the journey; never had so much of a moonwalk been allocated to simply getting from one place to another. Seen from the lander, the South Massif was a seemingly bland face of white-gray stone, but yesterday Schmitt had spotted outcrops of different rock types. No one doubted the trip would be worth it.

Tuesday, December 12
6:24 P.M., Houston time
5 days, 21 hours, 31 minutes Mission Elapsed Time

The ride to the South Massif would stand out in Gene Cernan's mind as one of the most exciting times of the mission. Not because of the geology; he didn't have much time to look at that because he was too busy trying to keep from driving into craters. Anxious to save time, he went as fast as he dared, speeding over blind hill crests without slowing down, ready to swerve if a pothole or rock came into view. For now, Cernan left the geological descriptions to Schmitt. It was remarkable, Cernan thought, how much his lunar module pilot could see from the Rover as it waltzed along, and how much he brought to bear on his observations. Schmitt was so familiar with samples from previous missions that he actually knew them by number.

7:37 P.M.

Seventy-three minutes after leaving the vicinity of the LM, having driven over the suspected landslide and onto a broad scarp, Cernan and Schmitt topped a last rise and descended into a broad trough at the base of the South Massif. Here, a crescent-shaped depression had been christened by Schmitt for the Norwegian Arctic explorer, and later statesman, Fridtjof Nansen. Now it would serve as the scene for an hour-long geologic assault. Had Schmitt and Cernan been able to look back onto the plains, they would have had a difficult time seeing their lunar module. But here in this broad, smooth hollow, *Challenger* and the rest of the valley were hidden from view; this was a place apart from the valley. Perched on this bright slope, Cernan and Schmitt stood at the end of Apollo's longest reach, for Nansen was about as far as two Rover-riding astronauts could travel and still have enough time for a meaningful exploration. It would have surprised no one to learn that Jack Schmitt had always felt the planners were too conservative when they calculated walkback limits, and that before the flight he had pushed, unsuccessfully, to have them relaxed. In any case, the thought of having to walk back to the LM was the furthest thing from his mind. If anything, he felt less at risk here than he had in 1957 and 1958, when he spent eighteen months in the fjords of western Norway working on his doctoral research. Working alone in difficult terrain is something most geologist consider downright foolhardly, but they do it anyway, especially when they are short on time, or have no money to pay a field assistant. In Norway, operating on a shoestring budget, Schmitt climbed alone, often making strenuous treks among sheer rock walls. He worked long into the Arctic summer evening, taking advantage of the midnight sun, because he wanted to get his money's worth out of the effort it took to get there. But Schmitt knew his own limits, and he saved the really difficult climbs until one of his colleagues from the States came to visit; the unsuspecting visitor found himself dragged off on a grueling ascent into the hill country. On the moon, Schmitt knew, he was far from alone, not just because Cernan was here, but because mission control was watching them like a hawk. If either man so much as adjusted the amount of cooling water from his backpack, Houston knew about it. And in his earphones, they heard the steady, even baritone of one Robert Allan Ridley Parker, astronomer-astronaut, Caltech Ph.D., and mission scientist for Apollo 17. Parker and Schmitt were more than a little alike. When he wanted to, the quiet, dark-haired astronomer could serve up the same blunt, needling verbiage Schmitt was famous for. The two scientists maintained a kind of disrespectful rapport, based on mutual insults, that is the coin of the realm at Caltech. The underlying message was, "God, you're incompetent; they really sank the standards when they let you out with a degree." Over the air, Parker maintained a virtual deadpan, but that didn't stop Schmitt from responding as usual. When Parker suggested he take some telephoto pictures of the Massifs "if they look interesting," Schmitt set him straight: "If they look interesting? If they look *interesting?!* Now, what kind of a thing is that to say?" Schmitt knew his friend was in no position to respond in kind, not with the world listening, and that, of course, gave Schmitt's barbs added bite. But in truth, Parker had not been silenced, because he'd inserted little personal touches into Cernan and Schmitt's cuff checklists. For example, at the conclusion of each section there was a message that read, "This is the end, not the beginning."

Nansen crater reminded Schmitt of an alpine valley above the timberline; the craters were smoothed by a blanket of dust that sparkled and threw back the sunlight like new snow. Resting on the surface were hundreds of blocks and fragments which the geologists suspected the landslide had brought down from the heights. He and Cernan had known they would find boulders here, because they had seen them on the orbital photographs. But what was their geologic context? Schmitt cast his gaze on the Massif's whitish face. On the high slope, perhaps a full mile above him, Schmitt could see a few thin, horizontal bands of color—bluish gray and tan-gray, the same colors as the rocks at his feet. There was more here than he and Cernan could possibly investigate in the fifty-odd minutes allotted; they would have to work fast. It was time, rather than the uniqueness of the lunar landscape, that made this fieldwork unlike any Schmitt had ever done. In Norway, he knew that if he had to, he'd scrape together the money to come back for the next field season. Here, the time pressure was monstrous; the drive for total efficiency ruled. Everything he did was for keeps.

8:07 P.M.

At last the mountains of the moon had yielded a rich harvest. Cernan and Schmitt had none of the frustrations that had plagued Scott and Irwin as they searched for boulders on Hadley Delta, or Young and Duke's hunt for the Descartes at Stone Mountain. The time allotted for Nansen was dwindling, raising the inevitable dilemma of whether to exploit a known harvest at the expense of a later, unknown one. That decision was left to the two men on the moon. Cernan decided to accept the back room's offer to take 10 minutes out of a later stop. When Cernan and Schmitt finally headed back to the Rover they had spent 63 minutes at Nansen—nothing compared to a terrestrial field stop, but one of the longest ever made on the moon. For their labors, the men had made an impressive haul; they would leave the South Massif with an entire saddle-bag full of rocks and soil, including a small white fragment, spotted by Schmitt in the middle of a gray boulder, that would prove to be nearly 4.5 billion years old, one of the oldest rocks ever brought back from the moon.

As the Rover pitched and rolled its way back down the mountain—a trip that would be just as memorable as the ride out—Schmitt told Parker he felt confident that the blue-gray and tan-gray rocks he and Cernan had collected could, by extrapolation, be linked to the two layers he'd spotted high on the Massif. Parker responded, "I'm reminded that extrapolation is the nature of our art." Hearing this admission of imprecision from a scientist, Gene Cernan uttered a slow, forced laugh.

9:20 P.M.
Science Operations Room
Manned Spacecraft Center

In the geology back room, Lee Silver studied a television monitor. The pictures from the moon had gotten progressively better with each mission, and the camera's controller, Ed Fendell, had honed his art to the point where he—and anyone else who was watching—could follow Cernan and Schmitt along in their work. Silver could not deny that for himself, Schmitt's presence on the moon gave Apollo 17 a special intensity. He had known Jack Schmitt's father and had met the boy when he was eleven years old. Almost a decade later, he'd lectured to Schmitt at Caltech. Now Schmitt, no longer Silver's student but his colleague, was at work on the surface of another planet.

With two J-missions behind him, Silver was, in his own words, the "old man" of this operation; for Apollo 17 he became the first and only scientist to serve not only on Bill Muehlberger's surface geology team but on the overnight planning team as well. He had not slept since Cernan and Schmitt landed, nor would he until they had left Taurus-Littrow. But Silver had been under pressure long before now. In the three years since he'd become involved with training Apollo astronauts and planning their explorations, the pace of events had been dizzying. The geology team had been designing traverses for Apollo 17 before they had even been able to analyze the rocks from 15 and 16. Silver had seen this phenomenon before; this was classic government science. A federal agency undertakes a massive scientific effort, but is forced by budgetary constraints to press on to new discoveries before the previous ones have been understood. Tied to NASA's mission schedule, the geology team did everything they could to keep up, but Silver, at least, hadn't been completely successful. Some of his Caltech colleagues had developed techniques so sensitive for probing lunar rocks that they could detect individual atoms, but Silver had not had the time or the resources to take advantage of them. And for all the scientists, the slow, painstaking analysis of the samples had lagged far behind the pace of lunar exploration. Some of those samples, Silver knew, contained information that would have affected the planning for each new mission, had there been time to learn of it. For example, no one had really analyzed the peculiar green rocks from Spur crater, which had been on earth for sixteen months. And when this mission was over the great enterprise called Apollo was shutting down, with so much of the moon left unexplored—and that was also classic government science.

If the exploration of the moon had put a professional strain on him and his colleagues, Silver understood that it had also exacted a high cost in personal terms, in strained marriages and disrupted families, for people throughout the program. Silver himself had been in a period of great turmoil. His responsibilities at Caltech were crushing; his personal life was rocked by crisis. One night at around 3 A.M. in front of the King's Inn on NASA Road One, Silver told Schmitt he couldn't keep this up, that he was pulling out of the training program. Schmitt became very upset, and Silver suddenly realized how much the man was depending on him. He decided to stay on, though he was never able to have the same involvement he'd had on Apollo 15.

That was the only time in the past sixteen months that Silver glimpsed a crack in Schmitt's stubborn armor. Schmitt himself would never admit he felt pressure, but there had been some subtle indications. Silver had seen Schmitt's demeanor change when he made it onto the 17 crew. There was less joking around than when he had been a backup crewman; this was serious business. And Silver knew about the pressure from the scientific community to get him on the flight; he also knew that Schmitt had won his seat at the expense of a very popular pilot-astronaut. Silver could not guess at how the other astronauts viewed the decision, but he did sense the scrutiny of the scientific community, who were watching Schmitt and waiting to see whether he would live up to his billing. Some of them, Silver was sure, were probably half-hoping he would screw up.

As far as Silver was concerned, Schmitt was doing a beautiful job. And Cernan was holding his own. Early on, some of the geologists had privately worried that Cernan, as a mission commander who enjoyed the limelight, might be reluctant to yield center stage to his lunar module pilot. They had underestimated Cernan's flexibility. Cernan was so accepting of the situation that Silver had felt it necessary to take him aside and encourage him not to let Schmitt be "his geologist," in effect, but to be more of a partner—which, having seen Cernan's powers of observation, Silver knew he could be. Watching the pair at work on the moon, Silver was impressed by how Cernan complemented Schmitt's focus on detail with a view of the big picture. But if, at times, Cernan almost seemed to be functioning as a field assistant—"Get your hammer," Schmitt told him at one point, "we're going to need it"—that was another measure of this commander's flexibility. The fact was that all through the moonwalks, Cernan had been making first-rate observations, calling Schmitt's attention to intricate fracture patterns in the rocks, changes in texture, subtle hints of bedding. They had become a well-honed field geology team.

Just now, Cernan and Schmitt were back on the valley floor, near a 1,500-foot-diameter crater named for *Doctor Zhivago*'s Lara. And Schmitt was doing what he did at every stop—looking around, sizing up what he saw, and relaying the information to the back room so that they could plan their goals accordingly.

"Bob, I'm at the east-southeast rim of a thirty-meter crater, in the light mantle, up on the scarp and maybe two hundred meters from the rim of Lara. . . . There's only about a half-centimeter of gray cover over very white material that forms the rim. . . ."

This was a new way of doing business, made possible by the fact that for the first time, there was a trained scientist on the moon. But it wasn't long before Schmitt began to lose the upper hand in his a *mano a mano* with the moon. There was too little time for detailed, systematic samples, but while Cernan hammered in a 32-inch-long core tube, Schmitt made his own solo investigations. He made a valiant effort to collect rocks by himself, holding his long-handled scoop in one hand and a Teflon baggie in the other. Every time he tried to set the scoop down, it fell over. Then he knocked over a big collection bag, scattering samples on the ground. Schmitt went down on hands and knees, rounded up the samples, and put them back in the bag. Then he leaned back on his heels to get his massive backpack over his feet, and launched himself back to a standing position—and dropped the collection bag in the process. Schmitt got down on one knee to retrieve the collection bag, but he stumbled and fell on his chest. At that point Bob Parker started calling Schmitt "Twinkletoes." Later, he radioed that the switchboard at the Manned Spacecraft Center was lighting up with calls from the Houston Ballet Foundation requesting Schmitt's services for the next season. Seizing the opportunity to audition, Schmitt did a graceful leap—and plummeted to the dust, laughing.

But that was a temporary setback. Minutes later, as Cernan and Schmitt headed for their next stop at a crater called Shorty, Silver and the rest of the back room were reminded how good their surrogates at Taurus-Littrow really were. At several points along the traverse, Cernan had stopped driving just long enough to allow Schmitt, still seated, to reach down with a special scoop, which had small bags attached, and collect a sample without getting off the Rover. One of these "Rover samples" had been planned while the men drove across the outer portion of the suspected landslide, which the geologists called by the noncommittal name "light mantle." But at the last minute, Muehlberger's team had sent word to cancel it in favor of more time at Shorty. Schmitt and Cernan decided to stop anyway—and it was a fortunate decision. That scoop of light-colored soil would realize one of the geologists' most cherished hopes for Apollo 17. All along, the geologists had suspected that the landslide—this sample helped prove that theory—had been triggered sometime in the distant past, by flying debris—ejecta from a distant impact—striking the South Massif. And the impact, they believed, was probably Tycho, 1,300 miles to the southwest. Analyzing the landslide material, the scientists would determine it had been lying on the valley floor for 109 million years, affording a compelling, if not airtight, time for Tycho's formation. In a stroke of incredible good luck, the trip to Taurus-Littrow appeared to have given the geologists one of the most important dates in lunar history. By exercising their own judgment as trained lunar field geologists, Cernan and Schmitt had beautifully shown their value in the exploration. And this moonwalk wasn't over.

10:18 P.M.

"Okay, Houston. Shorty is clearly a darker-rimmed crater. The inner wall is quite blocky. . . . And the impression I have of the mounds in the bottom is that they look like slump masses that may have come off the side." Jack Schmitt, about to explore the 360-foot crater, made his report to the back room. On the orbital photos, Shorty was ringed by a dark halo that stood out against the light mantle. The geologists suspected it was a volcanic explosion crater, perhaps one of the sources of the dark-mantle deposits. They knew it might well turn out to be just another impact crater, and that the halo was simply an ejecta blanket of dark soil. But the possibility that Shorty might at last reveal evidence of recent volcanism was simply too exciting to ignore.

As entertaining as Jack Schmitt's comedy of errors at Lara may have been for those watching on earth, it had greatly frustrated him. In part, he'd gotten into trouble because he'd had to work alone; he'd practiced a little solo sampling before the flight, but not nearly enough to become efficient. And also, he was getting tired. His hands and arms ached too much for him to keep his grip on things. On the drive to Shorty, he had told himself he would forgo any more attempts at solo work for a while. Now that he was here, his spirits were lifted by the sight of Shorty itself. The crater was big and deep, its walls littered with rocks, its floor a maze of rubble. The lighting was dramatic. But Schmitt knew he and Cernan would have to work fast; having spent an extra 10 minutes back at Nansen, they would have only half an hour here.

A large, intensely fractured boulder near the crater rim caught Schmitt's eye. Once Cernan was finished with his Rover chores, they would sample it; for now, Schmitt would take a panorama from the rim. He moved closer to the boulder and stopped, preparing to take his pictures. Instinctively, he scanned the area around his feet for details, and suddenly he saw that his boots had scuffed away the dust. Instead of the usual monotonous gray, he saw vivid, bright orange. For a moment he wondered whether it was an illusion. At an earlier stop, he'd seen a patch of color on the surface that turned out to be a reflection off some gold foil on the Rover. He lifted his outer visor partway and looked again. It was *real.* "There is *orange soil!*"

"Don't move until I see it," said Cernan, still working at the Rover. Privately, he wondered whether Schmitt had been out in the sun a little too long—until he turned around. "*Hey, it is!! I can see it from here!!*" Cernan's voice had risen a full octave.

"It's *orange,*" Schmitt said, as if he couldn't believe what he had found. It glowed like a highway sign in the glare of headlights.

Cernan wondered aloud, "How could there be orange soil on the moon?" Then he answered his own question: "It's been *oxidized.*" Yes, Schmitt agreed, it looked like the rusted soil you sometimes see in the desert. But if Cernan was right, what had done it? How, on a world devoid of water and air, could there be rust? The answer was volcanic gases. You can walk around the slopes of the Kilauea volcano in Hawaii, as Cernan and Schmitt had done in training, and see chunks of dark lava that have been turned bright orange, red, or yellow, by hot mineral-laden gases escaping from the depths. At that moment, in the back room, Lee Silver was practically jumping up and down with glee, saying, "That's it! That's the volcanic vent!" And Bill Muehlberger was saying, "Somebody unplug Silver."

Unaware of this ecstatic outburst, Schmitt immediately went to work. First he dug a trench into the orange, to find out the extent of the deposit, and he saw that the orange lay along a ellipse-shaped zone that ran parallel to the crater rim. Perfect; the same thing happens in Hawaii. The fractures that allow the gases to percolate upward often follow the perimeter of the volcanic craters, and as a result, so does the pattern of altered soil. Furthermore, the ellipse was divided into zones of color: The orange merged with yellow toward the edges, while in the center the soil was almost crimson. Such *zoning*, caused by different concentrations of gas, was another hallmark of volcanic action. "Man, if there ever was a—" Schmitt laughed. "I'm not going to say it." Yes, he was, even if he turned out to be wrong. "If I ever saw a classic alteration halo around a volcanic crater, this is it." When he arrived at Shorty, Schmitt had been all but certain that it was an impact crater; now there was a very real possibility that Shorty was volcanic. But unless he and Cernan could get an extension, they wouldn't have time to prove it. At the same moment, Lee Silver was telling Jim Lovell, "We need more time." But there wasn't any. The ever-tightening circle of oxygen consumption and walkback limit was closing in on Cernan and Schmitt. Unlike Nansen, the option for an extension at Shorty simply did not exist.

Cernan, finished with his Rover chores, at last joined Schmitt to collect samples. Schmitt dug his scoop into the orange and found it did not behave like dust; it broke apart in angular fragments. He dropped most of the first scoop before he could pour it into Cernan's sample bag. He told himself to slow down. A second try filled the bag with orange clumps. Then he captured the gray soil at the margins, for comparison. The next step—and perhaps the most important—was to find out how deep the deposit went, and what might be underneath it. The back room was already thinking about that; Parker passed up a request for a core tube. And at the same time, he informed the men that they had 20 minutes left.

Cernan drew his hammer. Schmitt watched as he pounded the core as hard as he could; the tube penetrated less than half an inch with each whack. "I'd offer to hit it," Schmitt said, "but I don't think I can, my hands are so tired." Cernan persisted and drove the core most of the way in. Thankfully the core came up without a struggle. "Even the core tube is red!" But not all of it. Schmitt was startled to see that the bottom of the tube was coated with dark purple-gray, almost black soil. For a moment, before capping it, the two men cradled the 3-foot core, holding in their hands one of Apollo's most exciting discoveries, as the scientists would realize when it and the other samples were on earth. Their impact would not be diminished one bit when it became clear that Schmitt had been wrong when he thought Shorty was a volcanic vent, that in fact it was a run-of-the-mill impact crater. The orange soil was anything but ordinary: It was made of tiny beads of glass that had once been molten droplets, poor in silica but rich in titanium and iron, propelled high into the lunar sky by a spectacular form of eruption called a fire fountain.

The origin of the fountain was a form of lava that contained dissolved volcanic gases. As it ascended from deep within the moon to the surface, the effect was that of shaking up a bottle of soda and then uncapping it: The gas rapidly came out of solution, propelling molten rock high into the lunar sky. Just as water pressure in a decorative fountain causes the liquid to break up into droplets, this so-called fire fountain was composed of an intensely hot spray. In the weak gravity, the droplets arced hundreds or perhaps thousands of feet through the vacuum. During their flight, they cooled into tiny glass spheres, which rained down on the valley of Taurus-Littrow. Different spheres cooled more slowly than others, affecting their final mineral content. Some of the beads were orange, some were crimson, not from oxidation, but due to a high titanium content. Others—many of them—were richer in iron, partially *devitrified* (that is, crystalline), and colored black. Those beads, which had startled Schmitt when he saw them at the bottom of the core tube, were the answer to the riddle of the dark mantle. The soils of the valley of Taurus-Littrow—and, by implication, the southeastern portion of Mare Serenitatis—were dark because they contained beads of dark, volcanic glass. But the fire fountains had not sprayed forth recently, as the geologists had hoped; the glass beads were 3.5 billion years old. Untold millions of years later, probably, the deposit had been covered over by a fresh lava flow, which protected the strange beads from being mixed with ordinary lunar soil by micrometeorite rain. They remained hidden for eons, until 19 million years ago, when the Shorty impact excavated them and brought them to the surface, where they awaited the footsteps of a geologist from earth. And the orange soil discovery would lead the geologists to analyze the beads of green glass from Apollo 15, which would also turn out to have formed by fire fountains. In both cases, the lava had come from hundreds of miles down in the lunar mantle, far deeper than any other Apollo samples. And on the surfaces of the beads, analysts found coatings of volatile elements—the first such elements found in any lunar samples—which were solid remnants of the original volcanic gases that had powered the eruptions. Knowing that these gases had existed within the moon would cause scientists to revise their models for its evolution.

The fact that Schmitt had been wrong when he suspected Shorty was a volcanic crater did not lessen the impact of his work. Despite the fact that Schmitt felt he had barely enough time to think, he and Cernan had performed so well that Muehlberger's planning team, after deliberating all night on whether to send the men back to Shorty the next day, decided that would not be necessary. The men had done everything needed to decipher Shorty's story. Schmitt even managed, as the remaining minutes ticked away, to gather pieces of the basalt boulder that had caught his eye to begin with. Meanwhile, Cernan snapped a spectacular series of panoramic pictures, and to his surprise, he spotted streaks of orange and dark gray on the crater walls. He was still trying to tell the back room what he saw, talking so fast that he was almost out of breath, when Parker cut him off: "You can talk about it when you get home."

5:12 A.M.

Cernan and Schmitt were bedding down for a well deserved rest when Joe Allen, taking the night shift in mission control, told them that the flight controllers were still marveling at the beautiful television pictures coming down from Taurus-Littrow. They were fascinated by the footprints and the Rover tracks, and they were speculating on what might someday disturb them, after Cernan and Schmitt had departed.

"That's an interesting thought, Joe," Cernan replied, "but I think we all know that somewhere, someday, someone will be here to disturb those tracks."

"No doubt about it, Geno," said Allen. They all knew that the young physicist in mission control would never have the chance himself. But Schmitt felt the need to stress that if it was too late for Joe Allen, there was still reason to be hopeful. "Don't be too pessimistic, Joe. I think it will happen."

"Oh, there's no doubt about that," Allen agreed. "But it's fun to think about what sort of device will ultimately disturb your tracks."

"Well," Schmitt said, "that device may look something like your little boy."

Allen laughed, thinking of his four-year-old son, David: "He'd make short work of them."

Cernan was in his hammock now, and Schmitt was about to get into his. They would have their eight hours rest, and then would come the final moonwalk. Schmitt took a last look at the earth through the overhead window, and rolled the shade closed. He told Allen with gentle irony, "Tomorrow we answer all the unanswered questions. Right?"

III: Witnesses to the Earthrise

Wednesday, December 13
El Lago, Texas

For days now, Jan Evans had ducked outside in hopes of seeing the moon, but instead she saw an impenetrable gray drizzle. Finally, on Wednesday the thirteenth, a cold front moved in and the sky cleared, and there it was, ripening toward full. Finally, she could see where Ron was. His moon wasn't the one she remembered from her Topeka, Kansas, childhood, on those winter nights during World War II when the whole town would turn off its lights during air raid drills. In her living room, her mother had draped a towel over the Philco radio with its big yellow dial. Outside, she could see the air raid warden patrolling the street. The trees were coated with ice, sparkling in the light of a full moon. That would always be her moon, not the one Ron was circling now. She would have loved to be a mouse in the corner, up there with him, if the thought hadn't scared her half to death. And Jan was quite content to listen to the squawk box that brought Ron to her. And she heard a lot of him, because Ron had decided to make a special effort to share his experience with the world as it was happening. Whether the world was listening was an open question; here at home the networks were giving Apollo 17 only spotty coverage at best. In New York, the sixth lunar landing made for only a brief interruption to the soap opera "Love Is a Many Splendored Thing," and the Johnny Mann Singers drew more viewers than live images of two explorers on the moon. But that was far from Jan's consciousness. Here in El Lago, with a house full of people—and a lawn full of reporters—she was having a ball following her husband's adventures. In between streams of technical jargon, she heard his excited chatter, his humming and whistling. She knew he was very busy, that the flight plan had him running from the moment he awoke until bedtime. On the way out to the moon, Ron had been on watch and had overslept for a solid hour, while mission control tried again and again to wake him up. But that was nothing new. Jan told everyone the story of how in the navy he'd been such a sound sleeper that his bunkmates on the carrier would leave a Baby Ben in a metal wash basin right next to his ear. But Ron wasn't oversleeping now, not with all that work to do. At least he was eating well, and that was a good thing, because if there was a chowhound on this crew, it was her husband. Before Gene and Jack had departed, there had been a small crisis when Ron lost his scissors somewhere in the command module, leaving him without any means of opening food bags for three days until they got back. They saved him by giving one of their pairs to him. But Jan had a laugh when she learned that on one of his solo days the flight plan hadn't left him any time for lunch.

After the Original 19 were selected, Ron had gone to Houston to look for a house, and she could still remember the day he called up and said, "You have to get an air conditioner for the car." She was afraid all this astronaut business had gone to his head—Air-conditioning for the car? On his salary?—but the next time he called, he was even more insistent: "I'm not kidding, you've got to get air-conditioning!" And when she and the children had trekked from San Diego, and across most of Texas, and arrived in tropical heat, she understood what he meant—air-conditioning was necessary to support human life. Ron was the last person who would let the title of astronaut go to his head.

The astronaut life had not been tough on her, the way it had on other wives. From the beginning, she considered it a luxury to have him home on weekends instead of gone for eight or nine months at a time, the way he had been when he was serving in Vietnam. And the risk of space flight were something she rarely thought about. She had seen the way NASA operated, and she trusted the system; she had come to feel that Ron had been in far greater danger flying missions over the jungle than he was now, a quarter-million miles from home. She knew Ron had always felt that being an astronaut he had gotten the best deal out of life; she also knew he had friends who had not been so lucky and were at this moment lying in North Vietnamese prisons. And she knew he had always felt a little guilty about that. But he had wanted to fly in space like nothing else she had ever known him to want. It had kept him going for six years, through the unglamorous support-crew duties of writing checklists, going to meetings, and setting switches before spacecraft tests; through the night shifts in mission control at the Capcom's mike. Now that he was up there, and his voice was coming to her from across that amazing distance, he sounded like the same old Ron she had met in high school—except that he was having the time of his life.

Jan didn't know the flight plan very well, but she knew the critical phases, things that were essential to carrying out the mission. She followed along when she could, and she got up when Ron got up, and went to bed when he did. She was grateful to Deke Slayton, who called every evening to review the day's events with her. "They made a great burn today," he would say. Or, "Things are going really well." Looking back later, she would say that Slayton was nothing less than the glue that held the astronaut families together.

Aboard the command module *America*, in lunar orbit

Most of the time, it didn't even feel like flying, except for once, when he lit up the SPS engine to change the orientation of his orbit, and he felt the sensation of actually flying a space ship; it felt like going full afterburner in a T-38. The rest of the time it was a lovely quiet. His ship was performing flawlessly, and he was proud of it—and what it stood for. He and his crewmates had named their command module *America*, and to Evans it was the perfect choice. The world should know it was the United States up here, that there was something else his country was giving to history besides a war, something uplifting. During training, Cernan had started calling him "Captain America." And if Evans was an unabashed patriot, then he wasn't afraid say, on the record, that he thought the Nixon administration had paid lip service to the space program without giving any real support, or that the politicians who opposed the program were being swayed by people who thought the money would be better spent on welfare—and he didn't hesitate to call those people kooks. Maybe that kind of outspokenness, almost unheard of among astronauts, came from having risked his neck for his country flying missions over Vietnam. But back then, he'd been just another unknown aviator.

And if he let his excitement hang out over the airwaves, that was just the way he liked to do things. Why not put aside the image of the cool space traveler? He knew people saw astronauts as reticent; he was changing that. When he wasn't enthusing to Houston, the onboard tape recorder caught him talking up a storm, to himself. "What happened to my grits? I lost my grits. There they are again. Anything you drop up here, it just disappears, flat disappears." He was trying to grow a beard for his son, Jon, but it itched so much he was tempted to shave it off. "Just be comfortable," Ken Mattingly in Houston said. "You got another week to go." It was a long mission. Even now, more than 6 days into it, he was only halfway through. And he was beginning to dislike his own company. "Man, I stink! Whew! Soap doesn't do any good. I'll be glad when the guys get back with the deodorant. Took all the deodorant with them to the surface. Outstanding stuff."

Evans made no secret of the fact that he would have liked to walk on the moon—but there was no doubt he meant it when he said that getting within 9 miles was good enough. He was doing his part, as Farouk El-Baz's last protégé in lunar orbit. And he thought about his crewmates and asked about their progress on the surface. Evans had known Cernan since they were both at the Naval Postgraduate School in Monterey, California. They'd applied for the astronaut program together, and when Cernan made it, they went out and got drunk together. It had been hard for them to lose Joe Engle, but when Schmitt came on the crew, they made room for him. Evans liked Schmitt well enough, and he gave him a lot of credit for hard work and persistence. "Dr. Rock," he called him. And Evans knew that you can take astronauts with the most diverse personalities and put them in the same spacecraft, and they will function as a team because of one thing: The mission comes first.

1:37 P.M., Houston time
6 days, 16 hours, 44 minutes Mission Elapsed Time

The phases of the moon and earth are complementary. When the moon is new, a full earth shines in the lunar night. As we see our satellite wax through crescent and first quarter to full, our own world wanes in the moon's sky until it is lost in blackness. On Wednesday, December 13, Gene Cernan pulled back the shade on *Challenger*'s rendezvous window and saw a blue-and-white crescent gleaming above the South Massif. He could barely make out a cloud system over the southeastern United States. In mission control, Capcom Gordon Fullerton confirmed that yes, that's what the satellite picture showed. How amazing, Cernan thought, to be standing on the moon, giving his home planet a weather report.

After a quick breakfast, he and Schmitt—by this time, a little weary—got ready for Apollo's final moonwalk. They were an hour late getting started, but there wasn't any particular need to rush.

At the very first mission review meeting Jack Schmitt had lobbied for a fourth moonwalk. He'd worked on the initial design studies for the J-missions, and he knew that four excursions had been part of those original plans, and now he had no qualms about telling the engineers what their hardware was capable of. On paper, he was right; you could squeeze in a fourth moonwalk, but it left you with very tight reserves of water and battery power in case of an emergency (like a blown water tank). The suits and backpacks hadn't been tested for the extra trip outside, though everyone suspected they would hold up. The planners spent the better part of a day with Schmitt, trying to talk him out of the idea, but to no avail; he knew he was right. Cernan just hung back and let Schmitt say his piece. Finally, Apollo spacecraft program manager Owen Morris asked Cernan for his own recommendation. "I think we can do it," Cernan said, "but I'll respect your judgment." Schmitt shot Cernan a dirty glance; that was the end of it.

And if Schmitt felt some lingering frustration at using Apollo to what he saw as less than its full capability, he no longer feared having to leave the moon without enough information. This day they would have a good seven hours on the surface, enough time to sample the North Massif, and the Sculptured Hills, and a couple of craters on the valley floor. And then, no more. So there they were, in their hard suits, moving like tin soldiers as they prepared to open the hatch. When Cernan emerged from the tiny cabin at around 4:30 P.M., he saw a small sign taped to the landing leg. He did not know who had put it there, but it must have been one of the checkout crew at the Cape. On each moonwalk, he'd paused to read it aloud as a kind of public prayer. Now he read it again: "Godspeed the crew of Apollo 17."

5:37 P.M.

The Rover headed north across a sun-drenched moonscape, past the giant crater Sherlock, and when it reached Turning Point Rock the men headed east toward their destination, a big, dark boulder perched on the side of the North Massif. They had all known the boulder would be there; it was visible on the Apollo 15 photographs, where it was a tiny dot trailed by a 1,500-foot-long furrow that recorded its plunge down the mountainside. As they got closer Cernan and Schmitt could see that it was actually five separate boulders; it had broken apart before coming to rest near the bottom of the Massif.

As always, the Rover gave them no hint of the severe slope. Cernan, on the uphill side, could barely raise his suited body out of his seat. Schmitt, meanwhile, thought for a moment that he might tumble down the hill. While Cernan struggled through his chores at the Rover, Schmitt climbed a few yards upslope and raised his outer visor, which was now badly scratched by dust, to peer at the wall of stone. It was just like the tan-gray rocks he'd seen yesterday at the South Massif, except that it had giant vesicles—holes where gas bubbles had once been—which meant that it had once been molten. But the holes weren't round, as you would see in a lava flow that cooled in place, but flattened ellipses, meaning that somehow, the rock had flowed before it had had a chance to solidify. Sometime later a great pressure had acted on the rock, for Schmitt could see fractures branching through it. It was difficult for him to tell much more. On earth, geologists must peer at their work through a variety of obscuring materials such as moss or lichen, or, in the desert, a smooth "varnish" of oxides created by the action of wind and rain. On the moon, rocks develop a brownish patina of impact glass from millions of years of micrometeorite rain. To get a good look at this rock, Schmitt knew, Cernan would have to put the hammer to it.

In the back room, Bill Muehlberger's team watched as Schmitt methodically examined the boulder. Every so often he would change his mind about what he saw, verbalizing the thought process that goes on in the mind of every geologist in the field. They knew that Schmitt had attacked complex problems like this one in Norway, and in Sudbury, Ontario, where the remains of a giant impact crater are preserved. Years later, Schmitt would say that his edge on the moon was that he was able to react to the millions of bits of visual data and sort out the most significant few, and do it within seconds. The mind, he would say, is not in a space suit.

Now the television camera caught him taking pictures, standing in one of the big hollows of the boulder track. With his golden outer visor raised, his face was clearly visible within the clear inner helmet, framed by his communications hat and darkened by a week-old beard. As he spoke to Cernan, something about a piece of gear, his mouth could be seen forming the words. Suddenly he was no longer a faceless, mirror-helmeted astronaut, but Harrison H. Schmitt, a geologist at the ultimate field site. He noticed the camera pointed at him, looked straight at it, and mugged a smile. Then, having spotted something new to investigate, he turned to his left and headed out of view.

• • •

Working together now, Cernan and Schmitt attacked the boulder, Cernan wielding hammer. It was ironic; Schmitt had helped design that hammer, but he couldn't hold it; the handle was too big for him to grip. The men had agreed in training that Cernan, with his massive hands, would be the mission's hammer bearer. The only thing Schmitt had to do was teach Cernan to hit the edge of a rock, which was much easier to chip, instead of pounding away, brute-force style, at the middle.

Schmitt made his way out of the shadow and around to the boulder's sunlit face, using his scoop like a walking stick. The more he looked, the more details he noticed. Within the tan-gray boulder were huge pieces of blue-gray, caught up like pebbles in the tar of a new road. They looked just like the blue-gray rocks they'd picked up at the South Massif. And within the blue-gray there were smaller, white chunks that looked like anorthosite, perhaps pieces of primordial crust. Cernan thought the entire boulder was one enormous breccia, but Schmitt wasn't so sure; the tan-gray rock had definitely been molten, and Schmitt thought it had formed when a pocket of magma caught up pieces of the blue-gray breccia within it. Whatever it was, this boulder had a story to tell, and if they were going to decipher it they would need as much time they could get. Parker informed them that they could have the lavish allotment of 1 hour and 20 minutes, but Schmitt was reluctant to take it all; it meant giving up another stop on the Massif later. After 36 minutes confronting the boulder Schmitt told Parker, "I've done the best I can."

But that would prove to be more than good enough. In the

months and years to come, analyses of the split boulder would testify to the violence of the impacts that formed the lunar basins. It turned out Cernan and Schmitt had both been right, in a sense; the boulder was one big breccia, but the tan-gray part had indeed been molten. It had melted because it had been subjected to forces almost beyond human comprehension. The split boulder was a kind of tableau, a geologic freeze-frame of the events that immediately followed the impact of the Serenitatis asteroid 3.8 billion years ago. Hurtling into the moon at perhaps 10 miles per second, the rocky intruder exploded with the power of billions of H-bombs. For perhaps a hundred miles around, huge volumes of the crust were pulverized or melted and flung outward in a spray of debris. Even as ejecta flew through space, shock waves contorted the lunar crust, pushing up rings of mountains like ripples on the surface of a pond; one of the rings became the Taurus Mountains. A vast sheet of molten rock sprayed out from the blast point, blanketing the newly formed mountains, sweeping up preexisting boulders in its broiling flow. Some of the boulders were themselves breccias that had recorded earlier cataclysms. Now more debris fleeing the growing basin piled on top of the newly formed Taurus Mountains. Within minutes, it was all over. The face of the moon was scarred by a crater 450 miles across. Over time the melt sheet cooled and solidified into a layer that was visible high on the Massifs.

A hundred million years later, *mare* lavas migrated upward through the fractured crust and erupted at the surface. In the Taurus-Littrow valley the last of these volcanic outpourings were accompanied by the fire fountains that created the dark mantle and the orange soil. Much, much later—“only” 800,000 years ago—a house-sized chunk from the North Massif was dislodged and tumbled down the mountainside to where two explorers now worked.

6:27 P.M.

The back room wanted a rake sample, so Schmitt headed back to the Rover; if they were going to have time for another stop on the mountain, he’d better work fast. Meanwhile, Cernan struggled up the hill so that he could shoot a documentary panorama that would include the boulder. At one point Schmitt happened to walk into the field of view, and the pictures, which would show him dwarfed by the dark mass of stone, would become some of the most famous of the Apollo program; Cernan would see them reproduced in countless books and magazine articles. Looking at them, he would also see a patch of soil on top of a low, rocky ledge that would forever bear the imprint of his hands where he had collected samples; he would wish that he had taken a moment to write his daughter Tracy's name in the dust.

From this height, for the first time, he could see all of Taurus-Littrow. Up here, he could spot all the places he and Schmitt had visited on the South Massif and the valley floor. To his left, beyond the flank of the North Massif, were the Sculptured Hills, where he and Schmitt would soon be headed. And way out on the plains, a tiny dot of titanium and Mylar. It occurred to Cernan that he had spent his adult life working farther and farther out on a limb, from flying jets to flying in earth orbit, then leaving his home world to come within five miles of another one, and finally returning to descend from the sky into this valley, getting outside, and, yesterday, driving miles into the distance, and over a hill, so that he could not even see his lunar module. During the sleep periods, lying in that tin can, that popped and creaked in the full-strength lunar sun, Cernan's thoughts were not about undoing the steps in that chain —getting off the moon, out of lunar orbit, and safely through the earth's atmosphere—but of what he was missing, what he could be doing instead of wasting precious hours by sleeping. As much as Schmitt, he understood how little three days was when you had a place like Taurus-Littrow to explore. And yet, they'd accomplished more than he ever expected. With the excited voice of a kid who's just climbed his first mountain Cernan said, "You know, Jack, when we finish . . . we will have covered this whole valley from corner to corner!"

Schmitt deadpanned, "That was the idea."

In the back room, Bill Muehlberger said, "It's only six-thirty. We've got a long night."

A surprised Dale Jackson said, "Is that what time of day it is? I'm all screwed up. I don't even know what day of the week it is."

And no wonder; everyone in the back room had been going full bore for three days. Even now, with a bonanza from the split boulder in hand, they worried that Cernan and Schmitt were cutting uncomfortably into the time reserved for the next stop. They told Jim Lovell to ask that Parker hurry them along. Lovell did so, then informed Muehlberger's team that a few extra minutes late in the moonwalk would have to be reserved for some chores at the ALSEP; he wanted to know which geology stop to take it out of. Jackson replied, "Take it out of Apollo 18."

7:49 P.M.

When historians review the scores of hours of Apollo videotapes, they will no doubt take note of Gene Cernan and Jack Schmitt's visit to the Sculptured Hills and what they did there, not with their hammer and scoop, but with themselves. To be sure, the geologic rewards were great; Schmitt discovered a small boulder which, along with some of the samples from Nansen, would reveal that the ancient crust of the moon was chemically more heterogeneous than theorists had suspected. But the exploration was just as revealing of the explorers themselves, because after three days of intense effort, they let themselves have a little fun. Having climbed high upslope to the boulder, Schmitt took an excuse to perform the geologist's time-honored ritual. He called out to Cernan, "Are you ready for this?" With one boot Schmitt shoved the rock into a lazy tumble that abruptly died. He chased after it, giving it a kick that sent dust flying but did little to move the rock. "Go! Roll!" he yelled. "Look, I would roll on this slope; why don't you?" And when he and Cernan had finished breaking off a sample, Schmitt gave a ballet-style leap and bounded downhill toward the Rover. Then, forsaking any trace of adulthood, he began hopping on two feet with arms outstretched, pretending to ski. He swung his body from side to side, sample bags flopping against his backpack. He made schussing noises. "I can't keep my edges! A little hard to get a good hip rotation!" A minute later Cernan came kangaroo-hopping down the hill after him—"*Wheee!*"—as if to affirm that the strongest emotion of being on the moon was simply elation.

10:36 P.M.

The replacement fender finally gave out during the drive back to the LM. Cernan said, "There's got to be a point where the dust just overtakes you, and everything mechanical quits moving." Lunar dust is abrasive—very abrasive, because of all the little rock fragments knocked off by micrometeorites. Cernan had only to look at the handle of the geology hammer; the rubber coating was worn through to bare metal. He and Schmitt had been religious about greasing the suit zippers and cleaning the wrist wrings, to keep the suits in good working order. And by God, they were holding up, despite the abuse heaped on them. Schmitt would never lose his amazement at that; just like almost everything else on Apollo, those suits worked better than anybody ever expected them to. It went back to the motivation of the people who built them, tested them, and flew them. It showed what can happen when people believe that the thing they are working on is the most important thing they will do in their lives, and they don't want to be responsible for screwing it up.

No doubt those suits could have gone further, and no doubt, so could Cernan and Schmitt—but not this day. The past six hours on the surface had been the most physically demanding of all their excursions. Unlike yesterday, when they'd spent as much as an hour riding the Rover between stops, today was fast paced, with much of the work on steep slopes. At the last geology stop, on the valley floor, Cernan and Schmitt had hit a wall; they worked not for the thrill of discovery but simply to get their work done. But at the last minute, with mission control hurrying them back to the Rover, Schmitt had discovered a layer of unusually bright white soil. "Come here, Gene," he said, "we can't leave this." They made the decision, on their own, to stay longer, and mission control let them. For a time, the discovery seemed to revive them, but when the geologists tried to argue for an extension, flight director Gerry Griffin said no; he'd heard Cernan and Schmitt complain about tired hands, and wouldn't take the risk of having them end up too worn out to handle an emergency. And that was the end of it.

Griffin had been right. Back at *Challenger*, amid the litter of three days of activity, Schmitt had helped pack up the rock boxes, and it was all he could do to get his hands to work. When he discovered an unused core tube, he drove it into the ground by hand, holding onto the lander for support; then he tossed this last lunar sample in with the others. Now, all that remained was for Cernan to drive the Rover to its final resting place, where it would televise the next day's liftoff. First, however, he and Cernan would attend to a few moments of ceremony. NASA was hosting young students from seventy different countries; they were watching the moonwalk from Houston. Before the flight Cernan had the idea to distribute pieces of a rock from Taurus-Littrow to museums around the world, and NASA planned to tie it in with the visit of these children. And so he and Cernan stood before the TV camera while Cernan, holding a rock, explained.

"It's a rock composed of many fragments, of many sizes, and many shapes, probably from all parts of the moon, perhaps billions of years old . . . that have grown together to become a cohesive rock, outlasting the nature of space, sort of living together in a very coherent, very peaceful manner. . . ." Cernan would look back wishing he'd had more time to prepare his remarks. ". . . We hope this will be a symbol of what our feelings are . . . and a symbol of mankind: that we can live in peace and harmony in the future.

"And now—let me bring this camera around—" Cernan grabbed the TV camera and pointed it at *Challenger*'s front landing gear, where a plaque commemorated the final lunar landing. Schmitt's contribution to that plaque had been to add, below the customary pictures of earth, a small picture of the moon, showing the places where men had landed. But Schmitt had not been successful in another hard-fought effort, to change its wording. He had wanted something to reflect what he still firmly believed, that even if no one came back to the moon for twenty years—let those explorers include Joe Allen's little boy and other children of his generation—the return was inevitable. Instead, the message spoke of finality. Now Cernan removed the cover and read the message.

"Here man completed his first explorations of the moon, December 1972 A.D. May the spirit of peace in which we came be reflected in the lives of all mankind." Schmitt listened to his commander's words. Within his suit he heard the reassuring whir of cooling pumps in his backpack, a sound he had almost come to take for granted. Now Cernan was saying, "This is our commemoration that will be here until someone like us, until some of you who are out there, who are the promise of the future, come back to read it again and to further the exploration and meaning of Apollo." Schmitt's fingers were raw within his gloves. His arms ached with the exertions of a full working day on the surface. And as he listened, he realized: *It's over.*

11:34 P.M.

After loading the rock boxes and other cargo into the LM—he had to climb up the ladder several times to pass the massive stowage bags to Schmitt—Gene Cernan stood alone on the moon, his space suit covered with grime. He was somewhat out of breath. Now he spoke, trying to cut through the fatigue, to imbue his words with energy, to speak for the history books. "Bob, this is Gene. As I take man's last steps from the surface, back home for some time to come—but we believe not too long into the future . . ." His words were interrupted by the sound of his own breathing. ". . . I believe history will record that America's challenge of today has forged man's destiny of tomorrow." As he spoke, Cernan took a last look at the bright and barren wilderness all around him. "And as we leave the moon at Taurus-Littrow, we leave as we came, and, God willing, as we shall return, with peace and hope for all mankind." He glanced at the earth, high in the southwestern sky, an unmoving, silent witness to a voyager about to begin a long journey back. Once more, he spoke: "Godspeed the crew of Apollo 17."

Friday, December 15
Late evening, Houston time
Aboard *America*, in lunar orbit

Gene Cernan would never have believed that he could feel so comfortable with his neck so far out. It had been more than a day since *Challenger* had blasted off from Taurus-Littrow, carrying him and

Schmitt into orbit and then, like riding on rails, back to a very happy Ron Evans. In the last phase of the rendezvous Cernan flew his craft in tight formation with the command module, as *Challenger* danced like a fighter plane under his control. It was his moment of exultation at a mission accomplished; it was his victory roll. With 60 percent of his maneuvering fuel still untouched Cernan would have loved to fly beautiful orbital-mechanics circles around Evans, but he did not. There could be no distractions. Minutes later he held the ascent stage still while Evans approached, ever closer, until the cabin filled with the welcome ripple-bang of twelve docking latches snapping shut. And Cernan felt that he had come home. Inside *America*, he couldn't believe how good it felt just to get out of his filthy pressure suit, strip off his long johns, and to wet a washcloth and wash the sweat and the dust off his body. Nothing would get his fingernails clean—the dust had penetrated deep into the quick; it would be weeks before it grew out.

No matter; Cernan felt like a new man, even though he and his crew were spending two more days in lunar orbit. During training, Cernan hadn't been in favor of these extra days in orbit, but Evans and Schmitt had been so enthusiastic about making more observations that he'd backed off. And now that he was here, he marveled at his own ease. He could settle in and soak up the view. When it came time to leave, then he would worry about the leaving.

Cernan was proud of Apollo 17. With 75 hours on the moon, 22 of them outside, and nearly 19 miles total distance covered, this mission had made its mark on the record books. Not a single major problem with the spacecraft; arguably the smoothest and most successful flight of the series. And he had accomplished the goal that he had set for himself almost a decade ago. He could say to himself what only five other men could say: He had steered a spaceship to a landing on the moon.

That evening, when *America* was on its 66th circuit of the moon, Jack Schmitt saw his first earthrise. Looking at that bright crescent, Schmitt sensed the same fragility his predecessors had seen. It had nothing to do with the earth's appearance, he would say later; it stemmed from its monumental isolation in the blackness. The earthrise brightened Schmitt's dark mood for a moment. Down on the surface, just before liftoff, he'd felt the first stirrings of sadness at the end of Apollo. Though he felt proud of his accomplishments on the flight, he had never seen his getting to the moon as a personal triumph; he'd always felt that he and Cernan were just lucky enough to implement the triumph of others. Apollo 17 represented the last acts of lunar exploration that would be done by members of their generation.

But the worst moment came later, just after the docking. The vibrations hadn't even died down when Gordon Fullerton called up and said, "I'd like to take a minute of your time here to read the following statement by the president of the United States of America." Schmitt was annoyed; they had work to do. But Fullerton went ahead and read the message.

"As the *Challenger* leaves the surface of the moon, we are conscious not of what we leave behind, but of what lies before us." Even in Fullerton's quiet, even tone, there was Nixon's cadence. "The dreams that draw humanity forward seem always to be redeemed, if we believe in them strongly enough and pursue them with diligence and courage." Ironic words, thought Schmitt, considering the fact that Apollo had been conceived under Kennedy and was ending under Nixon. Then Fullerton read the sentence that sank Schmitt's spirits. "This may be the last time in this century that men will walk on the moon, but space exploration will continue . . ." Schmitt barely heard what followed. He couldn't believe his ears—*the last time in this century?* He hated the words—hated them for their lack of vision. These words, from the leader of the nation! Even if Nixon really believed them, he didn't have to say so in a public statement, taking away the hopes of a generation of young people. Schmitt was furious that in the moment of triumph, he had been jolted out of the work of the mission to listen to a statement like that. He would fume silently about it for the rest of the flight.

Saturday, December 16
5:54 P.M., Houston time
9 days, 21 hours, 1 minute Mission Elapsed Time

When the SPS engine shut down, and a perfect Transearth Injection burn was behind them, Cernan's crew broke out the television camera. When *America* regained contact with earth, they broadcast a view of a huge and nearly full moon. Part of the far side was still visible, including Tsiolkovsky. One of the reasons Schmitt had pushed for a landing there, back in 1970, was that he had hoped it might reignite enough public interest to prompt NASA to reinstate Apollo 18 and 19. As he panned the TV camera across the face of the moon, he captured not only places he had hoped astronauts would go—the lunar poles, and Tycho—but some of the places where they had been: Tranquillity Base, and Hadley, the Descartes highlands, and a valley at the edge of the Sea of Serenity that would forever be known as Taurus-Littrow.

It was television that set Apollo apart as an exploratory venture. Schmitt could still remember the December night four years ago when he and engineer Jack Sevier had stayed up trying to figure out how Frank Borman's crew might televise a picture of the earth. They ended up telling them to use every filter in the photography kit together, taping them in front of the lens. The fix was a little inelegant, but it worked; that day, humanity saw its home on live TV for the first time. Eight months later, on a hot July night, the pictures were transmitted from a patch of dusty plains on the Sea of Tranquillity. Something extraordinary happened when those fuzzy black-and-white images appeared on television screens around the world: For the first time in the history of exploration, the human species—the developed world and the developing nations—participated in the moment along with the explorers themselves.

Now Schmitt was returning to that earth, having seen it from a great distance. The data held in his mind would be shared with his earthbound colleagues in a trailer at the space center in the days before the Christmas holiday. Then, for a time, he would enter into the debriefing of the self, and his thoughts would return to the sight of that small and lovely crescent beyond the alien shore. The ability to witness an earthrise, he would note, could happen only if humankind took the well-planned but still significant risk of sending three of its members into an environment utterly foreign to the one in which they had evolved—that is, lunar orbit—with no guarantee of getting those people back. That commitment, Schmitt would come to believe, marked a turning point in human evolution. Human beings had become a space-faring species. What they would do with that new status was still locked in the unknowable future, but Schmitt firmly believed that space exploration would dominate the future of humanity.

And what of that humanity, living into another day, unmindful of the leap it had made from the turning earth? In particular, what events were drawing attention in a country called the United States? In the headlines, Henry Kissinger had reported that serious problems had brought peace talks between the United States and North Vietnam to a halt. The United Mine Workers had elected a new president who pledged to clean up the union. A blizzard had dumped 28 inches of snow on northeastern Ohio, blocking highways and closing airports, with another foot expected before the storm moved east. In a Kansas City hospital, the man who had overseen the final step in the transformation of a nation of revolutionaries, then pioneers, and then immigrants, into a superpower, was battling lung congestion, heart problems, and kidney malfunction. Harry Truman was dying. And in a space ship called *America*, three men were coming home from the moon.

Sunday, December 17
2:37 P.M., Houston time
Mission Operations Control Room,
Manned Spacecraft Center

Stu Roosa sat next to the Capcom, looking at the big screen at the front of mission control. The image was stark, and for the uninitiated, like some of the visitors up in the viewing gallery, it must have looked very strange. It was mostly dark, except for a big white object that looked something like a snowman, or perhaps a mummy, lying on its stomach. But Roosa knew exactly what he was seeing.

The figure was Ron Evans. He was floating in the void 180,000 miles from earth, beginning his space walk to retrieve the photographic film from the side of the service module. Roosa listened to the transmissions from Apollo 17. He heard a whooshing sound—that was oxygen flowing into Evans's suit from the umbilical. And he heard Evans's exuberant voice: "Hot diggety dog!" There was nothing complicated about Ron Evans. He'd worked just as hard backing up Roosa for Apollo 14 as he had for his own flight. Now he was in his element. "Talk about being a spaceman, this is it!" On the screen, Roosa could see him waving. "Hello, Mom! Hello, Jan! Hi, Jon! How you doing? Hi, Jaime!"

There wasn't anyone in that control room who truly understood how Roosa felt, watching this. He had trained to make that spacewalk himself, dozens of times. With the lack of flights, Deke Slayton started putting experienced astronauts on the backup crews to save the extra time, money, and effort that would have been required to train rookies. When Roosa and Ed Mitchell came off Apollo 14, they were assigned to back up 16, with Fred Haise as the commander. Before Apollo 14, Mitchell had seen the assignment coming and had balked, but Slayton made it clear that he'd better go along or he wouldn't fly. But there was no resistance from Roosa. He'd wanted the assignment. There was always the chance that something would happen, and he'd get to go—though he would have had some very mixed feelings about flying if it meant Ken Mattingly were grounded again. Still, on the day Apollo 16 was launched he was standing with Fred Haise, and Haise was talking about how hard it was to train for more than a year, and get all cranked up to fly, and then just stand there and watch somebody else go instead. If Roosa said anything in response, it wasn't necessary. Anyone who'd ever done time on a backup crew knew what he was talking about. It was right around that time that the stamp affair broke, and in the shakeup that followed, Scott, Worden, and Irwin were taken off the Apollo 17 backup crew, and in their place Slayton assigned Roosa along with John Young and Charlie Duke, just back from their world tour. And Stu Roosa, veteran astronaut, found himself serving on two backup crews in a row. The three of them joked about being the only all-Southern crew. They grew mustaches, just for the hell of it, and they trained for a mission they knew they would probably never fly. By the time launch day rolled around Roosa figured he had more time in the command module simulator than any other astronaut. He was ready to go back to the moon—more than ready. His best friend Charlie Duke felt the same way, but he knew nothing was going to keep Jack Schmitt from getting to the moon, short of a broken leg—and even then, Duke figured, NASA would probably postpone the mission. Roosa had been more hopeful. He wanted a chance to do it again, and do it even better—if that was humanly possible. He'd take more photographs; he'd make more observations. And this time he wouldn't let so much of the experience get away from him. He'd be more relaxed; he'd have time to collect his thoughts. He'd *enjoy* it a little more. And he wanted to make that space walk, 180,000 miles from earth. But that experience, like a second trip to the moon, was now forever beyond his reach.

"Okay, put the old sun visor down now. I see what Charlie Duke meant. Man, it's dark out here . . . Whoops, come back here, little cassette . . . Okay, I'm coming back in."

• • •

A few days later, when Apollo 17 splashed down, the control room was so packed with people that it was hard to walk around. It seemed every flight director and flight controller who'd ever worked on Apollo had been drawn to this room to see the last mission end. And Roosa was there, seated next to the Capcom, along with Deke Slayton. He watched quietly as the command module descended under three perfect parachutes, while around him the room broke into applause. By the time Cernan, Evans, and Schmitt were being hoisted up to the helicopter, cigars were lit. Everyone was clapping and shaking hands. Minutes later the big screen showed Cernan and his crew on the carrier deck, making speeches.

There were probably as many emotions in that room as there were people. Chris Kraft would say later that he was absolutely relieved when he saw the parachutes; they were getting out of this risky business at just the right time, and with a great finish. One of the flight controllers would liken his own wistful sense of accomplishment to the feeling the architects of the pyramids must have had when the last one was finished. By nightfall there were splashdown parties in full swing all around the space-center communities, but there wasn't the same all-out feeling of celebration they'd had in 1969.

In 1973 Roosa was assigned to the Space Shuttle program, and for the next few years he went to meetings and saw the program fall further and further behind. By 1976 it was clear that the first flight would probably slip well beyond its 1978 target date. And Roosa knew he wasn't first in line; by the time he had a chance to fly again, it might be more than a decade after Apollo 14. He'd be a spacecraft commander, but he'd also be forty-seven years old. But if he was going to leave what Ron Evans called "the best job in the world," what would he do? He thought about going back to the air force, but he was told there was a lot of anti-astronaut sentiment there, because the stamp affair had been done by an all-air-force crew. So Roosa decided to leave NASA and try his hand at business. He'd been contacted by a corporation that wanted to expand its operations in the Middle East. They had offered him a vice president's position in Athens. It sounded challenging, and the kids would enjoy the change. But Roosa never really felt that he'd found something to replace what he did from 1966 to 1972. Someone had called it Mankind's Greatest Adventure; what could Roosa add to that? On the day he decided to leave, he called George Low, and they talked about his plans; Roosa said he was really going to miss NASA, but that it was time to move on. Then Low was quiet for a moment, and he said, "You know, there will never be another Apollo in anybody's life."

Epilogue: The Audiences of the Moon

In 1975, during a trip to Nepal with his wife, Joan, Stu Roosa visited a school to give a talk about his flight to the moon. Afterward, there were mysterious questions: "Who did you see?"

Roosa answered, "There is no one there." A murmur went through the place. Again the students asked what he had seen. Roosa was adamant: "There is nothing there. Not even wind. There is nothing."

Later, after the Roosas had gone, a teacher told the children, "You mustn't listen to him. He's wrong."

The Roosas were distressed when they learned that some Nepalese believe the spirits of their ancestors reside on the moon. Roosa had essentially told them there was no heaven. Joan wished the American government had briefed them better for the trip.

One day at the hotel, Joan went to get her hair done. The woman giving her a shampoo said, "You're married to the astronaut, aren't you?"

"Yes, I am."

"You know, you're married to a god." Joan laughed it off. But when she and Stuart went to that evening's gathering, all along the

mountain road children were kneeling with candles, in an act of reverence: a god had come to visit.

In the mid-1970s, a series of television advertisements for the American Express card featured notables whose faces were unknown to the public. Two of them were William Miller, who ran for vice president in 1964, and Mel Blanc, who did the voice of Bugs Bunny. And in one of the great ironies of our age, the producers chose a man who had been to the moon. On TVs across America, there was Pete Conrad, grinning his gap-toothed grin, saying, "Do you know me? I walked on the moon."

For Conrad, the ad changed things—for a little while. "Up until the time I did that," says Conrad, "I could go *anywhere,* and nobody knew me. Alright? I did that American Express commercial, and while that sumbitch was running, I couldn't take a leak—I remember trying to take a leak in Philadelphia, and a guy standing there says"—Conrad lurches into a crazed persona—"*I know you!*" Then the ad stopped running, and Conrad's sudden fame evaporated as quickly as it had come.

Today the moon voyagers are scattered across the country, and most of them are as anonymous as Pete Conrad. Many, like him, are corporate executives. And most agree with the credo Conrad stated one day, sitting in his office at McDonnell Douglas in Long Beach, California: "Don't ever look back. I'm serious. What's the point of looking back? I've already been there. How about some old guy sitting around telling you about how he played football for good ol' Yale University and was the world's greatest quarterback—that's all the bastard ever talks about. Okay?" To Conrad, the scene he has just described is so awful that no further explanation is necessary.

From time to time, however—usually for the benefit of an interviewer—Conrad does look back, and even among astronauts he is known as a great spinner of space yarns, or Lies and Sea Stories, as he calls them. But Conrad has kept the promise he made to himself on the day he was selected as an astronaut in 1962. Going to the moon, he says, hasn't changed him.

But that isn't what people want to hear, says Conrad. "They've got some preconceived notion that I should tell 'em I was frightened, or I was awe-inspired, or I saw the Lord—or . . . I don't know." The truth sounds unbelievable: when he was on the moon his strongest feeling was that it was the right place to be at the time. "That just shuts the door." So does the thought of a man who has been to the moon and who, today, does not look at it. "They *know* I'm *lying*. They say, 'Don't you go out and look at the moon?' And I honestly don't."

When Pete Conrad looks back on his spaceflight career, the high point isn't his lunar landing: it is the rescue of the Skylab space station in 1973. The station's outer shield, which protected it against heat and micrometeorites, was torn off during launch, taking one of its power-producing solar panels with it. Getting power to Skylab depended on freeing the remaining wing, which was lashed down by debris. Conrad's crew, who had been preparing for a month-long mission aboard Skylab, was now faced with carrying out a demanding repair. Arriving at the stricken station, they sweltered in desertlike heat for days until they could erect a makeshift sunshield. Two weeks into the flight, Conrad and his crewmate Joe Kerwin made a space walk and, with some difficulty, freed the stuck solar wing. Today, Conrad looks back with a healthy appreciation of the risks he took.

"My life was a lot further out on the line . . . on Skylab than it was on the moon," Conrad says. "That taxed me *personally,* put everything that I had spent my whole life . . . learning how to do, on the line. . . . Going to the moon was basically a nice, routine flight after the lightning. We didn't have any trouble after that. On Skylab we didn't know whether we [would leave] or stay for fourteen days." In 1978, Congress recognized the success by awarding Conrad the newly created Space Medal of Honor.

"Everybody thinks I got the Space Medal of Honor because I went to the moon. I say, 'No, it was for Skylab.' They say, 'Oh, Skylab. Yes. What was Skylab?' " It's the moon that people want to hear about, and like all his colleagues, when Conrad is introduced as one of the twenty-four men who went there, the question he is almost always asked is, What was it like? And he gives the neat, two-second answer he developed long ago: "*Super!* Really enjoyed it."

"Space changes nobody," says Stu Roosa flatly. To be sure, Roosa will talk about the impact of seeing the earth from deep space, and the lunar landscape. You can't see those things and not be affected. But it didn't change him. "You bring back from space what you bring into space," he says.

Roosa's words are ironic, considering the fact that one of the two men who rode with him on Apollo 14 says his own perspective, his life, even his very sense of being, were profoundly altered by the voyage. He is Ed Mitchell. Since 1971, Mitchell has tried to understand what happened to him inside the command module *Kitty Hawk* heading toward earth, when he saw the universe revealed in a flash of understanding. Now a freelance management consultant in Boca Raton, Florida, Mitchell talks with quiet intensity about his consciousness-changing experience in space, and the journey that followed.

When Mitchell came back from Apollo 14, the other astronauts heard about his attempt to transmit his thoughts through space. Some, he says, even came by his office to talk about it—when no one was watching. But none of them knew that Mitchell had returned to earth a changed man; even he did not really comprehend his experience.

"I didn't know what feelings were," Mitchell says of himself as a returned moon voyager. "People used to ask, 'What did it feel like to be on the moon?' I didn't know what it felt like! I could tell them what I did, and what I thought, but not how I felt. It pissed me off." But Mitchell realized that answering the question was a key to unraveling his revelation in space. He sought out two researchers in consciousness, Jean Houston and her husband, Bob Masters, and through them, relived the experience via hypnosis. From there, Mitchell began to gain understanding. He spent the next fifteen years piecing together a new truth.

In 1973, he founded the Institute of Noetic Sciences in Palo Alto, California, devoted to the scientific study of consciousness. Even after leaving daily operations at the institute in 1982, Mitchell continued his quest for scientific explanations of consciousness. He helped arrange experiments to test the claims of psychic Uri Geller, whose spoon-bending and other feats of telekinesis made him a talk-show celebrity. He has also delved into the work of such avant-garde scientists as physicist David Bohm, who proposed a kind of consciousness at the level of subatomic particles. Like Bohm, Mitchell believes that consciousness is an attribute of everything in existence, animate and inanimate.

Today, Mitchell believes that telepathy, clairvoyance, intuition, and other psychic phenomena have a single explanation: "It's just *information*." Information, Mitchell says, is the other half of the universe, the intangible complement to matter. An information field pervades space, like a cosmic data bank, that records, among other things, the accumulated experiences of all matter, including living organisms throughout history. Mitchell believes that a person who is sensitive to this information—which exists, he says, as a form of energy—can experience anything from clairvoyance to telekinesis. When he had a shift in consciousness on the way home from the moon, he says, he was tuning in to the data bank. Why didn't the other astronauts experience the same thing? They *did*, according to Mitchell, whether or not they express it. "The informational *input* is the same. But what is [actually received] is shaped by our level of awareness, and by the filter we call belief system." Mitchell believes he has found the resolution between science and spirituality that he hungered for as a teenager: "To me, divinity is the intelligence existing in the universe." That universe, he adds, is a learning, growing, changing organism, like the human beings who strive to understand it, and who have only begun to explore it.

"It is my joy to be with you tonight, to share the love of our Lord and Savior, Jesus Christ," Jim Irwin said in 1988, standing before a packed meeting hall in Colorado Springs. After a series of major heart attacks, he appeared thin and almost frail compared to his astronaut days. But his eyes were bright and penetrating, revealing an unmistakable inner strength and determination.

"I come to you as a former astronaut," Irwin continued. A wry smile crossed his face. "I feel like the ancient astronaut. The shrinking astronaut. But I'm glad that I'm still alive and that I can be here." This is how Jim Irwin spent the last two decades of his life, sharing his faith, testifying to the love of God that he discovered, he said, on the moon. Irwin embraced the task of recounting his experiences again and again, reciting stories he had long ago committed to memory. He talked about the white rock he spotted at Spur crater, presented on its own pedestal as if on an outstretched hand. Holding up a plastic model of the rock—there is almost nothing left of the real one, he would say, which had been cut up and sent to scientists around the world—he talked of the real impact of his trip to the moon, his realization that God is alive and available everywhere in the universe.

Back on earth, that awakening caused Irwin's life to change dramatically. Urged by his wife to heed this new spiritual call, supported by family friend and minister Bill Rittenhouse, Irwin left the astronaut corps and founded the High Flight ministry in 1972. The timing was significant; the stamp affair had just become news. By publicly admitting that he had made a mistake, Irwin took himself off the pedestal reserved for conquerors of the heavens; at the same time he began a life that could not be more different from the one he led as a test pilot. The quiet, introverted engineer now faced a life of public speaking, something he said he dreaded but grew to enjoy. One month he was in India, visiting polio-stricken children; another, he was witnessing in Eastern Europe. He gave his testimony in churches across the United States. Someone nicknamed him the moon missionary.

It was hard for some of Irwin's astronaut colleagues to understand his transformation; they would joke, "I don't know what happened to ol' Jim up there. . . ." Irwin had expected that response, but for his own part, he had no doubts about why his life took the path it did. "God had a plan for me, to leave the earth and to share the adventure with others, so that they can be lifted up." Going to the moon, he said, "prepared me for a role of greater service." To attend one of Irwin's appearances was to get a sense of how many lives he touched. In 1989, after sharing his faith on the campus of a small Baptist college near Worcester, Massachusetts, Irwin was mobbed by admirers and signed autographs for nearly an hour. That night the organizer of a dinner given in Irwin's honor by the local Hispanic community spoke of her dream to bring Irwin to Puerto Rico, so that the children there could hear how he reached for the stars, and know that they could do the same.

Irwin managed to find an element of exploration in his new life. Beginning in 1973 he participated in several expeditions to Turkey's Mount Ararat to search for the remains of Noah's Ark. In 1982, Irwin took a bad fall on Ararat's rocky slopes, almost losing his life. In the summer of 1986, near his Colorado Springs home, Irwin suffered a near-fatal cardiac arrest. Only weeks later he was recovering at home, pedaling an exercise bike and talking about going back to Turkey to continue his quest.

"I realize that my time is running out," he said calmly, looking off toward the mountains. "I want to take care of myself as best I can so I can stick around and be of help. I just hope that when the Lord's finished with me that I go quickly."

In the years that followed Irwin continued a grueling schedule of appearances, cutting down to two a day after his doctor warned him to slow down. In August 1991, Irwin died following a heart attack while in the Colorado mountains he loved.

If you talk to the moon voyagers, you will find that most of them think about the experience the way Bill Anders does. He doesn't describe the kind of life-changing impact that came to Jim Irwin and Ed Mitchell. He does say that the experience of being one of the first three men to circle the moon—and in particular, seeing the earthrise—broadened his perspective forever. It also made him something of an oracle, for a time. On the banquet circuit after Apollo 8, people would ask him and his crewmates about the chances for world peace and how the stock market was going to behave. "Pretty soon," Anders laughs, "it's easier to answer than to explain we don't know anything about it."

That wasn't the only thing Anders didn't know about when he left NASA. Like his colleagues, he was highly trained for the business of spaceflight—and untrained for real dollars-and-cents business. But he was clear about one thing: he didn't want to build a new career on his status as a moon voyager. It did open doors, however; in 1969, the Nixon White House appointed him as executive secretary of its National Aeronautics and Space Council. He went on to head the Nuclear Regulatory Commission, making use of his pre-NASA experience as a nuclear engineer. His performance in that job won him a stint as ambassador to Norway. In 1977, after eight years of government jobs in which he learned, among other things, "how to keep out of the way of Machiavelli's knife," Anders was recruited for an executive vice president's post at General Electric, which he called his "business boot camp" experience. In 1984 he was hired by Textron to run their aerospace and commercial divisions. Within a few years Anders was Textron's chief operating officer, earning more than a million dollars a year.

One day in 1987 Anders reflected on the challenges he and his colleagues faced in making second careers. "You're given unrealistic opportunities that you wouldn't have had otherwise, and measured by an unrealistic yardstick." The title of ex-astronaut may open some doors, Anders says, but once inside, "you really have to perform."

For Mike Collins, that meant a year as the assistant secretary of state for public affairs, a job that taught him, in his words, "how to operate in the strange, semihysterical environment" of official Washington. Thus prepared, he became the first director of the Smithsonian's new National Air and Space Museum, overseeing its design and construction. It opened ahead of schedule, a few days before the Bicentennial in 1976, and within budget—an almost unheard-of set of circumstances. Collins's hope that it would be the most exciting museum in the world was borne out by attendance figures: in three weeks it drew a million visitors; by the end of the year that figure had risen to 10 million.

For Jack Schmitt, who refuses to see his moon flight as the high point of his career—"The whole thing was a continuum for me," he says, "a steadily rising level of experience and no dropoff at the end"—recognition as a former astronaut helped him run for a Senate seat in 1976 against senior Democrat Joseph Montoya. "All the astronaut thing did was balance that ledger. He had seniority, I had astronaut. I could get as many speaking engagements as he could. But it didn't win the election. What won the election was, the Democrats were mad at Montoya."

Schmitt hoped he might bring "a different intellectual base to the discussion" of issues in the Senate, but his term was a frustrating and disillusioning experience. He felt like more of an outsider in the Senate than he had in the astronaut corps, he explains, because "I was a Republican, scientist, technologist. Interested in issues of the future, in addition to those of the present. And there wasn't anybody else there interested. There still isn't. . . . The Congress of the United States is not fertile ground for working those kinds of problems." Schmitt lost a reelection bid in 1982.

Meanwhile, for Bill Anders, success came at a price. He was on the road constantly; his only moments of relaxation were found aboard his sailboat, *Apogee*. "Money isn't everything," Anders said in 1989. "*Sailing* is everything." That year, when Anders was seriously thinking of retiring, he was recruited to head General Dynamics, one of the largest defense contractors in the country. By 1992, however, the Cold War was over, and Anders was in charge of guiding General Dynamics through the turbulent transition. After a controversial recovery effort that involved massive layoffs and a lucrative incentive program for the company managers, General Dynamics made a dramatic turnaround. GD's stock price more than quadrupled, and the company returned more than $2 billion to lenders and shareholders, a move that Anders says shocked competitors but won praise in the Pentagon and on Wall Street. Anders's own earnings for 1992 put him in *Fortune* magazine's list of the top ten highest paid CEO's. In 1993 he stepped down as CEO in what appeared to be retirement, moving to an island off the coast of Washington State. In reality, he is still serving as General Dynamics' chairman of the board, still traveling, still wishing he had more time for boating.

Sometimes, out on the water, he catches sight of the moon, and aside from a memory of feeling of disbelief—he wonders if the whole thing was a very realistic simulation—he is wryly aware that he doesn't see it the way most others do.

"Other people will say, 'Oh, what a beautiful moon.' I'll think to myself, If they could just see it up close. It appears beautiful because it's a long ways off, and you can't see it's really a very uninviting place." Anders concedes that he would probably feel differently if he'd been able to see the moon the way his twelve moonwalking colleagues had. It is still one of Anders's regrets: for all his satisfaction at having flown Apollo 8, he would rather have been the last man to walk on the moon than the first to go around it.

For many years Bill Anders noted the progress of the other astronauts as they tackled Life After the Moon, but he paid special attention to Frank Borman. Borman had left NASA with international fame, but had been determined not to coast on his celebrity. His plate was full with lucrative offers to join anything from aerospace giants to broadcasting companies. A few wanted him for PR purposes, to be the "company astronaut," and that was something he wanted to avoid. In the end he went to work for Eastern Airlines. The airline business, he felt, was something he could sink his teeth into. After seven years as a senior vice president Borman was named Eastern's top executive in 1976. His gruff management style—along with his aversion to the lush corporate life-style of his predecessors—ruffled more than a few feathers among his executives. But soon, after a long decline in the 1960s, Eastern was enjoying the most profitable years in its history. Then, in 1978, deregulation hit the airline industry, and Borman's troubles began to build. Beset by high labor costs and low employee morale, he found himself in a bitter power struggle with leaders of Eastern's unions that dragged on into the 1980s. The last traumatic confrontation came in early 1986, when Eastern's board accepted Frank Lorenzo's offer to buy the airline. In Borman's own words, it was his first failed mission. Says Anders, "I think he had a business problem that nobody could solve."

Meanwhile, Anders had been wishing for "some kind of Apollo 8 fraternity," and for a reunion with his former commander. For several years he'd marked the anniversary of the flight by sending his crewmates a telegram: LET'S GET TOGETHER. "After a while," Anders says, "I realized I wasn't even hearing back from

Borman. . . . I finally gave up." Anders couldn't have been more surprised when, one day in the summer of 1986, his secretary said, "Frank Borman is on the line." When Anders picked up the phone—"Bill? Frank."—Borman sounded as if he'd talked to him the day before. The call turned out to be for advice on an airplane engine Borman was buying, but about a year later Borman called again, just to say hello. "If you ever come out here," he told Anders, "love to see you."

"That phone call was the first time that Frank Borman and I have communicated when he's not under pressure," Anders said a few weeks later. He speculated that Borman, who had moved to Las Cruces, New Mexico, where his son ran a car dealership, had gotten over the sting of his defeat at Eastern. Said Anders, "He's probably mellowed."

In 1988 the Apollo 8 fraternity finally convened at the Aerospace Hall of Fame in San Diego, in time for the flight's twentieth anniversary. Seeing his crewmates together for the first time in sixteen years, Anders had a chance to renew the friendships—and settle an old score. For years, Frank Borman had maintained that he had taken the famous picture of the first earthrise; he even said he'd had to grab the camera away from Anders to do it. A few years later, just for fun, Jim Lovell threw his hat in the ring, insisting *he* took the picture. To settle the issue, Anders had a NASA photography expert research the matter, and the results seemed to be in Anders's favor. But that didn't stop Borman from sticking by his story. Fortunately for Anders, Apollo 8's onboard voice recorder issued the final, unequivocal verdict: Anders took the picture. In San Diego, during a slide presentation before a group of high school students, up came the earthrise photo.

"I'd better not comment on this one," Anders said. "Jim, why don't you comment?"

Lovell announced, "This picture has always been under contention . . . I think it's time now for Frank to have a public admission." Throughout the event, the earthrise picture became a running joke, and Borman took it all with good humor. Anders's suspicions about Borman were correct: he had indeed mellowed.

"I'm one of the few guys you may meet that's at peace with

himself," Borman says. "First of all, I'm lucky to be alive. I made it through some very traumatic times. I've got the same healthy, supportive wife I started out with—a real accomplishment in the NASA group. And I've got two boys that we have good relationships with." And Las Cruces, he says, is a place where no one cares that he was president of Eastern, or that he went to the moon.

At the reunion one evening, Borman entered a restaurant with his wife and some friends, and the maître d' stared at him for a long moment. Finally he said, "Didn't you used to be . . ."

"I used to be with an airline," Borman said. The young man didn't get the hint. Borman offered him another clue. "I used to be with NASA." Still no recognition. One of Borman's friends decided to end the suspense: "He used to be an astronaut." At last the maître d' realized he was meeting a man who had been to the moon. But before he could say anything Borman had the last word: "But now I'm just an old fart."

Years ago, Alan Shepard is said to have quipped, "Before I went to the moon, I was a rotten s.o.b. Now I'm just an s.o.b." To those who know him, that statement alone—the fact that Shepard admitted any change at all—speaks volumes.

"I suppose I'm a little nicer now than I used to be," Shepard says. That, he adds, has everything to do with getting to prove himself on Apollo 14. "Maybe it's because I wanted it so much, it was so important to me, and I did it, and I was able to relax and be a little more human."

It is Alan Shepard, ego still very much intact, who shows a surprising degree of balance when it comes to looking back. Of his space experience, he says, "You put it in a box, put it on the shelf, put a ribbon around it, and move on to something else." But he does not avoid opening the box, for example, when the moon is up. "I never look at the moon anymore without thinking about it, just saying, '*Wow.*' Only that. Then going on and doing whatever you were doing."

In person, he seems younger than his seventy years: tall, tanned, and fit, dressed casually but tastefully, he wears a gold chain around his neck bearing an Apollo 14 emblem. His brown hair, swept across his forehead, shows little gray. Alan Shepard is aging gracefully. And part of his maturing, he says, is the understanding that "there are things in life other than flying airplanes."

Today Shepard still serves as chairman of a consulting business called 7-14 Enterprises. He's on the board of directors of twelve or fifteen companies, including Kmart and Kwik Copy. And he is president of the Mercury 7 Foundation, created by the Original 7 to give scholarships to college students studying space and engineering. None of the moon voyagers save Neil Armstrong has as much name recognition, and Shepard is able to use his fame to raise money for the foundation. Though he is more often identified as the first American in space, he is also the first lunar golfer, and participates in lucrative celebrity tournaments. And it is for the foundation that Shepard is happy to look back.

"I don't mind going back and opening up the box again from time to time," Shepard says of his speaking engagements. "And I find it satisfying for me to do it and make some money for this foundation. To be able to share that experience with someone is fun, and at the same time create moneys for youngsters in college that want to get involved in scientific pursuits, I find that very satisfying."

A splendid moon, ripening toward full, graces the October dusk. As darkness settles on the town of Acton, Massachusetts, there is activity at the high school, and an air of expectancy. Cars stream into the parking lot, parents and children file into the large auditorium. This night they will hear the experiences of a man who walked on the moon. Even now, well before the presentation begins, he is seated in the front row, greeting some of his audience, signing autographs. Tall, white-haired, dark-suited, Gene Cernan has the larger-than-life look of a celebrity. That is something that has been reinforced by television; if you were watching in the weeks after the *Challenger* disaster, you probably saw him on "World News Tonight," or conversing with Ted Koppel from a giant "Nightline" screen. For more than twenty years Cernan has been making public appearances, and he still looks the way people expect an astronaut to look.

Cernan's audience this evening spans two generations. Many of the parents were in college when Cernan walked on the moon; most of them would not, if asked, have been able to identify him as the last to leave his footprints in the lunar dust. Nor would many of their children, who have grown up in an age in which space travel seems routine. For them, Apollo is a story in their history books. Tonight Cernan will talk to the parents, about the importance of education, about the need for America to find a goal in space to mobilize its technology and capture the imagination of its people. But first, he looks back. "I can take you to the surface of the moon so that you're standing next to me," Cernan likes to say, and standing on the stage he launches into a symphony of recollection. He tells of looking at the beauty of the sunlit earth, "surrounded by the blackest black you can conceive of. I can't show it to you, but I can tell you it exists, because I saw it with my own eyes." And the audience is transported by the words of a space traveler.

One afternoon the previous July, in his Houston office, Cernan talked about his life, his goals, and his identity as a moonwalker. Among the memorabilia on the walls are pictures of Cernan at Taurus-Littrow, standing before sun-drenched mountains and the American flag. Cernan has never felt the need to put his past behind him; in fact, he seems to have integrated it into his identity. And when he came back from the moon, Cernan had no intention of living in the past. After Apollo 17 he was assigned to other projects within NASA and then, in 1976, decided to strike out on his own. There was no letdown, he says: "You get ready, and you charge, and it's like having another mission." Over the years Cernan has been involved in a variety of business ventures: serving as vice president of an energy company in Houston, helping to start up a small airline in St. Louis, working as a marketing consultant for a Boston-based high-technology firm. In 1981 he created the Cernan Corporation, which channels his energies into aerospace consulting and other activities, including helping to create space exhibits for museums around the world. Like many of his former colleagues, he maintains a breakneck schedule, gone for weeks at a time, often home for only a day or two on the weekend. Like them, he shows no signs of slowing down. But even now, Cernan confesses, he has been unable to find a pursuit that really satisfies him. He says, "It's tough to find an encore."

But it isn't at all difficult to find people who still want to hear about what Cernan did more than two decades ago. And he is happy to answer. When Cernan talks about his moon experiences, not a hint of boredom dulls his words, which sometimes tumble so fast upon one another that he seems to speak without punctuation. And the people Cernan wants to reach most are children. All too aware how crucial education is to our nation's future, he says, Cernan takes every opportunity to use his own space experiences as inspiration. He tells students, "I urge you to dream—I did, and one day I found myself standing on the surface of the moon."

At the Acton high school, the children listen quietly while Cernan speaks and shows slides, but when he is done speaking to their parents, they have their own inquiries to make. A young girl asks, "How far can you jump on the moon?" Cernan lights up. "That's a good question. Does anyone know what gravity is?" Cernan had shown some nervousness before, talking to the adults, but it is gone now. For these children he describes the wonders of zero g, and how he opened a can of peaches and turned it upside down while he ate. He tells them mischievously, "If you want to learn to be a spaceman, go home tonight and try it." And there are more questions—How did you feel at liftoff? What color was the sky? Is it hot or cold in space? And Cernan answers them all, speaking well past the time anticipated, until a little girl asks the old standby: "What does it feel like when you're on the moon?"

Cernan smiles and bows his head for a moment. He begins to piece together an explanation. "You can move around very easily in one-sixth gravity, so it feels very comfortable. You're not warm, because your suit is air-conditioned with water." Then he interrupts himself and tries a different approach, one from the heart.

"I'll tell you what it feels like. It feels you're dreaming. You wonder when you're going to wake up. It's almost like your mom told you a wonderful story when you went to bed and, you know, sugar plums—it's like Santa Claus has already come. Being on the moon is like Santa Claus just gave you your wish."

Gene Cernan likes to point out a pair of photographs that share a frame on his office wall: one of himself at Taurus-Littrow, the other of Neil Armstrong at Tranquillity Base. The last man to walk on the moon and the first. They could not be more different in the way they have handled their positions; just as Cernan has embraced public attention, Armstrong has removed himself from it. Armstrong avoids public appearances and turns down most requests for interviews; he says little of his public life and nothing at all about his private life. And at most astronaut reunions, Armstrong is notable by his absence. In truth, Armstrong's response to the position that fate has given him was entirely in character.

In the fall of 1969, after the Apollo 11 postflight world tour, Armstrong returned to the Astronaut Office, but not for very long. He wanted to fly in space again, he says, "but when they asked me to move up to Washington, I guess that indicated to me that they had other thoughts of what I ought to be doing than flying." That turned out to be running NASA's aeronautics activities, a post that Armstrong held for a year. He left for one reason: he wanted to teach. In 1962 he had mentioned to the other astronauts that he planned to write an engineering textbook some day. No one was surprised when, in September 1971, Armstrong became an engineering professor at the University of Cincinnati.

In doing so, Armstrong left the mainstream behind and returned to the solitude of his native Ohio. He and his wife, Jan, purchased a dairy farm near the small city of Lebanon. To the rest of the world it seemed, as Mike Collins wrote in his memoir *Carrying the Fire*, that by leaving Washington, Armstrong had retreated to his castle and pulled up the drawbridge. At the mention of this, Armstrong laughs and says, "You know, those of us who live out in the hinterlands think that people that live *inside* the Beltway are the ones that have the problems."

Robert Hotz, formerly the editor of *Aviation Week* magazine and a longtime friend of Armstrong's, himself a farmer in rural

Maryland, says he completely understands what lured Armstrong away from NASA.

"Hell, you're in this high-tension world of aerospace. You get out on the farm. You look at the mountains across the valley, which are several million years old and are going to be there through the life of the planet. You understand that you're a short-term phenomenon, like the mosquitoes that come in the spring and fall. You get a perspective on yourself. You're getting back to the fundamentals of the planet. Neil feels that way, because we've talked about it, and so do I."

If Armstrong has kept a low profile, he has hardly been a recluse. In 1979 he left teaching for a variety of business activities. In 1986, he served as vice chairman of the presidential commission to investigate the *Challenger* disaster. He has appeared in advertisements for Chrysler, and has even hosted a cable TV documentary series on the history of flight. He gives assistance to historians who research Apollo. The common thread, in each case, is that Armstrong has met the world on his own terms, engaging only in activities he judges favorably. In short, Armstrong has handled the demands of his fame by rationing himself. While some of his colleagues, and others at NASA, wish he were a more visible spokesman for the agency and for the cause of space exploration, most of the others praise his approach. They say Armstrong was the ideal "choice" for the role of first man on the moon, as if it were an office to be filled. Were he more visible, they say, he would cheapen its currency.

Armstrong says he understands that fame is a direct result of media exposure, and that he has received a disproportionate share of it. Privately, he has said he can't understand why everyone focuses on the first lunar landing more than the other flights; after all, Apollo was a group effort. To him the title First Man on the Moon has little meaning, since in his mind the landing itself was the flight's most significant accomplishment—a feat he and Buzz Aldrin achieved at the same instant. But in general, the public and the media do not see it his way. Even other astronauts describe Armstrong's position as unique. However much they are besieged by phone calls and letters—from autograph seekers, space enthusiasts, documentary filmmakers, corporate executives—Armstrong gets more of it. One moonwalker says that when it comes to fame, if the astronauts were football players they would be a high school team, and Armstrong would be "the only guy in the NFL."

Perhaps the most famous picture of an astronaut is a photograph of Buzz Aldrin on the moon. He is facing the camera, standing at the edge of a small crater; in his mirrored visor is a tiny image of the photographer, Neil Armstrong, and the Sea of Tranquillity. If Armstrong's journey since that photograph was taken has been one of the most private, then Aldrin's has been among the most public and, at times, the most difficult.

Aldrin's troubles began shortly after he emerged from quarantine and began months of public appearances. Just as he had anticipated, he was uncomfortable with the spotlight and the role of PR spokesman for his agency. In 1971, after a period working on the Space Shuttle program, he decided to leave NASA to resume his air force career. Aldrin had hoped to be named commandant of cadets at the Air Force Academy; instead, after a ten-year absence from active duty, he was thrust into command of the test pilot school at Edwards. Aldrin, a non–test pilot with no managerial experience, found himself unable to perform up to his own high standards; his self-esteem was eroded. Within a year he was hospitalized for depression. Aldrin, who had spent his life striving for almost superhuman achievement, faced a devastating reality: he was sick and he needed help.

Two years later, he described his struggles against manic-depression and alcoholism in a confessional autobiography called *Return to Earth.* By going public, and in the process, debunking the myth of the perfect astronaut, Aldrin won himself both praise and criticism. But Aldrin was determined to tell the world what he went through, and to let others with similar problems know that they have company, even among men who walked on the moon. In doing so, he communicated his own humanity.

In the book, Aldrin charted a number of factors, including his own relentless drive for achievement, spurred by an intense and driven father. There had been other cases of depression in his family. Today Aldrin believes that given his family environment, coupled with his own personality, and possibly a genetic predisposition to the disease, he would probably have been destined for alcoholism late in life even if he had not gone to the moon. Instead, the crisis was accelerated by the double jeopardy that many returning moon voyagers faced: the sudden onslaught of world attention, followed almost immediately by the end of their astronaut careers. Aldrin, more than most of his colleagues, felt the loss: When Apollo 11 ended, so did his sense of purpose; that only made the hero's mantle harder to wear. And when he retired from the air force in 1972, he lost the ordered life he had thrived on since 1947, when he entered West Point. Without that structure, Aldrin says, his struggle only became more difficult. As he told an interviewer in 1993, "There I was—introverted, supersensitive, a perfectionist, concerned about what people thought of me. Jesus, it was a setup. No wonder I was in trouble."

By 1984 Aldrin's recovery was long completed. He had gone six years without a drink; his depression was under control. But somehow, a new purpose eluded him. The years at MIT and NASA, when he helped pioneer techniques for space rendezvous, were still a bright memory. Sometimes, looking into the shaving mirror, he told himself he would probably never experience that kind of creativity again.

But Aldrin was wrong. That year he again found inspiration in the intricate rhythms of orbital mechanics. This time, his innovation was the cycler, a spacecraft that, if placed on the proper trajectory, would circle continuously between the earth and Mars, ferrying astronauts and cargo with little need for fuel. Soon he was hard at work developing the cycler concept, along with ideas for an earth-orbit spaceport. By returning to the activity that gave him the most happiness—using his creative powers to advance space exploration—Aldrin had rediscovered his calling.

Today, happily remarried, trim and energetic, Aldrin is a one-man think tank, designing everything from new launch vehicles to scenarios for returning to the moon. But unlike thirty years ago, when Apollo was in full swing, he has found himself confronting a space program, its future clouded by uncertainty, that has not been ready for his innovations. Aldrin is undaunted. He spends his days networking with engineers, space advocates, and others who share his vision. As chairman of the National Space Society, he works to generate public support for future space activities. It isn't an easy way to make a living; Aldrin admits that he is probably too outspoken for most aerospace consulting work. "I don't want to tell the client what he wants to hear," he explained in 1992. "I want to tell them what I think." Instead of a livelihood, Aldrin says, his work is "an expensive hobby." As much as possible, he offsets his expenses by making personal appearances. And if his life still harbors frustrations and pitfalls—Aldrin says he has to be careful not to overwhelm himself with too many projects—then it has also given him great satisfaction. Today, with the struggles of the 1970s long behind him, he talks about his life with surprising openness and self-awareness. "I'm lucky," Aldrin says. "*I'm alive.* And . . . I'm a better person for having gone through that. I got a chance to redo my life."

For John Young, the man whom Lee Silver calls "the archetypical extraterrestrial," whether to remain an astronaut was never an issue. During NASA's long spaceflight hiatus of the 1970s, when his Apollo colleagues left to pursue other challenges, Young stayed where he was. By 1978 only a handful of veterans remained at the Johnson Space Center, where the astronaut corps was now infused with a new breed of young pilots, physicians, and scientists, men—and, for the first time, women—who had grown up with the reality of space travel. Some had been in grade school when Young copiloted the first Gemini flight. Now he was their chief, having taken over the office in 1974. At the time, the move had surprised a number of his colleagues who wondered how someone as quirky and difficult to know as Young would fare as a manager. No one questioned his dedication, his drive, or his engineering prowess. And he was still on flight status, training to command NASA's most important mission since the first moon landing: the maiden voyage of the reusable Space Shuttle.

On the day the shuttle program was approved by Congress back in 1972, Young was on the moon, and when he heard the news he said excitedly, "The country needs that shuttle mighty bad. You'll see." But making the shuttle a reality under the project's extremely tight budget taxed NASA's ingenuity. The shuttle orbiter would be a space plane the size of a DC-9, capable of flying 100 missions. To save money, designers were forced to scrap plans for a reusable, liquid-fueled booster. Instead the orbiter, mated to a fuel tank as big as a grain silo, would use its own liquid-fueled engines, with help from two powerful—and risky—solid rocket boosters. In all, the shuttle would be the most sophisticated flying machine ever built. Young focused his energies on the monumental technical difficulties of getting it ready for flight.

In April 1981, three years later than originally planned, the shuttle *Columbia* finally soared into orbit. When the TV camera was turned on, there was Young in the commander's seat, reading glasses on the bridge of his nose, smiling and saying, "The vehicle's performing like a champ." They had called Al Shepard the "Old Man" when he flew at age forty-seven; Young was fifty. A day later, after a reentry at twenty-five times the speed of sound, Young steered *Columbia* to a perfect landing on a desert runway at Edwards. Later, after the ground crews had arrived, Young emerged and bounded down the stairway to inspect his ship, punching the air with his fist like a relief pitcher who had just won the World Series. That day, Young told a crowd of well-wishers, "We're really not too far, the human race isn't, from going to the stars."

For a time it seemed that John Young's spaceflight career might be just as limitless. Late in 1983 he was back in orbit, this time as commander of a nine-day science mission called Spacelab 1. By January 1986, with twenty-four successful shuttle flights behind them, NASA was planning its most spectacular year since Apollo, with fifteen new launches. Young was to be part of it; he was unofficially slated for his seventh trip into space, this time to deploy NASA's scientific showpiece, the $1.5 billion Hubble Space Telescope. Friends say that he was as excited as a kid about the assignment. Then the shuttle *Challenger* exploded seventy-three seconds after launch, killing its seven-member crew, including schoolteacher Christa McAuliffe. The explosion shattered public perceptions about the seemingly routine nature of spaceflight, and about NASA's infallibility. Within the agency, amid shock and bewilderment, came the sudden realization that they had all been living too close to the edge. Young had bad dreams about the disaster for weeks.

In the aftermath, Young became an outspoken critic of his agency. One of his memos, citing an "awesome" list of potentially serious safety problems with the shuttle, was leaked to the press. NASA managers had repeatedly risked lives, it said, by letting the pressure of heavy launch schedules override safety concerns. And the agency was in danger of another accident with its plans for a fast-paced, post-*Challenger* launch schedule. Some colleagues said later that he was trying, in the only way he knew how, to protect his people. Feeling responsible, Young had strongly wanted to remain as chief astronaut until the shuttle was flying again. But in April 1987, in his twenty-fifth year as an astronaut, he was suddenly taken out of the Astronaut Office and made special assistant for engineering, operations, and safety to center director Aaron Cohen. Some, including Young, suspected that his criticisms had prompted the move, an idea that Cohen publicly denied. But the main reason seemed to be that it was simply time for him to move on. "I don't think *anybody* should stay in the same job for ten years," said one former high-level NASA official. "The problem is, you create legends bigger than life. I don't think it's good to create legends."

But John Young wasn't about to leave NASA. He told an interviewer, "I live the space program. I breathe it. I eat it. I sleep it. . . . I'm not willing to give it up as long as I can make a contribution to it." No one ever said Young's spaceflight career was over, and he continued to hope for another shuttle mission. And while he said he doesn't hold his two lunar voyages above any of his other flights—"You can't compare 'em"—Young maintained an enduring interest in the moon.

When the geologists held their annual Lunar and Planetary Science Conference, Young was always there, keeping up with new developments. In the mid-1980s he heard of the theory that still provides the best explanation for the origin of the moon: Soon after the earth formed, it was struck by an asteroid the size of Mars. So violent was this cataclysm that it vaporized parts of the earth's mantle and ejected the vapor into space, where it formed a disk of material that eventually coalesced to form the moon.

But at the Johnson Space Center, the moon had long faded from view. Apollo had receded into a nostalgic past, along with the ample space budgets of the 1960s. But in the early eighties a handful of space center scientists began to study scenarios for going back to the moon. Around the center their ideas were often greeted by skepticism or even ridicule, but John Young was one of their best allies, always encouraging, always interested. Meanwhile, outside NASA, interest was building in the far more audacious goal of sending humans to Mars. For these advocates the Red Planet was what the moon had been in 1961: mysterious, just beyond reach, awaiting the footsteps of human beings. To them, Mars was the goal NASA sorely needed to revitalize the space program and excite the American public. A return to the moon, they said, would only siphon off necessary resources for the Mars trip. The moon, they said, is boring.

"It's anything but boring," Young said quietly during an interview at the Johnson Space Center in 1989. "We don't even begin to understand it." When Young talks about the moon he speaks from the experience of confronting its unknowns, not only at the Descartes highlands but from an orbiting command module. He reminds us that it is a world with one-quarter the land area of the earth. "To think that twelve guys went there and we've figured it out, that's crazy."

Young also knows of the moon's enormous potential for astronomers. A telescope on the lunar surface would be free of the turbulence and light pollution that mars the view through the earth's atmosphere. All wavelengths of light would be accessible to it. On the lunar far side, radio telescopes would be free of any man-made interference. Telescopes of unprecedented size and resolving power could be built in the moon's one-sixth g. The late Harlan Smith of the University of Texas talked of building giant "nirvana telescopes" powerful enough to study the type in a newspaper on earth—or to see surface features in planets orbiting other stars.

Even more compelling, the moon may provide an answer to

the planet's pressing energy needs. Houston-based engineer David Criswell has designed solar power stations that could be set up on the moon and used to relay energy, in the form of microwaves, to receiving stations on earth. Manufactured from lunar materials, the first stations could produce 50 billion watts per year, Criswell says. Within forty years, if more were built, they could supply a whopping 20 trillion watts—the energy budget of the entire planet. The cost would be enormous, perhaps just shy of a trillion dollars. But selling power to the earth at a modest ten cents per kilowatt-hour, the venture could turn a profit after five years of operation.

Meanwhile, a group of scientists researching nuclear fusion have zeroed in on the moon's supply of helium-3, which was discovered in the Apollo soil samples in the early seventies. This isotope of helium is exceedingly rare on earth, but it is literally all over the moon; it is one of an assortment of gases deposited by the solar wind, the steady stream of charged particles emanating from the sun. A fusion reactor that used helium-3 as fuel, say the scientists, would be cleaner, safer, and more efficient than the systems now being developed. And since a metric ton of the gas could provide as much energy as about $3 billion worth of coal, it would be worth going to the moon to get it. Mining helium-3 would be a simple matter of scooping up moon dust, heating it, and collecting the gases that are driven off. And for the occupants of a lunar base, the process would also yield huge amounts of water, hydrogen, methane, and other gases valuable for survival and industry. In time, the scientists claim, helium-3 mining could sustain not only the earth's energy needs, but the life of an industrial community on the moon. It could even provide fuel for the Mars voyagers. Young says simply, "I think we should go put a base up there. That's what I think."

But in July 1989, as the twentieth anniversary of the first lunar landing approached, that seemed far from likely. By many accounts, NASA was only a shadow of the agency that had gone to the moon twenty years earlier. Young understood that a number of factors had contributed to the agency's decline, including a lack of direction from the top. What was necessary, he said, was for the president to give NASA, and the country, a new mission in space, to do what Kennedy had done twenty-eight years before.

On July 20, 1989, George Bush proposed the Space Exploration Initiative, a thirty-year effort that would include a permanent manned space station, a lunar base, and manned expeditions to Mars. And Young was full of enthusiasm. "I'd like to see it as a crash program," he said. He thought people would put their lives into it, the way they did in the old days.

By the end of 1992, the Space Exploration Initiative was dead, though it never officially disappeared from NASA's agenda. Two years in a row, the start-up funds had been voted down by Congress. In Houston, NASA's Office of Exploration, created to plan long-range programs, was disbanded. Meanwhile, space station *Freedom* was so mired in bureaucracy that Deke Slayton called it an "aerospace WPA." While outsiders questioned the station's value, NASA rewrote its purpose and reworked its design again and again. With $8 billion already spent, there was little more to show than a set of blueprints. By the following year, newly elected president Bill Clinton told NASA to cut costs and redesign the station yet again, then declared that it would become a joint program with the Russians. For a time in 1993, with the space station's fate still uncertain, the space center in Houston had become a place where, in the words of one thirty-year veteran, "people . . . are just holding down their jobs." By year's end, shuttle astronauts had successfully carried out a difficult repair mission in earth orbit to save the ailing Hubble Space Telescope. And for the first time in a long while, NASA's future looked brighter.

On a trip through Egypt in 1976, Stu Roosa and his wife visited a granite quarry near Aswan, where they saw an unfinished obelisk, perhaps thirty-five hundred years old. Had it been completed, this 1,100-ton monolith would have stood 137 feet high, the largest of its kind. But sometime before the artisans had finished their work, the stone cracked and they abandoned it where it is today, partially emerged from a rock slab on the valley floor. When Roosa thinks about Apollo, he often remembers the sight of that quarry in Aswan.

"I always thought Apollo was our unfinished obelisk," says Roosa. "It's like we started building this beautiful thing and then we quit." He shakes his head with a mixture of sadness and disbelief. "History will not be kind to us, because we were *stupid.*"

Today, at the NASA space centers in Houston and Florida, the Saturn V's for Apollo 18 and 19 lie on tourist stands, like unfinished obelisks, reminders of a time that seems now as remote as the moon itself. Across the distance of a quarter century, Apollo is an anomaly. There was a rare confluence of historical forces in 1961: A perceived threat to national prestige from the Soviet Union was met by a dynamic leader, John Kennedy, and economic prosperity allowed him to launch a massive effort to demonstrate America's capabilities. The moon was the ideal target—close enough to reach, audacious enough to capture the imagination.

Apollo happened so quickly that it all seems unreal. Eight years after Kennedy spoke to Congress, his challenge was met. "We couldn't *think* about it in that length of time today," says Ken Mattingly. "I tell all my friends, We could not go to the moon today. We *can not* do it."

Mattingly speaks out of frustration. For several years, as a consultant on the space station project, he saw firsthand the bureaucracy, the resistance to new ideas, and most important, the lack of national will. Of the station he says, "The damn country needs it but they don't know why!" Mastering the technology necessary to build a permanent workplace in earth orbit, he says, is essential preparation for tomorrow's far voyages.

"If you're ever going to Mars, you need that space station." Today, with the Apollo corporate memory nearly depleted from NASA and industry, Mattingly warns that we are in danger of forgetting how to explore space. "If you don't build things, you don't know *how* to build things. We can't handle a ten-year hiatus. There won't be anybody left. So is it a WPA? Maybe so, but it's a WPA with a purpose."

Today Mattingly works for General Dynamics' aerospace division in San Diego. It doesn't bother him that the nation lost interest in Apollo; that, he says, is a mark of progress—you have to accept the last step before you can take the next one. And if the glory days are behind us, Mattingly says, then that's where they should be.

"Apollo was the adolescence of space. Now we have to grow up and be adults." To Mattingly, that means the unglamorous but crucial work of making space pay its way, and he is trying to do just that. At General Dynamics Mattingly is working on revamping the venerable Atlas booster, a descendent of the rocket that put John Glenn in orbit in 1962. He hopes the Atlas will help open up cheaper access to space, so that the next generation can do the things he can only dream about. And when space begins to turn a profit, Mattingly says, then it will be time to explore again.

"We *will* go to Mars. And who knows what we'll find? Once again it will be the journey that is the true test, as much as what you learn when you get there." Mars comes no closer than 35 million miles; that trip, Mattingly says, is *really* leaving home. With present technology, the trip there will take six months. And don't expect the Mars voyagers to get the kind of attention that Mattingly's colleagues got: "When we send people to Mars, by the time they get there, and they call back, people are going to say, *Who?*" If Apollo was an anomaly, then the astronaut as national hero was equally so. And Mattingly stresses that he and the other moon voyagers were just symbols of the entire enterprise. They received the most attention, he says, but they didn't make the greatest contribution. Don't put astronauts on a pedestal: "There are extraordinary people *everywhere* in life . . . who are just as competent, just as cool, just as anything else you can mention. . . . All they need is an opportunity." Mattingly doesn't dwell on being one of the handful of men to go to the moon, though he says, "I wouldn't *trade* that for anything."

In Mattingly's mind what stands out most is what happened to him not in space but on earth. "It was being part of a team that was dedicated to something that transcended individual aspirations. That's what Apollo was. It was thousands of people who were willing to work day and night. . . . You can't imagine what that's like compared to an everyday experience."

It doesn't take space exploration to bring out the best in us, he says. If there is a lesson to be learned from Apollo, it is that we can do difficult things, when the objective is clearly defined, and when enough people and funds are dedicated to accomplishing it. From 1961 to 1972, the objective was as clear and inspiring as any you could ask for. You had only to go outside at night and look at it.

The indirect purpose . . . of all music played on the terraces of the audiences of the moon, seems to be to produce an agreement with reality.

—Wallace Stevens, *The Necessary Angel*

Even as he circled the moon, Alan Bean had promised himself to live his life the way he wanted to. In a sense, his vow was the opposite of Pete Conrad's. Bean was happy to let his moon trip influence him; he says it gave him the courage to drastically change his life. Today, the walls of Bean's condominium in suburban Houston are covered with mementos of his two spaceflights. But the main room of the house, the one where Bean spends most mornings, is the painting studio. Displayed around his easel are images Bean has created: Neil Armstrong, his gold visor reflecting the Sea of Tranquillity, unfurling an American flag in the vacuum; Ed Mitchell in midstride, map in hand, on his way to Cone crater; and Bean himself, poised in the footpad of the lunar module *Intrepid,* about to take his first step onto the lunar surface. Alan Bean has left spaceflight to record, as only he can, the experiences of the first human beings to visit another world. Nothing else, he says, could have lured him away from the Astronaut Office.

Bean did not leave abruptly. In July 1973, a month after Pete Conrad returned from Skylab, Bean took off for the station with his own crew. He didn't come home until the start of autumn. In their fifty-nine days in orbit Bean and his crewmates zapped the productivity meter so far off the scale that it was only halfway through the *next* Skylab mission that mission control stopped thinking there was something wrong with the third crew, whose performance was closer to normal. For Bean, like Pete Conrad, Skylab stands out above his moon flight—but not for the same reason. Circling the earth for weeks on end, Bean says, was less risky than going to the moon—but more demanding:

"I had a better personal feeling at the end of Skylab than I did at Apollo 12, *because* it took more self-discipline to work doing similar things day after day after day. On the moon mission it was different every day, and, I mean, anybody could do well on that, anybody that had been trained." In the end Bean is most proud of himself not for bravery, but productivity.

Then came the long wait for the Space Shuttle. Bean was getting more and more serious about painting, a hobby he had begun in 1962, in night classes while at Pax River. When the new crop of astronauts arrived Bean was in charge of training them, but occasionally he would take a couple of weeks off to try out the life of an artist. It could not have been more different from the life he'd led as a naval aviator and an astronaut, not only in substance but in image: jet pilots don't paint. But the more he painted, the more he realized he could do credit to the Apollo program in art. More than a century ago Frederic Remington captured the spirit of the Old West with his paintings of cowboys. In our time, Bean realized, one of the quintessential images is the moonwalker. In 1981, a few months after the first shuttle flight, Bean announced he was leaving NASA to become a full-time artist. His colleagues asked, "Are you sure you can make a living at that?" Bean wasn't sure, but he wasn't going to let that stop him. "If I can go to the moon," he told himself, "I can learn to be a good artist."

That happened with the aid of several Houston art teachers, especially a wildlife painter who became his mentor. The more Bean learned about painting, the more he realized that it shares an analytical quality with flying. Where he was once alert to the subtle change in pitch of a jet engine or the idiosyncrasies of an Apollo fuel cell, Bean now zeros in on the play of reflected light within a shadow, the subtle rainbow hues that hide on seemingly bland objects. Bean has brought his characteristic attention to detail to his new line of work. He is an expert on every variation of space suit, lunar module, and lunar roving vehicle. Near the easel sit the gloves Bean wore on space walks during his Skylab mission; now he uses them as a reference. It may take Bean more than a month to finish a painting; he is aiming for two hundred finished works over twenty years.

Bean has likened himself to a kind of tribal storyteller; his goal is to capture the moments that were important, or special, to the other astronauts. Sometimes he calls the astronaut he is painting to ask a question about the event. Sometimes, the others tell him their suggestions for new paintings. Bean's paintings have been exhibited in Houston galleries and have sold to private collectors and corporations. But of the customers Bean really wishes he could attract—the other astronauts—only one, shuttle astronaut Claude Nicollier, has bought one. He wryly admits they may not see $15,000 worth of art in one of his paintings. He wishes they did; he is sure they would enjoy owning their own lunar portrait more than they may realize. He seems wistful, saying this; he still sees in his colleagues the same loner tendencies that kept the Astronaut Office from feeling like a squadron. He says he wishes they could all network with each other and help each other; at the same time he knows that is against their nature.

Bean has thought about his fellow moon voyagers and the paths they have taken, and he believes there is a common thread. "I think that everyone who went to the moon came back more like they already were." And on the whole, his theory seems to hold up. Jim Irwin was a religious man before he went. Ed Mitchell was already interested in psychic phenomena. Neil Armstrong's retreat from the media spotlight, and Gene Cernan's acceptance of it, are entirely consistent with the people they were before Apollo. From Borman to Schmitt, Bean says, going to the moon only magnified tendencies that were already present.

And for Bean himself, going to the moon gave him the courage to choose a very different life. He did not know, when he made that promise to himself in lunar orbit, that it would lead to the painting studio. But looking back, he says, he was just following the call that was already inside him, and that, to quote the poet Robert Frost, has made all the difference. "I've been really happy since the moon trip," Bean says. "I feel blessed."

When Bean looks at the moon, he searches for the small bright dot that marks the place, out in the Ocean of Storms, where he and Pete Conrad walked. He wishes there would be some mystical feeling, but there never seems to be. He says he would love to go back, to see the moon with his new eyes, to find the subtle rainbows hiding in the gray-tan plains. If he could do it again, he'd put it all on hold for five minutes—the experiments, the sample gathering, the reporting—and just take it all in. *This is five minutes for me . . .*

• • •

Project Apollo remains the last great act this country has undertaken out of a sense of optimism, of looking forward to the future. That it came to fruition amid the upheaval of the sixties, alongside the carnage of the Vietnam War, only heightens the sense of irony and nostalgia, looking back twenty-five years later. By the time Apollo 11 landed, we were already a changed people; by the time of Apollo 17, we were irrevocably different from the nation we had been in 1961. It is the sense of purpose we felt then that seems as distant now as the moon itself. If NASA has lost direction, it is only because we have not chosen to give it one. Instead of letting the moon be the gateway to our future, we have let it become a brief chapter in our history. The irony is that in turning away from space exploration—whose progress is intimately linked to the future of mankind—we rob ourselves of the long-term vision we desperately need. Any society, if it is to flourish instead of merely survive, must strive to transcend its own limits. It is still as Kennedy said: Exploration, by virtue of difficulty, causes us to focus our abilities and make them better.

It is left to a future generation to return, to go about their work in the light of the earth, to pick up where Apollo left off. For a time it will still be magical to meet someone who has been to the moon, but gradually that mystique will fade, and a moon voyager will seem no more extraordinary than an explorer who has been to the Antarctic. And the moon will seem, as Jim Lovell likes to say, a little bit closer, only three days away.

But for the rest of this century, and probably for the rest of their lives, the men who have been to the moon still possess that special uniqueness that they are so quick to disavow. Until someone follows in their still-preserved footsteps, we are left to make the journey in our imaginations. It is a journey we can make, and should make, with our children: to look at the moon, and see Tranquillity, where Armstrong and Aldrin walked upon a stage one midsummer night, and the Ocean of Storms, and Fra Mauro, and Hadley, and Descartes, and Taurus-Littrow. To trace the paths of the orbiting command modules. To preserve not just the facts, the events, but the wonder.

In the end, when we confront the fact that human beings have been to the moon, we continue to have magic thoughts about the experience. We hold the magic, the awe, not the moon voyagers themselves. They flash quickly upon it and move on. The significant journey takes place not in their minds, but in ours.

Appendices

A: Astronaut Biographical Information

(source: Hawthorne, *Men and Women of Space*)

Edwin Eugene Aldrin, Jr. (legally changed to Buzz Aldrin, 1979)
born: January 20, 1930, Montclair, New Jersey
education: B.S., United States Military Academy, 1951; Sc.D., Massachusetts Institute of Technology, 1963
marriage: Joan A. Archer, 1954 (divorced); Beverly Van Zile, 1975 (divorced); Lois Driggs, 1987 (divorced)
children: James, 1955; Janice, 1957; Andrew, 1958
spaceflights: Gemini 12, Apollo 11
later activities: development of concepts for future space transportation systems.

William Alison Anders
born: October 17, 1933, Hong Kong; considers La Mesa, California, to be his hometown
education: B.S., United States Naval Academy, Annapolis, 1955; M.S. (nuclear engineering), Air Force Institute of Technology, 1962; completed Harvard Business School Advanced Management Program, 1979
marriage: Valerie Hoard, 1955
children: Alan, 1957; Glen, 1958; Gayle, 1960; Gregory, 1962; Eric, 1964; Diana, 1972
spaceflights: Apollo 8

later activities: Government (National Space Council, Atomic Energy Commissioner, Ambassador to Norway); business (aerospace, including CEO of General Dynamics)

Neil Alden Armstrong

born: August 5, 1930, Wapakoneta, Ohio
education: B.S. (aeronautical engineering), Purdue University, 1955
marriage: Janet Shearon, 1956 (divorced); Carol Knight, 1994
children: Eric, 1957; Karen, 1959 (deceased); Mark, 1963
spaceflights: Gemini 8, Apollo 11
later activities: Teaching; business (various); presidential commissions (including the Challenger investigation)
died: August 25, 2012, Cincinnati, Ohio, due to complications from heart surgery

Alan LaVem Bean

born: March 15, 1932, Wheeler, Texas; considered Forth Worth, Texas, to be his hometown
education: B.S. (aeronautical engineering), University of Texas at Austin, 1955; graduated from Navy Test Pilot School, Patuxent River, Maryland, 1960, and the School of Aviation Safety, University of Southern California, 1962
marriage: Sue Ragsdale, 1955 (divorced); Leslie Gombold, 1982
children: Clay, 1955; Amy, 1963
spaceflights: Apollo 12, Skylab 2
later activities: professional artist
died: May 26, 2018, Houston, Texas, following a sudden illness

Frank Borman

born: March 14, 1928, Gary, Indiana; grew up in Tucson, Arizona
education: B.S., United States Military Academy, 1950; M.S. (aeronautical engineering), California Institute of Technology, 1957; graduated from the Air Force Experimental Flight Test Pilot School, Edwards Air Force Base, California, 1960; graduated from the Air Force Aerospace Research Pilot School, 1961; completed Harvard Business School Advanced Management Program, 1970
marriage: Susan Bugbee, 1950
children: Frederick, 1951; Edwin, 1953
spaceflights: Gemini 7, Apollo 8
later activities: Business (commercial aviation, including president of Eastern Airlines; technology)

Roger Bruce Chaffee
born: February 15, 1935, Grand Rapids, Michigan
education: B.S. (aeronautical engineering), Purdue University, 1957
marriage: Martha Horn, 1957
children: Sheryl, 1958; Stephen, 1961
spaceflights: Apollo I (not flown)
died: January 27, 1967, Merritt Island, Florida, in a flash fire inside the Apollo 1 command module during a practice countdown

Eugene Andrew Cernan
born: March 14, 1934, Chicago, Illinois
education: B.S. (electrical engineering), Purdue University, 1956; M.S. (aeronautical engineering), Naval Postgraduate School, Monterey, California, 1964
marriage: Barbara Atchley, 1961 (divorced); Jan Nanna, 1987
children: Teresa ("Tracy"), 1963
spaceflights: Gemini 9, Apollo 10, Apollo 17
later activities: Business (aerospace management consulting)
died: January 16, 2017, Houston, Texas, following a lengthy illness

Michael Collins
born: October 31, 1930, Rome, Italy; considers Washington, D.C., to be his hometown
education: B.S. (military science), United States Military Academy, 1952; graduated from Aircraft Maintenance Officer School, 1958; graduated from Squadron Officer School, 1959; graduated from the Air Force Experimental Flight Test Pilot School, 1961; graduated from the Air Force Aerospace Research Pilot School, 1963; graduated from the Industrial College of the Armed Forces (by correspondence), 1971; attended Harvard Business School Advanced Management Program, 1974
marriage: Patricia Finnegan, 1957
children: Kathleen, 1959; Ann, 1961; Michael, 1963 (deceased)
spaceflights: Gemini 10, Apollo 11
later activities: Director, National Air and Space Museum; Undersecretary of the Smithsonian; business (aerospace); writing, painting

Charles Conrad, Jr.
born: June 2, 1930, Philadelphia, Pennsylvania
education: B.S. (aeronautical engineering), Princeton University, 1953; graduated from Naval Test Pilot School, 1958
marriage: Jane DuBose, 1953 (divorced); Nancy Fortner, 1990

children: Peter, 1954; Thomas, 1957; Andrew, 1959; Christopher, 1960 (deceased)
spaceflights: Gemini 5, Gemini 11, Apollo 12, Skylab 1
later activities: Business (aerospace, including vice president of the McDonnell Douglas Corporation)
died: July 8, 1999, Ojai, California, from injuries sustained in a motorcycle accident

Ronnie Walter Cunningham
born: March 16, 1932, Creston, Iowa; considers San Diego, California, his hometown
education: B.A. (physics), UCLA, 1960; M.A. (physics), UCLA, 1961; completed work for Ph.D., except for thesis, UCLA; completed Harvard Business School Advanced Management Program, 1974
marriage: Lo Irby, 1956 (divorced); Dorothy Vannerson, 1997
children: Brian, 1960; Kimberly, 1963
spaceflights: Apollo 7
later activities: Business (venture capital), public speaking

Charles Moss Duke, Jr.
born: October 3, 1935, Charlotte, North Carolina; considers Lancaster, South Carolina, to be his hometown
education: B.S. (naval sciences), United States Naval Academy, 1957; M.S. (aeronautics and astronautics), Massachusetts Institute of Technology, 1964; graduated from the Air Force Aerospace Research Pilot School, 1965
marriage: Dorothy Claiborne, 1963
children: Charles, 1965; Thomas, 1967
spaceflights: Apollo 16
later activities: Business (oil); Christian ministry

Donn Fulton Eisele
born: June 23, 1930, Columbus, Ohio
education: B.S., United States Naval Academy, 1952; M.S. (astronautics), Air Force Institute of Technology, 1960; graduated from the Air Force Aerospace Research Pilot School, 1962
marriage: Harriet Hamilton, (divorced); Susan Hearn, 1969
children: Melinda, 1954; Donn, 1956; Matthew, 1961 (deceased); Jon, 1964; Kristin, Andrew
spaceflights: Apollo 7

later activities: Director of Peace Corps, Thailand; business (various)
died: December 2, 1987, Tokyo, Japan, of a heart attack

Ronald Ellwin Evans
born: November 10, 1933, St. Francis, Kansas
education: B.S. (electrical engineering), University of Kansas, 1956; M.S. (aeronautical engineering), Naval Postgraduate School, 1964
marriage: Janet Pollom, 1957
children: Jaime, 1959; Jon, 1961
spaceflights: Apollo 17
later activities: Business (various); public speaking
died: April 7, 1990, Scottsdale, Arizona, of a heart attack

Richard Francis Gordon, Jr.
born: October 5, 1929, Seattle, Washington
education: B.S. (chemistry), University of Washington, 1951; graduated from the Naval Test Pilot School, 1957; graduate student in operations analysis at the Naval Postgraduate School, 1963
marriage: Barbara Field, 1953 (divorced); Linda Saunders, 1981 (deceased)
children: Carleen, 1954; Richard, 1955; Lawrence, 1957; Thomas, 1959; James, 1960 (deceased); Diane, 1961
spaceflights: Gemini 11, Apollo 12
later activities: Business (various)
died: November 6, 2017, San Marcos, California

Virgil Ivan Grissom
born: April 3, 1926, Mitchell, Indiana
education: B.S. (mechanical engineering), Purdue University, 1950; graduated from the Air Force Experimental Flight Test Pilot School, 1957
marriage: Betty Moore, 1945
children: Allan, 1950; Gary, 1953
spaceflights: Mercury-Redstone 4, Gemini 3, Apollo I (not flown)
died: January 27, 1967, Merritt Island, Florida, in a flash fire inside the Apollo 1 command module during a practice countdown

Fred Wallace Haise, Jr.
born: November 14, 1933, Biloxi, Mississippi
education: B.S. (aerospace engineering), University of Oklahoma,

1959; graduated from the Air Force Aerospace Research Pilot School, 1964
marriage: Mary Grant, 1954 (divorced); F. Patt Price, 1979
children: Mary, 1956; Frederick, 1958; Stephen, 1961; Thomas, 1970
spaceflights: Apollo 13
later activities: Business (aerospace)

James Benson Irwin
born: March 17, 1930, Pittsburgh, Pennsylvania
education: B.S. (naval science), United States Naval Academy, 1951; M.S. (aeronautical engineering), M.S. (instrumentation engineering), University of Michigan, both 1957; graduated from Squadron Officer School; graduated from the Air Command and Staff College; graduated from the Air Force Experimental Flight Test Pilot School, 1961; graduated from the Air Force Aerospace Research Pilot School, 1963
marriage: Mary Monroe, 1959
children: Joy, 1959; Jill, 1961; James, 1963; Jan, 1964; Joe, 1969 (adopted by the Irwins in 1973)
spaceflights: Apollo 15
later activities: Christian ministry
died: August 8, 1991, Glenwood Springs, Colorado, of a heart attack

James Arthur Lovell, Jr.
born: March 25, 1928, Cleveland, Ohio; considers Milwaukee, Wisconsin, to be his hometown
education: B.S., United States Naval Academy, 1952; graduated from the Naval Test Pilot School, 1958; graduated from the School of Aviation Safety, University of Southern California, 1961; completed Harvard Business School Advanced Management Program, 1971
marriage: Marilyn Gerlach, 1952
children: Barbara, 1953; James, 1955; Susan, 1958; Jeffrey, 1966
spaceflights: Gemini 7, Gemini 12, Apollo 8, Apollo 13
later activities: Business (telecommunications); public speaking

Thomas Kenneth Mattingly II
born: March 17, 1936, Chicago, Illinois
education: B.S. (aeronautical engineering), Auburn University, 1958; graduated from Air Force Aerospace Research Pilot School, 1966

marriage: Elizabeth Dailey, 1970 (separated; deceased, 1991)
children: Thomas, 1972
spaceflights: Apollo 16; Space Shuttle missions STS-4 and STS 51-C
later activities: Military space programs; business (aerospace)

James Alton McDivitt
born: June 10, 1929, Chicago, Illinois
education: B.S. (aeronautical engineering), University of Michigan, 1959; graduated from the Air Force Experimental Flight Test Pilot School, 1960; graduated from the Air Force Aerospace Research Pilot School, 1961
marriage: Patricia Haas, 1956 (divorced); Judith Odell, 1985
children: Michael, 1957; Ann, 1958; Patrick, 1960; Kathleen, 1966
spaceflights: Gemini 4, Apollo 9
later activities: Business (various)

Edgar Dean Mitchell
born: September 17, 1930, Hereford, Texas; considered Artesia, New Mexico, to be his hometown
education: B.S. (industrial management) Carnegie Institute of Technology (now Carnegie-Mellon University), 1952; B.S. (aeronautical engineering), Naval Postgraduate School, 1961; Sc.D. (aeronautics and astronautics), Massachusetts Institute of Technology, 1964; graduated from Air Force Aerospace Research Pilot School, 1966
marriage: Louise Randall, 1951 (divorced); Anita Rettig, 1973 (divorced); Sheilah Ledbetter, 1989 (divorced)
children: Karlyn, 1953; Elizabeth, 1959; Adam, 1984; Mitchell also adopted his second wife's children by a previous marriage Paul, 1963, and Mary, 1964.
spaceflights: Apollo 14
later activities: Founder, Institute of Noetic Sciences; business (consulting); public speaking
died: February 4, 2016, West Palm Beach, Florida

Stuart Allen Roosa
born: August 16, 1933, Durango, Colorado; grew up in Claremore, Oklahoma
education: B.S. (aeronautical engineering), University of Colorado, Boulder, 1960; graduated from the Air Command and Staff

College; graduated from the Air Force Aerospace Research Pilot School, 1965; completed Harvard Business School Advanced Management Program, 1973
marriage: Joan Barrett, 1957
children: Christopher, 1959; John, 1961; Stuart, 1962; Rosemary, 1963
spaceflights: Apollo 14
later activities: Business (president, Gulf Coast Coors)
died: December 12, 1994, Falls Church, Virginia, of complications from pancreatitis

Walter Marty Schirra, Jr.
born: March 12, 1923, Hackensack, New Jersey
education: B.S., United States Naval Academy, 1945; graduated from the Naval Test Pilot School, 1958
marriage: Josephine Fraser, 1946
children: Walter M., III, 1950; Suzanne, 1957
spacetlights: Mercury-Atlas 8, Gemini 6, Apollo 7
later activities: Business (various)
died: May 3, 2007, San Diego, California, of a heart attack

Harrison Hagan Schmitt
born: July 3, 1935, Santa Rita, New Mexico; grew up in Silver City, New Mexico
education: B.S. (geology), California Institute of Technology, 1957; studied at the University of Oslo, Norway, under a Fulbright Fellowship, 1957-58; Ph.D. (geology), Harvard University, 1964
marriage: Theresa Fitzgibbon, 1985
children: none
spaceflights: Apollo 17
later activities: Government (U.S. Senator); business (consulting in science, technology, public policy); teaching

Russell Louis Schweickart
born: October 25, 1935, Neptune, New Jersey
education: B.S. (aeronautical engineering), Massachusetts Institute of Technology, 1956; M.S. (aeronautics and astronautics), Massachusetts Institute of Technology, 1963
marriage: Clare Whitfield, 1958 (divorced); Nancy Ramsey, 1990
children: Vicki, 1959; Russell, 1960; Randolph, 1960; Elin, 1961; Diana, 1964
spaceflights: Apollo 9

later activities: Commissioner of energy, state of California; business (communications); founder and past president, Association of Space Explorers; public speaking (space development and global environmental issues); co-founder, B612 Foundation (for asteroid threat mitigation)

David Randolph Scott

born: June 6, 1932, Randolph Air Force Base, Texas
education: B.S. (military science), United States Military Academy, 1954; M.S. (aeronautics and astronautics), Massachusetts Institute of Technology, 1962; graduated from the Air Force Aerospace Research Pilot School, 1963
marriage: Lurton Ott, 1959
children: Tracy, 1961; Douglas, 1963
spaceflights: Gemini 8, Apollo 9, Apollo 15
later activities: Business (consulting in space technology); advisor on Apollo-related dramatic films and documentaries

Alan Bartlett Shepard, Jr.

born: November 18, 1923, East Derry, New Hampshire
education: B.S., United States Naval Academy, 1944; graduated from the Naval Test Pilot School, 1951; graduated from the Naval War College, Newport, Rhode Island, 1958
marriage: Louise Brewer, 1945
children: Laura, 1947; Juliana, 1951. The Shepards also raised a niece, Alice.
spaceflights: (Mercury-Redstone) MR-3; Apollo 14
later activities: Business (various); president of Mercury 7 Foundation (college scholarships in science and engineering)
died: July 21, 1998, Monterey, California, of leukemia

Thomas Patten Stafford

born: September 17, 1930, Weatherford, Oklahoma
education: B.S., United States Naval Academy, 1952; graduated from the Air Force Experimental Flight Test Pilot School, 1959
marriage: Faye Shoemaker, 1953 (divorced); Linda Dishman, 1988
children: Dionne, 1954; Karin, 1957
spaceflights: Gemini 6, Gemini 9, Apollo 10, Apollo-Soyuz Test Project (joint Soviet-American flight in 1975)
later activities: Business (consulting, aerospace, defense); advisor to NASA

John Leonard Swigert, Jr.
born: August 30, 1931, Denver, Colorado
education: B.S. (mechanical engineering), University of Colorado, 1953; M.S. (aerospace science), Rensselaer Polytechnic Institute, 1965; M.B.A. (business administration), University of Hartford, 1967
marriage: never married
children: none
spaceflights: Apollo 13
later activities: Government (elected to U.S. House of Representatives but died before taking office)
died: December 27, 1982, Georgetown University Hospital, Washington, D.C., of respiratory failure resulting from lung cancer

Edward Higgins White, II
born: November 14, 1930, San Antonio, Texas
education: B.S. (military science), United States Military Academy, 1952; M.S. (aeronautical engineering), University of Michigan, 1959; graduated from the Air Force Experimental Flight Test Pilot School, 1960
marriage: Patricia Finegan, 1953 (deceased 1983)
children: Edward H., III, 1953; Bonnie, 1956
spaceflights: Gemini 4, Apollo I (not flown)
died: January 27, 1967, Merritt Island, Florida, in a flash fire inside the Apollo 1 command module during a practice countdown

Alfred Merril Worden
born: February 7, 1932, Jackson, Michigan
education: B.S. (military science), United States Military Academy, 1955; M.S. (aeronautical and astronautical engineering), M.S. (instrumentation engineering), University of Michigan, 1963; graduated from the Instrument Pilots Instructor School, Randolph Air Force Base, Texas, 1963; graduated from the Empire Test Pilots School, Farnborough, England, 1965; graduated from the Air Force Aerospace Research Pilot School, 1965
marriage: Pamela Vander Beek, 1955 (divorced); Sandra Wilder, 1974 (divorced); Jill Hotchkiss, 1982
children: Merrill, 1958; Alison, 1960
spaceflights: Apollo 15
later activities: Business (technology development)

John Watts Young

born: September 24, 1930, San Francisco, California; considers Orlando, Florida, to be his hometown

education: B.S. (aeronautical engineering), Georgia Institute of Technology, 1952; graduated from the Naval Test Pilot School, 1959

marriage: Barbara White, 1955 (divorced); Susy Feldman, 1972

children: Sandy, 1957; John, 1959

spaceflights: Gemini 3, Gemini 10, Apollo 10, Apollo 16, (Space Shuttle) STS-1, STS-9/Spacelab 1

later activities: NASA administrative positions; space advocate

died: January 5, 2018, Houston, Texas, of complications from pneumonia

B: Persons Interviewed

(with affiliations during Apollo)

astronauts

Buzz Aldrin; Joseph P. Allen; William A. Anders; Neil A. Armstrong; Alan L. Bean; Frank Borman; Gerald P. Carr; Eugene A. Cernan; Michael Collins; Charles Conrad, Jr.; R. Walter Cunningham; Charles M. Duke, Jr.; Anthony W. England; Ronald E. Evans; Edward G. Gibson; Richard F. Gordon, Jr.; Fred W. Haise, Jr.; Karl G. Henize; James B. Irwin; Joseph P. Kerwin; James A. Lovell, Jr.; T. Kenneth Mattingly II; James A. McDivitt; Edgar D. Mitchell; Robert A. Parker; Stuart A. Roosa; Walter M. Schirra, Jr.; Harrison H. Schmitt; Russell L. Schweickart; David R. Scott; Alan B. Shepard, Jr.; Donald K. Slayton; Thomas P. Stafford; Alfred M. Worden; John W. Young

astronaut wives

Joan Aldrin; Valerie Anders; Susan Borman; JoAnn Carr; Jane Conrad; Dotty Duke; Jan Evans; Marilyn Lovell; Joan Roosa; Lurton Scott; Beth Williams

astronaut children

Jaime Evans; Tracy Scott

astronaut office manager, Kennedy Space Center

Charles Friedlander

Apollo program managers, Manned Spacecraft Center
George Abbey; Owen Morris

Apollo program managers, NASA Headquarters
Rocco Petrone; Samuel Phillips

crew systems engineers
David Ballard; Michael Brzezinski; John Covington; Raymond Zedekar

experiment scientists
Marcus Langseth; Jack Trombka

flight controllers
John Aaron; Steven Bales; Ronald Berry; Charles Deiterich; Robert Legler; John Llewellyn; Edward Pavelka

flight directors
Clifford Charlesworth; Peter Frank; Gerald Griffin; Eugene Kranz; Glynn Lunney

geologists
Robin Brett; Uel Clanton; John Dietrich; Michael Duke; Farouk El-Baz; Clifford Frondell; James Head; Friederich Hörz; Elbert King; Robert Laughon; Gary Lofgren; Harold Masursky; William Muehl-berger; William Phinney; Gerald Schaber; Eugene Shoemaker; Leon Silver; Gordon Swann; George Ulrich; Edward Wolfe

historians
Roger Bilstein; David Compton; Richard Hallion; Eric Jones; John Logsden; Alex Rowland

journalists
Howard Benedict; Robert Hotz; Roy Neal; Wayne Warga

mission planners
Rodney Rose; John Sevier; Howard Tindall

NASA administrators
Thomas Paine; James Beggs

NASA aircraft operations
Joseph Algranti

NASA managers, Headquarters
Chester Lee; Gerald Mossinghoff; Willis Shapley

NASA managers, Manned Spacecraft Center
Maxime Faget; Christopher Kraft

NASA public affairs officers
Jack Riley; Julian Scheer; Douglas Ward; Terry White

NASA X-15 and B-52 pilots
Johnnie Armstrong; William Dana; Milton Thompson

navy squadron pilots, VA44
Jack Keating; Jack Raider

North American Aviation
Joe Cuzzupoli; Charles Feltz; George Jeffs

simulator instructors and supervisors
Pleddy Baker; David Bragdon; Charles Floyd; Jay Honeycutt; Robert Pearson; Carl Shelley; David Strunk; Michael Wash

spacecraft engineers
Charles Mars

space suit technicians
Al Rochford; Troy Stewart

United States Information Agency
Simon Bourgin

C: Apollo Mission Data

Apollo 7 October 11–22, 1968
crew: Walter M. Schirra, Donn F. Eisele, R. Walter Cunningham
description: First manned earth-orbit test of the Apollo command and service modules.
mission duration: 10 days, 20 hours, 9 minutes
Apollo 8 December 21–27, 1968
crew: Frank Borman, James A. Lovell, Jr., William A. Anders
description: First manned flight around the moon. Apollo 8 orbited the moon 10 times on Christmas Eve, 1968.
time in lunar orbit: 20 hours, 7 minutes
mission duration: 6 days, 3 hours, 1 minute

Apollo 9 March 3–13, 1969
crew: James A. McDivitt, David R. Scott, Russell L. Schweickart

description: Earth-orbit test of the entire Apollo spacecraft. Included rendezvous maneuvers between the command module and lunar module, and a 38-minute space walk by Schweickart to test the lunar space suit and backpack.
spacecraft (command module, lunar module): *Gumdrop, Spider*
mission duration: 10 days, 1 hour, 1 minute

Apollo 10 May 18–26, 1969
crew: Thomas P. Stafford, John W. Young, Eugene A. Cernan
description: Dress rehearsal for the lunar landing.
spacecraft (command module, lunar module): *Charlie Brown, Snoopy*
time in lunar orbit: 2 days, 13 hours, 41 minutes
mission duration: 8 days, 0 hours, 3 minutes

Apollo 11 July 16–24, 1969
crew: Neil A. Armstrong, Michael Collins, Edwin E. Aldrin, Jr.
description: First lunar landing.
spacecraft (command module, lunar module): *Columbia, Eagle*
time in lunar orbit: 2 days, 11 hours, 34 minutes
lunar landing date, location: July 20, Sea of Tranquillity
time on lunar surface: 21 hours, 36 minutes
moonwalk duration: 2 hours, 31 minutes
pounds of samples collected: 47.7
mission duration: 8 days, 3 hours, 18 minutes

Apollo 12 November 14–24, 1969
crew: Charles Conrad, Jr., Richard F. Gordon, Jr., Alan L. Bean
description: Second lunar landing. Conrad and Bean were first to make a pinpoint lunar landing, touching down some 600 feet from the unmanned Surveyor 3 probe.
spacecraft (command module, lunar module): *Yankee Clipper, Intrepid*
time in lunar orbit: 3 days, 17 hours, 2 minutes
lunar landing date, location: November 19, Ocean of Storms
time on lunar surface: 1 day, 7 hours, 31 minutes
moonwalk durations: 1st: 3 hours, 56 minutes
2nd: 3 hours, 49 minutes
pounds of samples collected: 75.7
mission duration: 10 days, 4 hours, 36 minutes

Apollo 13 April 11–17, 1970

crew: James A. Lovell, Jr., John L. Swigert, Jr., Fred W. Haise, Jr.
description: Third lunar landing attempt; aborted following the explosion of an oxygen tank inside the service module.
spacecraft (command module, lunar module): *Odyssey*, *Aquarius*
mission duration: 5 days, 22 hours, 54 minutes

Apollo 14 January 31–February 9, 1971

crew: Alan B. Shepard, Jr., Stuart A. Roosa, Edgar D. Mitchell
description: Third lunar landing, and the first successful mission devoted entirely to scientific exploration of the moon.
spacecraft (command module, lunar module): *Kitty Hawk*, *Antares*
time in lunar orbit: 2 days, 18 hours, 40 minutes
lunar landing date, location: February 5, Fra Mauro
time on lunar surface: 1 day, 9 hours, 30 minutes
moonwalk durations: 1st: 4 hours, 47 minutes
2nd: 4 hours, 34 minutes
pounds of samples collected: 94.4
mission duration: 9 days, 0 hours, 2 minutes

Apollo 15 July 26–August 7, 1971

crew: David R. Scott, Alfred M. Worden, James B. Irwin
description: Fourth lunar landing. First of the extended scientific expeditions, called J-missions, featuring extended lunar stay time, long-duration backpacks, and the battery-powered Lunar Rover. The Apollo 15 service module was the first to be equipped with the new Scientific Instrument Module. Worden performed a 38–minute space walk to retrieve scientific film during the trip back to earth.
spacecraft (command module, lunar module): *Endeavour*, *Falcon*
time in lunar orbit: 6 days, 1 hour, 17 minutes
lunar landing date, location: July 30, Hadley-Apennine
time on lunar surface: 2 days, 18 hours, 54 minutes
moonwalk durations: standup: 33 minutes
1st: 6 hours, 32 minutes
2nd: 7 hours, 12 minutes
3rd: 4 hours, 49 minutes
pounds of samples collected: 169
mission duration: 12 days, 7 hours, 12 minutes

Apollo 16 April 16–27, 1972

crew: John W. Young, T. Kenneth Mattingly II, Charles M. Duke, Jr.
description: Fifth lunar landing; first exploration of the moon's central highlands. Included a space walk by Mattingly lasting 1 hour, 24 minutes.
spacecraft (command module, lunar module): *Casper, Orion*
time in lunar orbit: 5 days, 5 hours, 53 minutes
lunar landing date, location: April 20, Descartes highlands
time on lunar surface: 2 days, 23 hours, 2 minutes
moonwalk durations: 1st: 7 hours, 11 minutes
2nd: 7 hours, 23 minutes
3rd: 5 hours, 40 minutes
pounds of samples collected: 208.3
mission duration: 11 days, 1 hour, 51 minutes

Apollo 17 December 7–19, 1972

crew: Eugene A. Cernan, Ronald E. Evans, Harrison H. Schmitt
description: Sixth and final Apollo lunar landing. Schmitt became the first professional scientist to land on the moon. First U.S. manned night launch. Longest Apollo flight. Space walk by Evans lasted 1 hour, 7 minutes.
spacecraft (command module, lunar module): *America, Challenger*
time in lunar orbit: 6 days, 3 hours, 48 minutes
lunar landing date, location: December 11, Taurus-Littrow
time on lunar surface: 3 days, 2 hours, 59 minutes
moonwalk durations: 1st: 7 hours, 11 minutes
2nd: 7 hours, 36 minutes
3rd: 7 hours, 15 minutes
pounds of samples collected: 243.1
mission duration: 12 days, 13 hours, 51 minutes

Bibliography

A note about sources

My conversations with the astronauts, totaling hundreds of hours, are the heart of this book, but its framework is built from the extraordinary audio, video, and printed records of their missions. Every word that the astronauts transmitted to mission control was tape-recorded and transcribed. Many of their private conversations were captured by their spacecraft's onboard voice recorder. In the 1980s NASA declassified these tapes, which reveal thoughts and feelings—from irritation to awe—that the astronauts never shared with mission control. When the astronauts were back on earth, they reviewed their flights in detail; tapes and transcripts of these debriefings became another important source. Press kits, technical reports, and handbooks, prepared by NASA and Apollo contractors, provided other key data. Together, these records provided the means to reconstruct the events of the missions in detail, to form a framework on which to display the astronauts' experiences. Many illuminating perspectives came from the astronauts' wives and children, a number of whom participated in the project. Finally, I conducted dozens of background interviews with the people who planned the flights and monitored them from mission control, NASA managers, the scientists who trained the astronauts for their lunar explorations, and astronauts and engineers who worked in support roles before and during the missions.

I also benefited from access to the work of other researchers. At the NASA history office in Houston, David Compton shared many of the interviews he conducted for his own book on Apollo scientific exploration. At the history office of NASA Headquarters in Washington,

Lee Saegesser provided me with a number of unpublished sources, including interviews with astronauts and other Apollo participants conducted by writer Robert Sherrod during and after the program. Writer and filmmaker Al Reinart provided a tape of his interview with Apollo 13's Jack Swigert, who had died in 1982. Finally, Eric Jones's *Apollo 17 Lunar Surface Journal* was an essential source in writing the account of that mission.

Another body of source material exists in magazine and newspaper articles published at the time of the missions. Coverage in *Time, Newsweek, Life, National Geographic,* and *The New York Times* alerted me to a number of details to pursue in my conversations with the astronauts. The magazine *Aviation Week & Space Technology* was a helpful source for technical details.

Finally, a number of books deserve special mention. *Chariots for Apollo* by Brooks, Grimwood, and Swenson was an important reference on the Apollo spacecraft development, and many aspects of Apollo mission planning. *Apollo: The Race to the Moon* by Murray and Cox was an excellent source of information on the missions from the point of view of the flight controllers and mission planners. *First on the Moon* by Armstrong, Collins, and Aldrin provided a number of details on the Apollo 11 training, as well as biographical details for the three astronauts. Wilhelms's *To a Rocky Moon: A Geologist's History of Lunar Exploration* was an essential chronicle of Apollo's scientific evolution. And several astronauts' published accounts gave valuable perspectives on their personal experiences before and during their flights, and on the astronaut corps as a whole.

Aldrin, Buzz, and Malcolm McConnell. *Men from Earth.* 2nd ed. New York: Bantam Falcon Books, 1991.

Aldrin, Edwin E., Jr., with Wayne Warga. *Return to Earth.* New York: Random House, 1973.

Anderson, Frank W., Jr. *Orders of Magnitude: A History of* NACA *and* NASA, *1915–1976.* NASA History Series. Washington, D.C.: NASA, 1976.

Armstrong, Neil, Michael Collins, and Edwin E. Aldrin, Jr. *First on the Moon.* Boston: Little, Brown and Company, 1970.

Baker, David. *The History of Manned Spaceflight.* New Cavendish Books, 1981. Reprint. New York: Crown Publishers, 1982.

Beatty, J. Kelly, and Andrew Chaikin, eds. *The New Solar System.* 3d ed. Cambridge, Mass.: Sky Publishing; and Cambridge: Cambridge University Press, 1990.

Borman, Frank, with Robert J. Serling. *Countdown.* New York: Morrow, 1988.

Brooks, Courtney G., James M. Grimwood, and Loyd S. Swenson, Jr. *Chariots for Apollo: A History of Manned Lunar Spacecraft.* NASA SP-4205. Washington, D.C.: Government Printing Office, 1979.

Carpenter, M. Scott, L. Gordon Cooper, Jr., John H. Glenn, Jr., Virgil I. Grissom, Walter M. Schirra, Jr., Alan B. Shepard, Jr., and Donald K. Slayton. *We Seven.* New York: Simon & Schuster, 1962.

Collins, Michael. *Carrying the Fire: An Astronaut's Journeys.* New York: Farrar, Strauss & Giroux, 1974.

———. *Liftoff: The Story of America's Adventure in Space.* New York: Grove Press, 1988.

Compton, William David. *Living and Working in Space: A History of Skylab.* NASA SP-4208. Washington, D.C.: Government Printing Office.

———. *Where No Man Has Gone Before: A History of the Apollo Lunar Exploration Missions.* NASA SP-4214. Washington, D.C.: Government Printing Office, 1989.

Cooper, Henry S. F., Jr. *Apollo on the Moon.* New York: Dial, 1969.

———. *Moon Rocks.* New York: Dial, 1970.

———. *Thirteen: The Flight that Failed.* New York: Dial, 1973.

Cortright, Edgar M., ed. *Apollo Expeditions to the Moon.* NASA SP-350. Washington, D.C.: Government Printing Office, 1975.

Cunningham, Walter, with Mickey Herkowitz. *The All-American Boys.* New York: Macmillan, 1977.

Duke, Charlie and Dotty. *Moonwalker.* Nashville: Oliver-Nelson Books, 1990.

Gray, Mike. *Angle of Attack.* New York: W. W. Norton & Co., 1992.

Grissom, Betty, and Henry Still. *Starfall.* New York: Thomas Crowell, 1974.

Hacker, Barton C., and James M. Grimwood. *On the Shoulders of Titans: A History of Project Gemini.* NASA History Series. Washington, D.C.: NASA, 1977.

Hawthorne, Douglas B. *Men and Women of Space.* San Diego: Univelt, Inc., 1992.

Heiken, Grant, David Vaniman, and Bevan M. French. *Lunar Sourcebook: A User's Guide to the Moon.* New York: Cambridge University Press, 1991.

Irwin, James B., with William A. Emerson, Jr. *To Rule the Night.* Philadelphia: Holman (Lippincott), 1973.

Irwin, Mary, with Madalene Harris. *The Moon Is Not Enough.* Grand Rapids: The Zondervan Corporation, 1978.

King, Elbert A. *Moon Trip: A Personal Account of the Apollo Program and Its Science.* Houston: University of Houston Press, 1989.

Lay, Beirne, Jr. *Earthbound Astronauts: The Builders of Apollo-Saturn*. Englewood Cliffs, N.J.: Prentice-Hall, 1971.

Levine, Arnold S. *Managing NASA in the Apollo Era*. NASA SP-4102. Washington, D.C.: Government Printing Office, 1982.

Lewis, Richard S. *Appointment on the Moon*. New York: Ballantine, 1969.

———. *The Voyages of Apollo: The Exploration of the Moon*. New York Times Book Company, 1974.

MacKinnon, Douglas, and Joseph Baldanza. *Footprints*. Illustrated by Alan Bean. Washington, D.C.: Acropolis Books, 1989.

Mailer, Norman. *Of a Fire on the Moon*. Boston: Little, Brown, 1970.

Masursky, Harold, G. William Colton, and Farouk El-Baz, eds. *Apollo Over the Moon: A View from Orbit*. NASA SP-362. 1978.

Murray, Charles, and Catherine Bly Cox. *Apollo: The Race to the Moon*. New York: Simon & Schuster, 1989.

Mutch, Thomas A. *Geology of the Moon: A Stratigraphic View*. Princeton: Princeton University Press, 1970.

NASA. *Report of the Apollo 204 Review Board to the Administrator, National Aeronautics and Space Administration*. Washington, D.C.: Government Printing Office, April 4, 1967.

———. *The Apollo Spacecraft: A Chronology*. Vols. 1–4. Washington, D.C.: Government Printing Office, 1969–78.

NASA Lyndon B. Johnson Space Center. *Apollo 17 Preliminary Science Report*. NASA SP-330. Washington, D.C.: Government Printing Office, 1973.

———. *Biomedical Results of Apollo*. NASA SP-368. Washington, D.C.: Government Printing Office, 1975.

NASA Manned Spacecraft Center. *Analysis of Apollo 8 Photographs and Visual Observations*. NASA SP-201. Washington, D.C.: Government Printing Office, 1973.

———. *Apollo 11 Preliminary Science Report*. NASA SP-214. Washington, D.C.: Government Printing Office, 1969.

———. *Apollo 12 Preliminary Science Report*. NASA SP-235. Washington, D.C.: Government Printing Office, 1970.

———. *Analysis of Apollo 10 Photographs and Visual Observations*. NASA SP-232. Washington, D.C.: Government Printing Office, 1971a.

———. *Apollo 14 Preliminary Science Report*. NASA SP-272. Washington, D.C.: Government Printing Office, 1971b.

———. *Apollo 15 Preliminary Science Report*. NASA SP-289. Washington, D.C.: Government Printing Office, 1972a.

———. *Apollo 16 Preliminary Science Report*. NASA SP-315. Washington, D.C.: Government Printing Office, 1972b.

Newell, Homer E. *Beyond the Atmosphere: Early Years of Space Science*.

NASA SP-4211. Washington, D.C.: Government Printing Office, 1980.

Oberg, James E. *Red Star in Orbit.* New York: Random House, 1981.

O'Leary, Brian. *The Making of an Ex-Astronaut.* Boston: Houghton Mifflin, 1970.

Schirra, Walter M., Jr., with Richard N. Billings. *Schirra's Space.* Boston: Quinlan, 1988.

Surveyor Program [Office]. *Surveyor Program Results.* NASA SP-184. Washington, D.C.: Government Printing Office, 1969.

Taylor, Stuart Ross. *Lunar Science: A Post-Apollo View.* New York: Pergamon, 1975.

Ulrich, George E., Carroll Ann Hodges, and William R. Muehlberger. 1981. *Geology of the Apollo 16 Area, Central Lunar Highlands.* U.S. Geological Survey Professional Paper 1048. Washington, D.C.: Government Printing Office, 1981.

Wells, Helen T., Susan H. Whiteley, and Carrie E. Karegeannes. *Origins of NASA Names,* NASA SP-4402. Washington, D.C.: Government Printing Office, 1976.

Wilford, John N. *We Reach the Moon.* New York: Bantam Books, 1969.

Wilhelms, Don E. *The Geologic History of the Moon.* U.S. Geological Survey Professional Paper 1348. Washington, D.C.: Government Printing Office, 1987.

———. *To a Rocky Moon: A Geologist's History of Lunar Exploration.* Tucson and London: The University of Arizona Press, 1993.

Wolfe, Tom. *The Right Stuff.* New York: Farrar, Strauss & Giroux, 1979.

Worden, Alfred M. *Hello Earth: Greetings from* Endeavour. Los Angeles: Nash, 1974.

Author's Notes

Prologue

2 *Within days after Shepard's flight, he had made his decision:* Kennedy's decision is covered in detail in John Logsden's *The Decision to Go to the Moon: Project Apollo and the National Interest* (Chicago: University of Chicago Press, 1976) and Murray and Cox, *Apollo: The Race to the Moon.*

2 *the technological hurdles that would have to be cleared to build the Apollo spacecraft:* In 1960, among NASA's planned projects was a three-man spacecraft that would circumnavigate the moon. According to the NASA book *Origins of NASA Names,* by Ulrich et al., Abe Silverstein, NASA's Director of Spaceflight Development, proposed that it be called Apollo. The agency had already set a precedent by naming Project Mercury after a mythological figure, and Apollo had favorable connotations as the Greek god who pulled the sun across the sky in his golden, horse-drawn chariot each day. Apollo was also the god of archery, prophecy, poetry, and music. In 1961, following John Kennedy's challenge, Project Apollo became the name for the lunar landing program (*Origins of NASA Names,* p. 99).

4 *In 1959, Conrad had been one of sixty-nine young fliers:* Carpenter et al., *We Seven,* p. 6.

BOOK ONE

Chapter 1: "Fire in the Cockpit!"

13 Skepticism about Grissom and the Mercury hatch: In his book, *Schirra's Space* (p. 75), Wally Schirra points out that on his own Mercury mission he purposely blew the hatch after his spacecraft was lifted onto the deck of the recovery carrier. The recoil of the actuating handle was so violent that Schirra's hand was cut, even through his space suit glove. Grissom had not

sustained even a minor injury anywhere on his body—strongly supporting his claim.

13 *the command module simulator here at the Cape was a constant source of difficulty:* So many changes were being made to Grissom's spacecraft that the simultor crews could not keep up; the result was that the simulator never accurately reflected the precise configuration of the real command module. At the same time, the simulator's computer software underwent frequent changes, and often there were bugs which caused the simulator to break down (Murray and Cox, *Apollo: The Race to the Moon*, p. 186).

14 *Finally the problem was solved, and at 2:45 P.M.:* Sequence of events is from Report of the Apollo 204 Review Board, section 4.

14 *seated next to Roosa at the Stony console, Deke Slayton listened:* Helpful sources for Slayton's actions before, during, and after the fire were his witness statement before the Apollo 204 Review Board, made on February 8, 1967, and "The Ten Desperate Minutes," *Life*, April 21, 1967.

15 *Years later, he would still wonder whether or not he made the right choice:* Slayton had thought he might have been able to notice the spark thought to have triggered the fire, and to extinguish it before anything could happen. In hindsight, it seems a long shot, given the speed with which the fire spread.

16 *Block I . . . was never built to go to the moon:* Because Block I had no docking system, it could not link up with a lunar module, the spacecraft that would be used to land on the moon. It also lacked the proper guidance and communications systems for the lunar voyage.

16 *there would be only one Block I mission, Apollo 1:* Until December 1966, there had been two manned Block I flights scheduled: Grissom's, and a second earth-orbit test commanded by Wally Schirra. (At that time, Tom Stafford was on the backup crew for the second mission, along with Frank Borman and Mike Collins.) In December 1966, Schirra's flight was canceled on the grounds that it would be an unnecessary duplication of Grissom's. In its place, NASA planned the first Block II mission, with Jim McDivitt, Dave Scott, and Rusty Schweickart as the crew, and Stafford, Young, and Cernan as their backups. But on January 27, 1967, the Block II spacecraft had not yet been built.

16 *Wires were constantly being rerouted, black boxes replaced:* This information on the second Block I spacecraft is from Collins, *Carrying the Fire*, p. 256.

16 *The McDonnell Aircraft Corporation . . . had forged a harmonious relationship with NASA:* This was not the case at North American, whose relationship with NASA had been strained by personality conflicts from the beginning of Apollo. Charlie Frick, the man who preceded Joe Shea as head

of NASA's Apollo Spacecraft Program Office, was overbearing and even belligerent in his approach to North American. He and North American manager John Paup were at odds from the beginning. This conflict, and the events at North American leading up to the Apollo 1 fire, are described in Gray, *Angle of Attack.*

17 *But now Slayton heard another voice, clearly frantic:* One astronaut who listened repeatedly to the voice tapes from the fire says he never heard panic in Chaffee's voice; instead there was a sense that they were being rapidly engulfed and things were happening very fast. He adds, "You hear it time and again in [the test flight] business. . . . They may have known there wasn't any hope."

22 *Much of the once spotless cabin was covered with soot:* The appearance of the interior of the spacecraft is described in the Report of the Apollo 204 Review Board, section 5.

23 *"If there's ever a serious accident in the program, it's probably going to be me.":* Grissom and Still, *Starfall,* p. 172.

23 *"If we die, we want people to accept it":* Armstrong et al., *First on the Moon,* p. 49.

23 Weight and complexity of an oxygen/nitrogen mixture: Charlie Feltz, who was North American's chief engineer in the early days of Apollo, says that initially he pushed for a two-gas system precisely because pure oxygen was so hazardous, but that he was overruled by NASA. And according to Chris Kraft, the test-flight veteran who became one of NASA's central figures in Apollo, there were daunting technical obstacles to using an oxygen/nitrogen mixture. Instrumentation precise enough to monitor the composition of the mixture had not yet been developed.

23 *there was a spark inside Apollo 1:* For an in-depth discussion of the causes of the fire, see *Apollo: The Race to the Moon,* pp. 190–91 and 214.

24 *Velcro fasteners . . . exploded in a shower of fireballs:* At the time of the fire, there was ten times as much Velcro in the Apollo 1 cabin as originally specified, because the astronauts, who had always customized their spacecraft, wanted more of it (*Apollo: The Race to the Moon,* p. 214).

24 *But the managers . . . always had sound reasons for vetoing the change:* Slayton was one of several astronauts who told the author that even before the fire there were plans to change to a one-piece hatch. Furthermore, Charlie Feltz says that his original design called for an outward-opening hatch, but that NASA vetoed it in favor of the inward-opening design.

24 *Each pound of payload cost many times its own weight:* Adding weight to the command module was complicated by the fact that its designers had given it an offset center of gravity, as a means of generating lift—and

therefore, trajectory control—during reentry into the earth's atmosphere. Much of the spacecraft's weight was concentrated on the side away from the hatch. Making the hatch heavier required additional weight on the opposite side as ballast. Otherwise, the center of gravity would be shifted toward the middle of the command module, lessening its lift-generating ability.

24 *White and his backup, Dave Scott, used to practice opening the hatch for exercise:* The hatch weighed about 90 pounds, but lifting it was even harder because the astronaut had to reach back over his head.

25 *The greatest irony was that Gus Grissom:* The irony is even greater considering the fact that NASA managers vetoed the outward-opening hatch largely *because* they wanted to avoid a repeat of Grissom's Mercury incident.

Chapter 2: The Office

31 *Conrad ended up carrying Glenn's bags as well as his own:* This situation is described in Wolfe, *The Right Stuff,* p. 387.

32 *an extra $16,000 per year:* The amount of money for each man declined to around $10,000 when nineteen new astronauts were selected in 1966. Field Enterprises withdrew from the contract before the first Apollo flight. *Life* published its last astronaut-written stories in 1970, after the Apollo 13 mission.

The *Life* contract was a source of much controversy and brought NASA criticism from other news organizations. NASA administrator Jim Webb was against it. The astronauts, for their part, were grateful to have it: for their families, it offered a shield from the media; for the pilots themselves, it was an excuse to turn down innumerable requests for interviews. Says Walt Cunningham, "You can't imagine the pressure from the media chasing us for stories." The ethical question of whether the astronauts should be paid for their stories was settled, at least officially, when John Glenn spent a weekend with John Kennedy and his family at the presidential retreat at Hyannis Port, Massachusetts. Glenn argued that the astronauts deserved to be compensated for the invasion of their privacy that was required to satisfy the enormous public interest in them—which was over and above the news coverage of their missions. Kennedy agreed, as described in "Heroes, not of their own accord," a 1977 master's thesis by Perry Michael Whye, Iowa State University.

42 *announced that Gus Grissom was going to command Gemini 3 and that his copilot would be John Young:* According to other astronauts, the process of picking Grissom's copilot had a few twists and turns. In late February or early March 1963 Slayton called the astronauts into the briefing room at Ellington and told them that Al Shepard and Tom Stafford would fly

Gemini 3. Three weeks or a month later, Shepard was grounded. Everyone assumed Stafford would be named to fly with Grissom. But Frank Borman tells the story that he, not Stafford, was told he would be Grissom's copilot, and went over to Grissom's house to meet with him. The next thing he knew, Borman says, John Young had been named in his place. Evidently Grissom had stepped in and made his own choice.

44 *Al Shepard, who had been appointed chief of the Astronaut Office:* Just as Slayton's role was analogous to that of a wing commander, Shepard functioned much like a squadron commander, and reported directly to Slayton. While their duties overlapped, Slayton generally handled the astronauts' dealings with NASA management and with contractors. Shepard ran the internal affairs of the Astronaut Office, including scheduling travel for business and public relations, working out training schedules, and so on.

44 *In the Astronaut Office, Shepard usually kept a chilly distance from his troops:* Shepard's secretary kept on her desk a cube whose sides displayed cartoon faces with different expressions—smiling, scowling, blank, and so on. Depending on what mood her boss was in, she would orient the cube so that the appropriate face was visible to any astronaut walking past her desk. Thus warned, the pilots could, if necessary, perform an evasive maneuver.

45 *flying wasn't the basis of this competition:* The competition definitely spilled over into flying, however. On cross-country flights, if two teams of astronauts were flying two T-38's, they would race. If the conditions were just right, and if the pilots didn't waste any fuel taking off, they could make it all the way from L.A. to Houston without refueling. Even then, some astronauts landed with the last wisps of fuel burning in the engines.

47 *There were five others like him among the Fourteen:* Besides Cunningham and Schweickart, the other non-test pilots were Buzz Aldrin, Gene Cernan, and Roger Chaffee.

49 *paid little attention to these events:* However, civil rights had briefly become an issue within the astronaut corps in 1963. A black pilot named Ed Dwight, who had barely graduated from Chuck Yeager's space school, had applied for the astronaut program and been rejected. The astronauts were on a desert-survival course in Nevada when Slayton was summoned for a phone call from Washington. When he returned, he told the other astronauts that he had just spoken to Attorney General Robert Kennedy, who wanted NASA to accept Dwight. Slayton told the pilots, "I just spoke for all you guys. . . . I said if we had to take him and he wasn't qualified, then they'd have to find sixteen other people, because all of us would leave."

In his autobiography, Yeager wrote that during that time there were still relatively few black air force pilots, but that Dwight "sure as hell didn't represent the top of the talent pool. I had flown with outstanding [black] pilots like Emmett Hatch and Eddie Lavelle; but unfortunately, guys of their quality didn't apply for the [space school]. Dwight did." Yeager says that despite the fact that Dwight lacked the flying experience and engineering background to be admitted to his school, the White House pressured the air force to enroll him. "The only prejudice against Dwight," Yeager wrote, "was a conviction shared by all the instructors that he was not qualified to be in the school." With tutoring, Dwight graduated. But he was not selected as an astronaut. In 1967, the air force selected a black graduate of the space school, Robert Lawrence, for its Manned Orbiting Laboratory program. Shortly afterward, Lawrence was killed in the crash of his F-104 jet.

53 *They wryly called themselves the "Original 19":* The name was coined by John Young, who had served on the astronaut selection board in March 1966. (Collins, *Carrying the Fire,* p. 127.)

Chapter 3: First Around the Moon

I: The Decision

56 *Technical problems with the lunar module:* For a detailed account of the lunar module's development, see Brooks, et al., *Chariots for Apollo: A History of Manned Lunar Spacecraft.*

57 *It was the brainchild of George Low:* After the Fire, Low was named to replace Joe Shea as head of the Apollo Spacecraft Program Office.

58 *he and his deputy George Mueller:* Mueller pronounced his name "Miller."

58 *Webb . . . already knew his tenure would end when Johnson left office:* Webb went to see Johnson in September 1968 to discuss resigning; to his surprise, Johnson was ready to accept his resignation that day. Webb left NASA on October 7, only days before the first manned Apollo flight. (Murray and Cox, *Apollo: The Race to the Moon,* p. 323.)

61 *Bob Gilruth, the head of the Manned Spacecraft Center:* Gilruth, a pioneer from the early days of flight testing, was one of the giants of the manned space program. His close association with the Original 7—whom he sometimes called his "boys"—was said to be the source of much of the Mercury astronauts' power within NASA.

63 *Jim proudly proclaimed the "flight" a qualified success: Life,* September 27, 1963, p. 86a.

67 *a single meeting one August afternoon in the office of Chris Kraft:* In truth, this meeting could only lay out the basic objectives for the mission. The detailed planning of every aspect—from control of the spacecraft's trajectory to the astronauts' sleep schedules—was conducted in a series of

weekly Flight Operations Plans, chaired by engineer Rod Rose. For a definitive account of this and the other Apollo missions from the perspective of mission planners and controllers, including the personalities involved, see Murray and Cox, *Apollo: The Race to the Moon.*

70 *The translunar crossing of some 234,000 miles:* The moon's distance from earth varies from 221,500 miles to 252,700 miles. At the moment Apollo 8 went into lunar orbit it was 233,900 miles.

70 *mascons had to be understood:* Planners knew that the closer a spacecraft got to the moon, the greater the effects of any gravitational irregularities. The Lunar Orbiter probes had orbited at much greater altitudes, generally several hundred miles, than those planned for Apollo. The trajectory people could only guess how severely mascons might affect a command module 69 miles above the moon, or a lunar module at 50,000 feet, heading for a landing.

74 *Even when he wasn't in the simulator, Anders was learning his machine:* Anders was the systems expert, but to the press, he was also the rookie—something that irritated him no end. Borman and Lovell picked up on it, of course, which bothered him even more. Anders would only respond, wryly, "When it comes to going to the moon, everyone's a rookie."

75 *The odds of an astronaut's survival: Apollo: The Race to the Moon,* p. 102.

76 *After one test, Eisele quipped:* Baker, *The History of Manned Spaceflight,* p. 310.

77 *Whether another Soyuz was being readied for a circumlunar mission, no one knew:* In 1988 veteran cosmonaut Aleksey Leonov, the first man to walk in space, told the author that in the fall of 1968 he had been training to command the first circumlunar mission, with Oleg Makarov as his flight engineer. The men were to fly a spacecraft called Zond, which was a variant on the Soyuz design. For information on the Soviet plans see note for page 135 (Chapter 4, Part I).

77 *Borman cut him off:* According to Borman, there was talk of having one of the astronauts climb outside and do a spacewalk, with all the risks that entailed. Borman vetoed the idea.

77 *behind that macho, take-charge exterior was a very apprehensive astronaut:* As evidence that his behavior had nothing to do with apprehension about the risk of going to the moon, Borman says he was just as rigid and impatient after the Fire, during the recovery efforts at North American. For example, on those occasions when an astronaut appeared uninvited at Downey to champion some pet improvement for the command module, Borman sent him home. Finally he called Slayton in exasperation and said, "No more astronauts!" Borman's behavior raised more than a few hackles

in the Astronaut Office. But they couldn't afford to make a "gold-plated" command module, he said; they had no time to waste.

79 Lindbergh's visit: Some details are provided in "The Heron and the Astronaut," by Anne Morrow Lindbergh, *Life,* February 28, 1969, p. 19. Other information in Jim Lovell's letter to the author, 1990.

80 *most powerful thrust machine ever flown:* At this time, the Soviets were trying to develop two heavy-lift boosters: the G-1, which is said to have been slightly less powerful than the Saturn V, and the N-1, which would have been more powerful. Neither of these was successfully launched, however. See also note for page 135 (Chapter 4, part I).

II: A Hole in the Stars

85 *the launch pad's automatic sequencer took over:* The Terminal Countdown Automatic Sequencer, located within the mobile launch platform, was an electromechanical device that had no computer software.

89 *some 24,226 miles per hour, the speed necessary to reach the moon:* This was slightly less than the so-called *escape velocity* of 25,020 miles per hour, necessary for an astronaut to blast free of the earth's gravitational pull altogether. The Apollo astronauts did not need to go that fast; they needed only to go fast enough to reach the moon's sphere of gravitational influence.

92 *simply by floating around, an astronaut would push his vestibular system over the edge:* According to Jim Lovell, even on Gemini many astronauts "were nauseated at the beginning of their flights, but no one wanted to admit it." One of them was Gene Cernan, who told the author that he felt ill for the first day of his Gemini 9 mission. Lovell himself says he did not experience any discomfort on his two Gemini missions, but did feel some nausea when he first began to move around inside Apollo 8 soon after launch.

97 *chase down stray bits of vomit and feces with paper towels:* Today, Anders says he can't stand the smell of airline towelettes because it reminds him of this whole episode.

98–99 Borman's conversation about his illness: Most astronauts were reluctant to make public any details of their in-flight medical problems. This conversation is not recorded in the air-to-ground transcripts and was not available on tape from NASA. It was, however, described by Sam Phillips in his article "Apollo 8: A Most Fantastic Voyage," *National Geographic,* May 1969, p. 613. The exchange between Borman and Mike Collins is taken from the recording "To the Moon," produced by Time/Life records, 1969.

99 *he conversed with Capcom Ken Mattingly:* Mission controllers, including Capcoms, worked in eight-hour shifts. Mike Collins, Ken Mattingly, and Jerry Carr were the three main Capcoms for Apollo 8.

100 *When they turned the camera on the brilliant blue and white planet:* To the astronauts, the earth was some forty times brighter than the full moon appears to us.

101 Not much chance of that, *Anders thought:* Another source of Anders's worry was the newly designed, outward-opening side hatch. Every now and then he would glance warily at it. True, that hatch would have saved their lives in the event of a fire on the launch pad or some other emergency on the ground. But with the vacuum of space on the other side of it, a hatch that was easy to open just didn't give him a very warm feeling. Anders hoped the designers had chosen wisely; he would worry about that outward-opening hatch for the rest of the flight.

101 *Could Jules Verne have imagined the view from Apollo 8?:* As Lovell himself was well aware, there were eerie similarities between Apollo 8 and the flight described in Jules Verne's *From the Earth to the Moon* a century before. Verne's moon voyagers were three in number. Their names even bore a slight resemblance to those of the Apollo 8 crew: Barbicane, Nicholl, and Ardan. They were launched from Florida, in December, and recovered in the ocean. One significant difference was that Verne left his moon voyagers stranded in lunar orbit. After his readers protested, Verne brought his astronauts back to earth in the sequel, *Around the Moon.*

III: "In the Beginning . . ."

107 *its rendezvous with the moon at more than 5,000 miles per hour:* Apollo 8's speed relative to the moon reached 5,700 miles per hour at the time of engine ignition. But some 40 percent of that speed was due to the fact that the moon was racing along in the opposite direction at 2,300 miles per hour.

108 *"Alright, alright, come on," Borman said:* On the onboard voice tapes, there was about a minute's worth of conversation between Anders's "*Look* at that" and Borman's "Alright, alright, come on. . . ." During that time Borman tried to steer Anders back to the reading of the checklist, which he did for a short time, until his curiosity got the better of him and he had to take a second look. *"Fantastic,"* he said. "But you know, I still have trouble telling the holes from the bumps—" At which point, Borman cut him off.

110 *beach sand darkened by the cold embers of bonfires:* When Anders repeated that description in a TV transmission during the second orbit, he won himself a storm of hate mail from poets.

110 *"What does the ol' moon look like from sixty miles?"*: NASA figured distance in nautical miles (one nautical mile is equal to 1.15 statute miles). The figure heard in Apollo mission dialogue is not 69 miles but 60 nautical miles.

111 *it was still Arthur C. Clarke's moon:* The moon as it appears in *2001* arguably belongs as much to the film's producer, Stanley Kubrick, but it is Clarke that Anders mentions.

113 *Lovell, following the explorer's prerogative, had named them:* Anders went even further, giving names to dozens of craters on the far side. His choices were made to honor national leaders, NASA managers, and flight controllers. And he named three modest-sized craters for Borman, Lovell, and himself. He chose their locations near the boundary between the far side and the near side, so that when future astronauts flew over them they would be in radio contact with earth: "Roger, Houston. We're now over Anders . . ." Unfortunately, Anders's names were not approved by the International Astronomical Union, the recognized authority for all named features on the moon and planets. The IAU christened three craters for Borman, Lovell, and Anders, but to Anders's irritation they are located within a region of the farside that he and his crewmates never saw: it was in shadow during the flight.

115 *"How fast are they going now?"*: "Christmas cheers on the Apollo 8 home front," by Dora Jane Hamblin, *Life,* January 10, 1969, p. 790.

120 *with the help of a friend in Washington, Borman found something:* Borman sought the advice of a friend in the U.S. Information Agency, a man named Simon Bourgin who had been a traveling companion on the world tour that followed Gemini 7. Sometime later he got a letter from Bourgin with an idea for the broadcast, courtesy of a Washington newspaperman Bourgin knew. The newspaperman, Joe Laitin, had in turn gotten the idea from his wife.

121 *the moon's . . . appearance near lunar sunrise and sunset:* Apollo 8 orbited from east to west; this meant that, for example, the astronauts experienced sunrise over the part of the moon where sunset was occurring.

IV: "It's All Over but the Shouting"

126 *"Please be informed there is a Santa Claus"*: When he composed that line, Lovell was thinking of the turn-of-the-century newspaper columnist who reassured a little girl of the existence of St. Nick: "Yes, Virginia . . ."

126 *if they made a single mistake on the rest of the flight the brandy would get the blame:* Even without brandy, Lovell's fatigue got the better of him and he called up the wrong program on the computer, wiping out all its stored information about Apollo 8's orientation in space. An anxious half-

hour followed while Lovell worked to realign the command module's navigation platform. For a while afterwards, Borman worried that somehow the episode might have affected the part of the computer's memory that would handle the reentry maneuvers; Houston assured him there was nothing to worry about.

130 *Every so often what looked to be a fist-size chunk shot by:* Later, Anders would learn that those "chunks" had been mere pea-sized bits, each surrounded by a halo of glowing gas.

Chapter 4: "Before This Decade Is Out"

I: The Parlay

135 *the Soviets were now talking about missions in earth orbit:* In their congratulatory telegram to the Americans following Apollo 8's success, the Soviets went so far as to say that the moon race had never existed. Not only was this not true, but it is now clear that the Soviets continued to plan a moon program for some time after Apollo 8, even after it was clear they would lose the space race with the Americans. Their lunar landing attempt hinged on the giant N-1 booster, designed to deliver 10 million pounds of thrust. Four unmanned launches ended in disaster when the rocket exploded. The third of these occurred on July 3, 1969, only thirteen days before the scheduled launch of Apollo 11. The fourth took place late in 1972. For more information, see *Red Star in Orbit* by James Oberg; *Men from Earth* by Buzz Aldrin and Malcolm McConnell (second edition); "Russians Reveal Secrets of Mir, Buran, Lunar Landing Craft" by Craig Covault, in *Aviation Week & Space Technology*, February 10, 1992, pp. 38–39; and the PBS television *Nova* documentary, "The Russian Right Stuff."

148 Preliminary timeline for moonwalk, showing lunar module pilot out first: According to the NASA history *Chariots for Apollo*, such a checklist was written in 1964. Bill Anders also recalls that sometime before 1968 he and Al Bean were assigned to help write a preliminary checklist for the first moonwalk. "Almost as a joke," Anders says, on the slim chance that he might find himself on the first landing, he steered the writing of the checklist so that the words "LMP EGRESS" came before "CDR EGRESS"; in other words, that Anders would be out some number of minutes ahead of his commander. Later, Anders had a good laugh over the caper. Bean, who was just happy to be assigned to Apollo lunar activities, was oblivious to such maneuvering.

148 *Armstrong was anything but an exercise fanatic:* On that subject, there was a bit of mythology, a quote, often incorrectly attributed to Armstrong: "I believe the Lord gives us a finite number of heartbeats, and I'll be damned if I'm going to waste mine on exercise." In fact, Armstrong had once spoken those words, but he was quoting someone else, and he wasn't

endorsing that view. Nevertheless, as one astronaut said, "Neil was no jock."

148 *More than one astronaut remembers that Aldrin paid a visit, checklist in hand:* Today, Aldrin is understandably sensitive about this issue, which was emphasized by the media before, during, and after Apollo 11. In 1987 he told the author that before Apollo 11 his feelings about being first to walk on the moon had been "totally mixed." On the one hand, his dislike for publicity made him lean strongly the other way. And he knew it would be awkward if he received more attention after the flight than his commander, who would rightfully deserve accolades for flying the lunar landing. On the other hand, he said, for an aggressive person like himself, in the competitive arena of the Astronaut Office, it was only natural that he explore the possibility of being first—especially when there were operational grounds for doing so. Today, Aldrin maintains that the other astronauts have blown his efforts out of proportion. He had no more desire to be first on the moon, he says, than any of them.

149 *Mission planners had quietly come to the same conclusion in February:* Based on the author's interview with John Covington, who worked in the Crew Systems Division on timelines for the moonwalk, and on *Chariots for Apollo*, p. 322.

149 *to keep the first steps on another world free of any militarism:* Julian Scheer, who headed the public affairs office at NASA Headquarters in 1969, says that he felt strongly that a civilian, not a military officer, should be the one to represent NASA, a civilian agency, in making the first lunar footsteps. He voiced this opinion to Sam Phillips a few months before the launch. He adds, "It was nothing against Aldrin" personally.

In 1986, Phillips told the author that Armstrong's civilian status was a factor in the decision—which he said he left to Slayton and the other managers in Houston—but that he did not consider it an important factor. Asked whether NASA as an agency was consciously ruling against a military man as first on the moon, Phillips said, "No, absolutely not."

II: "We Is Down Among 'Em!"

150 One source used in the writing of this chapter was "Our Happy Moon Trip," *Life*, June 20, 1969.

150 *Apollo 10's new TV camera:* Color TV from space was Stafford's idea. He felt the public deserved a more vivid picture of spaceflight than the black-and-white pictures transmitted from Apollo 7 and 8. When he heard that engineers at Westinghouse were developing a spaceworthy color camera, he fought to have it ready in time for his mission.

151 *the sounds of Frank Sinatra:* In his recording of "Fly Me to the Moon," Sinatra changed the lyrics slightly. The official version, as supplied by Hampshire House Publishing Corp., is as follows:

Fly me to the moon and let me play among the stars
Let me see what spring is like on Jupiter and Mars
In other words, hold my hand
In other words, darling kiss me

151 *George Mueller . . . pushed for a landing on Apollo 10:* The idea for a dress rehearsal before the lunar landing attempt had been in the works since June 1967. By January 28, 1969, when Sam Phillips formally approved the mission for Apollo 10, planning was already well along. No one was surprised that Mueller was making such suggestions; according to one Apollo planner "Mueller was always floating new ideas. . . . He shot from both hips."

151 *Stafford's lunar module . . . was too heavy to land:* Originally Stafford had been assigned the first LM capable of making a landing, but in the wake of the delays that prompted the Apollo 8 decision, his lightweight LM went to Neil Armstrong, and Stafford took over Frank Borman's lander, which had been built before Grumman's super-weight-saving program took effect.

153 Vibrations during Apollo 10 Translunar Injection: According to Stafford, the vibrations were the result of a misadjusted pair of pressurization valves on the hydrogen tank within the Saturn's third stage. The valves were adjusted in such a way that as they vented excess pressure, they set up a kind of synchronized pulsing that increased in intensity. By readjusting the valves on later boosters, the problem was avoided.

155 *the astronauts called it "the lem":* This pronunciation is a holdover from the early 1960s, when the lander's official designation was lunar excursion module (LEM). When that was later shortened to lunar module (LM), the old pronunciation stuck. Although there were abbreviations for the command module (CM) and service module (SM), and for the joined pair (CSM), astronauts usually called the entire command ship "the command module."

156 *Once, a workman accidentally dropped a screwdriver inside the cabin:* Collins, *Carrying the Fire*, p. 324.

III: Down to the Wire

161 NACA's *High Speed Flight Station:* Pronounced "N-A-C-A," not "*naka*."

162 *a postflight party in full swing usually saw Armstrong at the piano:* The late Milt Thompson, a fellow NASA X-15 pilot, described an incident from one flight party that illustrates Armstrong's understated brand of humor. The bar was full of NACA and air force fliers, and Armstrong was at the piano. A group of men from the air force's missile-testing lab across the dry lake came in. In the minds of the fliers at Edwards, these non-pilots—who called themselves "Missileers"—were from the wrong side of the dry lake, so to speak. And yet, here they were, with their official badges pinned to their chests, looking pleased with themselves. One went over to Armstrong and requested a song. Armstrong glanced at the man, saw that he was wearing a Missileer badge instead of a set of wings. Without missing a note, Armstrong deadpanned, "Gee, I don't know any Missileer songs"—at which point the pilots in the bar laughed themselves onto the floor, while the ragtime rolled on. (For more information on the X-15 and its pilots, see *At the Edge of Space: The X-15 Flight Program*, by Milton O. Thompson, Smithsonian Institution Press, Washington, D.C., 1992.)

163 *"space is the frontier, and that's where I intend to go"*: When Armstrong made this statement he had already participated in a number of projects that anticipated his work in Apollo. In 1959 Armstrong and another NASA engineer were studying whether a manned booster could be flown into orbit. In the early 1960s, he participated in the initial development of the Lunar Landing Research Vehicle (LLRV). As an astronaut, Armstrong would use the LLRV and its descendant, the Lunar Landing Training Vehicle (LLTV), to practice flying the lunar module to a landing.

168 *Armstrong was all business:* Even in the simulator, however, Armstrong could display his sly wit. While training as the backup crew for Gemini 11, one day Armstrong and his copilot Bill Anders were practicing rendezvous maneuvers. Just minutes away from a critical rocket firing, Anders, busy with the onboard radar and the navigation charts, looked up from his work to find Armstrong asleep. When Anders tried to wake his commander, Armstrong looked at his watch, shot his copilot a narrow-eyed glance and went back to sleep."Okay," Anders thought, "I'll let him screw up." But then, at just at the right moment, Armstrong sprang awake, reached for the hand controller, and executed a flawless maneuver. Anders realized he'd been had. Later, when he queried Armstrong about the joke, Armstrong just laughed.

170 *one of his flight controllers nicknamed him General Savage:* The nickname was after the hero in the popular TV show "Twelve O'Clock High."

171 *These young men:* There were no women on the flight control teams for any Apollo mission; this was because there were few women engineers. For Apollo 11, a simulation instructor, Ann Accola, was one of a group of people working in a support "back room" to help establish the location of the lunar module from the astronauts' descriptions of the terrain. Today, women comprise almost a third of the flight control rosters (Gene Kranz letter to the author, 1993).

173 *Suddenly, it all went bad:* This simulation probably took place on June 27, which was a day full of simulated emergencies. According to the simulation log for Apollo 11, Armstrong and Aldrin spent a total of six hours in five separate simulated landings. Four of the five runs were aborts.

175 Collins's comments about Aldrin: From *Carrying the Fire*, p. 434.

175 *"I hated that probe, and was half convinced it hated me"*: *Carrying the Fire*, p. 339.

176 *During the spring of 1969 Collins felt the eyes of the world upon him:* Somehow, in the rush of activity, Collins found time to design a mission emblem, using an idea from Jim Lovell: an eagle coming in for a landing above a field of craters, with the earth suspended in the black sky beyond. There would be no names on the patch, only the words, "Apollo 11." The name *Eagle* was a natural choice for the lunar module, while Collins's command module would be called *Columbia*, a name that evoked not only national identity, but Jules Verne's mighty cannon, the Columbiad.

180 *Thomas Gold insisted that the moon was covered by a layer of fluffy powder dozens of feet thick*: One geologist recalled showing Gold a picture from one of the Surveyors and saying, "Tommy, look: They didn't sink!" Another geologist insisted that if Gold was an irritant, he was a beneficial one: by voicing his controversial theories, he forced the other scientists to examine their own ideas more thoroughly.

180 *if he wanted to build a sterilization machine, he would construct something like the surface of the moon*: The geologist was NASA's Elbert King. If anything, King and other geologists argued, a storage facility was necessary, not to keep the earth safe from "moon germs," but to protect the lunar samples from terrestrial contamination (King, *Moon Trip*, p. 61).

181 *Quarantine was something Armstrong and his crew would just as soon have done without, but they had no choice*: Mike Collins saw the contamination issue as a question of probability: a very small number (the chance of life on the moon) multiplied by a very large number (the implications if it did) still produces a finite number. On that basis, he and his crewmates accepted the need for quarantine as more than just a political necessity (*Carrying the Fire*, p. 317).

182 *"any recognizable disadvantage . . . the position I'm in"*: Frank and Susan Borman recalled that they tried to warn Armstrong and his crew about what lay ahead. Shortly after Armstrong, Aldrin, and Collins were named to the landing crew, the Bormans invited them and their wives for dinner. The Bormans had just returned from their Apollo 8 world tour, and they described with amazement how, in France, one official tried to give Borman a car and another offered Susan a fur coat, both of which they had politely but adamantly refused. "If you think that's bad," the Bormans

told the three men and their wives, "you're going to have all of Europe thrown at you, on sterling."

Chapter 5: The First Lunar Landing

I: The *Eagle* Has Landed

187 View of the moon in earthlight: Seen from earth at this time, the moon was a slender crescent. The rest of the near side, which was illuminated by earthlight, was also dimly visible. Inside Apollo 11, Armstrong and his crew saw the moon from a somewhat different angle. To them, the moon's sunlit portion was out of view. They could, however, see part of the lunar far side, which was in complete shadow; hence the "crescent of blackness."

190 *Gene Kranz and his team of flight controllers called the White Team:* The practice by flight directors of adopting colors began with the final Mercury flight and continued through Apollo. Each new flight director chose a new color for his team of controllers. When Kranz became a flight director in 1965, he chose white. The controllers that made up each flight director's team would change from one mission to the next, but thanks to their intensive training before each flight, they always functioned as a coherent whole (Murray and Cox, *Apollo: The Race to the Moon*, pp. 286, 288).

192 *"Dammit, we really did something"*: This speech is derived from a combination of the author's interview with Gene Kranz and the quote on p. 348 of *Apollo: The Race to the Moon*.

192 *suggested to Duke that* Eagle *yaw slightly to one side:* Because of *Eagle*'s face-down orientation, its antenna was not able to point in such a way as to enable clear transmissions to earth. By yawing the LM—that is, turning it slightly to one side—Armstrong and Aldrin improved the antenna's aim without affecting the pointing of their descent rocket.

194 *more like a simulation and less like history:* Kranz remembered that one controller did indeed comment, "It's just like a simulation." Kranz says, "I think several controllers relaxed with this comment; I know I certainly did. . . ."

194 *Kranz almost burst out laughing:* Kranz says Steve Bales "was so loud that he startled the entire room. He did not need a communications loop."

195 *He put the question to one of his back-room experts, Jack Garman:* The details of Bales's exchanges with Garman are described in *Apollo: The Race to the Moon*, p. 353.

195 *"I have too many things to do in my computation cycle"*: The reason for the computer overload, it was later determined, was the fact that *Eagle*'s rendezvous radar had been left on "automatic," as a precaution for an

emergency rendezvous with *Columbia.* But this meant that as *Eagle* descended, its computer was spending part of every computation cycle analyzing the signals from the rendezvous radar. By setting the rendezvous radar to "manual"—an instruction radioed to Armstrong and Aldrin only 30 minutes before their liftoff from the moon—the alarms were avoided during the ascent and subsequent rendezvous, a time when a healthy computer was even more critical than during the descent. See *Apollo: The Race to the Moon,* pp. 365–67.

II: Magnificent Desolation

202 *Inside, Joan was savoring relief*: The accounts of Joan Aldrin's activities on the day of the landing are based on the author's interview with her, as well as information in Armstrong et al., *First on the Moon,* and Mailer, *Of a Fire on the Moon.*

203 *She found the embrace of Buzz's uncle, Bob Moon:* Remarkable but true: Moon was the maiden name of Aldrin's maternal grandmother.

204 NASA *was still coping with a controversy stirred by the Genesis reading on Apollo 8:* According to a book by a former NASA public affairs officer, the Genesis reading brought "a shrill protest from agnostics who tried to convince the federal court that astronauts had no right to express religious sentiments in outer space. That backfired . . . when thousands of God-fearing people petitioned NASA to allow the astronauts freedom to do as they wished." Quoted in "Heroes, not of their own accord," a 1977 master's thesis by Perry Michael Whye, Iowa State University.

206 *in Armstrong's mind, it amounted to the first human contact with the moon:* For the landing, Armstrong gave some thought to quotes; before the flight he and Aldrin decided that if they reached the lunar surface they would use the call sign "Tranquillity Base"—"base" to connote exploration. They told only Charlie Duke, lest the first words from the moon take him by surprise—"Say again, Apollo 11?" And when it finally happened, Armstrong found himself adding quite spontaneously, "The *Eagle* has landed."

210 *The soil in the bag was almost black, like powdered graphite:* In retrospect, this should have come as no surprise; astronomers had long pointed out that the moon, which appears bright in the night sky, is actually a very dark object, reflecting on average only 7 percent of the sunlight striking it—a reflectivity comparable to that of asphalt.

212 *By international agreement no nation could claim the moon:* Equal access to the moon by all of humanity was a provision of the Space Treaty of 1967.

212 *Posing for Armstrong's camera:* There are no photographs of Edmund Hillary on the summit of Everest, simply because Tenzing did not know

how to use a camera, and as Hillary said, "Everest was no place to teach him." Coincidentally, there are no good pictures of Neil Armstrong on the moon. The only clear Hasselblad photo shows Armstrong with his back to the camera, working at *Eagle*'s equipment storage tray. He also shows up in somewhat fuzzier motion picture footage taken from the lander's cabin. Buzz Aldrin, who has been asked about this more times than he probably likes, has explained that it was not intentional; the way the timelines were worked out, Armstrong had the camera most of the time. During the brief periods when Aldrin did have the camera, he was focused on such operational tasks as photographing the lunar module and the surrounding terrain. In addition, Aldrin snapped the famous picture of his own footprint—not for the aesthetic merit of the image, but for the scientists who were interested in the mechanical properties of lunar dust.

213 *the images from the moon were like a window on a dream:* In the pages of science fiction there had been almost as many versions of the first moonwalk as there were science fiction writers, but to the author's knowledge only one included live television pictures of the event: Arthur Clarke's *Prelude to Space*, penned in 1947 (Arthur Clarke letter to the author, 1993).

216 *Armstrong realized there would not be nearly enough time:* In truth, Armstrong and Aldrin still had enough oxygen in their backpacks to stay out much longer than the time allotted, but the conservatism of this first landing dictated that they not use it.

217 Eagle *was nearly 200 feet away, looking like a scale model:* One geologist, watching the moonwalk at home, saw Armstrong run out of the field of view and thought, "Oh, my God, where is he going?" A number of observers were startled by Armstrong's apparent deviation from the flight plan. In truth, Armstrong did not violate any mission rules by running back to the crater. Mission planners had specified a maximum distance from the LM of 300 feet. However, they said, it was advisable, for conservatism's sake, that they remain within 100 feet.

217 *Armstrong had been gone for only about three minutes:* Aldrin had no idea of Armstrong's brief exploration. He was focused on his own work back at the LM, preparing the equipment for the documented sampling. The thought of running off to some interesting crater, he said years later, was not something he could allow himself to consider. In an extraordinary situation, he noted, it was useful to have a kind of tunnel vision: focus on the job at hand; don't deviate from the checklist; the better to avoid getting behind.

218 *But after two hours and thirty-one minutes:* The time of 2:31 refers to the interval between opening the hatch and closing the hatch. Armstrong was actually on the surface for a total of 2:13; Aldrin 1:42.

III: "Before This Decade Is Out"

219 *its lone occupant, Mike Collins:* Much of this section is drawn from *Carrying the Fire.*

222 *Moonlight flooded the cabin:* When Armstrong had settled into his makeshift bed inside *Eagle,* he realized another light was shining on his face. When he opened his eyes he was staring right at the earth, shining like a blue light bulb in the eyepiece of the navigation telescope on the other side of the cabin.

223 *there was no other way to think about it:* When Armstrong and Aldrin entered *Eagle* after the moonwalk they discovered a piece of plastic lying on the floor; it was the top of the circuit breaker used in arming the ascent engine. Without realizing it, Aldrin had broken it off with his backpack hours earlier, during the preparations for the moonwalk. No one was particularly concerned, simply because there were ways to work around the problem if the switch could not be used. But Aldrin managed to push the breaker in with a pencil, and in mission control, where controllers studied the telemetry from *Eagle,* it was clear that the fix had worked.

223 *"My secret terror for the last six months": Carrying the Fire,* pp. 411–12.

227 *Aldrin described the strange flashes:* The flashes were determined to be caused by high-energy cosmic rays entering the spacecraft and passing through the astronauts' eyes (*Biomedical Results of Apollo,* p. 355).

227 *That would come on a hot August night:* Armstrong didn't want any reporters, photographers, or other hoopla when the astronauts got out of the LRL. In a discussion with Chuck Berry on the day before the release, he jokingly threatened that he and his crew might feign illness: "I can't guarantee that the people won't limp or have contractions of some sort. If you want to take that kind of a chance, just have those cameras out there." He also said he wished he had a bottle of gentian violet so that he and his crew could paint little spots on their faces.

BOOK TWO

Chapter 6: Sailors on the Ocean of Storms

I: The Education of Alan Bean

232 *Apollo had given the country the technology to go to other worlds:* Sources for this section: John Logsden's unpublished manuscript on the transition from the Apollo to post-Apollo era (John Logsden, Center for Space Policy, George Washington University, Washington, D.C.); Levine, *Managing NASA in the Apollo Era;* the author's interviews with former

NASA administrators Tom Paine and James Beggs, and other NASA officials.

233 *In the LRL, five geologists:* The five were P. R. Bell and Elbert King of NASA's Lunar Receiving Laboratory, Ed Chao of the U.S. Geological Survey, Clifford Frondell of Harvard, and Robin Brett, formerly of USGS. See Cooper's *Moon Rocks* and Wilhelms's *To a Rocky Moon.*

238 *"Flight, try S-C-E to Aux":* As described on p. 375 of Murray and Cox, *Apollo: The Race to the Moon,* the Signal Condition Equipment was an electronics box that "performed an obscure role in translating the information from the sensors [onboard the spacecraft] into the signals that went to displays in the spacecraft and on the ground." When set to primary, the SCE would turn itself off under low-voltage conditions like those that occurred after the lightning strike. To turn it back on, the astronauts had to change the setting on the SCE to auxiliary. This resumed the flow of data to mission control.

239 *"Was that ever a sim they gave us!":* "Sim" was a commonly used abbreviation for simulation.

240 *"Look down there; those are campfires": Life,* December 19, 1969, p. 36.

242 *Conrad . . . took to calling him "Animal":* The nickname was taken from a character in the movie *Stalag 17,* about American soldiers in a German prisoner-of-war camp in World War II.

245 *For quite a while Bean felt like a minnow:* Looking back, Bean says, he shouldn't have blamed Shepard for his difficulties in the Astronaut Office. While Shepard did enjoy being ominous, he adds, it was unlikely that he was trying to harass Bean or any of the Fourteen. Bean says, "It's like the line about 'Don't tug on Superman's cape.' I was tugging." In particular, he was voicing opinions Shepard didn't agree with—and in retrospect, Bean says, those opinions were uninformed.

246 *the Apollo Applications Project, the space station planned for earth orbit in the 1970s:* The Apollo Applications Project was a key element in NASA's post-Apollo planning in the mid-1960s. As originally conceived by George Mueller, it would make use of Apollo hardware for an ambitious program of missions emphasizing space science and including space stations in earth orbit and long-duration visits to the lunar surface. One of Mueller's prime motivations was to preserve Apollo's 400,000-member government and industry team. But as NASA historian David Compton wrote, "Mueller faced a cruel paradox: the buildup of the Apollo industrial base left him no money to employ it effectively after the lunar landing." Before AAP could begin, its funding was slashed by Congress. The single surviving element, the

space station, was later renamed Skylab. See Compton, *Living and Working in Space: A History of Skylab.*

246 *Electronics Test, the boondocks of test flight:* The reason Electronics Test was considered the boondocks, Bean says, is because it was where the majority of test pilot graduates were assigned, and it would have been easy to get lost there.

251 *a young mathematician named Emil Schiesser made a breakthrough:* Described on p. 383 of *Apollo: The Race to the Moon.*

II: Shore Leave

254 *dancing their way to the moon:* To make the picture complete, imagine the three men wearing matching caps, which Conrad had broken out after the Translunar Injection burn and presented to his crew. They were made of white Beta-cloth and they were personalized just like the Corvettes, with "CDR," "CMP," and "LMP." Conrad's had a little Teflon propeller on top.

256 *He looked down at what seemed to be a string of small volcanoes:* In reality, they were probably irregularly shaped impact craters, which can resemble volcanic features.

257 *"Where do you want me to put you?":* Conrad's conversation with Dave Reed is described on p. 385 of *Apollo: The Race to the Moon.*

260 *The gauge that was supposed to display lateral motion seemed to be broken:* Only when he was back on earth would Conrad find out that the display was working fine; he had put *Intrepid* in a near-perfect vertical descent.

262 *"that may have been a small one for Neil, but it's a long one for me":* According to Conrad, he was never able to collect his five hundred dollars.

263 *the first full-fledged scientific station to be set up on another world:* The first ALSEP was almost doomed before it could start transmitting data. The station's power generator was fueled by a cask of plutonium. When Bean tried to remove the cask from its graphite storage case on the side of the lunar module, it wouldn't budge. Only with help from Conrad—who beat on the case with his hammer, so hard that the case cracked—did Bean finally pull the cask free. Powered by its nuclear-electric generator, the ALSEP was designed to relay data for years.

268 *Gordon was jumping up and down: Life,* December 19, 1969, p. 37.

269 *"Now's the time to think up all sorts of fancy prose":* Ibid.

III: In the Belly of the Snowman

271 *The right leg of his space suit . . . was slightly too short:* It was Conrad's own fault, though that was no comfort. About a week before launch, technicians had found an air leak where the right boot joined the leg. The suit was quickly flown back to its manufacturer, the International Latex Corporation in Dover, Delaware, where it was repaired and returned to the Cape. Conrad was then called in to have the suit refitted. He wanted to wear his water-cooled underwear for the fitting, because it had thick plastic tubes that ran along the shoulders. But the water-cooled garment had long ago been packed in the LM. Couldn't he wear his training set? Conrad asked. Absolutely not was the answer; nothing that wasn't pristine could be worn inside a flight-ready suit. So Conrad did the fitting in a pair of flight-qualified plain cotton long johns. He tried to leave enough room for the bulk of the plastic tubes, but he underestimated. He was only off by a quarter of an inch, but that was enough to make a difference.

274 *Each step launched him into the air for long seconds:* Bean notes that while it seemed as if he were going 10 feet with each step, in reality his strides were probably the same length as when he ran on the beach at the Cape. He says that in one-sixth g, his body was so light that he didn't have much traction, and that prevented him from pushing off very hard.

275 *Conrad, running ahead of him, managed to study the map even as he bounded along:* Despite Conrad's mastery of the "lunar lope," he became the first astronaut to fall down on the moon, while collecting samples during this moonwalk. He got up easily by grabbing Bean's hand. Most of the time, however, Conrad and Bean found they could avoid falling because in one-sixth g it happened so slowly that they could, by running, get their feet under themselves and stand up.

275 *As they ran . . . they were beginning to feel the strain of their adventure:* Some of the exertion was due to the stiffness of the waist joint of their suits, a problem that was remedied on the last three Apollo missions.

278 *While Conrad held the tool carrier, Bean rummaged among the samples:* Conrad and Bean don't agree about who did what. This version was reconstructed with help from Eric Jones of Los Alamos National Laboratory, who went over the caper with Conrad and Bean while researching his forthcoming Lunar Surface Journals.

282 *Where Bean's cold had come from, he had no idea:* Later it would turn out that both men had stuffy heads from zero gravity, which causes the fluids to migrate to the upper body.

283 *the sights had been the most spectacular of his life:* There were spectacular sights even on the relatively boring trip back to earth. On the last day of the mission the men became the first to witness an eclipse of the

sun by the earth. For over an hour, as *Yankee Clipper* flew through the earth's shadow, they saw their home planet ringed by a rainbow of sunlight shining through the atmosphere. In the middle of the darkened world they could see a faint glow; it was light from the full moon, shining on a midnight ocean.

Chapter 7: The Crown of an Astronaut's Career

I: A Change of Fortune

285 *two more Apollo flights were in jeopardy:* At this time, NASA was still planning to fly Apollo 18 and 19 but had postponed them until 1974, after the missions to the Apollo Applications (later Skylab) space station. See Compton, *Living and Working in Space: A History of Skylab,* p. 135.

285 *Bob Gilruth had privately called for an end to the moon landing program:* A number of people remember Gilruth expressing this sentiment. One of them is Chris Kraft. In 1989 Kraft recalled, "He said to me on several occasions [after the first lunar landing], 'We've done that. It's too risky. We're liable to lose one of these things, and we just can't afford to have that happen. . . . Why should we go anymore? We've got the rocks.' " Kraft added, "I personally didn't feel as strongly as Gilruth did about not flying again." Kraft believed Gilruth felt the way he did in part because he was near the end of his career, and age tends to temper the willingness to take risks. "He may even be right," Kraft said. "Experience is a great teacher."

301 *To convert the data from* Odyssey's *platform to* Aquarius's *frame of reference:* The work was complicated by a drastic change in reference. To visualize this, remember that the nose of the command module was docked to the lunar module's ceiling. An astronaut seated in the command module was oriented 90 degrees to his crewmates standing in the LM cabin. As a result, the three degrees of rotation possible for the joined spacecraft—roll, pitch, and yaw—looked different, depending on which of the two craft he was in. If *Odyssey* pitched up or down, then by Lovell and Haise's reckoning, so did *Aquarius.* But if *Odyssey* spun along its axis—which Swigert called roll—Lovell and Haise felt *Aquarius* yaw, that is, nod to one side. And when *Odyssey* wagged from side to side—a yaw maneuver—it looked like roll to Lovell and Haise.

II: The Moon Is a Harsh Mistress

304 *Haise and a navy pilot named Edgar Mitchell:* Ed Mitchell remembers that he and Haise became "great friends and great rivals" during their time at Grumman. They would take turns quizzing each other about esoteric details of the LM's design and operation. And they played one-upmanship any time one of them got a step closer to flying in space. When Mitchell

was named to Apollo 13, he appeared to have won the competition, but when the Apollo 13 and 14 crews were swapped, for reasons described in Chapter 8, Haise got the victory. Ironically, the victory placed Haise in a struggle for his life, while Mitchell spent hour after hour in the lunar module simulator, working out procedures for his friend to use aboard *Aquarius.*

304 *to perform in ways it had not been designed for:* The LM lifeboat idea dated back to 1961, before the preliminary design for the lander had even been drawn up. Grumman engineers had envisioned using the lander's engines to push a crippled command ship out of lunar orbit. Eight years later, training for Apollo 9, Gene Kranz's flight controllers were hit with a simulation in which Jim McDivitt and Rusty Schweickart were stranded in a lopsided orbit around the earth. They had to devise ways to stretch the LM's supplies to last 30 hours instead of 18, long enough for Dave Scott to rescue them. While that exercise foreshadowed the effort to save Lovell's crew, no one ever envisioned using the LM in precisely the way it was used on Apollo 13 (sources: the author's interviews with Gene Kranz; Murray and Cox, *Apollo: The Race to the Moon*, p. 423).

305 Power values for radio and environmental control system: These numbers were supplied by former flight controller Robert Legler. Specifically, 1.29 amps is for the S-band transceiver; 5.5 amps is the power needed to run one suit fan.

307 *Mattingly blushed and jingled the change in his pocket: Newsweek*, April 13, 1970, p. 63.

309 *Lovell could only imagine what Swigert was thinking now:* Apollo 13 was halfway to the moon before Swigert realized he had not filed his income taxes and that he would be quite unable to do so before the April 15 deadline. The subject came up as scientist-astronaut Joe Kerwin was reading the Sunday morning news: "Today's favorite pastime across the nation—Uh oh, have you guys completed your income tax?"

Swigert radioed, "How do I apply for an extension?" Mission control exploded with laughter. "It ain't too funny, things happened real fast down there and I do need an extension. I'm really serious . . ."

"You're breaking up the room down here," Kerwin said. A few minutes later he assured Swigert that there wouldn't be any problem: an automatic extension is granted to anyone who is out of the country at tax time.

310 *She'd asked some of the* NASA *people about the odds of saving the men:* For his own part, Chris Kraft says he would never have given Marilyn Lovell such a dismal forecast. He says that once the LM lifeboat plan was adopted he never had any doubts of getting Lovell's crew home. "It's not very romantic," Kraft says, "but it's true!"

312 *worse than they were willing to admit:* These doubts are based on some notes taken by Chris Kraft at one of the astronauts' post-flight debriefings.

The notes do not specifically identify the astronaut who made the comment. Today, Lovell (along with Haise) denies having had such thoughts, but it is possible that his feelings during the flight were different from those he had long afterward. And as Ken Mattingly points out, the flow of information from earth to Apollo 13 was imperfect enough to allow room for such thoughts—which he says would have been entirely natural—in the minds of Lovell and his crew.

III: The Chill of Space

328 *immediate problem of where to store urine:* If it became necessary—which it never did—Haise had even figured out a way to put urine to good use by transferring it into the LM's cooling system.

329 *another bit of ingenuity from Houston:* Another procedure devised in Houston by controller Robert Legler was especially critical: it allowed Lovell's crew to charge *Odyssey*'s batteries using *Aquarius*'s power.

332 *After three days of constant background noise from* Aquarius's *fans and pumps:* Fred Haise, wanting to preserve those sounds for posterity, recorded them with a hand-held tape recorder before *Aquarius* was jettisoned.

332 *A serious crack was another matter:* In Houston, some of the flight controllers had privately pondered the same thing and, like Haise, realized there was nothing they could do about it. See *Apollo: The Race to the Moon*, p. 443.

336 *As presidential adviser John Ehrlichman told historian John Logsden years later:* John Logsden's unpublished manuscript on the transition from the Apollo to post-Apollo era (John Logsden, Center for Space Policy, George Washington University, Washington, D.C.).

336 *After Apollo 13 some of Nixon's advisers:* Nixon told this to Stu Roosa when the Apollo 14 astronauts visited the White House in 1971.

Chapter 8: The Story of a Full-up Mission

I: Big Al Flies Again

337 *"Why don't you fix your little problem and light this candle?": Life*, May 12, 1961.

339 *John Glenn was so angry:* The author's conversation with writer Howard Benedict, 1992.

339 *In April, representatives of the President's Advisory Committee:* The space subcommittee of PSAC, which Shepard recalls as being composed primarily of medical specialists, was chaired by MIT scientist Jerome Wiesner, who had been against the manned space program from the beginning. Shepard interviews with author; Shepard interview with the JFK Library

Oral History Project, June 12, 1964 (JFK Library, Boston); Murray and Cox, *Apollo: The Race to the Moon*, pp. 66–67, 70–71.

343 *Shepard was on the way to becoming a millionaire:* Prior to Apollo 14 there were occasional rumors stating that Shepard was already a millionaire, but he denied them.

346 *"If you guys don't mind flying with an old retread"*: It was Pete Conrad who bestowed a new nickname on Shepard; it happened after the splashdown of Apollo 12, when Conrad, Gordon, and Bean arrived at Ellington Air Force Base in their quarantine trailer. Behind the glass, Conrad looked out at the well-wishers who had come to meet them, and now he saw the un-grounded Al Shepard. He grinned his gap-toothed grin, grabbed the PA microphone, and announced, "Here comes the rookie!" Shepard flashed Conrad a look that could have melted lead. Conrad was glad he'd said it. And it wasn't long before some began referring to Shepard, Roosa, and Mitchell as the "all-rookie crew."

347 *And as far as Deke Slayton was concerned, if Shepard wasn't qualified, no one was:* Slayton's confidence in Shepard is illustrated by the fact that Apollo 13 would have been NASA's last chance to make a lunar landing in 1969 if 11 and 12 had failed.

347 *"They ought to hire tiddlywinks players as astronauts"*: *Newsday*, February 5, 1971.

348 *the only one of the Original 7 to reach the moon:* This was contrary to the expectations of many observers, who had initially thought that one of the Original 7 would probably become the first man on the moon. In the mid-1960s, some of the geologists who trained the astronauts made four predictions of who would most likely command the first landing. Three of their picks were members of the Original 7: Wally Schirra, Scott Carpenter, and Gordon Cooper. (The fourth, and the only one who made it to the moon, was Pete Conrad.)

348 *Now he had been saved . . . by George Mueller:* By this time, however, Mueller was no longer at NASA, having left the agency at the end of 1969. (Levine, *Managing* NASA *in the Apollo Era*, p. 308.)

349 *At the end of August . . . canceled two more Apollo missions:* In a 1989 interview, Tom Paine, who had participated in the decision to cancel these missions, said that he had been faced with the dilemma of whether to continue exploring the moon at the expense of NASA's long-term future. True, the scientists were urging him not to trim any more missions, but to Paine Apollo seemed to be nearing a point of diminishing returns. Was it worth deferring future projects to land on the moon a few more times? Paine decided it was not. Under the circumstances, Paine felt, he had made the best of things.

It is clear that in addition to the budgetary and scientific issues, the risks involved in the lunar missions—to human lives and to the future of NASA—was an important factor in the decision, but it is difficult to know how much of a factor. In 1989 Chris Kraft said that he and his colleagues "put all that in a pot and stirred it up. And the fathers of NASA and leaders of Congress concluded: Stop after [Apollo]17 instead of [Apollo] 19."

For the scientists, the cancellations were a lesson in lost opportunities. Not until September 1970, after the decision had been made, did they voice protest. In the middle of that month thirty-nine lunar scientists wrote a letter to Congressman George P. Miller, the chairman of the House Committee on Science and Astronautics and a long-time supporter of the space program. Canceling the final three Apollo flights, they wrote, might cause the lunar science program to "fail in its chief purpose of reaching a new level of understanding" about the moon and about our own planet. Miller replied by stating that he and his committee had tried to get an additional $220 million for Apollo into the authorization bills for 1970 and 1971, but that "the Nixon administration, in realigning national priorities, has relegated the space program to a lesser role." However, he told the scientists, "Had your views on the Apollo program been as forcefully expressed to NASA and the Congress a year or more ago, this situation might have been prevented." See Compton, *Where No Man Has Gone Before: A History of the Apollo Lunar Exploration Missions*, p. 203. Other sources on the end of Apollo: Compton, *Living and Working in Space: A History of Skylab*; Levine, *Managing NASA in the Apollo Era*; John Logsden's unpublished manuscript on the transition from the Apollo to post-Apollo era. See also the author's article "Why Haven't We Been Back," *Air & Space/Smithsonian*, July 1989, pp. 90–97.

349 *Their husbands were playing around:* To be sure, a number of astronaut wives already knew of, or suspected, their husbands' extramarital affairs. One remarked in 1967 that she accepted a certain amount of infidelity the way she thought of speeding tickets: "It's bound to happen once in a while." That the astronauts' indiscretions weren't covered by the media is a reflection of their special place in 1960s America. As *Life*'s Dora Jane Hamblin wrote in 1977, "I think *Life* treated the men and their families with kid gloves. So did most of the rest of the press. These guys were heroes. . . . I knew, of course, about some very shaky marriages, some womanizing, some drinking, and never reported it. The guys wouldn't have let me, and neither would NASA" (unpublished article by former NASA public affairs officer Paul Haney; Hamblin letter to Perry Michael Whye, Iowa State University, January 1977).

350 *the modifications to the command and service modules:* To safeguard against a repeat of the Apollo 13 accident, engineers redesigned the service module's oxygen tanks and increased their number from two to three. They also installed valves between tanks to prevent any single rupture from de-

pleting more than one tank. In the command module, they added a contingency water storage system so that astronauts would have drinking water if the command module were disabled. In the lunar module, they added an extra battery for emergency power. *Apollo Program Summary Report*, 1975, NASA Lyndon B. Johnson Space Center (internal publication).

350 *Shepard sketched a design that showed an astronaut pin:* The astronaut insignia showed a three-tailed star ascending through a ring.

351 *when angered he was capable of outbursts of temper:* Once during a training exercise, Mitchell became frustrated with a balky experiment and shook it so hard that it broke (Associated Press news story, January 31, 1971, published in the *Washington Star*).

351 *Mitchell . . . seemed to be carrying the load for Shepard with the lander's systems:* Mitchell says Shepard leaned on him a great deal during training. By agreement with Shepard, Mitchell played the role of instructor pilot with his commander. At first, Mitchell coached Shepard through malfunction procedures; then, as Shepard caught on, Mitchell hung back. In the simulator one day a few months before launch, Mitchell turned to Shepard and said, "Okay, Boss. I believe you're ready to go."

352 *at age forty-seven:* At the age of forty-seven years and two months, Shepard was the oldest American to go into space up to that time. Soviet cosmonaut Georgiy Beregovoy was four months older when he made his Soyuz 9 flight in 1968 (Hawthorne, *Men and Women of Space*).

352 *Okay, buster, you volunteered for this thing:* Carpenter et al., *We Seven*, p. 195.

352 *a booster a hundred times more powerful:* Shepard's Redstone booster had 78,000 pounds of thrust; the first stage of the Saturn V had more than 7.5 million pounds of thrust.

II: To the Promised Land

353 *The culprit might be something as simple as a tiny piece of debris on the mechanism:* The cause of the problem was never determined. In their studies following the flight, engineers ruled out an ice particle as the culprit, but did list contamination or debris as possibilities.

356 *Mitchell was probably the only astronaut who missed the presence of psychologists:* MacKinnon and Baldanza, *Footprints*, pp. 91–92.

356–357 Mitchell's ESP experiment: The results of Mitchell's experiment do not bear easy interpretation. Because Apollo 14's launch happened forty minutes behind schedule, so did Mitchell's sleep periods on the way to and from the moon, throwing off the timing of his experiment. On earth, the test subjects were actually trying to "receive" forty minutes before Mitchell was trying to "send." Despite this fact, Mitchell says, the experiment produced useful results. When the subjects tried to imagine the picture that Mitchell was thinking of at any given moment, they were often wrong in their guesses—but they were wrong, Mitchell says, significantly more often than pure chance would have dictated. Mitchell says this suggests that their subconscious minds knew something was wrong (that is, the experiment wasn't happening according to plan) and, in a precognitive way, adjusted to the situation by giving wrong answers. The missed guesses, Mitchell says, led the press to call the experiment a failure. But to Mitchell, it demonstrated that psychic phenomena were indeed at work.

357 Beginning at 2 days, 6 hours, and 57 minutes into the flight, the Mission Elapsed Time clock was advanced 40 minutes, to make up for the 40-minute delay in Apollo 14's launch. Mission Elapsed Times of events in Chapter 8 reflect this change.

359 *Once more, with help from mission control, they had made a narrow escape:* As evidence of his and Shepard's determination to make the landing, Mitchell recalls that in simulations, even when the instructors made the LM's computer fail entirely, they had still landed. Mitchell would handle the yaw and roll while Shepard controlled pitch and rode the throttle. The simulator didn't "fly" very well that way—they made some awfully hairy approaches—but they got it down in one piece.

IV: The Climb

365 *scientists were divided into two main camps:* For a definitive account of the history of lunar science, see Wilhelms, *To a Rocky Moon: A Geologist's History of Lunar Exploration.*

368 *firmly perched on the rolling hills of Fra Mauro:* Just to be sure, Shepard and Mitchell rigged a plumb line from a piece of string tied to a handhold, to use as an attitude reference; the 8-ball and computer were turned off. *Antares* wasn't tipping over; the sound that awoke them proved to be a noisy valve.

368 *a mocking version of the Apollo 14 mission patch:* On the patch, the rendering of the Coyote is a satirical reference to Shepard's crew. For redheaded Stu Roosa, the Coyote has red fur. For Ed Mitchell, who was not big on exercise, he has a pot belly. And he has a long gray beard, for Old Man Shepard.

370 *With or without the MET, they would get there:* One other reason Mitchell wanted to reach the crater itself: he wanted to roll a rock into it. On geology field trips, rock-rolling was a hallowed ritual which, in its purest expression, saw two or three pilots lying on their backs, pushing against a massive boulder with their boots until it went crashing down a hillside among the cedar trees, or into the steaming throat of a volcano.

372 *The deepest rocks, and perhaps the most important, would lie near the crater's edge:* Cone crater lay atop a ridge that was, like the rest of the Fra Mauro hills, believed to consist of debris ejected from a giant impact crater called the Imbrium basin, 340 miles to the north. Finding the date of Imbrium's formation, and sampling the rocks it tore out of the lunar crust, was one of the geologists' top priorities. The rocks near Cone's rim were most likely to be true samples of Imbrium ejects, and therefore, potentially the most important.

372 *"Oh, let's give it a whirl!"*: The tone in Mitchell's voice was uncharacteristically unrestrained, and frustration wasn't the only reason. After Conrad and Bean broadcast lively chatter during their moonwalks, Mitchell's family and friends, anticipating his own flight, teased him for being too serious. He ought to follow Conrad's and Bean's example, they said. Mitchell says that he had that in the back of his mind as he and Shepard were climbing Cone crater: he was playing to the gallery just a bit.

376 *Ed Mitchell felt a weariness:* Mitchell's emotional state aside, he and his crewmates were physically spent. After Apollo 13, mission planners weren't about to let Shepard's crew linger in lunar orbit any longer than necessary, and as soon as Shepard and Mitchell had rejoined Roosa the three men had pushed through the procedures for jettisoning the LM, and then, the Transearth Injection. By the time they bedded down for the night it had been a twenty-two-hour day.

377 *a haul of rocks and photographs that made the geologists ecstatic:* Stu Roosa was told that his pictures of Descartes were so good that they actually exceeded the theoretical resolution of the lens.

377 *Shepard and Mitchell hadn't documented their finds as well as the scientists had expected:* "Letter from the Space Center," by Henry S. F. Cooper, *The New Yorker*, April 11, 1971, pp. 126–27.

378 *Shoemaker blasted* NASA *for doing a "completely miserable job": Newsday*, February 5, 1971.

BOOK THREE

Chapter 9: The Scientist

383 Schmitt's awful puns: Two examples: One of the tools used by astronauts on the moon is called gnomon; it's used as a reference in photographs of lunar rocks to indicate local vertical. One of Schmitt's favorite phrases was, "Gnomon is an island."

When Schmitt encountered Paul Gast, the Manned Spacecraft Center's chief geologist, in the hall he would say loudly, "I'm appalled! I'm aghast!" much to the irritation of the high-strung Gast.

Gene Cernan turned the tables on Schmitt one day when the pair were training for Apollo 17. Driving a mockup of the Lunar Rover across a formation called an alluvial fan, Cernan hit a bump that dislodged Schmitt, who wasn't wearing his seat belt. Cernan looked over at Schmitt, lying on the ground, and said into his radio, "Well, the Schmitt just hit the fan."

385 *Shoemaker had made the first detailed geologic map of part of the moon:* Before Shoemaker's map, a less detailed geologic map of the entire moon had been prepared by Robert Hackman and Arnold Mason of the USGS in Washington. It is unclear how much they influenced Shoemaker's work, or vice versa.

386 NASA *rejected all three:* Schmitt, by all signs in perfect health, was rejected simply because he'd had surgery to correct a congenitally malrotated colon.

389 *most of the pilots had little or no enthusiasm for geology classes:* There were some exceptions. Gordon Swann says he saw a fair amount of interest from Wally Schirra, Scott Carpenter, Gordon Cooper, and several others among the Old Heads. He credits his late colleague Dale Jackson, one of the astronauts' early instructors, with this achievement. Some of the rookies, like Roger Chaffee, saw geology as their ticket to the lunar surface, and they worked hard at it; Chaffee was among the most promising students in the office.

390 *When Bill Anders wanted extra preparation:* More often than not, Schmitt's meetings with Anders took place in the sauna in the astronaut gym—not only because that was one of the few opportunities Anders could fit into his hectic schedule, but also, Schmitt suspected, to keep Borman from finding out.

391 *Armstrong turned in an excellent performance on the moon:* In particular, geologists have praised the variety of the rocks Armstrong collected during his hurried minutes on the moon.

392 *Schmitt called on his friends at Harvard and Caltech for help:* Shoemaker had become the chairman of Caltech's Division of Geological Sciences in January 1969.

392 *Silver made some pivotal refinements to the method:* Silver's method, which applies to rocks between 10 million and 4.5 billion years old, has become the standard means of dating the oldest rocks. The oldest rocks yet discovered on earth are between 3.8 and 4 billion years old.

392 *Silver, with his grad student Mike Duke:* Duke was one of the geologists who applied for the scientist-astronaut selection in 1965.

397 *Schmitt became a simulator hound:* The flight directors, whom Schmitt had gotten to know better than he knew many of the astronauts, were

happy to have him act as the astronaut for their own training runs, something the pilots rarely offered to do.

Chapter 10: A Fire to Be Lighted

402 *the first of the so-called J-missions:* At one time, NASA planned a series of I-missions, which were to involve extensive scientific observations from lunar orbit, without a landing, but these missions were never developed.

402 *a place called Hadley:* To the geologists, it was officially known as Hadley-Apennine.

402 *the lava plains that formed the valley floor:* Strictly speaking, these lava plains are not part of Mare Imbrium, but are instead part of an isolated patch of *mare* called Palus Putredinis, the Marsh of Decay.

404 *Jim Irwin . . . the only person who could have gone to the moon with him:* USGS geologist Jerry Schaber, who participated in briefings with Scott and Irwin, remembers that Scott tried to look after his lunar module pilot. After one hefty dinner at the crew quarters Lew Hartzell brought out a coconut cream pie. When Irwin reached for it Scott said, "Ah, Jim, you better watch it. Tomorrow's our [preflight] weigh-in." Irwin looked sad; Scott promptly took a helping of pie. He didn't have any problem keeping his weight down, Schaber said, but apparently Irwin did.

405 *this chunky white rock, called anorthosite:* On earth, anorthosite formed many times during the Precambrian. Not all of it is white; due to minute impurities it may be black or green. In rare cases—particularly in one locality in northern Labrador—it exhibits a spectacular play of color called Labradorescence.

407 *The geologists could hardly believe the resistance:* Gordon Swann says he had to fight to get a geologic hammer onto the missions. The engineers asked him why the astronauts needed a hammer to break rocks; couldn't they just pick up small ones? Swann explained that the big rocks might be important samples. Then the engineers told Swann they were afraid that if an astronaut hammered on a boulder, flying rock chips would damage his space suit. Swann pointed out that in all his years in the field, his own clothing had never been pierced by a flying rock chip. Tests using a space-suited test subject dispelled the worry. But in the end, Swann says, the only reason the geologists were able to get a hammer onto the Apollo 11 stowage list was that the astronauts would need it to hammer in core tubes. When it came time for Apollo 12, Swann says, the geologists assumed there wouldn't be any more problems with the hammer. But the engineers said, "You already flew that experiment. Why do you need to fly it again?" Bill Muehlberger got the same question when he wanted Apollo 16 astronauts John Young and Charlie Duke to take a 500mm lens to the moon—after all, it had already been done on Apollo 15.

407 *they depended on the astronauts' words:* Every so often the listeners would hear a description of a very strange "rock," and as they listened they would realize they were hearing a description of a beer can, or a cowpie.

Chapter 11: To the Mountains of the Moon

I: "Exploration at Its Greatest"

413 *"the* Falcon *is on the Plain at Hadley":* The Plain was a name chosen by Scott as a reference to the parade ground at West Point.

418 *followed by Irwin's grin-and-bear-it laughter:* After the flight, Irwin received a plaque from the engineers who built the Rover, bearing a pair of opaque welder's goggles and the inscription: "To be worn in case you are ever again riding the Rover on the moon with Dave Scott driving."

419 *he'd even driven over it in simulations:* A specially built Rover simulator, using the same video camera and terrain model used in the lunar module simulator, allowed the men to practice driving across Hadley.

421 *how long the rock had been sitting there:* Analyses showed that the rock had been there for 500 million years.

II: High Point

424 *after the last of a regular series of evening geology briefings at the Cape:* The evening briefings were Scott's idea. He was hungry for more preparation from the geologists, and he scheduled the briefings when the day of meetings and simulator runs was over, when they would be free from interruptions. The geologists would come to the crew quarters for one of Lew Hartzell's hearty dinners, then they would go to work. Any idea pertaining to the geologic exploration of the Apollo 15 landing site was up for discussion. At one point, a young geologist named Jim Head, who worked with Farouk El-Baz at Bellcomm, was leading a discussion about Hadley Rille. Scott asked, "If Jim and I were able to drive down to the bottom of the rille, would that be useful?" Head could barely contain himself—the bottom might be teeming with clues to the rille's origin—and he answered, "Oh, *yeah*—Jesus, you could spend the rest of your *life* down there!" When everyone realized what he had unwittingly said, the room exploded in laughter.

428 *Allen combined a youthful enthusiasm with a keen scientific mind:* He was also an accomplished flier. In pilot training, Allen had proved himself by winning the highest awards for formation flying, instrument flying, and acrobatics.

434 *One of them, based on declassified spy-satellite technology:* The driving force behind getting this camera, called the Panoramic Camera, onto Apollo was a Survey geologist named Hal Masursky. Masursky's detailed,

engaging lectures about the moon made him one of the astronauts' favorites.

436 *Either there was no life out there, or the cosmos must be teeming with it:* Of all Worden's experiences on the flight, the view of the universe he had during lunar night appears to have affected him the most. He says it led him to wonder whether it was possible that human beings are descendants of alien beings who visited earth in the distant past.

III: The Spirit of Galileo

440 *"the lesson of the moon's antiquity and changelessness":* Wilhelms, *To a Rocky Moon: A Geologist's History of Lunar Exploration,* p. 280.

442 *He reached into a pocket on his suit and pulled out a falcon feather:* A friend of Scott's at the air force academy in Colorado Springs had arranged it; it had been shed by the falcon that was the academy's mascot.

443 *Centuries before, the story was told, Galileo Galilei:* Historians doubt that Galileo actually performed the experiment. Instead, they say, he may have solved the problem by sheer power of reasoning. However, he may indeed have dropped objects from the Leaning of Tower of Pisa for a different purpose: to evaluate the effects of air resistance on falling bodies of various shapes and sizes.

443 *a small aluminum figure, a stylized representation of a fallen astronaut:* Scott had commissioned the sculpture from a Dutch artist named Paul van Hoeydonck, whom he met in New York.

IV: The Final Selection

445 *Scott pulled out a stack of first-day covers:* The U.S. Postal Service issued a new stamp during the flight of Apollo 15. Ten years had passed since Alan Shepard's Mercury flight, and the stamp commemorated a decade of achievement by Americans in space. On it, two astronauts rode a Rover across a bright moonscape. A special first-day cover, hand-canceled by Scott in a brief ceremony on the lunar surface, was brought back to earth for display. When Scott's crew arrived on the carrier *Okinawa,* they arranged to buy copies of the new stamp, and affixed them to their own envelopes, then had them canceled at the ship's post office (Justice Department internal memo, December 6, 1978).

448 *"One of the most brilliant missions in space science ever flown":* Compton, *Living and Working in Space: A History of Skylab,* p. 240.

449 *When Cernan crashed his helicopter in the Banana River during training:* About a week before the launch of Apollo 14, Cernan was flying a helicopter over the Banana River, enjoying the clear air and the smooth, mirrorlike water—so smooth, in fact, that he misjudged his altitude and

crashed into the river. The chopper exploded in flames, and Cernan had to dive into the water to escape being burned to death. He arrived back at the crew quarters as Shepard was having breakfast. Shepard looked up, astonished to see Cernan standing there, soaking wet in his scorched flight suit. Cernan, who during training had joked about hoping Shepard would break his leg so he could take his place, said, simply, "You win."

450 *In his mind, only one thing mattered:* Slayton said in 1989 that Engle's performance on the backup crew of Apollo 14 wasn't the issue in the Apollo 17 crew selection; neither was the pressure from the scientific community.

451 *"Yes, sir, I'll do the best job I can":* Schmitt's memory is unclear, but he thinks that the phone call was probably from James Fletcher, who had become the NASA administrator in 1971.

451 *explaining to his kids that he wasn't going to the moon: New York Times,* August 20, 1971. Engle's children were twelve-year-old Laurie and nine-year-old Jon.

Chapter 12: The Unexpected Moon

I: Luna Incognita

456 *a high point of 69 miles and a low point of 9 miles—the so-called descent orbit:* From Apollo 14 on, the astronauts used the SPS engine to place the joined command module and lunar module into the descent orbit, a maneuver previously performed by the lunar module alone, using its descent engine. This was done in order to save the LM's fuel for the Powered Descent, allowing the lander to carry more payload to the lunar surface.

459 *Young and Duke would have to take a wave-off:* Wave-off is a term used in aviation when a pilot is told by ground controllers to delay a landing attempt or to land elsewhere.

462 *"You're right; we* wouldn't *have let you land!":* While Mattingly recalls that McDivitt was serious, that seems extremely unlikely, given the amount of data available to him and others in Houston and at Downey. McDivitt does not remember the conversation but says he must have been pulling Mattingly's leg.

465 NASA *wanted the geologic objectives to be as carefully planned as any other aspect of the missions:* Gordon Swann recalls that NASA engineers were always asking him and the other geologists for a "voice plan"—in other words, they wanted to know what the astronauts would say as they collected rocks. Swann tried to explain to them that there wasn't any way to *know* what the astronauts would say, since no one could predict with certainty what they would find. If he could write up a voice plan, he told them, he wouldn't need to send the astronauts to the moon.

III: ". . . Or Wherever Geologists Go"

491 *the KC-135, a converted cargo plane:* Anyone who has worked in the KC-135 knows that it can be a difficult experience. After each period of zero g, the plane pulls out of its dive, and weightlessness is replaced by an onslaught of twice-normal gravity. For an astronaut in a space suit—especially one that might be less than adequately cooled—it was even more unpleasant. In a training session of 40 parabolas or more, it was all some of the pilots could do to keep from getting sick—not something they wanted to happen inside a pressure suit. No wonder the KC-135 was nicknamed the Vomit Comet.

493 *"science advances most when its predictions prove wrong":* Wilhelms, *To a Rocky Moon: A Geologist's History of Lunar Exploration*, p. 284.

Chapter 13: The Last Men on the Moon

I: Sunrise at Midnight

495 *Born in Liberia, he'd been taken aboard a slave ship at the age of twelve:* See "A Conversation with the Nation's Oldest Citizen," by Grover Lewis, *Rolling Stone*, February 1, 1973, pp. 22–26.

497 *determined to see their wings clipped:* A number of Apollo astronauts say that some of the Original 7 alienated people in the space center with their arrogance. As one example, one of the Fourteen described what he called the "Schirra effect": In meetings with the engineers, Schirra would take a strong position on some aspect of spacecraft design, and no one could talk him out of it, even when it was clear that he was wrong. Finally, he would say, "If you'd been there"—that is, in space—"you'd know." Then he would walk out. Another astronaut, from the Nine, says Alan Shepard's involvement with banks and other lucrative business ventures generated resentment among some NASA managers.

Such history aside, it is clear that the stamp affair angered many at NASA. The transcripts of the congressional hearings have never been published, but one NASA official who was there remembers that Connecticut senator Lowell Weicker verbally assaulted NASA managers. And according to one astronaut, when Chris Kraft came back to Houston from the hearings, he told a room full of astronauts, "The [astronauts] went in there and were treated like heroes. I went in there and I was totally humiliated. That is never going to happen to me again."

499 *The sequencer, aware of its own error:* What was the cause of the problem that caused a $450 million lunar mission, and the efforts of thousands of people, to grind temporarily to a halt? As discovered later by technicians, it was a single defective diode in a printed circuit card within the sequencer. *Apollo 17 Mission Report* (NASA internal publication), p. 14–1; Baker, *The History of Manned Spaceflight*, p. 438.

506 *Schmitt had first raised the idea late in the spring of 1970:* Schmitt's efforts in 1970 focused on the last four landing missions, Apollo 16–19. The geologists were already pushing for a landing at Tycho for Apollo 16. For Apollo 17, Schmitt was suggesting Mare Orientale, the giant impact basin that straddles the boundary between the near and far sides. For 18, he proposed a landing at the lunar north pole, where scientists hoped permanently shadowed craters might contain deposits of volatile elements, now in the form of ice, that had once escaped from the lunar interior. Then, Apollo 19 would go to Tsiolkovsky. The "lunar mafia"—Schmitt's name for his flight-controller friends—had tracked down some Tiros communications satellites, which TRW would have sold to NASA for about $80 million (details from the *Apollo 17 Lunar Surface Journal,* edited by Eric Jones, in press).

509 At 2 days, 17 hours into the flight, the Mission Elapsed Time clock was advanced 2 hours and 40 minutes, to make up for the delay in Apollo 17's launch. Mission Elapsed Times of events in Chapter 13 reflect this change.

509 Challenger *crept down the last 25 feet:* In truth, the word "crept" is only appropriate for the very last portion of the descent. At one point when *Challenger* was very close to the surface, Schmitt told Cernan his descent rate was "a little high"—mild words, but the slight intensity of Schmitt's voice hints at his concern. Today Schmitt says, with a laugh, "Gene had an aggressive streak in his flying."

509 *to make his own tracks in ancient dust:* Schmitt's first step on the moon was almost his downfall. When he stepped off the footpad, his boot came down on the sloping side of a boulder that was partly covered by beads of impact-created glass—like tiny ball bearings—and his leg went out from under him. Still gripping the ladder, he managed to avoid the fall.

511 *though only he could see it, he said,* "Look at the light": Only Cernan could see the glow, because he had the only window in the boost protective cover.

511 *Cernan . . . would talk about them for the rest of his life:* Schmitt points out that he has also been talking about those sights ever since Apollo 17. "I can't believe people are still interested," Schmitt says, "but they are."

II: Apollo at the Limit

516 *once inside—if a student is clever enough to get in:* According to Schmitt, the students always are.

518 *an optical illusion caused at certain angles of illumination:* Today, most lunar scientists believe that the grooves Scott and Irwin saw were indeed an illusion due to lighting effects.

519 *Anxious to save time, he went as fast as he dared:* On level ground, going flat out, the Rover hit 12 kilometers per hour, about half the speed of a world-class marathon runner.

519 *the planners were too conservative when they calculated walkback limits:* Based on Shepard and Mitchell's journeys at Fra Mauro, it was assumed that two astronauts forced to return to the LM on foot would achieve a speed of about one and two-thirds miles per hour. Since no one had ever been forced to put these calculations to the test—fortunately—Schmitt had been unable to offer a better estimate.

525 *the geologists suspected that the landslide:* Schmitt says that, in reality, a better term is avalanche, which refers to a slide that is fluid in character. In the case of the light mantle, Schmitt says, gases implanted in the soil by the solar wind would have allowed grains of dust to slide past one another. He believes the gases probably dissipated as the material descended, so that the avalanche became a landslide.

525 *Cernan and Schmitt headed for . . . a crater called Shorty:* Shorty had been named for a character in San Francisco author Richard Brautigan's novel *Trout Fishing in America.*

528 *now there was a very real possibility that Shorty was volcanic:* Another possibility, in Schmitt's mind, was that Shorty was an impact crater that had fractured the crust, allowing volcanic gases to escape to the surface.

528 *The ever-tightening circle of oxygen consumption and walkback limit:* As the J-missions progressed, the walkback limit was observed with less and less conservatism. Cernan and Schmitt were significantly closer to theirs at Shorty than any astronauts before them had been.

530 *hundreds of miles down in the lunar mantle:* Lee Silver has estimated that the magma that produced the fire fountains may have originated some 300 miles below the surface, over a quarter of the way to the moon's center.

III: Witnesses to the Earthrise

535 *getting within 9 miles was good enough:* Like the previous three crews, Evans and his crewmates had spent about a day in the descent orbit, with a low point of 50,000 feet.

535 *Farouk El-Baz's last protégé in lunar orbit:* Evans was told about the orange soil discovery, and the next time he was over the landing site he could see patches of orange, subtle but definitely there. Soon he was spotting it in other places, especially along the southwest border of Mare Serenitatis. Why hadn't he seen this earlier? After the mission he and El-Baz agreed: if you know what you are looking for, you find it.

536 *He did not know who had put it there:* Cernan has dropped hints that it may have been himself who put the sign on *Challenger*'s front landing leg. Schmitt, meanwhile, doesn't remember seeing it at all and says it probably never existed: it was just something Cernan decided to say over the radio, and Schmitt played along.

539 *the violence of the impacts that formed the lunar basins:* Apollo helped reveal that the time prior to about 4 billion years ago was an extremely violent one, both on the earth and moon. Schmitt suggests that the impacts of giant asteroids, which also pounded the young earth, may have played a role in the origin of life on our own planet. Perhaps, Schmitt says, chemical components of these asteroids, along with the tremendous energy of their impacts, helped create the conditions necessary for life to arise.

546 *He hated the words:* In 1988 Schmitt said, "I'm madder now than I was then. Nixon was right."

548 *with no guarantee of getting those people back:* Schmitt points out that in the strict sense, the risk for him and his crewmates circling the moon was no greater—and perhaps less—than for people in a submarine in the deep ocean. The greatest impact, he says, was psychological: "You're *no longer of the earth.*" He likens it to the first sailors who left sight of land. Like most of his other perceptions about his flight, this one came about in retrospect—at the time, he says, he was too busy to think about such things.

Epilogue: The Audiences of the Moon

556 *Roosa's words are ironic:* There is nothing unusual about the fact that Roosa did not know that his own crewmate had had a life-changing experience on the flight. Most of the astronauts have had little idea of what their colleagues took away from their lunar voyages. In part, this is because astronauts rarely asked each other about any aspects of their flights besides the technical ones. Even when they did, it wasn't necessarily a rich exchange. Dave Scott remembers that during a party after Apollo 14, he asked Alan Shepard what it had been like on the moon. Shepard answered in one word: "*Spectacular.*" And that, Scott says, was all he needed to hear; he could see the rest in Shepard's eyes.

558 *there is almost nothing left of the real one:* According to NASA's lunar sample curator, Jim Gooding, about half the Genesis Rock is still kept inside the lunar sample vault in Houston. However, it has been broken up into small fragments by scientists searching for minerals that could be used to establish a more accurate age for the rock. But the Genesis Rock is an unusual example. Gooding points out that despite the large number of ongoing studies of lunar material—NASA sends out more than a thousand prepared samples to scientists each year—researchers are able to conduct

their analyses using tiny amounts of rock or soil. As a result, most of the Apollo sample collection has not been touched. Some 74 percent by weight remains pristine within the Houston facility. Another 14 percent is kept in a special reserve storage facility in San Antonio. At least a tiny portion of every lunar sample has been studied, except for two core tubes collected on Apollo 16 and 17. For more information on the lunar sample facility, see the author's article, "Pieces of the Sky," *Sky & Telescope*, April 1982, pp. 344–49.

558 *By publicly admitting he had made a mistake, Irwin took himself off the pedestal:* In interviews with the author, Irwin confessed some resentment that the other astronauts did not come to the defense of him and his crewmates. Like Scott and Worden, he maintained that previous astronauts had done similar things, but that none would admit it. The author was not able to validate this claim in off-the-record conversations with other astronauts.

In 1983, Al Worden filed suit against NASA to return the first-day covers the agency had confiscated from him and his crewmates in 1972. In an out-of-court settlement, the Justice Department directed NASA to return the covers, and according to Worden, one official told him that the Apollo 15 crew had committed no wrongdoing by their actions regarding the first-day covers before and after their flight.

562 *he would rather have been the last man to walk on the moon:* Many astronauts have shared Anders's regret at not having walked on the moon. For Apollo 13's Fred Haise, the disappointment was compounded by the experience of a failed mission. But Haise's commander has a different attitude. If Jim Lovell could pick the flights he would like to be on, even with clear hindsight, one of them would be Apollo 13. It doesn't take anything extraordinary to do what is expected of you, Lovell says, but fighting for his life 200,000 miles from home tested him in ways that even a lunar landing wouldn't have. Says Lovell, "Apollo 13 was a test pilot's mission." He regrets that the Society of Experimental Test Pilots never recognized him and his crew for their performances. And when Congress awarded the Space Medal of Honor to Pete Conrad, Neil Armstrong, and a handful of other astronauts, Lovell was disappointed once more. A medal of honor, he points out, is given for action above and beyond the call of duty. But today, he says, "I don't hold a grudge any longer."

Apollo 13 has also had special meaning for Ken Mattingly. If people expect that he is glad for the twist of fate that kept him from the hardships Jim Lovell and his crew endured, glad to have a successful flight on his résumé instead of a failed one, then they are wrong.

"I have personally prospered by not being onboard," Mattingly says. "But I wish I had been. That's where I belonged. If it had been a success, if it had all worked out well, I wouldn't feel the way I do. But it meant more adversity than you would expect, and as a result I feel like somehow . . . I didn't do my share." He adds, "Does that make any sense? Hell, no."

563 *the famous picture of the first earthrise:* Here is Borman's earthrise story, as he told it to the author in March 1988:

"I'm looking over the lunar horizon, and there's the earth coming up. And I'm saying, 'Bill, take that picture! Get that one!' He says, 'I can't.' 'Why not?' 'I don't have enough film. All my film's allocated for scientific'—I said, 'Bill, you're full of baloney; that is the only picture that anybody will remember from this goddamned flight! None of your volcanoes and craters—Take that picture!' He said, 'No.' So I took the camera and took the goddamned picture. That's the truth of the story. And it's probably on the transcripts too. Did you read it?"

563 *the final, unequivocal verdict: Anders took the picture:* The author broke this news to Borman in March 1988. Susan Borman told her husband he had an apology to make (to Anders). Borman, laughing, said, "I'm not gonna change my story!"

564 *"I've got the same . . . wife I started out with":* Of the twenty-one astronauts who were married when they went to the moon, eleven later divorced.

567 *When Cernan talks about his moon experiences, not a hint of boredom dulls his words:* Rusty Schweickart says when he first heard Cernan talk about his experiences, at a public appearance in Australia in the mid-1970s, he was astounded. "Gene gave the most absolutely *eloquent, touching, personal* account of the experience of spaceflight and how it has shaped his life—I mean, I sat there next to him *absolutely blown away!* . . . I never thought Gene had that in him. Let alone the willingness to say anything about it."

568 The photograph of Neil Armstrong at Tranquillity Base: There are only a handful of still photographs of Armstrong on the moon; these were taken by Aldrin as part of documentary photography of the lunar module and the moon itself. The photograph in Cernan's office shows Armstrong collecting the contingency sample; it is an enlargement made from 16mm movie film.

571 *"No wonder I was in trouble":* In John Preston, "Buzz: The man who fell to Earth," *The Press* (Christchurch, New Zealand), January 15, 1994.

574 *the best explanation for the origin of the moon:* As described on pp. 353–54 of *To a Rocky Moon* by Don Wilhelms, this theory, proposed by A. G. W. Cameron and William K. Hartmann, accounts for the moon's similarity to the earth's mantle as well as its lack of water and volatile elements, which were driven off by the heat of the impact and escaped into space. While there are still problems to be worked out with the theory,

it does appear to explain not only the moon's composition, but the nature of its orbit, which the scientists say would be a direct result of the events that followed the collision.

575–576 *an answer to the planet's pressing energy needs:* See the author's article, "Shoot for the Moon." *Air & Space/Smithsonian,* December 1991/January 1992, pp. 42–51.

579 *that trip, Mattingly says, is* really *leaving home:* There are leavings, and then there are leavings. Fred Haise says, "I'd like to be the first guy to go to another star, and [watch as] our sun goes away."

Index

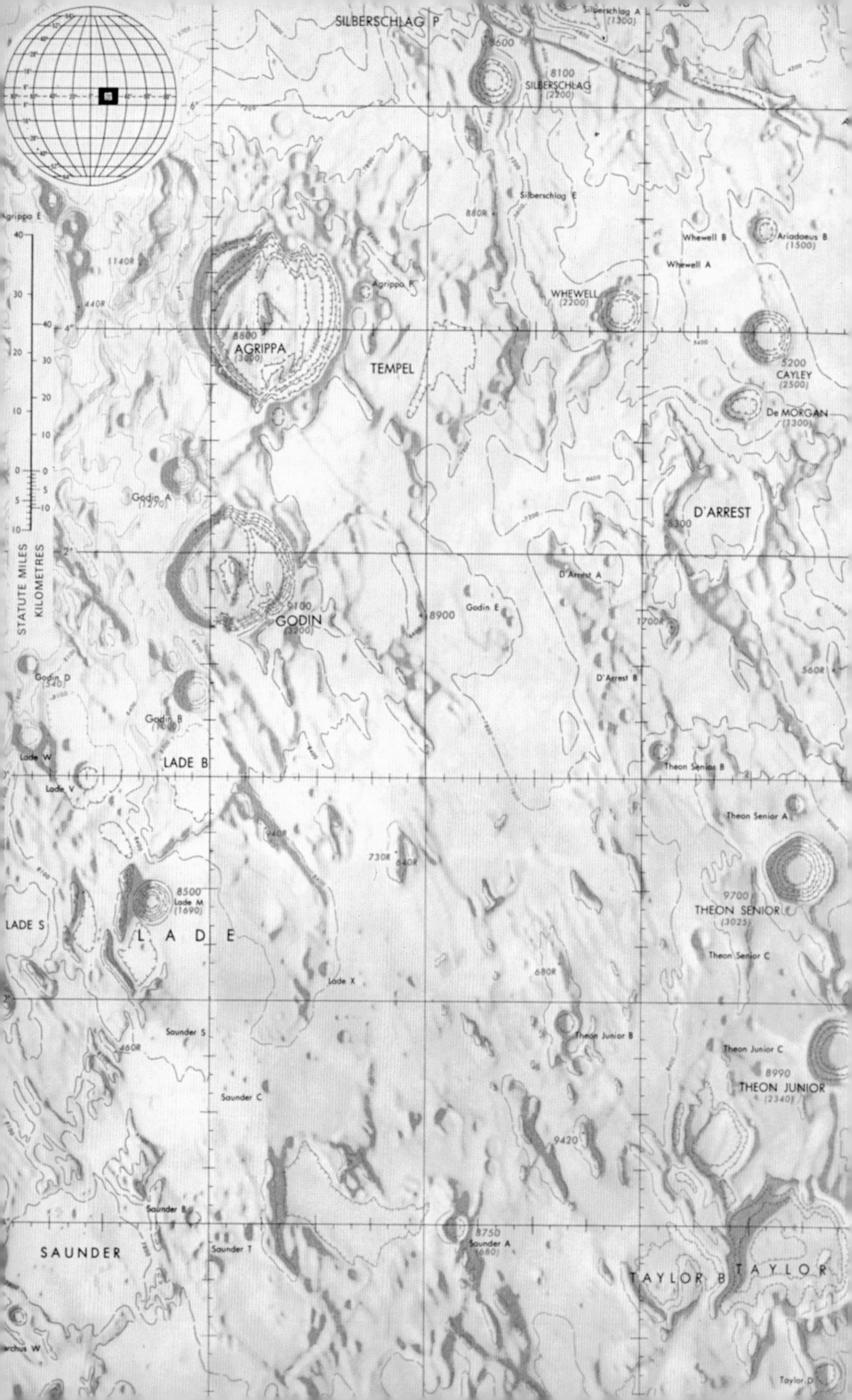
SILBERSCHLAG P
Silberschlag A
(1300)
8600
8100
SILBERSCHLAG
(2200)
Silberschlag E
880R
Agrippa E
1140R
440R
Whewell B
Ariadaeus B
(1500)
Whewell A
Agrippa F
WHEWELL
(2200)
8800
AGRIPPA
(3000)
TEMPEL
5200
CAYLEY
(2500)
De MORGAN
(1300)
STATUTE MILES
KILOMETRES
Godin A
(1270)
D'ARREST
8300
D'Arrest A
8100
GODIN
(3200)
8900
Godin E
1700R
560R
Godin D
(540)
D'Arrest B
Godin B
LADE B
Lade W
Lade V
Theon Senior B
Theon Senior A
940R
730R
640R
8500
Lade M
(1690)
9700
THEON SENIOR
(3025)
LADE S
L A D E
Theon Senior C
680R
Lade X
Saunder S
Theon Junior B
460R
Theon Junior C
8990
THEON JUNIOR
(2340)
Saunder C
9420
Saunder B
8750
Saunder A
(680)
SAUNDER
Saunder T
TAYLOR B
TAYLOR
Taylor D